Design Theory and Methods using CAD/CAE

Design Theory and Methods using CAD/CAE

The Computer Aided Engineering Design Series

Kuang-Hua Chang

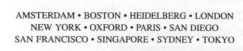

AMSTERDAM • BOSTON • HEIDELBERG • LONDON
NEW YORK • OXFORD • PARIS • SAN DIEGO
SAN FRANCISCO • SINGAPORE • SYDNEY • TOKYO

ELSEVIER

Academic Press is an imprint of Elsevier

Academic Press is an imprint of Elsevier
32 Jamestown Road, London NW1 7BY, UK
525 B Street, Suite 1800, San Diego, CA 92101-4495, USA
225 Wyman Street, Waltham, MA 02451, USA
The Boulevard, Langford Lane, Kidlington, Oxford OX5 1GB, UK

First published 2015

Notices

Knowledge and best practice in this field are constantly changing. As new research and experience broaden our understanding, changes in research methods, professional practices, or medical treatment may become necessary.

Practitioners and researchers must always rely on their own experience and knowledge in evaluating and using any information, methods, compounds, or experiments described herein. In using such information or methods they should be mindful of their own safety and the safety of others, including parties for whom they have a professional responsibility.

To the fullest extent of the law, neither the Publisher nor the authors, contributors, or editors, assume any liability for any injury and/or damage to persons or property as a matter of products liability, negligence or otherwise, or from any use or operation of any methods, products, instructions, or ideas contained in the material herein.

ISBN: 978-0-12-398512-5

British Library Cataloguing-in-Publication Data
A catalogue record for this book is available from the British Library

Library of Congress Cataloging-in-Publication Data
A catalog record for this book is available from the Library of Congress

For information on all Academic Press publications
visit our website at http://store.elsevier.com

Typeset by TNQ Books and Journals
www.tnq.co.in

To my best friends in Oklahoma, James and Qin-Fang Chen (陳壽祥、周秦芳), *Tony and Freda Chen* (陳列、賴慧慈), *Fred and Binro Lee* (李南海、何濱洛), *Victor and Jill Chen* (陳功、簡璞珍), *and Jaffee Wu* (吳繼凱). *You are a blessing sent from above to guide and support me. I am truly honored and grateful for the precious friendships and many years of happiness you all have brought to me.*

Contents

Preface

The conventional product development process employs a design–build–test philosophy. The sequentially executed product development process often results in a prolonged lead time and an elevated product cost. The e-Design paradigm presented in the *Computer Aided Engineering Design* series employs IT-enabled technology, including computer-aided design, engineering, and manufacturing (CAD/CAE/CAM) tools, as well as advanced prototyping technology to support product design from concept to detailed designs, and ultimately manufacturing. This e-Design approach employs virtual prototyping (VP) technology to support a cross-functional team in analyzing product performance, reliability, and manufacturing costs early in the product development stage and in conducting quantitative trade-offs for design decision making. Physical prototypes of the product design are then produced using rapid prototyping (RP) technique mainly for design verification. The e-Design approach holds potential for shortening the overall product development cycle, improving product quality, and reducing product cost. The *Computer Aided Engineering Design* series intends to provide readers with a comprehensive coverage of essential elements for understanding and practicing the e-Design paradigm in support of product design, including design method and process, and computer-based tools and technology. The book series consists of four modules: *Product Design Modeling Using CAD/CAE*, *Product Performance Evaluation Using CAD/CAE*, *Product Manufacturing and Cost Estimating Using CAD/CAE*, and *Design Theory and Methods Using CAD/CAE*. The *Product Design Modeling Using CAD/CAE* book discusses virtual mockup of the product that is first created in the CAD environment. The critical design parameterization that converts the product solid model into parametric representation, enabling the search for better designs, is an indispensable element of practicing the e-Design paradigm, especially in the detailed design stage. The second book, *Product Performance Evaluation Using CAD/CAE*, focuses on applying numerous computer-aided engineering (CAE) technologies and software tools to support evaluation of product performance, including structural analysis, fatigue and fracture, rigid body kinematics and dynamics, and failure probability prediction and reliability analysis. The third book, *Product Manufacturing and Cost Estimating Using CAD/CAE*, introduces computer-aided manufacturing (CAM) technology to support manufacturing simulations and process planning, RP technology and computer numerical control (CNC) machining for fast product prototyping, as well as manufacturing cost estimate that can be incorporated into product cost calculations. The product performance, reliability, and cost calculated can then be brought together to the cross-functional team for design trade-offs based on quantitative engineering data obtained from simulations. Design trade-off is one of the key topics included in the fourth book, *Design Theory and Methods Using CAD/CAE*. In addition to conventional design optimization methods, the fourth book discusses decision theory, utility theory, and decision-based design. Simple examples are included to help readers understand the fundamentals of concepts and methods introduced in this book series.

In addition to the discussion on design principles, methods, and processes, this book series offers detailed review on the commercial off-the-shelf software tools for the support of modeling, simulations, manufacturing, and product data management and data exchanges. Tutorial style lessons on using commercial software tools are provided together with project-based exercises. Two suites of engineering software are covered: they are Pro/ENGINEER-based, including Pro/MECHANICA Structure, Pro/ENGINEER Mechanism Design, and Pro/MFG; and SolidWorks-based, including

SolidWorks Simulation, SolidWorks Motion, and CAMWorks. These tutorial lessons are designed to help readers gain hands-on experiences to practice the e-Design paradigm.

The book you are reading, *Design Theory and Methods Using CAD/CAE*, is the fourth (last) module of the *Computer Aided Engineering Design* series. The objective of the *Design Theory and Methods* book is to provide readers with fundamental understanding in product design theory and methods, and apply the theory and methods to support engineering design applications in the context of e-Design. In Chapter 1, a brief introduction to the e-Design paradigm and tool environment will be given. Following this introduction, important topics in design theory and methods, including decision methods and theory in engineering design, design optimization, structural design sensitivity analysis, as well as multi-objective design optimization will be discussed.

Chapter 2 focuses on decision making for engineering design, in which conventional decision methods and decision theory, as well as decision-based design developed recently are discussed. The conventional methods, such as decision tree and decision table, have been widely employed by industry in support of design decision making. On the other hand, decision theory offers scientific and theoretical basis for design decision making, which gained attentions of researchers in recent years. This chapter offers a short review on popular decision methods, design theory, as well as the application of the theory to support engineering design. This chapter serves as a prelude to chapters that follow.

Chapter 3 discusses design optimization, which is one of the mainstream methods in engineering design. We discuss linear and nonlinear programming and offer mathematical basis for design problem formulation and solutions. We include both gradient-based and non-gradient approaches for solving optimization problems. In this chapter, readers should see clearly the limitations of the non-gradient approaches in terms of the computational efforts of the design problems, especially, large-scale problems. The gradient-based approaches are more suitable to the typical problems in the context of e-Design. We focus on single-objective optimization that serves the basis to understand multiobjective optimization to be discussed in Chapter 5 that is much relevant to practical applications. We address issues involved in dealing with practical engineering design problems and discuss an interactive design approach, including design trade-off and what-if study, which is more suitable for support of large-scale design problems. We offer case studies to illustrate practical applications of the methods discussed and a brief review on software tools that are commercially available for support of various types of optimization problems.

Chapter 4 provides a brief discussion on the sensitivity analysis, that is, gradient calculations that are essential for design using the gradient-based methods. In this chapter, we narrow our focus on structural problems in hope to introduce basic concept and methods in gradient calculations. We include in this chapter popular topics, such as sizing, shape, and topology designs. We also offer case studies to illustrate practical applications of the methods discussed. Some aspect of the ideas and methods on gradient calculations for structural problems can be extended to support other engineering disciplines; for example, motion and machining time. Two case studies are presented to illustrate a practical scenario that involves integration of topology and shape optimization and an advanced topic that supports shape design for crack propagation at the atomistic level using multiscale simulations.

In Chapter 5, we introduce multiobjective design optimization concept and methods. We start with simple examples to illustrate the concept and introduce Pareto optimality. We then discuss major solution techniques categorized by the articulation of preferences. We also include multiobjective genetic algorithms that gain popularity in recent years. In addition, we revisit decision-based design using both utility theory and game theory introduced in Chapter 3. We make a few comments on the decision-based design approach from the context of multiobjective optimization. We include a discussion on software tools that offer readers knowledge on existing tools for adoption and further investigation. We also include two advanced topics, reliability-based design optimization and design optimization for product manufacturing cost.

In addition to theories and methods, two companion projects are included: *Project S5 Design with SolidWorks* and *Project P5 Design with Pro/ENGINEER*. These projects offer tutorial lessons that should help readers to learn and be able to use the respective software tools for support of practical design applications. We include two examples in each projects, design optimization of a cantilever beam, and multidisciplinary design optimization for a single-piston engine. Example files needed for going through the tutorial lessons are available for download at the book's companion site. The goal of the projects is to help readers become confident and competent in using CAD/CAE/CAM and optimization tools for creating adequate product design model and adopt effective solution techniques in carrying out product design tasks.

This *Design Theory and Methods* book should serve well for a half semester (8 weeks) instruction in engineering colleges of general universities. Typically, a three-hour lecture and one-hour laboratory exercise per week are desired. This book (and the book series) aims at providing engineering senior and first-year graduate students a comprehensive reference to learn advanced technology in support of engineering design using IT-enabled technology. Typical engineering courses that the book serves include Computer-Aided Design, Engineering Design, Integrated Product and Process Development, Concurrent Engineering, Design and Manufacturing, Modern Product Design, Computer-Aided Engineering, as well as Senior Capstone Design. In addition to classroom instruction, this book should support practicing engineers who wish to learn more about the e-Design paradigm at their own pace.

About the Author

Dr. Kuang-Hua Chang is a *David Ross Boyd Professor* and *Williams Companies Foundation Presidential Professor* at the University of Oklahoma (OU), Norman, Oklahoma. He received his diploma in Mechanical Engineering from the National Taipei Institute of Technology, Taiwan, in 1980; and M.S. and Ph.D. degrees in Mechanical Engineering from the University of Iowa, Iowa City, Iowa, in 1987 and 1990, respectively. Since then, he joined the Center for Computer-Aided Design (CCAD) at Iowa as a Research Scientist and CAE Technical Area Manager. In 1997, he joined OU. He teaches mechanical design and manufacturing, in addition to conducting research in computer-aided modeling and simulation for design and manufacturing of mechanical systems.

His work has been published in 7 books and more than 150 articles in international journals and conference proceedings. He has also served as technical consultants to US industry and foreign companies, including LG-Electronics, Seagate Technology, and so forth. Dr. Chang received numerous awards for his teaching and research in the past few years, including the Williams Companies Foundation presidential professorship in 2005 for *meeting the highest standards of excellence in scholarship and teaching*, OU Regents Award for Superior Accomplishment in Research and Creative Activity in 2004, OU BP AMOCO Foundation Good Teaching Award in 2002, and OU Regents Award for Superior Teaching in 2010. He is a five-time recipient of CoE Alumni Teaching Award, given to top teachers in CoE. His research paper was given a Best Paper Award at the iCEER-2005 iNEER Conference for Engineering Education and Research in 2005. In 2006, he was awarded a Ralph R. Teetor Educational Award by SAE *in recognition of significant contributions to teaching, research and student development*. Dr. Chang was honored by the OKC Mayor's Committee on Disability Concerns with the 2009 Don Davis Award, which is *the highest honor granted in public recognition of extraordinarily meritorious service which has substantially advanced opportunities for people with disabilities by removing social, attitudinal, & environmental barriers in the greater Oklahoma City area.* In 2013, Dr. Chang was named David Ross Boyd Professor, one of the highest honors at the University of Oklahoma, for *having consistently demonstrated outstanding teaching, guidance, and leadership for students in an academic discipline or in an interdisciplinary program within the University.*

About the Cover Page

The cover page shows the solid model of an airplane torque tube in computer that is one of the three similar tubes located in the front leading edge of the wing, three on each side. These torques tubes were re-engineered for enhanced product reliability and reduced manufacturing lead-time. The problems are twofold. First, the current magnesium tubes designed decades ago suffer poor corrosion-resistance, requiring frequent repairs. Second, magnesium tubes are made by casting, which is extremely uneconomical when used for producing a very small quantity required for the remaining aircraft fleet. Lead-times were excessive and the cost was extremely high for acquiring the tubes. Involving the original equipment manufacturer (OEM) in re-engineering the torque tubes has also proven to be cost prohibitive. Therefore, without the original technical data package, sample tubes are first measured for critical geometric dimensions using a coordinate measurement machine (CMM). The measurement data are employed for constructing parametric solid models using a CAD system, in this case, Pro/ENGINEER. Once the parametric solid model is available, the product and process re-engineering activities are conducted concurrently. In re-engineering the tubes, strength analyses are conducted for both magnesium and aluminum solid models. In order to reduce the weight of the aluminum tube while maintaining its strength, the tube geometry is changed using a design optimization method (see advanced topic section in Chapter 5 for more details). A sample aluminum tube has been machined at OU and delivered to an Air Force Depot for form, fit checking, and material strength test. The aluminum tube is both stronger and more corrosion-resistant than the magnesium tube it will replace. Machining the tubes using CNC is more cost effective (50% cost reduction), and, more importantly, the manufacturing lead-time is reduced from about 18 months to just one week.

Acknowledgment

I would like to first thank Mr Joseph P. Hayton for recognizing the need of such an engineering design book series that offers knowledge in modern engineering design principles, methods, and tools to mechanical engineering students. His enthusiasm in moving the book project forward and eventually publishing the book series is highly appreciated. Mr Hayton's colleagues at Elsevier, Ms Lisa Jones and her production team, and Ms Chelsea Johnson, have made significant contributions in transforming the original manuscripts into a well-organized and professionally polished book that is suitable and presentable to our readers.

I am grateful to my current and former graduate students Dr. Yunxiang Wang, Dr. Mangesh Edke, Dr. Qunli Sun, Dr. Sung-Hwan Joo, Dr. Xiaoming Yu, Dr. Hsiu-Ying Hwang, Mr. Trey Wheeler, Mr Iulian Grindeanu, Mr. Tyler Bunting, Mr. David Gibson, Mr. Chienchih Chen, Mr. Tim Long, Mr. Poh-Soong Tang, and Mr. Javier Silver, for their excellent efforts in conducting research on numerous aspects of engineering design. Ideas and results that came out of their research have been largely incorporated into this book. Their dedication to the research in developing computer-aided approaches for support of product design modeling is acknowledged and is highly appreciated.

INTRODUCTION TO e-DESIGN

CHAPTER OUTLINE

Conventional product development employs a design-build-test philosophy. The sequentially executed development process often results in prolonged lead times and elevated product costs. The proposed e-Design paradigm employs IT-enabled technology for product design, including virtual prototyping (VP) to support a cross-functional team in analyzing product performance, reliability, and manufacturing costs early in product development, and in making quantitative trade-offs for design decision making. Physical prototypes of the product design are then produced using the rapid prototyping (RP) technique and computer numerical control (CNC) to support design verification and functional prototyping, respectively.

e-Design holds potential for shortening the overall product development cycle, improving product quality, and reducing product costs. It offers three concepts and methods for product development:

- Bringing product performance, quality, and manufacturing costs together early in design for consideration.
- Supporting design decision making based on quantitative product performance data.
- Incorporating physical prototyping techniques to support design verification and functional prototyping.

1.1 INTRODUCTION

A conventional product development process that is usually conducted sequentially suffers the problem of the *design paradox* (Ullman 1992). This refers to the dichotomy or mismatch between the design engineer's knowledge about the product and the number of decisions to be made (flexibility) throughout the product development cycle (see Figure 1.1). Major design decisions are usually made in the early design stage when the product is not very well understood. Consequently, engineering changes are frequently requested in later product development stages, when product design evolves and is better understood, to correct decisions made earlier.

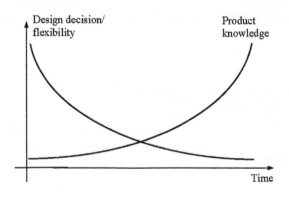

FIGURE 1.1

The design paradox.

Conventional product development is a design-build-test process. Product performance and reliability assessments depend heavily on physical tests, which involve fabricating functional prototypes of the product and usually lengthy and expensive physical tests. Fabricating prototypes usually involves manufacturing process planning and fixtures and tooling for a very small amount of production. The process can be expensive and lengthy, especially when a design change is requested to correct problems found in physical tests.

In conventional product development, design and manufacturing tend to be disjointed. Often, manufacturability of a product is not considered in design. Manufacturing issues usually appear when the design is finalized and tests are completed. Design defects related to manufacturing in process planning or production are usually found too late to be corrected. Consequently, more manufacturing procedures are necessary for production, resulting in elevated product cost.

With this highly structured and sequential process, the product development cycle tends to be extended, cost is elevated, and product quality is often compromised to avoid further delay. Costs and the number of engineering change requests (ECRs) throughout the product development cycle are often proportional according to the pattern shown in Figure 1.2. It is reported that only 8% of the total product budget is spent for design; however, in the early stage, design determines 80% of the lifetime cost of the product (Anderson 1990). Realistically, today's industries will not survive worldwide competition unless they introduce new products of better quality, at lower cost, and with shorter lead times. Many approaches and concepts have been proposed over the years, all with a common goal—to shorten the product development cycle, improve product quality, and reduce product cost.

A number of proposed approaches are along the lines of virtual prototyping (Lee 1999), which is a simulation-based method that helps engineers understand product behavior and make design decisions in a virtual environment. The virtual environment is a computational framework in which the geometric and physical properties of products are accurately simulated and represented. A number of successful virtual prototypes have been reported, such as Boeing's 777 jetliner, General Motors' locomotive engine, Chrysler's automotive interior design, and the Stockholm Metro's Car 2000 (Lee 1999). In addition to virtual prototyping, the concurrent engineering (CE) concept and methodology have been studied and developed with emphasis on subjects such as product life cycle design, design for X-abilities (DFX), integrated product and process development (IPPD), and Six Sigma (Prasad 1996).

Although significant research has been conducted in improving the product development process and successful stories have been reported, industry at large is not taking advantage of new product development paradigms. The main reason is that small and mid-size companies cannot afford to

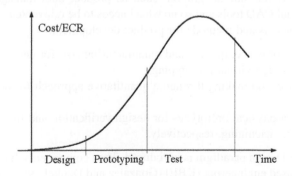

FIGURE 1.2

Cost/ECR versus time in a conventional design cycle.

develop an in-house computer tool environment like those of Boeing and the Big-Three automakers. On the other hand, commercial software tools are not tailored to meet the specific needs of individual companies; they often lack proper engineering capabilities to support specific product development needs, and most of them are not properly integrated. Therefore, companies are using commercial tools to support segments of their product development without employing the new design paradigms to their full advantage.

The e-Design paradigm does not supersede any of the approaches discussed. Rather, it is simply a realization of concurrent engineering through virtual and physical prototyping with a systematic and quantitative method for design decision making. Moreover, e-Design specializes in performance and reliability assessment and improvement of complex, large-scale, compute-intensive mechanical systems. The paradigm also uses design for manufacturability (DFM), design for manufacturing and assembly (DFMA), and manufacturing cost estimates through virtual manufacturing process planning and simulation for design considerations.

The objective of this chapter is to present an overview of the e-Design paradigm and the sample tool environment that supports a cross-functional team in simulating and designing mechanical products concurrently in the early design stage. In turn, better-quality products can be designed and manufactured at lower cost. With intensive knowledge of the product gained from simulations, better design decisions can be made, breaking the aforementioned design paradox. With the advancement of computer simulations, more hardware tests can be replaced by computer simulations, thus reducing cost and shortening product development time. The desirable cost and ECR distributions throughout the product development cycle shown in Figure 1.3 can be achieved through the e-Design paradigm.

A typical e-Design software environment can be built using a combination of existing computer-aided design (CAD), computer-aided engineering (CAE), and computer-aided manufacturing (CAM) as the base, and integrating discipline-specific software tools that are commercially available for specific simulation tasks. The main technique in building the e-Design environment is tool integration. Tool integration techniques, including product data models, wrappers, engineering views, and design process management, have been developed (Tsai et al. 1995) and are described in *Design Theory and Methods using CAD/CAE*, a book in The Computer Aided Engineering Design Series. This integrated e-Design tool environment allows small and mid-size companies to conduct efficient product development using the e-Design paradigm. The tool environment is flexible so that additional engineering tools can be incorporated with a lesser effort.

In addition, the basis for tool integration, such as product data management (PDM), is well established in commercial CAD tools and so no wheel needs to be reinvented. The e-Design paradigm employs three main concepts and methods for product development:

- Bringing product performance, quality, and manufacturing cost for design considerations in the early design stage through virtual prototyping.
- Supporting design decision making through a quantitative approach for both concept and detail designs.
- Incorporating product physical prototypes for design verification and functional tests via rapid prototyping and CNC machining, respectively.

In this chapter, the e-Design paradigm is introduced. Then components that make up the paradigm, including knowledge-based engineering (KBE) (Gonzalez and Dankel 1993), virtual prototyping, and

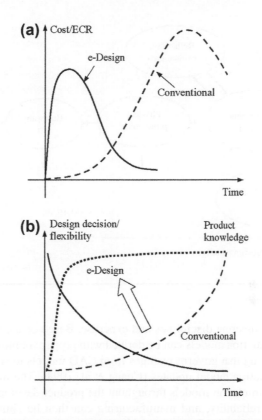

FIGURE 1.3

(a) Cost/ECR versus e-Design cycle time; (b) product knowledge versus e-Design cycle time.

physical prototyping, are briefly presented. Designs of a simple airplane engine and a high-mobility multipurpose wheeled vehicle (HMMWV) are briefly discussed to illustrate the e-Design paradigm. Details of modeling and simulation are provided in later chapters.

1.2 THE e-DESIGN PARADIGM

As shown in Figure 1.4, in e-Design, a product design concept is first realized in solid model form by design engineers using CAD tools. The initial product is often established based on the designer's experience and legacy data of previous product lines. It is highly desirable to capture and organize designer experience and legacy data to support decision making in a discrete form so as to realize an initial concept. The KBE (Gonzalez and Dankel 1993) that computerizes knowledge about specific product domains to support design engineers in arriving at a solution to a design problem supports the concept design. In addition, a KBE system integrated with a CAD tool may directly generate a solid model of the concept design that directly serves downstream design and manufacturing simulations.

With the product solid model represented in CAD, simulations for product performance, reliability, and manufacturing can be conducted. The product development tasks and the cross-functional team are

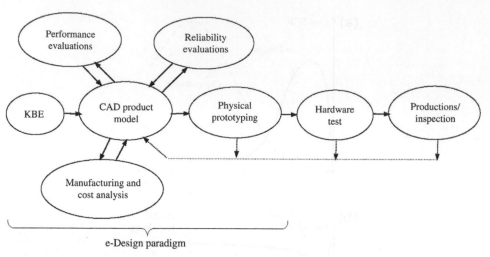

FIGURE 1.4

The e-Design paradigm.

organized according to engineering disciplines and expertise. Based on a centralized computer-aided design product model, simulation models can be derived with proper simplifications and assumptions. However, a one-way mapping that governs changes from CAD models to simulation models must be established for rapid simulation model updates (Chang et al. 1998). The mapping maintains consistency between CAD and simulation models throughout the product development cycle.

Product performance, reliability, and manufacturing can then be simulated concurrently. Performance, quality, and costs obtained from multidisciplinary simulations are brought together for review by the cross-functional team. Design variables—including geometric dimensions and material properties of the product CAD models that significantly influence performance, quality, and cost—can be identified by the cross-functional team in the CAD product model. These key performance, quality, and cost measures, as well as design variables, constitute a product design model. With such a model, a systematic design approach, including a parametric study for concept design and a trade-off study for detail design, can be conducted to improve the product with a minimum number of design iterations.

The product designed in the virtual environment can then be fabricated using rapid prototyping machines for physical prototypes directly from product CAD solid models, without tooling and process planning. The physical prototypes support the cross-functional team for design verification and assembly checking. Change requests that are made at this point can be accommodated in the virtual environment without high cost and delay.

The physics-based simulation technology potentially minimizes the need for product hardware tests. Because substantial modeling and simulations are performed, unexpected design defects encountered during the hardware tests are reduced, thus minimizing the feedback loop for design modifications. Moreover, the production process is smooth since the manufacturing process has been planned and simulated. Potential manufacturing-related problems will have been largely addressed in earlier stages.

A number of commercial CAD systems provide a suite of integrated CAD/CAE/CAM capabilities (e.g., Pro/ENGINEER and SolidWorks®). Other CAD systems, including CATIA® and NX, support one or more aspects of the engineering analysis. In addition, third-party software companies have made significant efforts in connecting their capabilities to CAD systems. As a representative example, CAE and CAM software companies worked with SolidWorks and integrated their software into SolidWorks environments such as CAMWorks®. Each individual tool is seamlessly integrated into SolidWorks.

In this book, Pro/ENGINEER and SolidWorks, with a built-in suite of CAE/CAM modules, are employed as the base for the e-Design environment. In addition to their superior solid modeling capability based on parametric technology (Zeid 1991), Pro/MECHANICA® and SolidWorks Simulation support simulations of nominal engineering, including structural and thermal problems. Mechanism Design of Pro/ENGINEER and SolidWorks Motion support motion simulation of mechanical systems. Moreover, CAM capabilities implemented in CAD, such as Pro/MFG (Parametric Technology Corp., www.ptc.com), and CAMWorks, provide an excellent basis for manufacturing process planning and simulations. Additional CAD/CAE/CAM tools introduced to support modeling and simulation of broader engineering problems encountered in general mechanical systems can be developed and added to the tool environment as needed.

1.3 VIRTUAL PROTOTYPING

Virtual prototyping is the backbone of the e-Design paradigm. As presented in this chapter, VP consists of constructing a parametric product model in CAD, conducting product performance simulations and reliability evaluations using CAE software, and carrying out manufacturing simulations and cost estimates using CAM software. Product modeling and simulations using integrated CAD/CAE/CAM software are the basic and common activities involved in virtual prototyping. However, a systematic design method, including parametric study and design trade-offs, is indispensable for design decision making.

1.3.1 PARAMETERIZED CAD PRODUCT MODEL

A parametric product model in CAD is essential to the e-Design paradigm. The product model evolves to a higher-fidelity level from concept to detail design stages (Chang et al. 1998). In the concept design stage, a considerable portion of the product may contain non-CAD data. For example, when the gross motion of the mechanical system is sought, the non-CAD data may include engine, tires, or transmission if a ground vehicle is being designed. Engineering characteristics of the non-CAD parts and assemblies are usually described by engineering parameters, physics laws, or mathematical equations. This non-CAD representation is often added to the product model in the concept design stage for a complete product model. As the design evolves, non-CAD parts and assemblies are refined into solid-model forms for subsystem and component designs as well as for manufacturing process planning.

A primary challenge in conducting product performance simulations is generating simulation models and maintaining consistency between CAD and simulation models through mapping. Challenges involved in model generation and in structural and dynamic simulations are discussed next, in which an airplane engine model in the detail design stage, as shown in Figure 1.5, is used for illustration.

(a) (b)

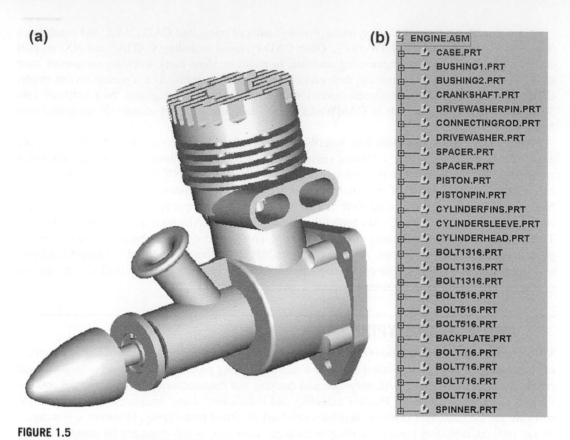

FIGURE 1.5

Airplane engine model: (a) CAD model and (b) model tree.

1.3.1.1 Parameterized product model

A parameterized product model defined in CAD allows design engineers to conveniently explore design alternatives for support of product design. The CAD product model is parameterized by defining dimensions that govern the geometry of parts through geometric features and by establishing relations between dimensions within and across parts. Through dimensions and relations, changes can be made simply by modifying a few dimensional values. Changes are propagated automatically throughout the mechanical product following the dimensions and relations. A single-piston airplane engine with a change in its bore diameter is shown in Figure 1.6, so as illustrating change propagation through parametric dimensions and relationships. More in-depth discussion of the modeling and parameterization of the engine example can be found in *Product Design Modeling using CAD/CAE*, a book in The Computer Aided Engineering Design Series.

1.3.1.2 Analysis models

For product structural analysis, finite element analysis (FEA) is often employed. In addition to structural geometry, loads, boundary conditions, and material properties can be conveniently defined in the CAD model. Most CAD tools are equipped with fully automatic mesh generation capability. This capability is convenient but often leads to large FEA models with some geometric discrepancy at the

(a)

(b)

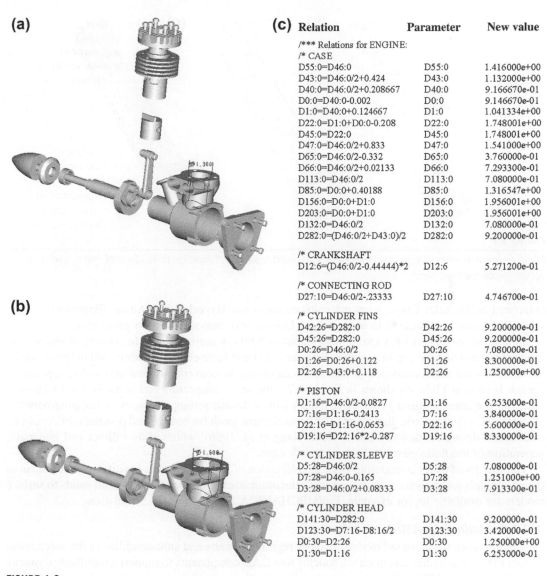

(c)

Relation	Parameter	New value
/*** Relations for ENGINE:		
/* CASE		
D55:0=D46:0	D55:0	1.416000e+00
D43:0=D46:0/2+0.424	D43:0	1.132000e+00
D40:0=D46:0/2+0.208667	D40:0	9.166670e-01
D0:0=D40:0-0.002	D0:0	9.146670e-01
D1:0=D40:0+0.124667	D1:0	1.041334e+00
D22:0=D1:0+D0:0-0.208	D22:0	1.748001e+00
D45:0=D22:0	D45:0	1.748001e+00
D47:0=D46:0/2+0.833	D47:0	1.541000e+00
D65:0=D46:0/2-0.332	D65:0	3.760000e-01
D66:0=D46:0/2+0.02133	D66:0	7.293300e-01
D113:0=D46:0/2	D113:0	7.080000e-01
D85:0=D0:0+0.40188	D85:0	1.316547e+00
D156:0=D0:0+D1:0	D156:0	1.956001e+00
D203:0=D0:0+D1:0	D203:0	1.956001e+00
D132:0=D46:0/2	D132:0	7.080000e-01
D282:0=(D46:0/2+D43:0)/2	D282:0	9.200000e-01
/* CRANKSHAFT		
D12:6=(D46:0/2-0.44444)*2	D12:6	5.271200e-01
/* CONNECTING ROD		
D27:10=D46:0/2-.23333	D27:10	4.746700e-01
/* CYLINDER FINS		
D42:26=D282:0	D42:26	9.200000e-01
D45:26=D282:0	D45:26	9.200000e-01
D0:26=D46:0/2	D0:26	7.080000e-01
D1:26=D0:26+0.122	D1:26	8.300000e-01
D2:26=D43:0+0.118	D2:26	1.250000e+00
/* PISTON		
D1:16=D46:0/2-0.0827	D1:16	6.253000e-01
D7:16=D1:16-0.2413	D7:16	3.840000e-01
D22:16=D1:16-0.0653	D22:16	5.600000e-01
D19:16=D22:16*2-0.287	D19:16	8.330000e-01
/* CYLINDER SLEEVE		
D5:28=D46:0/2	D5:28	7.080000e-01
D7:28=D46:0-0.165	D7:28	1.251000e+00
D3:28=D46:0/2+0.08333	D3:28	7.913300e-01
/* CYLINDER HEAD		
D141:30=D282:0	D141:30	9.200000e-01
D123:30=D7:16-D8:16/2	D123:30	3.420000e-01
D0:30=D2:26	D0:30	1.250000e+00
D1:30=D1:16	D1:30	6.253000e-01

FIGURE 1.6

Design change propagation: (a) bore diameter = 1.3 in.; (b) bore diameter changed to 1.6 in.; (c) relations of geometric dimensions.

part boundary. Plus, triangular and tetrahedral elements are often the only elements supported. An engine connecting rod example meshed using Pro/MESH (part of Pro/MECHANICA) with default mesh parameters is shown in Figure 1.7. The FEA model consists of 1,270 nodes and 4,800 tetrahedron elements, yet it still reveals discrepancy to the true CAD geometry. Moreover, mesh distortion due to large deformation of the structure, such as hyperelastic problems, often causes FEA to abort prematurely. Semiautomatic mesh generation is more realistic; therefore, tools such as MSC/Patran®

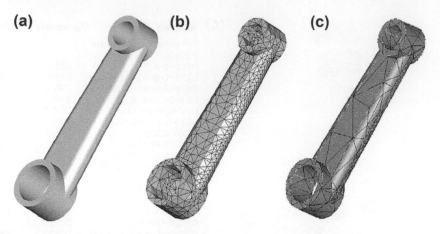

FIGURE 1.7

Finite element meshes of a connecting rod: (a) CAD solid model, (b) h-version finite element mesh, and (c) p-version finite element mesh.

(MacNeal-Schwendler Corp., www.mscsoftware.com) and HyperMesh® (Altair® Engineering, Inc., www.altair.com) are essential to support the e-Design environment for mesh generation.

In general, p-version FEA (Szabó and Babuška 1991) is more suitable for structural analysis in terms of minimizing the gap in geometry between CAD and finite element models, and in lessening the tendency toward mesh distortion. It also offers capability in convergence analysis that is superior to regular h-version FEA. As shown in Figure 1.7c, the same connecting rod is meshed with 568 tetrahedron p-elements, using Pro/MECHANICA with a default setting. A one-way mapping between changes in CAD geometric dimensions and finite element mesh for both h- and p-version FEAs can be established through a design velocity field (Haug et al. 1986), which allows direct and automatic generation of the finite element mesh of new designs.

Another issue worth considering is the simplification of 3D solid models to surface (shell) or curve (beam) models for analysis. Capabilities that semiautomatically convert 3D thin-shell solids to surface models are available in, for example, Pro/MECHANICA and SolidWorks Simulation.

1.3.1.3 Motion simulation models

Generating motion simulation models involves regrouping parts and subassemblies of the mechanical system in CAD as bodies and often introducing non-CAD components to support a multibody dynamic simulation (Haug 1989). Engineers must define the joints or force connections between bodies, including joint type and reference coordinates. Mass properties of each body are computed by CAD with the material properties specified. Integration between Mechanism Design and Pro/ENGINEER, as well as between SolidWorks Motion (Chang 2008) and SolidWorks, is seamless. Design changes made in geometric dimensions propagate to the motion model directly. In addition, simulation tools, such as Dynamic Analysis and Design Systems (DADS) (LMS, www.lmsintl.com/DADS) and communication and data systems integration, are also integrated with CAD with proper parametric mapping from CAD to simulation models that support parametric study. As an example, the motion inside an airplane engine is modeled as a slider-crank mechanism in Mechanism Design, as shown in Figure 1.8.

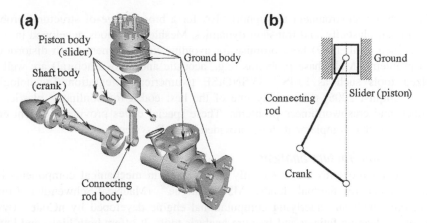

FIGURE 1.8

Engine motion model: (a) definition and (b) schematic view.

A common mistake made in creating motion simulation models is selecting improper joints to connect bodies. Introducing improper joints creates an invalid or inaccurate model that does not simulate the true behavior of the mechanical system. Intelligent modeling capability that automatically specifies joints in accordance with assembly relations defined between parts and subassemblies in solid models is available in, for example, SolidWorks Motion.

1.3.2 PRODUCT PERFORMANCE ANALYSIS

As mentioned earlier, product performance evaluation using physics-based simulation in the computer environment is usually called, in a narrow sense, virtual prototyping, or VP. With the advancement of simulation technology, more engineering questions can be answered realistically through simulations, thus minimizing the needs for physical tests. However, some key questions cannot be answered for sophisticated engineering problems—for example, the crashworthiness of ground vehicles. Although VP will probably never replace hardware tests completely, the savings it achieves for less sophisticated problems is significant and beneficial.

1.3.2.1 Motion analysis

System motion simulations include workspace analysis (kinematics), rigid- and flexible-body dynamics, and inverse dynamic analysis. Mechanism Design and SolidWorks Motion, based on theoretical work (Kane and Levinson 1985), mainly support kinematics and rigid-body simulations for mechanical systems. They do not properly support mechanical system simulation such as a vehicle moving on a user-defined terrain. General-purpose dynamic simulation tools, such as DADS (www.lsmintl.com) or Adams® (www.mscsoftware.com), are more desirable for simulation of general mechanical systems.

1.3.2.2 Structural analysis

Pro/MECHANICA supports linear static, vibration, buckling, fatigue, and other such analyses, using p-version FEA. General-purpose finite element codes, such as MSC/Nastran® (MacNeal-Schwendler Corp., www.mscsoftware.com) and ANSYS® (ANSYS Analysis Systems, Inc., www.ansys.com) are ideal for the e-Design environment to support FEA for a broad range of structural problems—for example, nonlinear, plasticity, and transient dynamics. Meshless methods developed in recent years (for example, Chen et al. 1997) hold promise for avoiding finite element mesh distortion in large-deformation problems. Multiphase problems (e.g., acoustic and aero-structural) are well supported by specialized tools such as LMS® SYSNOISE (Numerical Integration Technologies 1998). LS-DYNA® (Hallquist 2006) is currently one of the best codes for nonlinear, plastic, dynamics, friction-contact, and crashworthiness problems. These special codes provide excellent engineering analysis capabilities that complement those provided in CAD systems.

1.3.2.3 Fatigue and fracture analysis

Fatigue and fracture problems are commonly encountered in mechanical components because of repeated mechanical or thermal loads. MSC Fatigue® (MacNeal-Schwendler Corp., www.mscsoftware.com), with an underlying computational engine developed by nCode® (www.ncode.com) is one of the leading fatigue and fracture analysis tools. It offers both high- and low-cycle fatigue analyses. A critical plane approach is available in MSC Fatigue for prediction of fatigue life due to general multiaxial loads.

Note that the recently developed extended finite element method (XFEM) supports fracture propagation without re-meshing (Moës et al. 2002). XFEM was recently integrated in ABAQUS®. Also note that additional capabilities, such as thermal analysis, computational fluid dynamics (CFD) and combustion, can be added to meet specific needs in analyzing mechanical products. Integration of additional engineering disciplines are briefly discussed in Section 1.3.4.

1.3.2.4 Product reliability evaluations

Product reliability evaluations in the e-Design environment focus on the probability of specific failure events (or failure mode). The failure event corresponds to a product performance measure, such as the fatigue life of a mechanical component. For the reliability analysis of a single failure event, the failure event or failure function is defined as (Madsen et al. 1986)

$$g(X) = \psi^u - \psi(X) \tag{1.1}$$

where

ψ is a product performance measure
ψ^u is the upper bound (or design requirement) of the product performance
X is a vector of random variables.

When product performance does not meet the requirement—that is, when $\psi^u \leq \psi(X)$, the event fails. Therefore, the probability of failure P_f of the particular event $g(X) \leq 0$ is

$$P_f = P[g(X) \leq 0] \tag{1.2}$$

where $P[\bullet]$ is the probability of event $\bullet$.

Given the joint probability density function $f_X(x)$ of the random variables X, the probability of failure for a single event of a mechanical component can be expressed as

$$P_f = P[g(X) \leq 0] = \int \int_{g(X) \leq 0} \cdots \int f_X(x) dx. \tag{1.3}$$

The probability of failure in Eq. 1.3 is commonly evaluated using the Monte Carlo method or the first- or second-order reliability method (FORM or SORM) (Wu and Wirsching 1984, Yu et al. 1998).

Once the probabilities of several failure events in subsystems or components are computed, system reliability can be obtained by, for example, fault-tree analysis (Ertas and Jones 1993). No general-purpose software tool for reliability analysis of general mechanical systems is commercially available yet. Numerical evaluation of stochastic structures under stress (NESSUS®) (www.nessus.swri.org), which is currently in development can be a good candidate for incorporation into the e-Design environment. With the probability of failure, critical quality design criteria, such as mean time between failure (MTBF), can be computed (Ertas and Jones 1993).

Two main challenges exist in reliability analysis: One, realistic distribution data are difficult to acquire and often are not available in the early stage; and two, failure probability computations are often expensive. The first challenge may be alleviated by employing legacy data from previous product lines. Approximation techniques (e.g., Yu et al. 1998) can be employed to make the computation affordable even for an individual failure event within a mechanical component.

1.3.3 PRODUCT VIRTUAL MANUFACTURING

Virtual manufacturing addresses issues of design for manufacturability (DFM) (Prasad 1996) and design for manufacturing and assembly (DFMA) (Boothroyd et al. 1994) early in product development. In the e-Design paradigm, DFM and DFMA are performed by conducting virtual manufacturing and assembly using, for example, Pro/MFG. DFM and DFMA of the product are verified through animations of the virtual manufacturing and assembly process.

Pro/MFG is a Pro/ENGINEER module supporting the virtual machining process, including milling, drilling, and turning. By incorporating part design and also defining workpieces, workcells, fixtures, cutting tools, and cutting parameters, Pro/MFG automatically generates a toolpath (see Figure 1.9a), which simulates the machining process (Figure 1.9b), calculates machining time, and produces cutter location (CL) data. The CL data can be post-processed for CNC codes. In addition, casting, sheet metal, molding, and welding can be simulated using Pro/CASTING, Pro/SHEETME-TAL, Pro/MOLD, and Pro/WELDING, respectively.

With such virtual manufacturing process planning and animation, manufacturability of the product design can, to some extent, be verified. The DFMA tool (Boothroyd et al. 1994) developed by Boothroyd Dewhurst, Inc., assists the cross-functional team in quantifying product assembly time and labor costs. It also challenges the team to simplify product structure, thereby reducing product as well as assembly costs.

One of the limitations in using virtual manufacturing tools (e.g., Pro/MFG) is that chip formation (Fang and Jawahir 1996), a primary consideration in computer numerical control (CNC), is not incorporated into the simulation. In addition, machining parameters, such as power consumption, machining temperature, and tool life, which contribute to manufacturing costs are not yet simulated.

1.3.4 TOOL INTEGRATION

Techniques developed to support tool integration (Chang et al. 1998) include parameterized product data models, engineering views, tool wrappers, and design process management. Parameterized

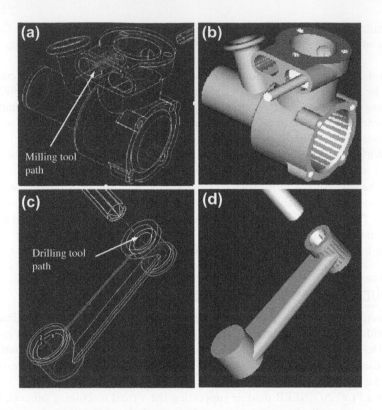

FIGURE 1.9

Virtual machining process: (a) engine case—milling toolpath; (b) milling simulation; (c) connecting rod—drilling toolpath; (d) drilling simulation.

product data models represent engineering data that are needed for conducting virtual prototyping of the mechanical system. The main sources of the product data model are CAD and non-CAD models. The product data model evolves throughout the product development cycle as illustrated in Figure 1.10.

Engineering views allow engineers from various disciplines to view the product from their own technical perspectives. Through engineering views, engineers create simulation models that are consistent with the product model by simplifying the CAD representation, as needed adding non-CAD product representation and mapping. Tool wrappers provide two-way data translation and transmission between engineering tools and the product data model. Design process management provides the team leader with a tool to monitor and manage the design process. When a new tool of an existing discipline, for example ANSYS for structural FEA, is to be integrated, a wrapper for it must be developed. Three main tasks must be carried out when a new engineering discipline, say computational fluid dynamics (CFD), is added to the environment. First, the product data model must be extended to include engineering data needed to support CFD. Second, engineering views must be added to allow design engineers to generate CFD models. Finally, wrappers must be developed for specific CFD tools.

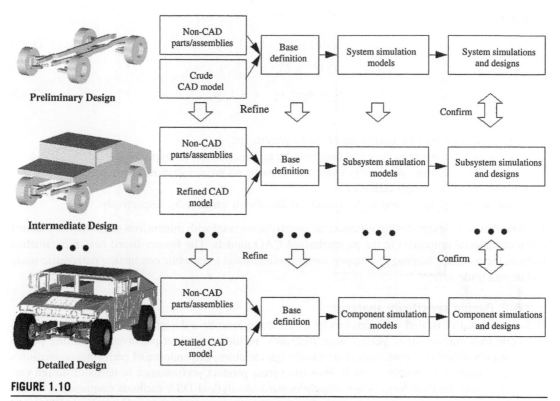

FIGURE 1.10

Hierarchical product models evolved through the e-Design process.

1.3.5 DESIGN DECISION MAKING

Product performance, reliability, and manufacturing cost that are evaluated using simulations can be brought to the cross-functional team for review. Product performance and reliability are checked against product specifications that have been defined and have evolved from the beginning of the product development process. Manufacturing cost derived from the virtual manufacturing simulations can be added to product cost. The cross-functional team must address areas of concern identified in product performance, reliability, and manufacturability, and it must identify a set of design variables that influence these areas. Design modifications can then be conducted. In the past, quality functional deployment (QFD) (Ertas and Jones 1993) was largely employed in design modification to assign qualitative weighting factors to product performance and design changes. e-Design employs a systematic and quantitative approach to design modifications (for example, Yu et al. 1997).

1.3.5.1 Design problem formulation

Before a design can be improved, design problems must be defined. A design problem is often presented in a mathematical form, typically as

$$\text{Minimize } \varphi(\boldsymbol{b}) \tag{1.4a}$$

Subject to

$$\psi_i(\boldsymbol{b}) \leq \psi_i^u \quad i = 1, \, m \tag{1.4b}$$

$$P_{f_j}(\boldsymbol{b}) \leq P_{f_j}^u \quad j = 1, \, n \tag{1.4c}$$

$$b_k^l \leq b_k \leq b_k^u \quad k = 1, \, p \tag{1.4d}$$

where

$\varphi(\boldsymbol{b})$ is the objective (or cost) function to be minimized
$\psi_i(\boldsymbol{b})$ is the ith constraint function that must be no greater than its upper bound ψ_i^u
$P_{f_j}(\boldsymbol{b})$ is the jth failure probability index that must be no greater than its upper bound $P_{f_j}^u$
$\boldsymbol{b}$ is the vector of design variables
b_k^l and b_k^u are the lower and upper bounds of the design variable b_k, respectively.

Note that in e-Design design variables are usually associated with dimensions of geometric features and part material properties in the parameterized CAD models. The feature-based design parameters serve as the common language to support the cross-functional team while conducting parametric study and design trade-offs.

1.3.5.2 Design sensitivity analysis

Before quantitative design decisions can be made, there must be a design sensitivity analysis (DSA) that computes derivatives of performance measures, including product performance, failure probability, and manufacturing cost, with respect to design variables. Dependence of performance measures on design variables is usually implicit. How to express product performance in terms of design variables in a mathematical form is not straightforward. Analytical DSA methods combined with numerical computations have been developed mainly for structural responses (Haug et al. 1986) and fatigue and fracture (Chang et al. 1997). DSA for failure probability with respect to both deterministic and random variables has also been developed (Yu et al. 1997). In addition, DSA and optimization using meshless methods have been developed for large-deformation problems (Grindeanu et al. 1999). More details about the analytical DSA for structural responses also referred to Haug et al. (1985).

For problems such as motion and manufacturing cost, where premature or no analytical DSA capability is available, the finite difference method is the only choice. The finite difference method is expressed in the following equation:

$$\frac{\partial \psi}{\partial b_j} \approx \frac{\psi(\boldsymbol{b} + \Delta b_j) - \psi(\boldsymbol{b})}{\Delta b_j} \tag{1.5}$$

where Δb_j is a perturbation in the jth design variable. With sensitivity information, parametric study and design trade-offs can be conducted for design improvements at the concept and detail stages, respectively.

1.3.5.3 Parametric study

A parametric study that perturbs design variables in the product design model to explore various design alternatives can effectively support product concept designs. The parametric study is simple and easy to perform as long as the mapping between CAD and simulation models has been established. The mapping supports fast simulation model generation for performance analyses. It also supports DSA using the

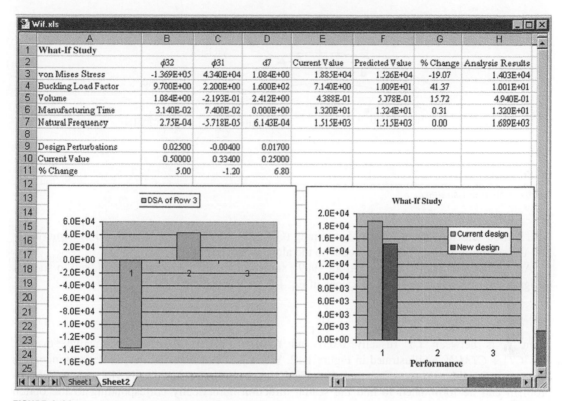

FIGURE 1.11

Spreadsheet for parametric study and design trade-offs.

finite difference method. The parametric study is possible for concept design because the number of design variables to perturb is usually small. A spreadsheet with a proper formula defined among cells is well suited to support the parametric study. The use of Microsoft Excel is illustrated in Figure 1.11.

1.3.5.4 Design trade-off analysis

With design trade-off analysis, the design engineer can find the most appropriate design search direction for the design problem formulated in Eq. 1.4, using four possible algorithms:

- Reduce cost.
- Correct constraint neglecting cost.
- Correct constraint with a constant cost.
- Correct constraint with a cost increment.

As a general rule, the first algorithm, reduce cost, can be chosen when the design is feasible; in other words, all constraint functions are within the desired limits. When the design is infeasible, generally one may start with the third algorithm, correct constraint with a constant cost. If the design remains infeasible, the fourth algorithm, correct constraint with a cost increment—say 10%—may be appropriate. If a feasible design is still not found, the second algorithm, correct constraint neglecting

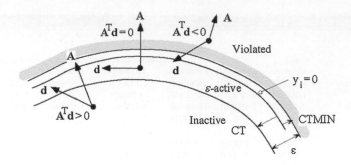

FIGURE 1.12

ε-active constraint strategy.

cost, can be selected. A quadratic programming (QP) subproblem can be formulated to numerically find the search direction that corresponds to the algorithm selected.

An ε-active constraint strategy (Arora 1989), shown in Figure 1.12, can be employed to support design trade-offs. The constraint functions in Eq. 1.4 are normalized by

$$y_i = \frac{\psi_i}{\psi_i^u} - 1 \le 0, \quad i = 1, m \tag{1.6}$$

when y_i is between *CT* (usually 0.03) and *CTMIN* (usually 0.005), it is active—that is, $\varepsilon = |CT| + CTMIN$, as illustrated in Figure 1.12. When y_i is less than *CT*, the constraint function is inactive or feasible. When y_i is larger than *CTMIN*, the constraint function is violated. A QP subproblem can be formulated to find the search direction numerically corresponding to the option selected. For example, the QP subproblem for the first algorithm (cost reduction) can be formulated as

$$\text{Minimize} \quad \mathbf{c}^T\mathbf{d} + 0.5\,\mathbf{d}^T\mathbf{d}$$

$$\text{Subject} \quad \mathbf{A}^T\mathbf{d} \le \mathbf{y} \tag{1.7}$$

$$\mathbf{b}^L - \mathbf{b}^{(k)} \le \mathbf{d} \le \mathbf{b}^U - \mathbf{b}^{(k)}$$

where

$$\mathbf{c} = [c_1, c_2, ..., c_{n1+n2}]^T, \quad c_i = \partial\varphi/\partial b_i$$

$\mathbf{d}$ is the search direction to be determined.

$$A_{ij} = \partial P_{y_i}/\partial b_j; \quad \mathbf{b} = [b_1, b_2, ...b_n]^T$$

k is the current design iteration.

The objective of the design trade-off algorithm is to find the optimal search direction $\mathbf{d}$ under a given circumstance. Details are discussed in *Design Theory and Methods using CAD/CAE*, a book in The Computer Aided Engineering Design Series.

1.3.5.5 What-if study

After the search direction $\mathbf{d}$ is found, a number of step sizes α can be used to perturb the design along the direction $\mathbf{d}$. Objective and constraint function values, represented as ψ_i, at a perturbed design

$b + \alpha d$ can be approximated using the first-order sensitivity information of the functions by Taylor series expansion about the current design b without going through simulations; that is,

$$\psi_i(b + \alpha d) \approx \psi_i(b) + \frac{\partial \psi_i}{\partial b} \alpha d. \tag{1.8}$$

Note that since there is no analysis involved, the what-if study can be carried out very efficiently. This allows the design engineer to explore design alternatives more effectively.

Once a satisfactory design is identified, after trying out different step sizes α in an approximation sense, the design model can be updated to the new design and then simulations of the new design can be conducted. Equation 1.8 also supports parametric study, in which the design perturbation δb is determined by engineers based on sensitivity information. To ensure a reasonably accurate function prediction using Eq. 1.8, the step sizes must be small so that the perturbation $\partial \psi_i / (\partial b)(\alpha d)$ is, as a rule of thumb, less than 10% of the function value $\psi_i(b)$.

1.4 PHYSICAL PROTOTYPING

In general, two techniques are suitable for fabricating physical prototypes of the product in the design process: rapid prototyping (RP) and computer numerical control (CNC) machining. RP systems, based on solid freeform fabrication (SFF) technology (Jacobs 1994), fabricate physical prototypes of the structure for design verification. The CNC machining fabricates functional parts as well as the mold or die for mass production of the product.

1.4.1 RAPID PROTOTYPING

The Solid Freeform Fabrication (SFF) technology, also called Rapid Prototyping (RP), is an additive process that employs a layer-building technique based on horizontal cross-sectional data from a 3D CAD model. Beginning with the bottom-most cross-section of the CAD model, the rapid prototyping machine creates a thin layer of material by slicing the model into so-called 2½ D layers. The system then creates an additional layer on top of the first based on the next higher cross-section. The process repeats until the part is completely built. It is illustrated using an engine case in the example shown in Figure 1.13. Rapid prototyping systems are capable of creating parts with small internal cavities and complex geometry.

Most important, SFF follows the same layering process for any given 3D CAD models, so it requires neither tooling nor manufacturing process planning for prototyping, as required by conventional manufacturing methods. Based on CAD solid models, the SFF technique fabricates physical prototypes of the product in a short turnaround time for design verification. It also supports tooling for product manufacturing, such as mold or die fabrications, through, for example, investment casting (Kalpakjian 1992).

Note that there are various types of SFF systems commercially available, such as the SLA® 7000 and Sinterstation® by 3D Systems (Figures 1.14a and 1.14b). In this chapter, the Dimension 1200 sst® machine (www.stratasys.com), as shown in Figure 1.14c, is presented. More details about it as well as other RP systems will be discussed in *Product Manufacturing and Cost Estimating using CAD/CAE*, a book in The Computer Aided Engineering Design Series.

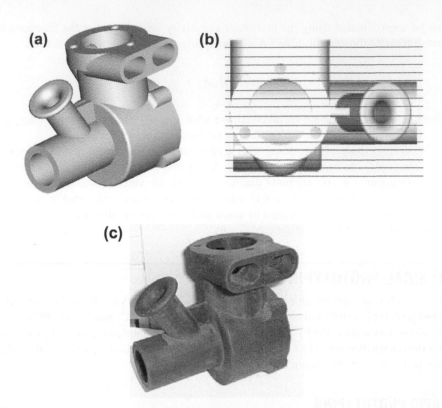

FIGURE 1.13

SFF: layered manufacturing: (a) 3D CAD model, (b) 2-1/2D slicing, and (c) physical model.

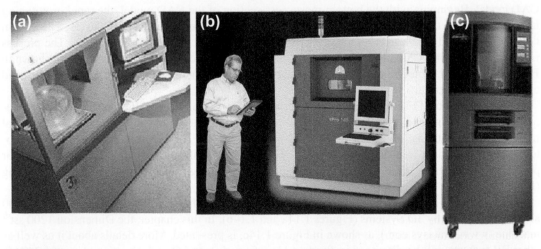

FIGURE 1.14

Commercial RP systems: (a) 3D systems' SLA 7000, (b) Sinterstation 2500, (c) Stratasys inc.'s dimension 1200 sst.

Sources: (b) 3D Systems Corporation, USA; (c) Stratasys Ltd.

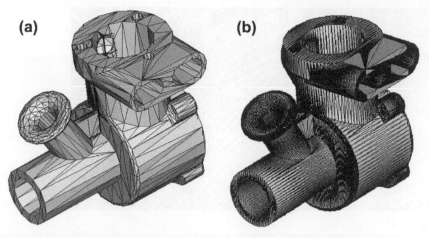

FIGURE 1.15

STL engine case models: (a) coarse and (b) refined.

The CAD solid model of the product is first converted into a stereolithographic (STL) format (Chua and Leong 1998), which is a faceted boundary representation uniformly accepted by the industry. Both the coarse and refined STL models of an engine case are shown in Figure 1.15. Even though the STL model is an approximation of the true CAD geometry, increasing the number of triangles can minimize the geometric error effectively. This can be achieved by specifying a smaller chord length, which is defined as the maximum distance between the true geometric boundary and the neighboring edge of the triangle. The faceted representation is then sliced into a series of 2D sections along a prespecified direction. The slicing software is SFF-system dependent.

The Dimension 1200 sst employs fused deposition manufacturing (FDM) technology. Acrylonitrile butadiene styrene (ABS) materials are softened (by elevating temperature), squeezed through a nozzle on the print heads, and laid on the substrate as build and support materials, respectively, following the 2D contours sliced from the 3D solid model (Figure 1.16). Note that various crosshatch options are available in CatalystEX® software (www.dimensionprinting.com), which comes with the rapid prototyping system.

The physical prototypes are mainly for the cross-functional team to verify the product design and check the assembly. However, they can also be used for discussion with marketing personnel to develop marketing ideas. In addition, the prototypes can be given to potential customers for feedback, thus bringing customers into the design loop early in product development.

1.4.2 CNC MACHINING

The machining operations of virtual manufacturing, such as milling, turning, and drilling, allow designers to plan the machining process, generate the machining toolpath, visualize and simulate machining operations, and estimate machining time. Moreover, the toolpath generated can be converted into CNC codes (M-codes and G-codes) (Chang et al. 1998, McMahon and Browne 1998) to fabricate functional parts as well as a die or mold for production.

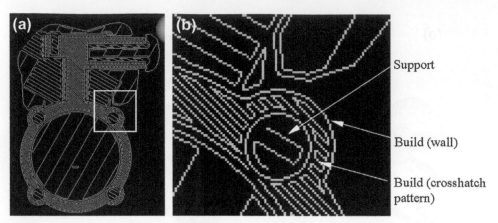

FIGURE 1.16

Crosshatch pattern of a typical cut-out layer: (a) overall and (b) enlarged.

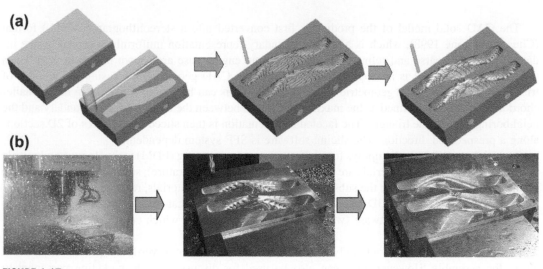

FIGURE 1.17

Cover die machining: (a) virtual and (b) CNC.

For example, the cover die of a mechanical part is machined from an 8 in. × 5.25 in. × 2 in. steel block, as shown in Figure 1.17a. The cutter location data files generated from virtual machining are post-processed into machine control data (MCD)—that is, G- and M-codes, for CNC machining, using post-processor UNCX01.P11 in Pro/MFG. In addition to volume milling and contour surface milling, drilling operations are conducted to create the waterlines. A 3-axis CNC mill, HAAS VF-series (HAAS Automation, Inc. 1996), is employed for fabricating the die for casting the mechanical part (Figure 1.17b).

1.5 EXAMPLE: SIMPLE AIRPLANE ENGINE

A single-piston, two-stroke, spark-ignition airplane engine (shown in Figure 1.5) is employed to illustrate the e-Design paradigm and tool environment. The cross-functional team is asked to develop a new model of the engine with a 30% increment in both maximum torque and horsepower at 1,215 rpm. The design of the new engine will be carried out at two interrelated levels: system and component. At the system level, the performance measure is the power output; at the component level, the structural integrity and manufacturing cost of each component are analyzed for improvement. Note that only a very brief discussion is provided in this introductory chapter. The computation and modeling details are discussed in later chapters and *Product Design Modeling using CAD/CAE*, a book in The Computer Aided Engineering Design Series.

1.5.1 SYSTEM-LEVEL DESIGN

Power is proportional to the rotational speed of the crankshaft (N), the swept volume (V_s), and the brake mean effective pressure (P_b) (Taylor 1985):

$$W_b = P_b \, V_s \, N. \tag{1.9}$$

The effective pressure P_b applied on top of the piston depends on, among other factors, the swept volume and the rotational speed of the crankshaft. The pressure is limited by the integrity of the engine structure.

Design variables at the system level include bore diameter (d46:0) and stroke, defined as the distance between the top face of the piston at the bottom and top dead-center positions. In the CAD model, the stroke is defined as the sum of the crank offset length (d6:6) and the connecting rod length (d0:10), as shown in Figure 1.18. To achieve the requirement for system performance, these three design variables are modified as listed in Table 1.1. The design variable values were calculated following theory and practice for internal combustion engines (Taylor 1985). Details of the computation can be found in Silva (2000).

The solid models of the entire engine are automatically updated and properly assembled using the parametric relations established earlier (refer to Figure 1.6b). The change causes P_b to increase from 140 to 180 lbs, so the peak load increases from 400 to 600 lbs. The load magnitude and path applied to the major load-carrying components, such as the connecting rod and crankshaft, are therefore altered. Results from motion analysis show that the system performs well kinematically. Reaction forces applied to the major load-carrying components are computed—for example, for the connecting rod shown in Figure 1.19. The change also affects manufacturing time for some components.

1.5.2 COMPONENT-LEVEL DESIGN

Structural performance is evaluated and redesigned to meet the requirements. In addition, virtual manufacturing is conducted for components with significant design changes. Build materials (volume) and manufacturing times constitute a significant portion of the product cost. In this section, the design of the connecting rod is presented to demonstrate the design decision-making method discussed.

Because of the increased load transmitted through the piston and the increased stroke length, the connecting rod can experience buckling failure during combustion. In addition, because changes in stroke length, stiffness, and mass vary, the natural frequency of the rod may be different. Moreover,

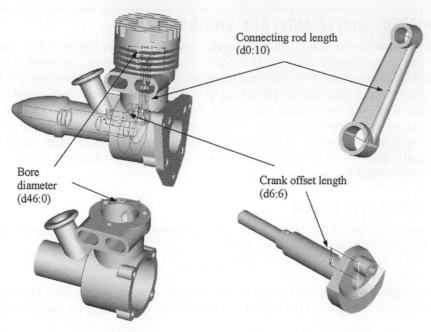

FIGURE 1.18

Engine assembly with design variables at the system level.

Table 1.1 Changes in Design Variables at the System Level

Design Variable	Current Value (in.)	New Value (in.)	Change (in.)	% Change
Bore diameter (d46:0)	1.416	1.6	0.164	11.6
Crank length (d6:6)	0.5833	0.72	0.1567	26.9
Connecting rod length (d0:10)	2.25	2.49	0.24	10.7

load is repeatedly applied to the connecting rod, potentially leading to fatigue failure. Structural FEA are conducted to evaluate performance. In addition, virtual manufacturing is carried out to determine the machining cost of the rod.

Because of the increment of the connecting rod length (d0:10) and the magnitude of the external load applied (see Figure 1.20), the rod's maximum von Mises stress increases from 13,600 to 18,850 psi and the buckling load factor decreases from 33 to 7. The first natural frequency is 1,515 Hz. The machining time estimated for the connecting rod is 13.2 minutes using hole-drilling and face-milling operations (shown earlier in Figure 1.9d).

1.5.3 DESIGN TRADE-OFF

The design trade-off method discussed in Section 1.3.5 is applied to the components, with significant changes resulting from the system-level design. Only the design trade-off conducted for the connecting rod is discussed.

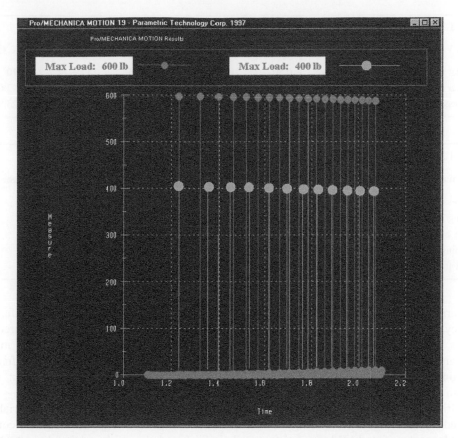

FIGURE 1.19

Dynamic load applied to the connecting rod.

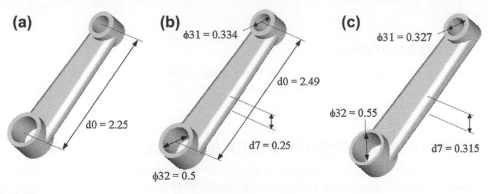

FIGURE 1.20

Engine connecting rod: (a) original design; (b) changes at the system level; (c) changes at the component level.

Table 1.2 Changes in Design Variables at the Component Level

Design Variable	Current Value (in.)	New Value (in.)	% Change
Diameter of the large hole (ϕ32)	0.50	0.55	10
Diameter of the small hole (ϕ31)	0.334	0.32728	−2.01
Thickness (d7)	0.25	0.31484	25.9

Table 1.3 Changes in Performance Measures at the Component Level

Performance Measure	Current Value	New Value	% Change
VM stress	18.9 ksi	10.5 ksi	−44.4
Buckling load factor	7.1	14.2	100
Volume	0.438813 in.^{3}	0.5488 in.^{3}	25.1
Machining time	13.2 min	13.2 min	0
Natural frequency	1515 Hz	1840 Hz	21.5

Performance measures for the connecting rod, including buckling load factor, fatigue life, natural frequency, volume, and machining costs (time), are brought together for design trade-off. Three design variables, ϕ32, ϕ31, and d7, are identified, as shown in Figure 1.20b. The objective is to minimize volume and manufacturing time subject to maximum allowable von Mises stress, operating frequency, and minimum allowable buckling load factor. The engine is designed to work at 21 kHz, and the minimum allowable buckling load factor for the connecting rod is assumed to be 10.

Sensitivity coefficients for performance and cost measures with respect to design variables are calculated (refer to Figure 1.11) using the finite difference method. Design trade-offs are conducted followed by a what-if study. When a satisfactory design is found, the solid model of the rod is updated for performance evaluation and virtual manufacturing. This process is repeated twice when all the requirements are met. The design change is summarized in Tables 1.2 and 1.3, which show that the machining time is maintained and a small volume increment is needed to achieve the required performance.

1.5.4 RAPID PROTOTYPING

When the design is finalized through virtual prototyping, rapid prototyping is used to fabricate a physical prototype of the engine, as shown in Figure 1.21. The prototype can be used for design verification as well as tolerance and assembly checking.

1.6 EXAMPLE: HIGH-MOBILITY MULTIPURPOSE WHEELED VEHICLE

The overall objective of the high-mobility multipurpose wheeled vehicle (HMMWV) design is to ensure that the vehicle's suspension is durable and reliable after accommodating an additional armor loading of 2,900 lb. A design scenario using a hierarchical product model (see Figure 1.10) that evolves during the design process is presented in this section.

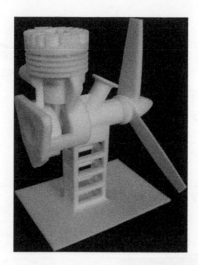

FIGURE 1.21

Physical prototypes of engine parts.

In the preliminary design stage, vehicle motion is simulated and design changes are performed to improve the vehicle's gross motion. At this stage, the dynamic behavior of the HMMWV's suspension is simulated and designed. The specific objectives of the preliminary design are to avoid the problem of metal-to-metal contact in the shock absorber due to added armor load, and to improve the driver's comfort by reducing vertical acceleration at the HMMWV driver's seat.

By modifying the spring constant to improve the HMMWV suspension design at the preliminary design stage, the load path generated in HMMWV dynamics simulation is affected in the suspension unit. In the detail design stage, the objective is to assess and redesign the durability, reliability, and structural performance of selected suspension components affected by the added armor load that result in changes in load path and load magnitude.

Note that only a very brief discussion is provided in this introductory chapter. The computation and modeling details are discussed in later chapters.

1.6.1 HIERARCHICAL PRODUCT MODEL

In this particular case, a hierarchical product model is employed to support the HMMWV's design. In all models, nonsuspension parts, such as instrument panel, seats, and lights, are not modeled. Important vehicle components, such as engine and transmission, are modeled using engineering parameters without depending on CAD representation. A low-fidelity CAD model consisting of 18 parts (Figure 1.22) is created using Pro/ENGINEER to support the preliminary design. This model has accurate joint definition and fairly accurate mass property, but less accurate geometry. The goal of the low-fidelity model is to support vehicle dynamic simulation. It is created using substantially less effort compared to that required for the detailed model.

The detailed product model, consisting of more than 200 parts and assemblies (Figure 1.23), is created to support the detail design of suspension components. The detailed model is derived from the preliminary model by (1) breaking an entity into more parts and assemblies (e.g., the gear hub

FIGURE 1.22

HMMWV CAD model for preliminary design.

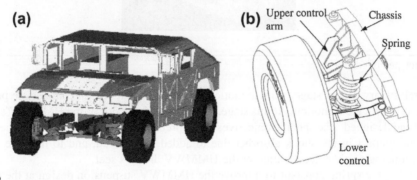

FIGURE 1.23

HMMWV CAD model for detail design.

assembly, shown in Figure 1.24) to simulate and design detailed parts, and (2) refining the geometry of mechanical components to support structural FEA (e.g., the lower control arm, shown in Figure 1.25).

1.6.2 PRELIMINARY DESIGN

The HMMWV is driven repeatedly on a virtual proving ground, as shown in Figure 1.26, with a constant speed of 20 MPH for a period of 23 seconds. A dynamic simulation model, shown in Figure 1.27, is first derived from the low-fidelity CAD solid model of the HMMWV (refer to Figure 1.22). A more in-depth discussion of the HMMWV vehicle dynamic model is provided in Chapter 3.

Using DADS, severe metal-to-metal contact is identified within the shock absorber, caused by the added armor load and rough driving conditions, as shown in Figure 1.28. The spring constant is adjusted to avoid any contact problems; it is increased in proportion to the mass increment of the added armor to maintain the vehicle's natural frequency. This design change not only eliminates the contact problem (see Figure 1.28) but also reduces the amplitude of vertical acceleration at the driver's seat, which improves driving comfort (see Figure 1.29). However, the change alters the load path in the components of the suspension subsystem—for example, the shock absorber force acting on the control arm increases about 75%, as shown in Figure 1.30.

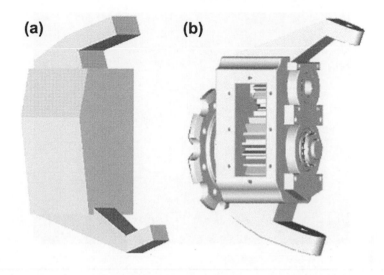

FIGURE 1.24

HMMWV gear hub assembly models: (a) preliminary and (b) detailed.

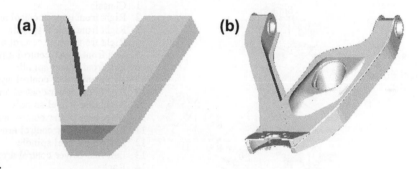

FIGURE 1.25

HMMWV lower control arm models: (a) preliminary and (b) detailed.

1.6.3 DETAIL DESIGN

Simulations are carried out for fatigue, vibration, and buckling of the lower control arm (Figure 1.30); reliability of gears in the gear hub assembly (refer to Figure 1.24b); the spring of the shock absorber (see Figure 1.23); and the bearings of the control arm (see Figure 1.30).

Using ANSYS, the first natural frequency of the lower control arm is obtained as 64 Hz, which is far away from vehicle vibration frequency, eliminating concern about resonance. The buckling load factor is analyzed using the peak load at time 10.05 seconds in the 23-second simulation period. The result shows that the control arm will not buckle even under the most severe load. Therefore, the current design is acceptable as far as buckling and resonance of the lower control arm are concerned.

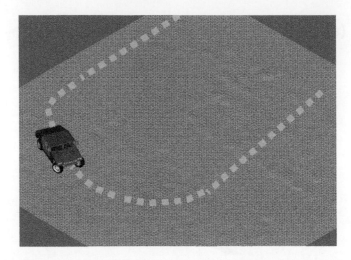

FIGURE 1.26

HMMWV dynamic simulation.

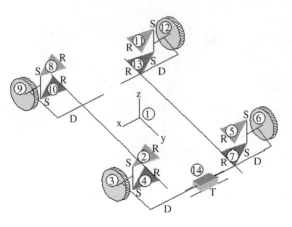

Body
1 Chassis
2 Right front upper control arm
3 Right front wheel spindle
4 Right front lower control arm
5 Left front upper control arm
6 Left front wheel spindle
7 Left front lower control arm
8 Right rear upper control arm
9 Right rear wheel spindle
10 Right rear lower control arm
11 Left rear upper control arm
12 Left rear wheel spindle
13 Left rear lower control arm
14 Rack

Joint types

R: Revolute joint
T: Translational joint
S: Spherical joint
D: Distance constraint

FIGURE 1.27

HMMWV dynamic model.

Results obtained from fatigue analyses show that fatigue life (crack initiation) of the lower control arm degrades significantly—for example, from 6.61E+09 to 1.79E+07 blocks (one block is 20 seconds) at critical areas (see Figure 1.31b)—because of the additional armor load and change of load path. Therefore, the design must be altered to improve control arm durability. Reliability of the bearing, gear, and spring at a 99% fatigue failure rate is 2.18E+07, 3.36E+06, and 1.27E+02 blocks, respectively. Note that the fatigue life of the spring at the required reliability is not desirable.

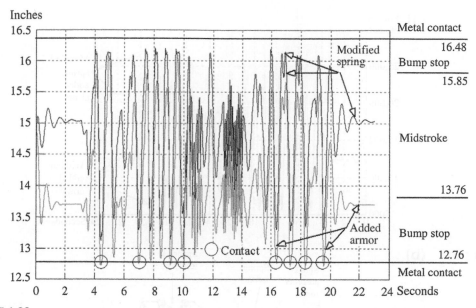

FIGURE 1.28

Shock absorber operation distance (in inches).

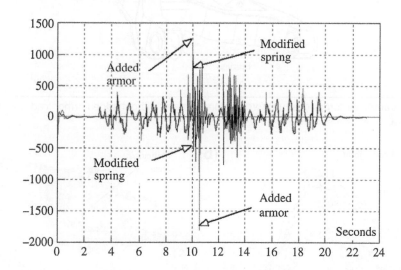

FIGURE 1.29

HMMWV driver seat vertical accelerations (in./sec^2).

1.6.4 DESIGN TRADE-OFF

Eleven design parameters, including geometric dimensions (d1 and d2 in Figure 1.32a), material property (cyclic strength coefficient K' of the lower control arm), and thickness of the control arm sheet metal (t1 to t7 in Figure 1.32b) are defined to support design modification.

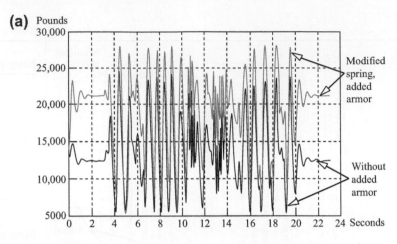

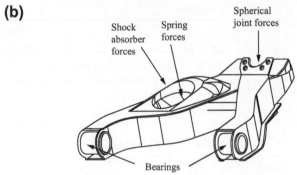

FIGURE 1.30

History of shock absorber forces (lbs): (a) force history with and without added armor load, (b) locations of force application.

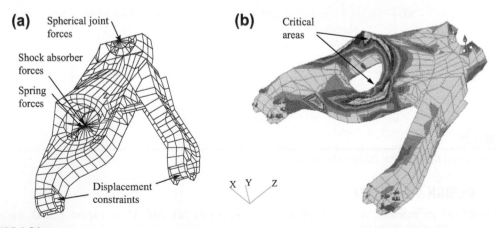

FIGURE 1.31

HMMWV lower control arm models: (a) finite element and (b) fatigue life prediction.

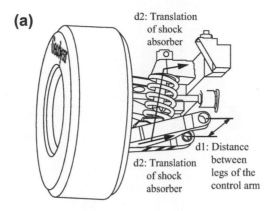

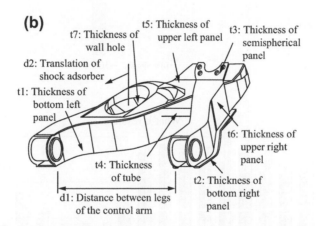

FIGURE 1.32

Design parameters defined for the control arm: (a) suspension geometric dimensions and (b) thickness dimensions.

A global design trade-off that involves changes in more than one component is conducted first. Geometric design parameters d1 and d2 are modified to reduce loads applied to the control arm, bearing, spring, and gears in the gear hub so that the durability and reliability of these components can be improved. Changes in d1 and d2 affect not only the lower control arm but also the upper control arm and the chassis frame. Sensitivity coefficients of loads at discretized time steps (a total of 10 selected time steps) with respect to parameters d1 and d2 are calculated using a finite difference method. Sensitivity coefficients can be displayed in bar charts (see Figure 1.33a) to guide design modifications. A what-if study is carried out with a design perturbation of 0.6 and 0.3 in. for d1 and d2, respectively, to obtain a reduction in loads. An example of the what-if results is shown in Figure 1.33b.

A local design trade-off that involves design parameters of a single component is carried out for the lower control arm. Thickness design parameters t1 to t7 and the material design parameter K' are modified to increase the control arm's fatigue life. Fatigue life at ten nodes of its finite element model

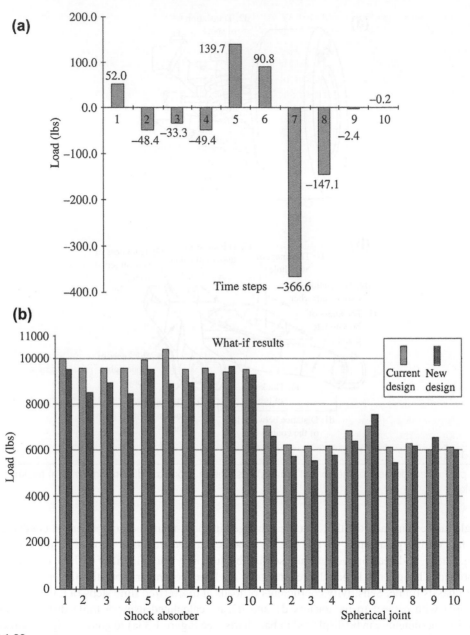

FIGURE 1.33

Sensitivity of load on the spherical joint of control arm w.r.t d2 at 10 time steps (a) design sensitivity display and (b) what-if study.

in the critical area is measured. Sensitivity coefficients of control arm fatigue life at these nodes with respect to the thickness and material parameters are calculated. A design trade-off method using a QP algorithm is employed because of the large number of design parameters and performance measures involved. An improved design obtained shows that with a 0.6% weight increment, fatigue life at the critical area increases about ten times: from 1.79 E+07 to 1.68 E+08 blocks.

A dynamic simulation is performed again with the detailed model and modified design to ensure that the metal contact problem, encountered in the preliminary design stage, is eliminated as a result of model refinement and design changes in the detail design stage. The global design trade-off reduces the load applied to the shock absorber spring. This reduction significantly increases the spring fatigue life to the desired level.

1.7 SUMMARY

In this chapter, the e-Design paradigm and software tool environment were discussed. The e-Design paradigm employs virtual prototyping for product design and rapid prototyping and computer numerical control (CNC) for fabricating physical prototypes of a design for design verification and functional tests. The e-Design paradigm offers three unique features:

- The VP technique, which simulates product performance, reliability, and manufacturing costs; and brings these measures to design.
- A systematic and quantitative method for design decision making for the parameterized product in solid model forms.
- RP and CNC for fabricating prototypes of the design that verify product design and bring marketing personnel and potential customers into the design loop.

The e-Design approach holds potential for shortening the overall product development cycle, improving product quality, and reducing product costs. With intensive knowledge of the product gained from simulations, better design decisions can be made, thereby overcoming what is known as the *design paradox*. With the advancement of computer simulations, more hardware tests can be replaced by them, reducing cost and shortening product development time. Manufacturing-related issues can be largely addressed through virtual manufacturing in early design stages. Moreover, manufacturing process planning conducted in virtual manufacturing streamlines the production process.

QUESTIONS AND EXERCISES

1.1. In this assignment, you are asked to search and review articles (such as in *Mechanical Engineering* magazine) that document successful stories in industry that involve employing the e-Design paradigm and/or employing CAD/CAE/CAM technology for product design.
- Briefly summarize the company's history and its main products.
- Briefly summarize the approach and process that the company adopted for product development in the past.
- Why must the company make changes? List a few factors.
- Which approach and process does the company currently employ?
- What is the impact of the changes to the company?
- In which journal, magazine, or website was the article published?

1.2. In this chapter we briefly discussed rapid prototyping technology and the Dimension 1200 sst machine. The sst uses fused deposition manufacturing technology for support of layer manufacturing. Search and review articles to understand the FDM technology and machines that employ such technology other than the Dimension series.

REFERENCES

Anderson, D.M., 1990. Design for Manufacturability: Optimizing Cost, Quality, and Time to Market. CIM Press.

Arora, J.S., 1989. Introduction to Optimal Design. McGraw-Hill.

Boothroyd, G., Dewhurst, P., Knight, W., 1994. Product Design for Manufacturing and Assembly. Marcel Dekker.

Chang, K.H., Choi, K.K., Wang, J., Tsai, C.S., Hardee, E., 1998. A multi-level product model for simulation-based design of mechanical systems. Concurrent Engineering Research and Application (CERA) Journal 6 (2), 131–144.

Chang, K.H., Yu, X., Choi, K.K., 1997. Shape design sensitivity analysis and optimization for structural durability. International Journal of Numerical Methods in Engineering 40, 1719–1743.

Chang, T.C., Wysk, R.A., Wang, H.P., 1998. Computer-Aided Manufacturing, second ed. Prentice Hall.

Chen, J.S., Pan, C., Wu, T.C., 1997. Large deformation analysis of rubber based on a reproducing kernel particle method. Computational Mechanics 19, 153–168.

Chua, C.K., Leong, K.F., 1998. Rapid Prototyping: Principles and Applications in Manufacturing. John Wiley.

Ertas, A., Jones, J.C., 1993. The Engineering Design Process. John Wiley.

Fang, X.D., Jawahir, I.S., 1996. A hybrid algorithm for predicting chip form/chip breakability in machining. International Journal of Machine Tools and Manufacture 36 (10), 1093–1107.

Gonzalez, A.J., Dankel, D.D., 1993. The Engineering of Knowledge-Based Systems, Theory and Practice. Prentice Hall.

Grindeanu, I., Choi, K.K., Chen, J.S., Chang, K.H., 1999. Design sensitivity analysis and optimization of hyperelastic structures using a meshless method. AIAA Journal 37 (8), 990–997.

HAAS Automation Inc., 1996. VF Series Operations Manual.

Hallquist, J.O., 2006. LS-DYNA3D Theory Manual. Livermore Software Technology Corp.

Haug, E.J., Choi, K.K., Komkov, V., 1986. Design Sensitivity Analysis of Structural Systems. Academic Press.

Haug, E.J., 1989. Computer-Aided Kinematics and Dynamics of Mechanical Systems, vol. I: Basic Methods. Allyn and Bacon.

Jacobs, P.F., 1994. StereoLithography and Other RP&M Technologies. ASME Press.

Kalpakjian, S., 1992. Manufacturing Engineering and Technology, second ed. Addison-Wesley.

Kane, T.R., Levinson, D.A., 1985. Dynamics: Theory and Applications. McGraw-Hill.

Lee, W., 1999. Principles of CAD/CAM/CAE Systems. Addison-Wesley Longman.

Madsen, H.O., Krenk, S., Lind, N.C., 1986. Methods of Structural Safety. Prentice Hall.

McMahon, C., Browne, J., 1998. CADCAM, second ed. Addison-Wesley.

Moës, N., Gravouil, A., Belytschko, T., 2002. Nonplanar 3D crack growth by the extended finite element and level sets. Part I: Mechanical model. International Journal for Numerical Methods in Engineering 53 (11), 2549–2568.

Numerical Integration Technologies, 1998. SYSNOISE 5.0.

Prasad, B., 1996. Concurrent Engineering Fundamentals, vols. I and II: Integrated Product and Process Organization. Prentice Hall.

Silva, J., 2000. Concurrent Design and Manufacturing for Mechanical Systems. MS thesis. University of Oklahoma.

Szabó, B., Babuška, I., 1991. Finite Element Analysis. John Wiley.

Taylor, C., 1985. The Thermal-Combustion Engine in Theory and Practice, vol. I: Thermodynamics, Fluid Flow, Performance, second ed. MIT Press.

Tsai, C.S., Chang, K.H., Wang, J., 1995. Integration infrastructure for a simulation-based design environment, Proceedings, Computers in Engineering Conference and the Engineering Data Symposium. ASME Design Theory and Methodology Conference.

Ullman, D.G., 1992. The Mechanical Design Process. McGraw-Hill.

Wu, Y.T., Wirsching, P.H., 1984. Advanced reliability method for fatigue analysis. Journal of Engineering Mechanics 110, 536–563.

Yu, X., Chang, K.H., Choi, K.K., 1998. Probabilistic structural durability prediction. AIAA Journal 36 (4), 628–637.

Yu, X., Choi, K.K., Chang, K.H., 1997. A mixed design approach for probabilistic structural durability. Journal of Structural Optimization 14, 81–90.

Zeid, I., 1991. CAD/CAM Theory and Practice. McGraw-Hill.

SOURCES

Adams: www.mscsoftware.com

ANSYS: www.ansys.com

CAMWorks: www.camworks.com

CatalystEX: www.dimensionprinting.com

LMS DADS: http://lsmintl.com

Dimension sst: www.stratasys.com

HAAS VF-Series: www.haascnc.com

HyperMesh: www.altairhyperworks.com

LS-DYNA: www.lstc.com

MSC/Nastran, MSC/Patran: www.mscsoftware.com

nCode: www.ncode.com

NESSUS: www.nessus.swri.org

Pro/ENGINEER, Pro/MECHANICA, Pro/MFG, Pro/SHEETMETAL, Pro/WELDING, etc.: www.ptc.com

SLA-7000, Sinterstation: www.3dsystems.com

SolidWorks Motion, SolidWorks Simulation: www.solidworks.com

SYSNOISE 5.0: www.lmsintl.com/SYSNOISE

DECISIONS IN ENGINEERING DESIGN

CHAPTER OUTLINE

Engineering design is the process by which engineers' intellect, creativity, and knowledge are translated into useful engineering products that satisfy particular functional requirements and meet engineering specifications while complying with all constraints. The traditional design approach has been one of deterministic problem-solving, typically involving efforts to meet functional requirements subject to various technical specifications and economic constraints, among others. In general, engineering design is a loosely structured, open-ended activity that includes problem definition, representation, performance evaluation, and decision making.

A number of approaches have been proposed to organize, guide, and facilitate the design process. The main objective is seeking a logical and rigorous means to aid in developing a satisfactory design, or one that is acceptable to the customer or user of the product. All approaches heavily involve decision making, which is integral to the engineering design process and is an important element in nearly all phases of design. In fact, it is fair to state that the center of all approaches is decision making.

In this chapter, we define decision making as the process of identifying and choosing alternatives from the set of possible alternatives. Alternatives are developed based on certain criteria and requirements. In the meantime, the preferences of the decision maker are incorporated to sufficiently reduce uncertainty about alternatives, which helps to achieve the desired goals and ensure a high-quality decision.

Various methods are commonly used to aid designers in decision making, such as a decision matrix (Voland 2004), a decision tree (Eatas and Jones 1996), quality function deployment (Akao 2004), and so forth. These methods are generally ad hoc and incorporate relatively high levels of subjective judgment. An additional set of methods address variability, quality, and uncertainty in the design process, such as the Taguchi method (Lochner and Matar 1990), Six Sigma (Park and Antony 2008), Design for X (Huang 1996), and so on. These tools are more analytical and are typically coupled to the processes used to produce products. Design theories also exist, such as Suh's axiomatic design (Suh 1990), which are less widely used but offer more rigorous analytical bases. Finally, certain other methods are used primarily in the fields of management science and economics, such as utility and game theory, which are being explored in the current research for feasibility and applicability to support decision making in engineering design.

In the late 1990s, the National Science Foundation (NSF) initiated a series of studies to determine research priorities in engineering design by examining industry and education needs and to formulate recommendations for the NSF's Engineering Design Program (NSF 1996). The NSF funds an online decision-based design open workshop to engage design theory researchers in a dialogue to establish a common foundation for research and educational endeavors. The NSF also sponsored Gordon Research Conferences in 1998 and 2000 on theoretical foundations to examine theories and techniques for decision making under conditions of risk, uncertainty, and conflicting human values. Such studies promoted research in design theory and led to the development of decision theories and decision-based design, including the application of utility theory and game theory to support design decision making. For those who are interested in pursuing research topics in design theory or decision-based design, please refer to Lewis et al. (2006) for excellent discussions.

This chapter is essentially a prelude to the broad subject of design theory and methods, in which we view engineering design as a decision-making process and recognize the substantial role that decision theory can play in design. Therefore, we start by discussing basic decision methods and theory in this chapter to provide readers with an overview and broad understanding of design decision making. With such an understanding, we extend the discussion in later chapters to more practical methods and tools, such as design optimization, that are widely employed by engineers for the support of design decision making. More specifically, this chapter aims to (1) introduce basic decision theory and methods that aid readers in applying them for general decision making, (2) provide basic concepts of utility and game theories that were explored in recently to aid engineering designs, and (3) use simple design examples to illustrate the concepts and methods for applying the theories to support practical engineering design problems.

2.1 **INTRODUCTION**

Design is a process involving constant decision making. In an engineering design context, the role of decision making can be defined in several ways. The decision process is influenced by sets of conditions or contexts; some are controllable, such as business context, and some are uncontrollable, such as market and economy conditions. The business context represents the long-term view of the company and is in general largely in the control of the company. Decisions such as capital investments, product lines and upgrades, and product marketing strategy are determined by the company. However, some aspects of business contexts, such as market share (which is influenced by competing products), are somewhat uncontrollable. Also, the state of the economy and market demands are not controlled by the company. Correctly assessing the context for making a decision is important because it dictates the level of effort and long-term impact. Decisions with long-term impacts often are irreversible after implementation; therefore, the decision maker must seriously analyze the context and impact of alternatives before arriving at a decision. A large number of short-term incremental decisions can, however, be made relatively risk free in general.

Whether the conditions are controllable or not, there is always uncertainty involved in decision making. Narrowing the focus to product design, the uncertainty largely comes from the inputs, such as the completeness of and variation in product requirements and constraints established by the customers. Closing with the customer is an iterative process, in which reconciling the customer's needs with the developer's design capabilities require collaboration and experience with the product. Decisions made in earlier stages must be re-evaluated from time to time and adjustments need to be made in response to factors or events of high uncertainty that were not predictable or controllable throughout the design process.

In addition to the uncertainty involved in decision making, the designer's preference in choosing one alternative over another plays an important role in design, especially in dealing with multi-objective design problems, in which a designer is juggling competing objectives. In some cases, design decisions are made by design groups of a product development team, in which decisions are made to maximize their respective objectives that could be mutually competing or even conflicting.

In general in product development, decisions are made at different levels under different kinds of scenarios. At a high level, decisions are made for scenarios such as team organization, product cost, work breakdown, and suppliers. At mid-level, a decision involves issues such as design requirements, material selection, subsystems and components, and the manufacturing process. At a low level, a designer determines design objectives, geometric shape and dimensions of the individual components, and so forth. There are many methods that aid in decision making. Some of these methods developed decades ago are more ad hoc and incorporated relatively high levels of subjective judgment, such as decision matrices, in which weighting factors that significantly impact the decision are assigned by the designer. When these methods are used, they are generally applied to support more significant project decision making at a higher level. Methods developed more recently involve rigorous theory and mathematical frameworks in decision making, such as using utility theory.

In this chapter, we intend to provide a basic introduction on both conventional methods and rigorous decision theory. We start by introducing conventional methods that are commonly employed and are easily found in design textbooks, including the decision matrix and decision tree, in Section 2.2. These methods are deterministic and rely on subjective judgment. Although we start with conventional methods, one of our focuses in this chapter is to discuss rigorous decision theory. In Section 2.3, we introduce decision theory, in which decision making is formulated and solved mathematically. In Sections 2.4 and 2.5, we discuss utility theory and game theory, respectively, which have been explored and adapted to support engineering design recently. In Section 2.6, we use simple design examples to illustrate the concept and steps in applying utility and game theories to solve simple design examples. These examples demonstrate the application of modern decision theories to support engineering design.

2.2 CONVENTIONAL METHODS

Numerous methods developed decades ago are commonly employed to aid decision making. We introduce representative and popular methods that in general support decisions at high and mid-levels, including the decision matrix and decision tree. Readers are referred elsewhere (Akao 2004; Lochner and Matar 1990; Park and Antony 2008) for other popular methods, such as quality function deployment, the Taguchi method, and Six Sigma.

2.2.1 DECISION MATRIX METHOD

Decision matrix techniques are used to define attributes, weigh them, and appropriately sum the weighted attributes to give a relative ranking among design alternatives. Note that, in practice, attributes are weighted as numeric figure based on a prescribed ranking system for individual design alternatives. In some contexts, such as design optimization, attributes are also called design objectives, which are to be maximized or minimized, or constraint functions, which must be kept within limits. In general, attributes are also referred to as design criteria or decision criteria.

A decision matrix consists of rows and columns that allow the evaluation of alternatives relative to various decision criteria. We use a material selection of airplane torque tubes shown in Figure 2.1 as an example to illustrate the method. The torque tubes are located in the front leading edge of the airplane wing, three on each side. The tubes are being redesigned to address concerns raised by the maintenance depot. The problems are twofold. First, the current magnesium tubes have poor corrosion resistance, requiring frequent repairs. Second, magnesium tubes are currently made by casting, which is extremely uneconomical when only a small quantity is required for the maintenance of the remaining aircraft fleet. Lead times were excessive and the cost was extremely high for acquiring the tubes. Therefore, it is desirable to manufacture the tubes using a machining process instead of casting. The

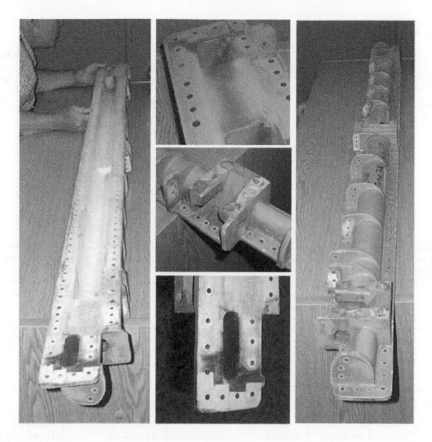

FIGURE 2.1

A sample magnesium torque tube with severe corrosion.

goal is to choose the best possible material to replace the magnesium tubes in order to enhance product reliability and reduce manufacturing lead time, among other design criteria.

Before creating a decision matrix, a set of material options (or alternatives) and their properties relevant to the decision criteria must be identified. In this case, the general properties and cost of five common materials for aerospace applications are collected and listed in Table 2.1.

A typical decision matrix is then constructed, in which a set of design criteria, including strength, weight, machinability, corrosion resistance, and material cost, are listed in the first row of the decision matrix, as shown in Table 2.2. Note that machining cost is implicitly incorporated into the machinability criterion. We adopt a rating system of 1–5, with 1 being the worst and 5 the best, to assign a rating factor (R_f) to each material under individual criteria. Individual rating factors are assigned by referring to the properties in Table 2.1 and scaling them roughly proportionally. As shown in Table 2.2, the tallied score, the so-called decision factor (D_f), indicates that titanium Ti-6A1-4V is the best choice. However, the cost of using titanium may be too high to justify the needs. According to Table 2.2, the next choice is magnesium, which unfortunately is the material we want to replace to begin with.

Table 2.1 Properties and Cost of Material Options

Material	Yield Strength (ksi)	Elongation (%)	Vickers Hardness (HV)	Density lb/in^3	Machinability Index[a]	Galvanic Corrosion[b]	Relative Cost per Weight[c]
Magnesium AZ31B-H24	26	12	50	0.064	50	−1.6	1.8
Titanium Ti-6A1-4V	140	8	360	0.16	510	0	18
Stainless steel 430	40	20	260	0.28	165	−0.5	1
Aluminum 7075-T6	73	11	150	0.101	100	−0.8	1.4
Aluminum 2024-T4	47	19	120	0.1	110	−0.8	1.4

[a]*SAE 1117 free-machining steel has an index of 100. Higher index numbers mean that some machining operations may be more expensive to perform compared to 1117 steel.*
[b]*The Galvanic Series of Metals can be used to determine the likelihood of a galvanic reaction, and galvanic corrosion or bimetallic corrosion, between two different metals in a seawater environment. Data were taken from Atlas Steel Technical Note No. 7 "Galvanic Corrosion," with 0.3 for graphite being the best and −1.6 for magnesium being the worst.*
[c]*Data were extracted from www.roymech.co.uk/Useful_Tables/Matter/Costs.html.*

Table 2.2 Decision Matrix for Material Selection of Airplane Torque Tubes

Material	Strength	Weight	Machinability	Corrosion Resistance	Material Cost	Score
Magnesium AZ31B-H24	1	5	5	1	2	14
Titanium Ti-6A1-4V	5	3	1	5	1	15
Stainless steel 430	2	1	2	3	5	13
Aluminum 7075-T6	3	2	1.5	2	2	10.5
Aluminum 2024-T4	2	2	1.5	2	2	9.5

Why does the decision matrix lead to the choice that is either too expensive or comes back to the one that we want to avoid at the first place? Certainly, one of the main issues is rating factors we assigned. For example, why is the strength factor of aluminum 7075-T6 equal to 3? Can it be 3.5? A slight adjustment in the rating factors could lead to different results. Another issue is that we assume that all design criteria have the same degree of importance. In general, they do not. In this example, corrosion and cost are considered to be the two most important criteria. In addition, weight is another important consideration for parts installed on aircrafts. Therefore, some sort of weighting scale is normally assigned to account for this variation. For this example, weighting factors 1–5 are assigned to

Table 2.3 Decision Matrix with Weighting Factors						
Material	Strength	Weight	Machinability	Corrosion Resistance	Material Cost	Score
Weighting Factor	(1)	(2)	(4)	(5)	(4)	
Magnesium AZ31B-H24	1	5	5	1	2	44
Titanium Ti-6A1-4V	5	3	1	5	1	44
Stainless steel 430	2	1	2	3	5	47
Aluminum 7075-T6	3	2	1.5	2	2	31
Aluminum 2024-T4	2	2	1.5	2	2	30

individual design criteria, with corrosion resistance being 5, machinability and material cost that contribute largely to the overall cost being 4, and weight being 2, as shown in Table 2.3.

Each weighting factor (W_f) is multiplied by the corresponding rating factor (R_f) for each material option, producing the decision factor (D_f):

$$D_f = W_f \times R_f \qquad (2.1)$$

These decision factors are then summed for each option and stored under the Score column. With the weighting factors, the best choice is revealed to be steel, although it is only marginally better than magnesium and titanium.

The decision matrix can be an important and useful tool to aid in design decision making, but the designer should be mindful that nothing takes the place of common sense and good judgment. For tools like the decision matrix to be viable, rating factors must be assigned as objectively as possible, and weighting factors must be determined to reflect the priority among design criteria. The method should help make a proper decision, rather than dictate the decision. The final decision should not be made solely on the results of the decision matrix. The value of a decision matrix is that it forces us to view the various alternatives in a careful and thoughtful manner. It is important to realize that these factors have a certain built-in uncertainty and subjectivity that may result in erroneous conclusions as to the best choice among options. A good designer will maintain a questioning attitude, always seeking further confirmation that the decision was correct as the design process evolves.

One major shortfall of the decision matrix method is that it does not take uncertainty into consideration. Uncertainty, such as the probability of the aircrafts exposed to highly corrosive operating environments, could impact the decision had the probability is known or reliably predicted in advance. Moreover, personal judgment is highly involved, especially in assigning rating factors R_f and weighting factors W_f. Different designers may come to a different design decision for the same design problem. Moreover, a designer's preference could play an important role in decision making, but it is not captured in any form.

2.2.2 DECISION TREE METHOD

The decision tree method is another way to evaluate different alternatives. This method is often used in evaluating business investment decisions, considering the outcome of possible future decisions, including the effect of uncertainties. The strength of the method is that it allows an evaluation of the benefits of present and future profits against the investment. It is a useful technique when a decision must be made in succession into the future.

A decision tree is represented graphically using four elements:

- Branches: straight lines (——) that terminate at each end with one of three types of nodes
- Decision nodes: depicted as squares (□)
- Event (or chance) nodes: depicted as circles (○)
- Payoff nodes: depicted as price tags (⬠)

Every decision tree is a collection of branches connected to each other by nodes. A tree is constructed from the left end starting with a decision node. Branches emanating from a decision node represent individual options. The right end of each branch must terminate in one of the three types of nodes: decision, event (or chance), or payoff. Note that the event is also called the state of nature, which is in general out of the control of a designer.

To illustrate the method, we consider the decision tree shown in Figure 2.2, which is concerned with deciding whether the original equipment manufacturer (OEM) in the earthmoving equipment

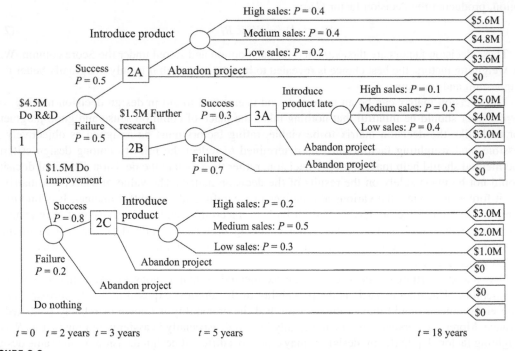

FIGURE 2.2

Decision tree for an OEM project.

business should carry out research and development for a new product or simply improve an existing product to capture a potential market niche in the next 15–20 years. The OEM has extensive experience in developing earthmoving equipment, such as the backhoe. However, the OEM has no direct experience with the new product, which involves contour crafting that prints a house in days instead of months while drastically reducing material and energy consumption (see, e.g., www.contourcrafting.org). With preliminary research completed, it was found that a $4.5 million investment is required up-front to develop a new product. On the other hand, a $1.5 million investment is needed for improving an existing product. The decision for the OEM to make is do nothing, improve a current product that supports contour crafting, or develop brand-new contour crafting equipment.

As stated above, in Figure 2.2, a decision point in the decision tree is depicted by a square, and circles designate chance events that are outside the control of the decision maker. The length of line between nodes in the decision tree is not necessarily scaled with time, although the tree does depict precedence relations.

The first decision is whether to proceed with the $4.5 million project for a new product, spend $1.5 million for improving an existing product, or not do anything. If the OEM decides to proceed with the $4.5 million investment, the director of engineering department estimates that at the end of the third year, there is a 50% chance of being ready to introduce the product. If the product is introduced to the market, it is estimated to have a life of 15 years. It is also estimated by the marketing department that the probabilities of high, medium, and low sales are 40%, 40%, and 20%, respectively. The payoffs of each are estimated as $5.6 million, $4.8 million, and $3.6 million, respectively.

If the development fails to create the new product within the given time and under budget, it is estimated that an additional $1.5 million would allow the team to complete the work in an additional 2 years. The probability of successfully completing the project at the end of 5 years is 30%. Due to the delay in bringing the product to market, it is estimated by the marketing department that the probabilities of high, medium, and low sales are 10%, 50%, and 40%, respectively, due to more severe competition in the market. The payoffs of each are estimated as $5.0 million, $4.0 million, and $3.0 million, respectively, which are less than those of the case of successful product in 3 years due to a shorter overall product lifespan. If a product cannot be completed in 5 years, the project will have to be abandoned because there will be too much competition.

The other option is to improve an existing product in 2 years and introduce the product to the market 1 year earlier. This improved product will support roughly 90% of the functions envisioned in the new product. The probability of successfully completing the project at the end of 2 years is higher (80%) because the OEM has adequate knowledge and technical expertise to do so. However, because the improved product does not provide complete functions as required for contour crafting, the sales are less optimistic. The high, medium, and low sales are estimated as 20%, 50%, and 30%, respectively, with their respective payoffs to be $3.0 million, $2.0 million, and $1.0 million. If the improved product is not ready in 2 years, the management believes there is no point to continue because the advantage of introducing the product early to the market is no longer viable.

A decision tree that incorporates the relevant information (the best information that the OEM is able to offer to the decision maker) is sketched in Figure 2.2. With the tree in place, a best possible decision can be made.

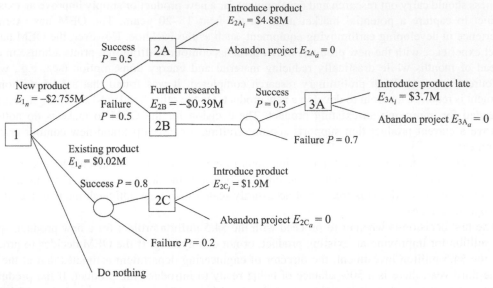

FIGURE 2.3

Solutions to the decision tree.

The best place to start is from the end of the branches and work backward, as illustrated in Figure 2.3. We define an expected value as in Eq. 2.2 to measure the profit or loss of individual decisions. The expected value for a decision is defined as

$$E = \sum_i P_i C_i - \sum_j I_j \tag{2.2}$$

where P_i and C_i are respectively the probability of individual event and its payoff, and I_j is the investment. The expected values are calculated at each decision point. For example, the expected values for the decision at 2A to introduce product or abandon project in 3 years are, respectively,

$$E_{2A_i} = 0.4(\$5.6M) + 0.4(\$4.8M) + 0.2(\$3.6M) = \$4.88M$$

$$E_{2A_a} = 0$$

The expected values at decision point 3A—that is, to introduce product late or abandon the project—are, respectively,

$$E_{3A_i} = 0.1(\$5.0M) + 0.5(\$4.0M) + 0.4(\$3.0M) = \$3.7M$$

$$E_{3A_a} = 0$$

Then, at the decision point 2B, the expected value for the decision of adding $1.5 million and introducing the product to the market at the end of the fifth year is

$$E_{2B} = [0.3(\$3.7) + 0.7(0)] - \$1.5M = -\$0.39M$$

Thus, carrying out the analysis for the delayed project back to 2B shows that to continue the project beyond that point results in a negative expected value. The proper decision, therefore, is to abandon the project if it is not successful in the first 3 years to cut the losses.

Now, we calculate the expected values for the option of improving an existing product. At the decision point 2C, the expected values of introducing the product or abandoning the project are, respectively,

$$E_{2C_i} = 0.2(\$3.0M) + 0.5(\$2.0M) + 0.3(\$1.0M) = \$1.9M$$

$$E_{2C_a} = 0$$

Finally, the expected value for the on-time project in introducing a new product at decision point 1 is

$$E_{1_n} = 0.5(\$4.88M) + 0.5(-\$0.39M) - \$5M = -\$2.755M$$

in which the large negative value is due to either the expected payoffs are too modest or the cost for the new product development is too great to be warranted by the payoff.

The expected value for a successful project in improving an existing product at decision point 1 is

$$E_{1_e} = 0.8(\$1.9M) - \$1.5M = \$0.02M$$

indicating a small margin of profit can be expected. Therefore, based on the estimates of payoffs, probabilities, and costs, the OEM should proceed with the option of improving an existing product, instead of developing a new product.

As illustrated in this example, decision tree is a useful and effective tool in support of business-type decision making. However, the expected values that the final decision is made upon are dependent on the estimates of payoffs, probabilities, and costs. These estimates are highly uncertain to say the least, although the engineering and marketing departments might have completed adequate research in coming up with these estimates. For example, if a competitor introduces a similar product to the market, the probabilities of sales and payoffs are mostly negatively impacted. Furthermore, the impact of later decisions (and the information generated by them) on earlier decisions is important in the effective use of decision trees. It is important for the decision tree to be regarded as a dynamic device, one in which new (and better) information is integrated as it becomes available.

Although we bring in probability into the decision tree and conduct so-called decision making under risk, the probabilities are assumed to be known in advance. This is not the case for many decision-making scenarios in practice. In the following sections, we introduce decision theory, which provides a rational and more rigorous framework for the support of decision making. We also introduce theory and methods that help decision making under numerous situations that are more general and practical.

2.3 BASICS OF DECISION THEORY

Decision theory provides a rational framework for choosing between alternative courses of action when the consequences resulting from this choice are imperfectly known (North 1968). Decision theory has been applied more to business management situations than to engineering design decisions.

The purpose of this section is to acquaint the reader with the basic concepts of decision theory, including elements of a decision, decision criteria, decision under uncertainty, and attitude toward risk. Later, in Sections 2.4 and 2.5, we discuss two major theories extended from the basics of decision theory—utility theory and game theory. These theories were explored for support of engineering design in recent years. We also present examples to illustrate the concepts and steps of applying these theories for simple and yet practical engineering design problems in Section 2.6.

In this section, we start with a simple example that involves a decision for buying a new or used car. The major criteria for making such a decision are cost and reliability. The car buyer wants to buy a reliable car with lesser cost that lasts for 10 years without major problems, such as engine overhaul or transmission repairs. To simplify our discussion, we assume the car buyer is interested only in one specific car model of new or used. Buying a new car will cost more at the beginning, but the probability of having major problems in 10 years is less. On the other hand, buying a used car may be cheaper up-front, but the probability of encountering major problems in the 10-year period is higher, which may increase the overall cost at the end. We use this example to illustrate terminologies and basic concepts to facilitate later discussions.

2.3.1 ELEMENTS OF A DECISION

There are three basic elements to any decision: course of actions, the state of nature, and the payoffs. Courses of action (a_i), also called acts or alternatives, are the choices available, which are under the control of decision maker—for example, the choice of buying a new or used car, as discussed. States of nature (ϕ_j), also called events, is an exhaustive list of possible future events. The decision maker has no direct control over the occurrence of a particular event. The car buyer does not know whether the car he or she buys will encounter major problems in the next 10 years. Payoff (v), also called outcome, quantifies effectiveness associated with specified combination of a course of action and state of nature. Payoff is often arranged in a summary table.

Returning to the car-buying example, let us say that the buyer did some research and was referred to online consumer reports, which provided basic information about the maintenance data and major repair costs of the specific car model in which the buyer is interested. Based on the information the buyer came up with, the estimates of the overall cost of new and used cars for a 10-year period, including purchase price, expenses (gas), regular maintenance (oil change, brake pads, air filters, and so on), and repair cost (including major repairs), are listed in Table 2.4. This is an example of a payoff table. In general, a payoff table can be generalized and formulated such as that in Table 2.5, where a_i is the ith alternate course of action, ϕ_j is the jth state of nature, and v is the payoff or value associated with specified combination of a course of action and state of nature.

Table 2.4 Cost of Different Options and Events (Payoff Table)

States of Nature (Events)	Courses of Action	
	New Car (N)	Used Car (U)
Without major problems (W)	$25,000	$15,000
With major problems (F)	$35,000	$30,000

Table 2.5 Generic Payoff Table

States of Nature (Events)	Courses of Action			
	a_1	a_2	...	a_m
ϕ_1	$v(a_1, \phi_1)$	$v(a_2, \phi_1)$	...	$v(a_m, \phi_1)$
ϕ_2	$v(a_1, \phi_2)$	$v(a_2, \phi_2)$	...	$v(a_m, \phi_2)$
...	...	...	...	...
ϕ_n	$v(a_1, \phi_n)$	$v(a_2, \phi_n)$	...	$v(a_m, \phi_n)$

In many cases, we can make better decisions if we establish probabilities for the states of nature. These probabilities may be based on historical data or subjective estimates, which are less desirable. They are often referred to as states of knowledge, which describe the degree of certainty that can be associated with the respective states of nature. The state of knowledge affects significantly how the decision is made, which is discussed in the following subsections.

2.3.2 DECISION-MAKING MODELS

Decision-making models usually are classified with respect to the state of knowledge. Depending on the state of knowledge, decisions are made under certainty, under risk, under uncertainty, and under conflict.

Decision under certainty implies that each action results in a known outcome that will occur with a probability of 100%. Action refers to a choice of alternatives. For example, a designer may choose one of the five materials listed in Table 2.1. The outcome of an action is assumed to be known. The design matrix method discussed in Section 2.2.1, which assumes definite rating factors and weighting factors, is a typical example of decision making under certainty.

Decision under risk assumes that each action results in known outcomes that will occur with a known probability, which is less than 100%. The decision tree discussed in Section 2.2.2 supports decision making with assigned probabilities for events that branch out at chance nodes. In the OEM example discussed in Section 2.2.2, the director of the engineering department estimated that, at the end of the third year, there was a 50% chance of the new product being ready to enter the market, if the OEM decided to proceed with the $4.5 million investment. Although such decision making is risky, the outcomes and probabilities are assumed to be known.

Decision under uncertainty implies that the probabilities that affect the outcomes usually are not known with much confidence. This presents a more realistic situation. Therefore, the decision maker should concentrate on achieving the best estimate of those probabilities. A formal theory of decision making that takes uncertainty into consideration is the utility theory, which will be discussed in Section 2.4.

Decision under conflict refers to a situation where decisions are made by more than one person or teams simultaneously or sequentially, and each decision maker or team is trying to maximize one's own objectives. This type of decision is supported by game theory, which will be briefly discussed in Section 2.5.

The feasibility of applying utility theory and game theory to support engineering design was explored and documented in details (e.g., Lewis et al. 2006). We use simple design examples to illustrate the concept in Section 2.6.

Before we move on to the utility theory and game theory, we introduce a few more basic concepts on the decision theory. We focus on two decision models: decision under risk and decision under uncertainty. We use the simple car-buying example to illustrate the decision models. In addition, we introduce utility functions that characterize the decision maker's attitude toward risk, which is essential to understanding the utility theory to be introduced later.

2.3.3 DECISION UNDER RISK

In this section, we revisit the decision under risk, in which the probabilities that affect the outcomes usually are assumed to be known. We introduce more rigorous treatment to the decision-making model, in which we revisit the car-buying example for illustration.

To facilitate our discussion, we assign N and U to the options of new and used car, respectively; and W and F to the events of car works well without major repairs and car fails due to major problems, respectively. The buyer also consulted with the dealership, and the historical data suggest that there is an 80% chance that a new car does not encounter major problems in 10 years. The historical data also suggest that the probability is 50% for a used car. At the time, this was the best possible estimate that the buyer was able to attain. Next, we construct a decision tree to aid the buyer in making a decision. The payoff table of Table 2.4 is expanded by incorporating the probabilities and expected values. In Table 2.6, the notation $P(W|N)$ stands for the probability of new car (N) that works (W) well in 10 years. Similarly, $P(F|U)$ stands for the probability of used car (U) that fails (F) in 10 years (i.e., encountering major problems), and so on.

Assuming these historical data are reliable, the payoff table and the probability estimates can be combined to arrive at the expected payoff of individual decisions. The expected payoff is also called the expected monetary value (EMV) in decision theory. The calculations are summarized as follows:

$$E(a_i) = \sum_j P(\phi_j) v(a_i, \phi_j) \tag{2.3}$$

where $E(a_i)$ is the EMV of event a_i.

Table 2.6 Cost of Different Options and Events (Payoff Table)

States of Nature (Events)	Courses of Action					
	New Car (N)			Used Car (U)		
	Probability	Payoff	Expected Value	Probability	Payoff	Expected Value
Without major problems (W)	$P(W|N) = 0.8$	$25,000	$20,000	$P(W|U) = 0.5$	$15,000	$7500
With major problems (F)	$P(F|N) = 0.2$	$35,000	$7000	$P(F|U) = 0.5$	$30,000	$15,000
			$27,000			$22,500

Hence, for the event of buying a new car, the expected payoff is

$$E(N) = \sum_j P(\phi_j)v(N, \phi_j) = P(W|N)v(N, W) + P(F|N)v(N, F)$$
$$= 0.8(\$25,000) + 0.2(\$35,000) = \$27,000$$

Similarly, for the event of buying a used car, the expected payoff is $E(U) = \$22,500$, as shown in Table 2.6. Therefore, based on the expected payoff, buying a used car presents a better option.

The car-buying example can be represented in a decision tree similar to that of Section 2.2.2. The decision tree for the car-buying example is shown in Figure 2.4.

As discussed in Section 2.2.2, the general approach to solving a decision tree is to move backward through the tree (from right to left) until we reach the originating decision node. We select a payoff node and move left to trace the branch to encounter the next node. If the next node is a chance node, we calculate the expected value (E) of all nodes connected immediately to the right of the encountered node by using Eq. 2.3. In this example, we calculate the expected values for the options of buying a new car $E(N)$ and buying a used car $E(U)$, respectively.

We enter these values to their respective chance nodes, and then move left to encounter a decision node—in this case, the originating decision node, as shown in Figure 2.4b. At the decision node, we select the branch that leads to the best value. In this case, $22,500$ is the best value, representing a lesser overall cost. At this point, we reach a decision of buying a used car.

One of the difficulties in using the decision tree method is coming up with the probabilities of the uncertain event occurring. In the car-buying example, how certain is the event that the probability of a used car without major problem is $P(W|U) = 50\%$? In this case, we do not have to agonize too much over the accuracy of the estimate because we can easily test to see if the decision to buy a used car is sensitive to the probability estimated. We first let the probability be $P(W|U) = x$. Then the

(a)   **(b)**

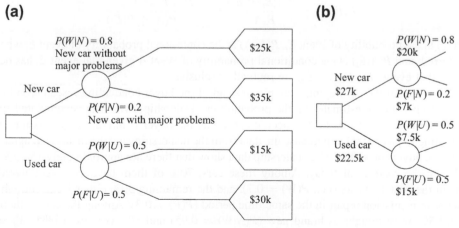

FIGURE 2.4

Decision tree for a car buyer: (a) starting tree and (b) solution tree.

probability of a used car having major problems is $P(F|U) = 1 - x$. The expected value of buying a used car is

$$E(U) = x(\$15k) + (1 - x)(\$30k) = \$30k - \$15k \ x$$

If we equate the two expected values of used and new car $E(U) = E(N) = \$27k$, we have $x = 0.2$. In other words, as long as the probability of a used car without major problems is greater than 20%, buying a used car is still a good choice.

2.3.4 DECISION UNDER UNCERTAINTY

In this section, we discuss decisions under uncertainty, in which the probabilities that affect the outcomes usually are not known with much confidence. This is the most general and practical decision model. The essence of this decision model is improving the confidence level of the knowledge of the states.

We introduce the Bayesian approach to decision making, in which the improved estimation of probabilities is achieved by using Bayes' theorem, which is a means of revising prior estimates of probabilities on the basis of new information. Certainly, it is logical, when faced with making a decision with uncertainty present, to try to remove the elements of uncertainty (or minimize their impact) by gathering more information about the nature of the events. The more knowledge we possess, the less the uncertainty. However, there usually are real limits of cost and time to achieve complete knowledge. The Bayesian approach is a very satisfactory compromise by which we combine additional knowledge with our initial estimation (prior probabilities) to form revised probabilities (posterior probabilities) that, in turn, provide us with a revised basis on which to make our decision.

Bayes' theorem was discussed in Chapter 5 of *Product Performance Evaluation Using CAD/CAE* (Chang 2013), one of the book modules of the Computer Aided Engineering Design series. Readers are encouraged to read that chapter to gain a basic understanding on the concepts of probability. Here, we simply state the theorem below:

$$P(E_i|A) = \frac{P(A|E_i)P(E_i)}{P(A)} = \frac{P(A|E_i)P(E_i)}{\sum_{i=1}^{m} P(A|E_i)P(E_i)} \tag{2.4}$$

where $P(E_i)$ is the probability of event E_i, $P(E_i|A)$ is the conditional probability of event E_i when event A has occurred, and $P(A|E_i)$ is the conditional probability of event A given that event E_i has occurred. We assume that events E_1, E_2, ..., E_m are mutually exclusive.

Now let us revisit the car-buying example. Instead of the buyer conducting research on his or her own to come up with the estimates on the probabilities, dealership owns data accumulated for many years through offering services to the specific car model the buyer is interested. We assume that the buyer is willing to pay \$1000 to purchase the data from the dealership, and the dealership agrees to sell the data to the buyer for \$1000. The dealership data show that there are approximately 150,000 cars of the model on the road as of today. Among these cars, 70% of them have not encountered major problems in the past 10 years (i.e., $P(W) = 0.7$), and the remaining 30% either went through engine overhauls or transmission repair in the same time period ($P(F) = 0.3$). Among the cars without major problems, 95% were bought as brand new ($P(N|W) = 0.95$) and 5% were used ($P(U|W) = 0.05$). Among the cars with major problems, 15% of them were bought as brand new car ($P(N|F) = 0.15$) and 85% were used ($P(U|F) = 0.85$). These data are illustrated in Figure 2.5 as a probability tree.

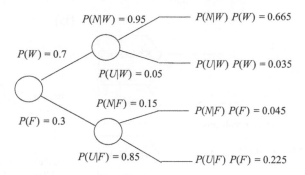

FIGURE 2.5

Probability tree for car failure data.

As illustrated in the probability tree, the probabilities of new car with and without major problems can be calculated using Eq. 2.4 as, respectively,

$$P(F|N) = \frac{P(N|F)P(F)}{P(N|W)P(W) + P(N|F)P(F)} = \frac{0.045}{0.665 + 0.045} = 0.0634$$

$$P(W|N) = \frac{P(N|W)P(W)}{P(N|W)P(W) + P(N|F)P(F)} = \frac{0.665}{0.665 + 0.045} = 0.937$$

and the probabilities of used car with and without major problem can be calculated using Eq. 2.4 as, respectively,

$$P(F|U) = \frac{P(U|F)P(F)}{P(U|W)P(W) + P(U|F)P(F)} = \frac{0.225}{0.225 + 0.035} = 0.865$$

$$P(W|U) = \frac{P(U|W)P(W)}{P(U|W)P(W) + P(U|F)P(F)} = \frac{0.035}{0.225 + 0.035} = 0.135$$

With the improved probability data, we revisit the decision tree shown in Figure 2.4. Following the probability data shown in Figure 2.5, the expected values of the chance nodes on top (buying new car) and bottom (buying used car), as shown in Figure 2.6, become

$$E(N) = 0.937(\$26k) + 0.0634(\$36k) = \$26.64k$$

and

$$E(U) = 0.135(\$16k) + 0.865(\$31k) = \$28.98k$$

in which the $1000 expense of data purchasing has been added to the payoffs. The results show that buying a new car is a better and economical option. Thus, by using posterior probability, we increased our knowledge of the probability that leads to a decision of buying a new car in contrast to the earlier decision.

Using posterior probability, we increased our knowledge of the probability that could help us arrive at a better decision. However, in our discussion so far, we assume the payoff of individual

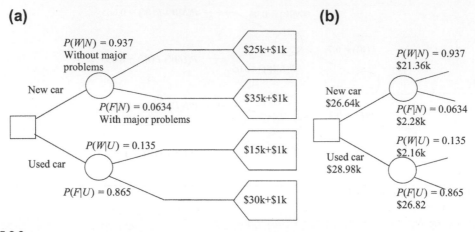

(a)

$P(W|N) = 0.937$
Without major problems

New car

$P(F|N) = 0.0634$
With major problems

$P(W|U) = 0.135$

Used car

$P(F|U) = 0.865$

$25k+$1k

$35k+$1k

$15k+$1k

$30k+$1k

(b)

$P(W|N) = 0.937$
$21.36k$

New car
$26.64k$

$P(F|N) = 0.0634$
$2.28k$

$P(W|U) = 0.135$
$2.16k$

Used car
$28.98k$

$P(F|U) = 0.865$
26.82

FIGURE 2.6

Decision tree for a car buyer with posterior probability: (a) starting tree and (b) solution tree.

events is precisely known. The question is: Does driving a new car without encountering major problem for 10 years really cost $25,000? Similarly, does a failed used car really cost $30,000? In reality, we do not have precise knowledge in these outcomes. In addition, a person might prefer buying a new car to getting a used car because he or she feels safer driving a new car. After all, the decision is being made by a human being. Personal preferences that could play an important role in decision making have not been taken into consideration. We address these issues next.

2.4 UTILITY THEORY

In Sections 2.3.3 and 2.3.4, we used an expected value rule to support decision making under risk and uncertainty. In many situations, it is highly desired that a decision maker's preference is incorporated. Utility theory offers rigorous mathematical framework, within which we are able to examine the preferences of individuals and incorporate them into decision making. In this section, we discuss utility theory from a design perspective. We discuss basics of the theory, including assumptions that lead to axioms of the theory, the utility functions that capture an individual's attitude toward risk, and the construction of utility functions for single and multiple attributes. In the context of design, the attributes are design criteria or design objectives to attain.

2.4.1 BASIC ASSUMPTIONS

There are four assumptions that lead to basic axioms of the utility theory. The first and perhaps the biggest assumption to be made is that any two possible outcomes resulting from a decision can be compared. Given any two possible outcomes, the decision maker can say which one he or she prefers. In some cases, the decision maker can say that they are equally desirable or undesirable. A reasonable extension of the existence of one's preference among outcomes is that the preference is transitive; that is, if one prefers A to B and B to C, then it follows that one prefers A to C.

FIGURE 2.7

Lottery diagram.

The second assumption, originated by von Neumann and Morgenstern (von Neumann and Morgenstern 1947), forms the core of modern utility theory. This assumption states that one can assign preferences in the same manner to lotteries involving prizes as one can to the prizes themselves. The lottery gives the probability of one getting prize A is p, and the probability that one gets prize B is $1 - p$. Such a lottery is donated as $(p, A; 1 - p, B)$, as represented in Figure 2.7, in which p is between 0 and 1. In utility theory, uncertainty is modeled through lotteries.

Now suppose one is asked to state his or her preferences for prize A, prize B, and a lottery of the above type $(p, A; 1 - p, B)$. Let us assume one prefers prize A to prize B. Then, based on von Neumann and Morgenstern, one would prefer prize A to the lottery $(p, A; 1 - p, B)$ because there is a probability $1 - p$ that one would be getting the inferior prize B of the lottery. One would also prefer the lottery $(p, A; 1 - p, B)$ to prize B for all probabilities p between 0 and 1. In other words, one would rather have the preferred prize A than the lottery, and one would rather have the lottery than the inferior prize B. Furthermore, it seems logical that, given a choice between two lotteries involving prizes A and B, one would choose the lottery with the higher probability of getting the preferred prize A. That is, one prefers lottery $(p, A; 1 - p, B)$ to $(p', A; 1 - p', B)$ if and only if p is greater than p'.

The third assumption is that one's preferences are not affected by the way in which the uncertainty is resolved, bit by bit, or all at once. To illustrate the assumption, let us consider a compound lottery—a lottery in which at least one of the prizes is not an outcome but another lottery among outcomes. For example, consider the lottery $(p, A; 1 - p, (p', B; 1 - p', C))$, as depicted in Figure 2.8a. According to the third assumption, one can decompose a compound lottery by multiplying the probability of the lottery prize in the first lottery by the probabilities of individual prizes in the second lottery. One should be indifferent between $(p, A; 1 - p, (p', B; 1 - p', C))$ and $(p, A; (1 - p)p', B; (1 - p)(1 - p'), C)$, as depicted in Figure 2.8b.

The fourth assumption is continuity. Consider three prizes: A, B, and C. One prefers A to C, and C to B. We shall assert that there must exist some probability p so that one is indifferent to receiving prize C or the lottery $(p, A; 1 - p, B)$ between A and B. C is called the certain equivalent of the lottery $(p, A;$

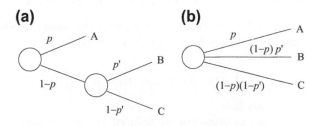

FIGURE 2.8

Lottery diagrams: (a) compound lottery and (b) equivalent simple lottery.

$1 - p$, B). In other words, if prize A is preferred to prize C and C is preferred to prize B, for some p between 0 and 1, there exists a lottery $(p, A; 1 - p, B)$ such that one is indifferent between this lottery and prize C.

2.4.2 UTILITY AXIOMS

We now summarize the assumptions we have made into the following axioms. We have prizes or outcomes A, B, and C from a decision. We use the following notations:

$\succ$ means "is preferred to," for example, A $\succ$ B means A is preferred to B.

$\sim$ means "is indifferent to," for example, A $\sim$ B means the decision maker is indifferent between A and B.

There are six axioms that serve the basis of the utility theory. They are orderability, transitivity, continuity, substitutability, monotonicity, and decomposability, which are defined as follows.

Orderability: Given any two prizes or outcomes, a rational person prefers one of them, else the two as equally preferable. Mathematically, the axiom is written as

$$(A \succ B) \vee (B \succ A) \vee (A \sim B) \tag{2.5}$$

in which "$\vee$" means "or." Equations (2.5) reads A is preferred to B, or B is preferred to A, or A and B are indifferent.

Transitivity: Preferences can be established between prizes and lotteries in an unambiguous fashion. The preferences are transitive; that is, given any three prizes or outcomes A, B, and C, if one prefers A to B and prefers B to C, one must prefer A to C. Mathematically, we have

$$(A \succ B) \wedge (B \succ C) \Rightarrow (A \succ C) \tag{2.6}$$

in which "$\wedge$" means "and," and "$\Rightarrow$" means "implies."

Continuity: If A $\succ$ C $\succ$ B, there exists a real number p with $0 < p < 1$ such that C $\sim (p, A; 1 - p, B)$. That is, it makes no difference to the decision maker whether C or the lottery $(p, A; 1 - p, B)$ is offered to him or her as a prize. Mathematically, we have

$$(A \succ C \succ B) \Rightarrow \exists \ p \ \in (0, 1) | (p, \ A; \ 1 - p, \ B) \sim C \tag{2.7}$$

in which "$\exists$" means "there exists" and "$|$" means "such that."

Substitutability: If one is indifferent between two lotteries, A and B, then there is a more complex lottery in which A can be substituted with B. Mathematically, we have

$$(A \sim B) \Rightarrow (p, \ A; \ (1 - p), \ C) \sim (p, \ B; \ (1 - p), \ C) \tag{2.8}$$

Monotonicity: If one prefers A to B, then one must prefer the lottery in which A occurs with a higher probability; in other words, if A $\succ$ B, then $(p, A; 1 - p, B) \succ (p', A; 1 - p', B)$ if and only if $p > p'$. Mathematically, we have

$$(A \succ B) \Rightarrow (p > p' \Leftrightarrow (p, \ A; \ 1 - p, \ B) \succ (p', \ A; \ 1 - p', \ B)) \tag{2.9}$$

in which "$\Leftrightarrow$" means "if and only if."

Decomposability: Compound lotteries can be reduced to simpler lotteries using the laws of probability; that is,

$$(p, \ A; \ 1 - p, \ (p', \ B; \ 1 - p', \ C)) \sim (p, \ A; \ (1 - p)p', \ B; \ (1 - p) \ (1 - p'), \ C) \tag{2.10}$$

2.4.3 UTILITY FUNCTIONS

If a decision maker obeys the axioms of the utility theory, there is a concise mathematical representation possible for preferences: a utility function $u(\cdot)$ that assigns a number to each lottery or prize. The utility function has the following properties:

$$u(A) > u(B) \Leftrightarrow A \succ B; \quad \text{and} \quad u(A) = u(B) \Leftrightarrow A \sim B \tag{2.11}$$

and

$$C \sim (p, \ A; \ 1-p, \ B) \Rightarrow u(C) = p \ u(A) + (1-p) \ u(B) \tag{2.12}$$

which implies that the utility of a lottery is the mathematical expectation of the utility of the prizes. It is this "expected value" property that makes a utility function useful because it allows complicated lotteries to be evaluated quite easily.

It is important to realize that all the utility function does is offering a means of consistently describing the decision maker's preferences through a scale of real numbers, provided that these preferences are consistent with the first four previously mentioned assumptions. The utility function is no more than a means to logical deduction based on given preferences. The preferences come first and the utility function is only a convenient means of describing them.

2.4.4 ATTITUDE TOWARD RISK

In Section 2.4.3, we assume a decision maker makes the decisions in the presence of uncertainty by maximizing its expected utility. In many situations, this rule does not adequately model the choices most people actually make. One famous example is the St. Petersburg lottery or St. Petersburg paradox, which is related to probability and decision theory in economics.

The St. Petersburg paradox, first published by Daniel Bernoulli in 1738 (Sommer 1954), is a situation where a naive decision criterion that takes only the expected value into account predicts a course of action that presumably no actual person would be willing to take. The paradox is illustrated as follows. A casino offers a game of chance for a single player in which a fair coin is tossed at each stage. The pot starts at \$1 and is doubled every time a head appears. The first time a tail appears, the game ends and the player wins whatever is in the pot. Thus, the player wins \$1 if a tail appears on the first toss, \$2 if a head appears on the first toss and a tail on the second, \$4 if a head appears on the first two tosses and a tail on the third, \$8 if a head appears on the first three tosses and a tail on the fourth, and so on. In short, the player wins \$$2^n$, where n heads are tossed before the first tail appears. What would be a fair price to pay the casino for entering the game?

To answer this, we need to consider what would be the average payoff. With probability 1/2, the player wins \$1; with probability 1/4, the player wins \$2; with probability 1/8, the player wins \$4, with probability $1/(2^n)$, the player wins \$$2^{n-1}$. The expected value is thus

$$E = \frac{1}{2} \times \$1 + \frac{1}{4} \times \$2 + \frac{1}{8} \times \$4 + \cdots \frac{1}{2^n} \times \$2^{n-1} + \cdots = \sum_{n=1}^{\infty} \frac{1}{2^n} \times \$2^{n-1} = \sum_{n=1}^{\infty} \frac{1}{2} = \infty$$

Assuming the game can continue as long as the coin toss results in heads and in particular that the casino has unlimited resources, this sum grows without bound and so the expected win for repeated play is an infinite amount of money. Considering nothing but the expectation value of the net change in one's monetary wealth, one should therefore play the game at any price if offered the opportunity. Contrary to this outcome, very few people are willing to pay large amounts of money to play this game. In fact, few of us would pay even $25 to enter such a game (Martin 2004). A hypothesized reason is that people perceive the risk associated with the game and consequently alter their behavior.

Bernoulli formalized this discrepancy between expected value and the behavior of individuals in terms of utility as the expected utility hypothesis: individuals make decisions with respect to investments in order to maximize expected utility (Sommer 1954). The expected utility hypothesis is a description of human behavior.

In general, utility is the measure of satisfaction or value that the decision maker associates with each outcome. In practice, very often we establish a relationship between monetary outcomes and their utility, which provides the basis for formulating a maximum expected utility rule for decision making. This relationship is created in a form of utility function $u(\$)$ that assigns a numerical value of utility between 0 (least preferred) and 1 (most preferred) for each monetary outcome.

We use the following example for illustration. A person named Jeff is deciding to gamble in a casino. Jeff is given a $200 as a welcome gift. If Jeff chooses to play a game, in which there is 20% probability of winning $5000 and 80% of losing $1000, he will have to first pay the $200 gift back to the casino in order to play the game. If he chooses not to play the game, he keeps the $200.

A decision tree shown in Figure 2.9a depicted the situation, in which Jeff is choosing between Option A for not playing the game (walk away with $200) and Option B for a chance to win $5000 and risk losing $1000. The expected values for options A and B are, respectively,

$$E(A) = \$200, \text{ and}$$

$$E(B) = 0.2(\$5000) + 0.8(-\$1000) = \$200$$

Thus, an expected value decision maker would be indifferent regarding A and B. In fact, more people may prefer Option A than B since A guarantees a $200 gift; although B offers a chance of winning $5000, the risk of losing $1000 is too high to justify it.

Certainly, a higher probability of winning (p) or a different dollar amount of the welcome gift, designated as x, may alter Jeff's decision. If we generalize the decision illustrated in Figure 2.9a by not specifying a numerical value for p or x, the modified decision tree is shown in Figure 2.9b.

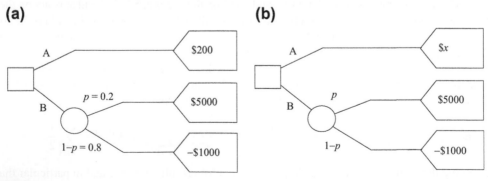

FIGURE 2.9

Decision tree for (a) outcomes A and B and (b) modified tree with p and x unspecified.

Now, if for a given set of values for x and p, we conclude that options A and B are equally attractive, implying that their utilities are equal (i.e., $u(A) = u(B)$) or in terms of p and x, we have

$$u(\$x) = p \ u(\$5000) + (1 - p) \ u(-\$1000) \tag{2.13}$$

To construct the utility function for this problem, we examine a series of at least three decision problems. We consider a different value of x and ask what value of p is needed to make options A and B equally attractive—that is, $u(A) = u(B)$.

If $x = \$5000$, regardless of an individual's attitude toward risk, the only rational response is $p = 1$. That is, there is no gain or loss to either play or not play the game. It is then logical to assign utility 1 to $u(\$5000)$ as the most preferred outcome. On the other hand, if $x = -\$1000$, regardless of an individual's attitude toward risk, the only rational response is $p = 0$. That is, there is no gain or loss to either play or not play the game. It is then logical to assign utility 0 to $u(-\$1000)$ as the least preferred outcome.

Now, let us say the welcome gift given to Jeff is $200. The question is how high the wining probability has to go in order to attract Jeff to play the game. Certainly, this percentage is different for different individuals because an individual's attitude toward risk is different. Let us say that when the winning percentage goes up to 40%, Jeff is indifferent between the two options; that is, he can go either way. To Jeff, the preference to the two options is identical. Therefore, $u(\$200) = 0.4$. We may go over a similar exercise to obtain more utilities for different x values. Eventually, a utility function that represents Jeff's attitude toward risk can be constructed.

Instead of going over more exercises, we plot the dollar amounts and utilities of the three set of values as three squares on graph formed by $ and $u(\$)$ as abscissa and ordinate, respectively, as shown in Figure 2.10. If we assume the utility function is monotonic, a curve that passes through the three points shown in Figure 2.10 is defined as the utility function, which is concave. A concave utility function reflects the risk-averse (or risk-avoiding) nature of the decision maker, implying that the decision maker is more conservative. Note that the assumption of using a monotonic function is logical because a conservative decision maker will most likely stay conservative under circumstances, in

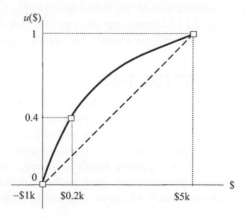

FIGURE 2.10

Utility function constructed for the example problem.

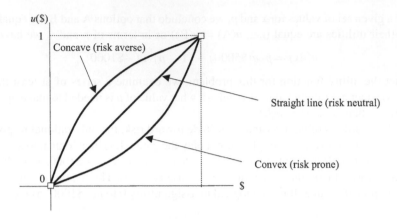

FIGURE 2.11

General form of risk-averse, risk-neutral, and risk-prone utility functions.

which the utility curve that represents decision maker's attitude toward risk is always above the straight line, representing that of a risk-neutral decision maker. It takes a larger utility value of $u = 0.4$ for Jeff, a conservative decision maker, than a risk-neutral person ($u = 0.2$) to enter the game.

Once a utility function is constructed, like that in Figure 2.10, we are able to predict that when Jeff is offered a welcome gift of, for example $2000, the casino must increase the probability of winning $5000 to much higher than 50% in order to attract him to play the game. That is, for a given $x, we can find p using the utility function, and vice versa. If the utility function accurately captures Jeff's attitude toward risk, Jeff may hire an agent or implement computer software to make a decision for him.

We mentioned that the concave utility function shown in Figure 2.10 reflects the risk-averse nature of the decision maker. A utility in the form of a straight line, as shown in Figure 2.11, indicates that the utility of any outcome is proportional to the dollar value of that outcome, which reflects those who make decisions following the expected value rule. We called these people risk-neutral decision makers. The convex curve shown in Figure 2.11 indicates the attitude of a risk-prone decision maker, who would choose a larger but uncertain benefit over a small, but certain, benefit.

2.4.5 CONSTRUCTION OF UTILITY FUNCTIONS

As discussed above, one way to construct a utility function is to acquire client's (e.g., Jeff in the example of playing casino game) responses to a series of questions. In many cases, this approach may not be practical due to numerous factors, such as a client's availability. A more common approach to construct utility functions, which is not data intensive, is to build a mathematical model for the utility function by prescribing the parametric in a generic family of curves, which are monotonic. Many functions reveal the characteristics of monotonicity, such as the exponential function $y = e^x - 1$, $y = 1 - 2^{-x}$, and so forth. One of such family of curves commonly adopted is

$$u(s) = \frac{1 - e^{-rs}}{1 - e^{-r}} \tag{2.14}$$

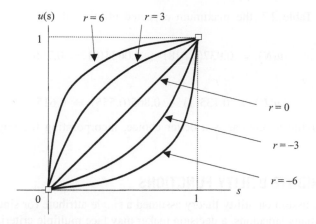

FIGURE 2.12

Mathematical model for the family of utility functions.

where s is a normalized form of the outcome x, defined as

$$s = \frac{x - x_{worse}}{x_{best} - x_{worse}} \tag{2.15}$$

where x_{best} and x_{worst} are the most preferred and the least preferred outcomes, respectively. When $x = x_{worst}$, $s = 0$; and $x = x_{best}$, $s = 1$. Also, when $r = 0$, $u(s) = s$, which is a straight line connecting $(0,0)$ and $(1,1)$, as shown in Figure 2.12, representing risk neutral. Note that $u(s) = s$ is obtained by applying L'Hôpital's rule, which we learned in Calculus. When $r > 0$, the utility function is concave, representing risk-averse behavior; when $r < 0$, the utility function is convex, representing risk-prone behavior.

Next, we apply the decision-making rule based on the maximum expected utility to the car-buying example. Recall the decision tree shown in Figure 2.6a. First, we identify the best and worst outcomes as $x_{best} = \$16,000$ and $x_{worst} = \$36,000$, respectively. From Eq. 2.15, we calculate the normalized parameter s corresponding to each monetary outcome, as shown in Table 2.7. We assume that the utility function defined in Eq. 2.14 with $r = 3$ accurately reflects the attitude toward risk of the buyer, indicating the car buyer is risk averse in spending money, like most of us. The utilities of individual outcomes can then be calculated using Eq. 2.14 and they are shown in Table 2.7. Using the utility

Table 2.7 The Normalized Parameter and Utilities		
x	s	u
$16,000	1	1.000
$26,000	0.5	0.818
$31,000	0.25	0.555
$36,000	0	0

values calculated in Table 2.7, the maximum expected utilities of buying new and used car are, respectively,

$$u(N) = 0.937(0.818) + 0.0634(0) = 0.766$$

and

$$u(U) = 0.135(1) + 0.865(0.555) = 0.615$$

which shows that buying a new car is a better choice, incorporating the buyer's attitude toward risk.

2.4.6 MULTIATTRIBUTE UTILITY FUNCTIONS

To this point, our discussion on utility theory assumed a single attribute (or single design criteria, or single objective). In many situations, a decision maker may face multiple criteria. For example, a car buyer wants a car with a long expected lifespan and a low price. In general, expensive cars last longer, implying that the criteria may be in conflict for some cases. Moreover, lifespan and purchase price are different measures and are difficult to compare directly.

In general, utility functions can be constructed to convert numerical attribute scales to utility unit scales, which allow direct comparison of diverse measures. When decision making involves a single attribute, the utility function constructed for the attribute is called a single attribute utility (SAU) function—for example, the function defined in Eq. 2.14. In this section, we introduce multiple attribute utility (MAU) functions and a structured methodology designed to handle the trade-offs among multiple objectives.

The mathematical combination of the SAU functions through certain scaling techniques yields a MAU function, which provides a utility function for the overall design with all attributes considered simultaneously. The scaling techniques reflect the decision maker's preferences on the attributes. For the formulation of the MAU function, additive and multiplicative formulations are commonly considered.

2.4.6.1 Additive MAU functions

The main advantage of the additive utility function is its relative simplicity. The additive form, defined by Eq. 2.16, allows for no preference interactions among attributes. Thus, the change in utility caused by a problem in one attribute does not depend on whether there are any problems in other attributes.

For a set of outcomes $x_1, x_2, \ldots, x_m$ on their respective m attributes, its combined utility using additive form is computed as

$$u(x_1, x_2, \ldots, x_m) = k_1 u_1(x_1) + k_2 u_2(x_2) + \cdots k_m u_m(x_m) = \sum_{i=1}^{m} k_i u_i(x_i) \qquad (2.16)$$

where $k_i > 0$ is the ith scaling factor and $\sum_{i=1}^{m} k_i = 1$, and u_i is the ith utility function. Note that k_i specifies the willingness of the decision maker to make trade-offs between different attributes. Recall that each utility function is normalized such that $u_i \in [0,1]$. Because the sum of all the scaling constants is 1, the value of MAU function defined in Eq. 2.16 is also between 0 and 1; that is, $u(x_1, x_2, \ldots, x_m) \in [0,1]$.

Table 2.8 Prices and Lifespans on the Three Candidate Cars

Attributes	Courses of Action		
	Car A	Car B	Car C
Price	$17,000	$10,000	$8000
Lifespan (years)	12	9	6

Table 2.9 Utility Values Set for the Two Attributes on the Three Candidate Cars

Attributes	Courses of Action		
	Car A	Car B	Car C
u_P	0	0.778	1
u_L	1	0.5	0

For example, a car buyer wants to buy a car with a long expected lifespan and a low price. The buyer narrowed down his or her choices to three alternatives: car A is a relatively expensive sedan with a reputation for longevity, car B is known for its reliability, and car C is a relatively inexpensive domestic automobile. The buyer has done some research and evaluated these three cars on both attributes, as shown in Table 2.8.

We set u_P and u_L for utilities of price and lifespan, respectively. Based on Table 2.8, we set $u_P(C) = u_P(\$8000) = 1$ for car C; $u_P(A) = u_P(\$17,000) = 0$ for car A. We assume the buyer is risk neutral on both price and lifespan; that is, $r = 0$ in the utility function defined in Eq. 2.14; that is, $u(s) = s$. Hence, $u_P(B) = u_P(\$10,000) = 0.778$ for car B. Similarly, we have $u_L(A) = u_L(12) = 1$; $u_L(C) = u_L(6) = 0$; hence, $u_L(B) = u_L(9) = 0.5$, as summarized in Table 2.9.

We assume that price is a lot more important factor than the lifespan of the car; therefore, we assign $k_P = 0.75$ and $k_L = 0.25$. Then, the utilities for the three alternatives are, respectively,

$$u(A) = k_P \ u_P(A) + k_L \ u_L(A) = 0.75 \ (0) + 0.25 \ (1) = 0.25$$

$$u(B) = k_P \ u_P(B) + k_L \ u_L(B) = 0.75 \ (0.78) + 0.25 \ (0.5) = 0.709$$

$$u(C) = k_P \ u_P(C) + k_L \ u_L(C) = 0.75 \ (1) + 0.25 \ (0) = 0.75$$

Example 2.1 offers more detailed illustration in the following section.

2.4.6.2 Multiplicative MAU functions

The other commonly employed MAU functions are multiplicative, which support preference interactions among attributes. A multiplicative MAU function is defined in Eq. 2.17:

$$u(x_1, x_2, \ldots, x_m) = \frac{1}{K} \left\{ \prod_{i=1}^{m} [1 + K k_i u_i(x_i)] - 1 \right\} \tag{2.17}$$

EXAMPLE 2.1

A different car buyer wants to buy a car with a long expected lifespan and a low price and has narrowed his or her choices down to the same three alternatives cars A, B, and C, with price and lifespan information listed in Table 2.8. The car buyer is risk averse in purchase price and is neutral in lifespan. We assume a utility function defined in Eq. 2.14 with $r = 3$, which characterizes the buyer's attitude toward risk. Calculate the combined utilities for the three alternatives using the additive form. We assume that price is equally as important as the lifespan of the car; therefore, $k_P = 0.5$ and $k_L = 0.5$.

Solutions

Because the buyer is risk neutral on lifespan, we have $u_L(A) = u_L(12) = 1$; $u_L(C) = u_L(6) = 0$; hence, $u_L(B) = u_L(9) = 0.5$. The buyer is risk averse with $r = 3$. From Eq. 2.15, we have

$$s_P = \frac{x - x_{worse}}{x_{best} - x_{worse}} = \frac{x - \$17,000}{\$8000 - \$17,000} = -\frac{x - \$17,000}{\$9000}$$

Hence, $s_P(B) = s_P(\$10,000) = 0.778$. Then, from the utility function defined in Eq. 2.14 with $r = 3$, we have

$$u_P(B) = u(s_P(B)) = \frac{1 - e^{-rs_P(B)}}{1 - e^{-r}} = \frac{1 - e^{-3(0.778)}}{1 - e^{-3}} = 0.950$$

Then, the utilities for the three alternatives are, respectively,

$$u(A) = k_P\ u_P(A) + k_L\ u_L(A) = 0.5\ (0) + 0.5\ (1) = 0.5$$

$$u(B) = k_P\ u_P(B) + k_L\ u_L(B) = 0.5\ (0.950) + 0.5\ (0.5) = 0.725$$

$$u(C) = k_P\ u_P(C) + k_L\ u_L(C) = 0.5\ (1) + 0.5\ (0) = 0.5$$

K is a new parameter called the composite scaling constant, which satisfies the following equation:

$$1 + K = \prod_{i=1}^{m}(1 + Kk_i) \tag{2.18}$$

If the scaling factor is $K = 0$, indicating no interacting of attribute preference, this formulation is equivalent to its additive form, assuming the sum of all the scaling factors k_i is 1; that is, $\sum_{i=1}^{m} k_i = 1$.

It is shown in (Hyman 2003; Shtub et al. 1994) that if $\sum_{i=1}^{m} k_i > 1$, then $-1 < K < 0$, and if $\sum_{i=1}^{m} k_i < 1$, then $K > 0$. Examples 2.2 and 2.3 provides a few more details on the multiplicative MAU function.

EXAMPLE 2.2

For $m = 2$ and $k_1 + k_2 = 1$, $k_1 > 0$ and $k_2 > 0$, show that the scaling factor defined in Eq. 2.18 is $K = 0$, and the multiplicative MAU function u defined in Eq. 2.17 becomes a linear MAU function.

Solutions

From Eq. 2.18, we have

$$1 + K = \prod_{i=1}^{2}(1 + Kk_i) = (1 + Kk_1)(1 + Kk_2) = 1 + K(k_1 + k_2) + K^2 k_1 k_2$$

Because $k_1 + k_2 = 1$, the above equation becomes $1 + K = 1 + K + K^2 k_1 k_2$. Because $k_1 > 0$ and $k_2 > 0$, we have $K = 0$. Hence,

$$u(x_1, x_2, \ldots, x_m) = \frac{1}{K}\left\{\prod_{i=1}^{m}[1 + Kk_i u_i(x_i)] - 1\right\} = \frac{1}{K}\{(1 + Kk_1 u_1)(1 + Kk_2 u_2) - 1\}$$

$$= \frac{1}{K}\{K(k_1 u_1 + k_2 u_2) + K^2 k_1 u_1 k_2 u_2\} = k_1 u_1 + k_2 u_2$$

EXAMPLE 2.3

Continue with Example 2.1 but use a multiplicative MAU function with $k_P = 0.5$ and $k_L = 0.25$.

Solutions

First, from Eq. 2.18, we have

$$1 + K = \prod_{i=1}^{2}(1 + Kk_i) = 1 + K(k_P + k_L) + K^2 k_P k_L = 1 + 0.75K + 0.125K^2$$

Hence $K = 2$.

Then, the utilities for the three alternatives are, respectively,

$$u(A) = \frac{1}{K}\{(1 + Kk_P u_P(A))(1 + Kk_L u_L(A)) - 1\} = \frac{1}{2}\{(1 + 2(0.5)(0))(1 + 2(0.25)(1)) - 1\} = 0.25$$

$$u(B) = \frac{1}{K}\{(1 + Kk_P u_P(B))(1 + Kk_L u_L(B)) - 1\} = \frac{1}{2}\{(1 + 2(0.5)(0.950))(1 + 2(0.25)(0.5)) - 1\} = 0.719$$

$$u(C) = \frac{1}{K}\{(1 + Kk_P u_P(C))(1 + Kk_L u_L(C)) - 1\} = \frac{1}{2}\{(1 + 2(0.5)(1))(1 + 2(0.25)(0)) - 1\} = 0.5$$

2.5 GAME THEORY

The design theory we discussed so far exclusively focused on a situation in which a single decision maker needs to find a best possible decision among alternatives. A decision needed to be made by one decision maker to produce maximum payoffs in single or multiple attributes (or criteria) under a set of constraints. In many situations, however, the payoff of a decision made by an individual depends not only on what he or she does but also on the outcome of the decisions or choices that other individuals make. In engineering design, design decisions are often not made by a single designer or a single design team. Instead, multiple designers or multiple design teams are working on the product design and are involved in design decision making, with each designer or team being responsible for one of more design objectives and/or subsystems. For example, a structural engineer focuses on maximizing the strength and durability of a load-bearing component, while the goal of a manufacturing engineer is to produce the component with least cost in less time. Design decisions made by the structural engineer in determining the geometric shape of the component may affect the cost and time of manufacturing the component and vice versa.

In practice, some of the design decisions are made simultaneously at a specific time of a design phase, and some are made in sequence throughout the design process. With several designers (or design teams) each with his or her own objectives, the nature of the design decisions can take several paths and the overall design may not be desired. This is because that a single designer or team can theoretically do better and his or her decision could dominate, hence largely determining the performance of the overall product, which may or may not be desired.

Game theory applied to this situation provides a method for understanding and perhaps guiding the design decision making. Game theory is a study of strategic decision making. More formally, it is the study of mathematical models of conflict and cooperation between intelligent, rational decision makers. As such, game theory could serve as an important and effective management tool for use in the situations of multiple designers or multiple design teams that are decentralized.

Game theory is mainly used in economics, political science, and psychology, as well as logic and biology. It was recently explored for engineering design. Our purpose of discussing game theory and later employing the theory as a design tool in Section 2.6.2 is not to offer a complete solution for design decision making but to present a plausible idea that has been explored for its feasibility for support of engineering design. As of now, applying game theory as a design tool is still an open research topic that requires much effort to convert the theory into practical tools in the near future.

In the following, we first discuss in Section 2.5.1 the elements of a game and the strategy to determine an equilibrium (or a rational solution) to the game; that is, a stable state in which either one outcome occurs or a set of outcomes occur with known probability. Then, in Section 2.5.2, we present a two-person matrix game and discuss Nash equilibrium for the game. The information presented in these two subsections should provide enough information for readers to proceed with two other kinds of games, sequential games and cooperative games, which are discussed in Sections 2.5.4 and 2.5.5, respectively. To minimize the complexity in mathematical discussion, we assume for most cases two-player games. Readers are referred to the literature (e.g., Aliprantis and Chakrabarti 2006; Hazelrigg 1996) for a more thorough and in-depth discussion of the subject.

2.5.1 ELEMENTS OF A GAME

We start by introducing a well-known game in the literature: the prisoner's dilemma game or prisoner's game. Thereafter, we introduce basic elements in a game by referring to the prisoner's game as an example. With this basic understanding, we proceed with formulating a game mathematically.

In the prisoner's game, two criminals are apprehended by the police, who have strong evidence that they are guilty of the crime, but only weak evidence that they are guilty of a second. Without a confession by one criminal or the other, a conviction for the second crime cannot be obtained. Upon capture, the criminals are immediately separated for interrogation. They are both told that existing evidence will convict them of the first crime, and for that they will each serve 5 years in jail. However, they are offered the opportunity to confess to the second crime. If one confesses and the other one does not, the confessor will have his sentence reduced to 3 years, while the other will have his sentence lengthened to 10 years. But if both confess, both their sentences will be lengthened to 8 years. Certainly, the criminals desire to serve less time in jail. Therefore, a payoff table with negative of the time served in jail is constructed in Table 2.10. The first numeric figure in the pair denotes the payoff for prisoner A and the second for prisoner B.

If the prisoners could hold each other to binding agreements, both would be better off if neither confessed; in this case, each gets 5 years. This is a so-called cooperative game because both players cooperate to maximize their collective payoffs. Acting alone, the better strategy for each is to confess. This is a so-called noncooperative game. If prisoner A chooses to confess, B gets 10 years if he withholds or 8 years if he confesses. If A chooses to withhold, B gets 5 years if he withholds and only 3 years if he confesses. In short, no matter what A does, B is better off by confessing. We say that the

Table 2.10 A Payoff Table for the Prisoner's Dilemma Game			
		Prisoner B	
		Confess	Withhold
Prisoner A	Confess	−8, −8	−3, −10
	Withhold	−10, −3	−5, −5

strategy of confessing is a strictly dominant strategy for B. The same applies to A. Thus, both prisoners confess, and both are worse off for their actions.

In the absence of any communication or any coordination scheme, rational players are expected to play their strictly dominant strategies because a strictly dominant strategy gives a player an unequivocally higher payoff. A solution to the prisoner's dilemma game can therefore end up being (confess, confess). This is the solution using strictly dominant strategies. The solution (confess, confess) is called the Nash equilibrium, which will be discussed further shortly.

In this prisoner's game—and any game in general—the following elements are present:

1. There are two (or more) participants (or players). In this example, there are two criminals, A and B.
2. Each player has a set of alternative choices, and a player can take action on the available choices. In this example, each criminal can choose to confess or withhold. We denote by a_i a choice that player i can make. We refer to the set of choices available to the player i as player i's action set, $A_i = \{a_i\}$.
3. Each player develops a strategy to play the game. A strategy s_i is a rule to tell player i which available actions to choose, subject to available information, each time he or she takes an action. In this example, the strategy set for criminal B that maximizes payoffs (provided that information of A's action is available) includes: if A confesses, B confesses; if A withholds, B confesses. The strategies available to player i are referred to as the player's strategy set or strategy space $S_i = \{s_i\}$. For all n players in a game, we define the strategy combination or strategy profile $s = (s_1, s_2, ..., s_n)$. For the prisoner's game, the strategy combination is $s = (s_A, s_B)$.
4. For each outcome, there is a payoff that each player gets. In the case of the prisoner's game, the payoffs are given by the pairs (a, b) for each outcome, as shown in Table 2.10.

The above are the essential ingredients that constitute what is called a game in strategic form (or normal form). Note that, as seen in (2) and (3), we use action and strategy interchangeably. In general, a strategic form game consists of a set of players; for each player, there is a strategy set; and for each outcome (or strategy combination) of the game, there is a payoff for each player. When a game is presented in normal form, it is presumed that each player acts simultaneously or, at least, without knowing the actions of the other. If players have some information about the choices of other players, the game is usually presented in extensive form (or tree form). We discuss the extensive form of a sequential game in Section 2.5.3.

Next, we formulate the game mathematically and discuss solution techniques. We focus on two-player games in a matrix form first.

2.5.2 TWO-PERSON MATRIX GAMES

A matrix game, such as the prisoner's dilemma game, is a two-player game such that:

1. Player A has a finite strategy set S_A with m elements; that is, $S_A = \{s_{A1}, s_{A2}, ..., s_{Am}\}$. In the prisoner's dilemma game, $S_A = \{s_{A1} = \text{confess}, s_{A2} = \text{withhold}\}$.
2. Player B has a finite strategy set S_B with n elements; that is, $S_B = \{s_{B1}, s_{B2}, ..., s_{Bn}\}$. In the prisoner's dilemma game, $S_B = \{s_{B1} = \text{confess}, s_{B2} = \text{withhold}\}$.
3. Denoting s_A an element of set S_A (i.e., $s_A \in S_A$) and s_B an element of set S_B (i.e., $s_B \in S_B$), the payoffs of the players are measured by utility functions $u_A(s_A, s_B) \in R$ and $u_B(s_A, s_B) \in R$ (R is a

Table 2.11 The Payoff Table of the Two-Person Matrix Game

Player A	Strategy	Player B			
		s_{B1}	s_{B2}	...	s_{Bn}
	s_{A1}	(a_{11}, b_{11})	(a_{12}, b_{12})	...	(a_{1n}, b_{1n})
	s_{A2}	(a_{21}, b_{21})	(a_{22}, b_{22})	...	(a_{2n}, b_{2n})
	...	...	...	...	...
	s_{Am}	(a_{m1}, b_{m1})	(a_{m2}, b_{m2})	...	(a_{mn}, b_{mn})

real number) of the outcomes, in which $s = (s_A, s_B) \in S_A \times S_B$. For example, in the prisoner's game, $u_A(s_{A1} = \text{confess}, s_{B2} = \text{withhold}) = -3$, and $u_B(s_{A1} = \text{confess}, s_{B2} = \text{withhold}) = -10$. $s = (s_A, s_B)$ is called a strategy combination or a strategy profile.

Mathematically, the matrix game formulated above can be described as follows: At a certain time, player A chooses a strategy $s_A \in S_A$. Simultaneously and independently, player B chooses a strategy $s_B \in S_B$. Once this is done or once a strategy profile $\mathbf{s} = (s_A, s_B)$ is determined, each player receives the respective payoff $a_{ij} = u_A(s_{Ai}, s_{Bj})$ and $b_{ij} = u_B(s_{Ai}, s_{Bj})$. The payoffs can be arranged in the form of an $m \times n$ matrix shown in Table 2.11.

The idea of a solution of a game is usually identified by the concept of the Nash equilibrium, defined next.

A pair of strategies $(s_A^*, s_B^*) \in S_A \times S_B$ is a Nash equilibrium of a matrix game if

$$1. \quad u_A(s_A^*, s_B^*) \geq u_A(s_A, s_B^*) \text{ for each } s_A \in S_A, \text{ and} \tag{2.19}$$

$$2. \quad u_B(s_A^*, s_B^*) \geq u_B(s_A^*, s_B) \text{ for each } s_B \in S_B \tag{2.20}$$

in which the strictly dominant strategies are followed. In fact, a Nash equilibrium is an outcome of the game from which none of the players have an incentive to deviate. This is because it is optimal for a player to choose the Nash equilibrium strategy. In this sense, a Nash equilibrium has a property that is self-enforcing. Example 2.4 provides more illustration.

Finding the Nash equilibrium by verifying Eqs 2.19 and 2.20 for all possible strategies may not be practical, especially when there are more than two players and each player is given a large strategy set. There are numerous ways to find the Nash equilibrium. In this subsection, we present the approach of a necessary condition learned in Calculus followed by Example 2.5 for illustration. Later, in Section 2.6.2, we employ a graphical approach for a pressure vessel design problem.

We assume the strategy sets are open intervals of real numbers and the payoff functions are differentiable. In this case, a necessary condition for a strategy set $(s_1^*, s_2^*, \ldots, s_n^*)$ of an n-player game to be a Nash equilibrium of the game is

$$\frac{\partial u_i(s_1^*, s_2^*, \ldots, s_n^*)}{\partial s_i} = 0 \ , \quad i = 0, \ n \tag{2.21}$$

When the system of equations of Eq. 2.21 has a unique solution, then it is the only possible Nash equilibrium of the game. In many cases, the Nash equilibrium is not unique. In some situation, other factors need to be considered to verify that the solution is the Nash equilibrium of the game—for example, second derivatives of the payoff functions with respect to the strategy set.

EXAMPLE 2.4

Verify that in the prisoner's dilemma example, the pair of strategies (confess, confess) is a Nash equilibrium and (withhold, withhold) is not.

Solutions

In the prisoner's dilemma example, the strategy sets for players A and B are, respectively, $S_A = \{s_{A1} = \text{confess}, s_{A2} = \text{withhold}\}$ and $S_B = \{s_{B1} = \text{confess}, s_{B2} = \text{withhold}\}$, and $m = n = 2$. The pair of strategies ($s_A^* = s_{A1} = \text{confess}, s_B^* = s_{B1} = \text{confess}$) is a Nash equilibrium because, for player A,

1. $u_A(s_A^*, s_B^*) \geq u_A(s_A, s_B^*)$ for each $s_A \in S_A$; that is, $u_A(s_A^* = s_{A1} = \text{confess}, s_B^* = s_{B1} = \text{confess}) = -8$ and $u_A(s_{A1} = \text{confess}, s_B^* = s_{B1} = \text{confess}) = -8$ and $u_A(s_{A2} = \text{withhold}, s_B^* = s_{B1} = \text{confess}) = -10$.
   For player B,
2. $u_B(s_A^*, s_B^*) \geq u_B(s_A^*, s_B)$ for each $s_B \in S_B$; that is, $u_B(s_A^* = s_{A1} = \text{confess}, s_B^* = s_{B1} = \text{confess}) = -8$ and $u_B(s_{A1}^* = \text{confess}, s_B = s_{B1} = \text{confess}) = -8$ and $u_B(s_{A1}^* = \text{confess}, s_B = s_{B2} = \text{withhold}) = -10$.

   Note that if both players withhold, each receives a 5-year sentence, which seems to be a better solution to the situation. Is it a Nash equilibrium? Let us take a look. For player A,

1. $u_A(s_A^* = s_{A1} = \text{withhold}, s_B^* = s_{B1} = \text{withhold}) = -5$, and $u_A(s_{A1} = \text{withhold}, s_B^* = s_{B1} = \text{withhold}) = -5$ and $u_A(s_{A2} = \text{confess}, s_B^* = s_{B1} = \text{withhold}) = -3$. Therefore, $u_A(s_A^*, s_B^*) \geq u_A(s_A, s_B^*)$ does not hold.
   For player B,
2. $u_B(s_A^* = s_{A1} = \text{withhold}, s_B^* = s_{B1} = \text{withhold}) = -5$, and $u_B(s_{A1}^* = \text{withhold}, s_B = s_{B1} = \text{withhold}) = -5$ and $u_B(s_{A1}^* = \text{withhold}, s_B = s_{B2} = \text{confess}) = -3$. Therefore, $u_B(s_A^*, s_B^*) \geq u_B(s_A^*, s_B)$ does not hold.

   According to Eqs 2.19 and 2.20, (withhold, withhold) is not a Nash equilibrium and is not self-enforcing. This is because that there is no binding agreement between the prisoners, and a player receives a longer sentence (10 years) if he chooses to withhold while the other player confesses.

The next example is extracted from Aliprantis and Chakrabarti (2006) and is called the Cournot duopoly model, initially analyzed by French mathematician Augustin Cournot (Example 2.5).

2.5.3 SEQUENTIAL GAMES

A sequential game involves multiple players who do not make decisions simultaneously, and one player's decision affects the outcomes and decisions of other players. A sequential game is represented by a game tree (also called the extensive form) with players moving sequentially. We assume information is perfectly known to a player at the time of decision making.

Consider a simple two-person sequential game of two stages with perfect information, whose game tree is shown in Figure 2.13. Here, each vertex (or node) represents a point of choice for a player. The player is specified by an alphabet listed by the vertex. The arrow out of the vertex represents a possible action for that player. The payoffs are specified at the terminal nodes placed at bottom of the tree. Two-stage implies that decisions are made in two stages, one by each player.

In the game shown in Figure 2.13, there are two players. Player A moves first and chooses either L1 or R1. Player B sees player A's move and then chooses L2 or R2, or L3 or R3 depending on player A's move. Suppose that player A chooses L1; then, node 2 is reached; if player B chooses L2, in which terminal node 4 is reached, then player A gets 2 and player B gets 1. If player A chooses R1, then node 3 is reached and if player B chooses R3, then node 7 is reached; then, player A gets 1 and player B gets 0. Now, we are ready to define a sequential game of two players.

A tree T is said to be a two-player game tree if the following are true:

1. Each nonterminal node of the tree is owned by exactly one player. For example, in Figure 2.13, node 1 is owned by player A, and nodes 2 and 3 are owned by player B.

EXAMPLE 2.5

Two firms, firm 1 and firm 2, produce an identical product of amount q_1 and q_2, respectively. The price per unit of the product in the market is determined by $p(q) = A - q$, where $q = q_1 + q_2$, and A is a fixed number. The price equation shows that the per-unit price of the product is decreasing with an increasing total production amount $q = q_1 + q_2$. Assume that the total cost for firm i of producing the output q_i is $C_i = c_i q_i$, where c_i is cost per unit, a positive number. What are the optimal outputs q_1 and q_2 that each firm should produce in order to maximize profit? Note that the profit of each firm depends on the output of the other firm. They choose their production quantities independently and simultaneously. There is no communication or binding agreement between the firms in regard to the production quantities.

Solutions

This problem can be modeled as a two-player matrix game that is noncooperative. It is reasonable to think of the Nash equilibrium as a self-enforced solution. We find the Nash equilibrium of the game using the first-order derivative test of Eq. 2.21.

We first formulate an equation describing profit as function of quantities. For firms 1 and 2, we have, respectively,

$$u_1(q_1, q_2) = p(q)\ q_1 - c_1 q_1 = (A - q_1 - q_2)\ q_1 - c_1 q_1 = -(q_1)^2 + (A - q_2 - c_1)\ q_1 \qquad (2.22)$$

and

$$u_2(q_1, q_2) = p(q)\ q_2 - c_2 q_2 = (A - q_1 - q_2)\ q_2 - c_2 q_2 = -(q_2)^2 + (A - q_1 - c_2)\ q_2 \qquad (2.23)$$

Taking derivatives of Eq. 2.22 with respect to q_1 and Eq. 2.23 with respect to q_2, we have

$$\frac{\partial u_1(q_1, q_2)}{\partial q_1} = -2q_1 - q_2 + A - c_1 = 0 \qquad (2.24)$$

$$\frac{\partial u_2(q_1, q_2)}{\partial q_2} = -q_1 - 2q_2 + A - c_2 = 0 \qquad (2.25)$$

Solving Eqs 2.24 and 2.25, we have

$$q_1^* = \frac{A + c_2 - 2c_1}{3} \qquad (2.26)$$

$$q_2^* = \frac{A + c_1 - 2c_2}{3} \qquad (2.27)$$

Note that if $A > c_1 + c_2$, then $q_1^* > 0$ and $q_2^* > 0$, implying that both firms produce a positive output at the Nash equilibrium.

If we assume that $A = 200$ and $c_1 = c_2 = 2$, then from Eq. 2.26, $q_1^* = 66$, and from Eq. 2.27, $q_2^* = 66$. Then, from Eqs 2.22 and 2.23, $u_1 = u_2 = 4356$. Is this the best payoff for each firm?

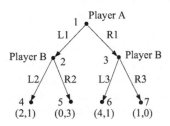

FIGURE 2.13

A game tree for a sequential game.

2. At each terminal node v of the tree, a payoff vector is assigned; that is, $\mathbf{u}(v) = (u_A(v), u_B(v))$. For example, at terminal node 4, $\mathbf{u}(4) = (2,1)$, as shown in Figure 2.13.

The above definition for a sequential game of two players can be easily extended to that of n players.

A strategy for a player i in a sequential game consists of the choices that the player is going to make at the nodes he or she owns. As illustrated in Figure 2.13, a strategy for a player in a sequential game is a complete plan of how to play the game, prescribing the choices at every node owned by the player. For example, in Figure 2.13, at node 2, two choices are available for player B: L2 and R2. In other words, a player's strategy will indicate the choices that the player has planned to make a priori (i.e., before the game starts). A strategy profile for a two-person sequential game is then represented as $\mathbf{s} = (s_A, s_B)$, where each s_A and s_B is a strategy for players A and B, respectively. For example, as shown in Figure 2.13, S_A can be L1 or R1, and S_B is a function from {node 2, node 3} to $\{s_{B21} = L2,$ $s_{B22} = R2; s_{B31} = L3, s_{B32} = R3\}$ with the feasibility restriction that, from node 2, player B can only choose L2 or R2 with a similar restriction on choices from node 3. In other words, the choice of player B depends on the choice made by player A according to the prescribed game tree. Note that a strategy profile uniquely determines the terminal node v that is reached. The payoff (or utility) of each player is a function u_A and u_B of the strategy profile (s_A, s_B).

A solution of a sequential game is also understood to be a Nash equilibrium. In a two-player sequential game, a strategy profile (s_A^*, s_B^*) is said to be a Nash equilibrium if, for each player, we have

$$1. \quad u_A(s_A^*, s_B^*) \geq u_A(s_A, s_B^*) \text{ for each } s_A \in S_A, \text{ and} \tag{2.28}$$

$$2. \quad u_B(s_A^*, s_B^*) \geq u_B(s_A^*, s_B) \text{ for each } s_B \in S_B \tag{2.29}$$

In other words, a Nash equilibrium is a strategy profile (s_A^*, s_B^*) such that player A cannot improve his payoff by changing his strategy if the player B does not change, and vice versa. Note that Eqs 2.28 and 2.29 are identical to those of Eqs 2.19 and 2.20 in a strategy form game.

A backward induction method is often employed to solve for the Nash equilibrium of a sequential game. Backward induction is the process of reasoning backward in time, from the end of a problem or situation, to determine a sequence of optimal actions. It proceeds by first considering the last time a decision might be made and choosing what to do in any situation at that time. Using this information, one can then determine what to do at the second-to-last time of decision. This process continues backward until one has determined the best action for every possible situation at every point in time. We illustrate the use of the method in the following example (Example 2.6).

EXAMPLE 2.6

Find the Nash equilibrium for the sequential game shown in Figure 2.13.

Solutions

We use the backward induction and conditions shown in Eqs 2.28 and 2.29 to find the Nash equilibrium.

We first assume $s_A^* = s_{A1} = L1$, then $s_B^* = s_{B2} = R2$ because $u_B(L1, R2) = 3 \geq \{u_B(L1, R2) = 3, u_B(L1, L2) = 1\}$. If $s_A^* = s_{A2} = R1$, then $s_B^* = s_{B2} = L3$ because $u_B(R1, L3) = 1 \geq \{u_B(R1, L3) = 1, u_B(R1, R3) = 0\}$.

Now, we go backward and use Eq. 2.28 to find s_A^* from the two strategies (L1, R2) and (R1, L3). We found that $s_A^* = s_{A2} = R1$ because $u_A(R1, L3) = 4 \geq \{u_A(R1, L3) = 4, u_A(L1, R2) = 3\}$. Hence the Nash equilibrium is found at $(s_A^*, s_B^*) = (R1, L3)$—that is, the path $1 \rightarrow 3 \rightarrow 6$. If player B does not change its strategy $(s_B^* = L3)$, player A cannot improve its payoff by changing its strategy—for example, from R1 to L1. Similarly, if player A does not change its strategy $(s_A^* = R1)$, player B cannot improve its payoff by changing its strategy—for example, from L3 to R3.

The next example (Example 2.7) is again extracted from Aliprantis and Chakrabarti (2006), called the Stackelberg duopoly model, which is slightly modified from the example of the Cournot duopoly model to illustrate a few more details of the sequential game.

EXAMPLE 2.7

Continue with Example 2.5, except that firm 1 chooses a production quantity $q_1 \geq 0$. Firm 2 observes q_1 and then chooses its own production quantity q_2. Find the optimal production quantities q_1 and q_2 each firm should produce in order to maximize profit.

Solutions

This problem can be formulated as a two-person sequential game of two stages and with perfect information. We use backward induction again for this problem. We first find the output q_2^* of firm 2 that maximizes firm 2's profit given the output q_1 of firm 1. That is, $q_2^* = q_2^*(q_1)$. The profit of firm 2 can be formulated as

$$u_2(q_1, q_2^*) = \max_{q_2 \geq 0} u_2(q_1, q_2) = \max_{q_2 \geq 0} [-(q_2)^2 + (A - q_1 - c_2)q_2] \tag{2.30}$$

Taking the first and second derivatives of Eq. 2.30 with respect to q_2, we have

$$\frac{\partial u_2}{\partial q_2} = -2q_2 + (A - q_1 - c_2) \tag{2.31}$$

and

$$\frac{\partial^2 u_2}{\partial q_2^2} = -2 < 0 \tag{2.32}$$

Solving for q_2 from Eq. 2.31, we have

$$q_2^* = q_2^*(q_1) = \frac{A - q_1 - c_2}{2} \tag{2.33}$$

which gives maximum profit u_2 for firm 2, provided $q_1 < A - c_2$ (so that $q_2^* > 0$).

Firm 1 should now anticipate that firm 2 will choose q_2^* if firm 1 chooses q_1. Therefore, firm 1 will want to choose q_1 to maximize its profit, defined as

$$u_1(q_1, q_2^*) = -(q_1)^2 + q_1(A - q_2^* - c_1) = -(q_1)^2 + q_1(A - \frac{A - q_1 - c_2}{2} - c_1) = \frac{1}{2}[-(q_1)^2 + q_1(A + c_2 - 2c_1)] \tag{2.34}$$

subject to $q_1 \geq 0$. Taking again the first and second derivatives of Eq. 2.34 with respect to q_1, we have

$$\frac{\partial u_1}{\partial q_1} = -q_1 + \frac{A + c_2 - 2c_1}{2} \tag{2.35}$$

and

$$\frac{\partial^2 u_1}{\partial q_1^2} = -1 < 0 \tag{2.36}$$

Therefore, from Eq. 2.35, we have

$$q_1^* = \frac{A + c_2 - 2c_1}{2} \tag{2.37}$$

which gives the maximum profit u_1 for firm 1. Substituting Eq. 2.37 to Eq. 2.33, we have

$$q_2^* = \frac{A - \frac{A + c_2 - 2c_1}{2} - c_2}{2} = \frac{A + 2c_1 - 3c_2}{4} \tag{2.38}$$

If we assume $A = 200$ and $c_1 = c_2 = 2$, then from Eq. 2.37, $q_1^* = 99$. From Eq. 2.38, $q_2^* = 49.5$. Then, $u_1 = 4900.5$ and $u_2 = 2450.25$.

2.5.4 COOPERATIVE GAMES

Now we go back to the prisoner's game. We have observed that both criminals in the prisoner's game are better off if they both withhold. This is the basis for one solution concept in cooperative games.

First, we define a criterion to rank outcomes from the point of view of the group of players as a whole. We can say that one outcome is better than another if at least one player is better off and no one is worse off. For example, in the prisoner's game, (withhold, withhold) is better than (confess, confess). This is called the Pareto criterion, after the Italian economist and mechanical engineer, Vilfredo Pareto. If an outcome cannot be improved, then we say that the outcome is Pareto optimal—that is, optimal in terms of the Pareto criterion. In the prisoner's game, (withhold, withhold) is Pareto optimal because no one can be made better off without making the other prisoner worse off. In fact, (confess, withhold) is also Pareto optimal because neither player is better off without making the other player worse off.

Mathematically, the Pareto solution for a two-person game can be defined as follows:

$$A \text{ strategy profile } \mathbf{s}^* = (s_A^*, \ s_B^*) \in S_A \times S_B \text{ is a Pareto optimal iff (if and only if) } \nexists \text{ (there does}$$
$$\text{not exist) another strategy profile } \mathbf{s} \in S_A \times S_B \text{ such that } \mathbf{u}(\mathbf{s}^*) \geq \mathbf{u}(\mathbf{s})$$
$$\text{and at least } u_A(\mathbf{s}^*) > u_A(\mathbf{s}) \text{ or } u_B(\mathbf{s}^*) > u_B(\mathbf{s}).$$

$$(2.39)$$

It is obvious that $\mathbf{s}^* = $ (withhold, withhold) satisfies Eq. 2.39.

If there were a unique Pareto optimal outcome for a cooperative game, that would seem to be a good solution concept. However, there is not. For example, in the prisoner's game, $\mathbf{u}$(confess, withhold) $= (-3, -10)$ is also a Pareto solution (see Example 2.8). In general, there are infinitely many Pareto optima for any fairly complicated game. Pareto optimal and solution techniques will be formally introduced in Chapter 5 under the subject of multiobjective optimization. In general, Pareto optimal or Pareto solutions are better understood in the so-called criterion space. We use the Cournot duopoly model to illustrate Pareto optimal in Example 2.9.

2.6 DESIGN EXAMPLES

We discussed two major theories—utility theory and game theory—which were explored to support engineering design. In this section, we use two engineering design examples to illustrate the details. We first discuss using utility theory as a design tool and apply the design tool to solve a simple cantilever beam example. Then, we use game theory as a design tool to solve a pressure vessel design problem, which is formulated as noncooperative, sequential, and cooperative games for illustration. Both examples are purposely made simple so that the focus stays on concept illustration.

2.6.1 UTILITY THEORY AS A DESIGN TOOL

In this subsection, we illustrate a few details in employing utility theory to support multiobjective engineering designs. We use a simple cantilever beam example of two design attributes (or design objectives) to solve for two design problems, unconstrained and constrained.

EXAMPLE 2.8

Verify that (confess, withhold) is a Pareto solution and (confess, confess) is not, using Eq. 2.39.

Solutions

We first sketch the outcome or payoff in the so-called criterion space as shown below.

It is obvious that the prisoners are better off when their respective payoffs move closer to the origin O. As discussed in Section 2.5.2, the Nash equilibrium for a noncooperative game is point α, $\mathbf{u}_\alpha = (-8,-8)$. Point γ is a Pareto solution as discussed. We are interested in finding out if point β is a Pareto solution. At point β (confess, withhold), $\mathbf{u}_\beta = (-3,-10)$. Now, we check if point β is a Pareto solution using Eq. 2.39.

We take point α, $\mathbf{u}_\alpha = (-8,-8)$. Is $\mathbf{u}_\beta = (-3,-10) \geq \mathbf{u}_\alpha(-8,-8)$? The answer is no. Next, we take point γ, $\mathbf{u}_\gamma = (-5,-5)$. Is $\mathbf{u}_\beta = (-3,-10) \geq \mathbf{u}_\gamma(-5,-5)$? The answer is no. Then, we take the final point ζ. $\mathbf{u}_\zeta = (-10,-3)$. Is $\mathbf{u}_\beta = (-3,-10) \geq \mathbf{u}_\zeta = (-10,-3)$? The answer is no. Therefore, there is no other point such that $\mathbf{u}_\beta \geq \mathbf{u}(s)$; from Eq. 2.39, $\mathbf{u}_\beta = (-3,-10)$ is Pareto optimal.

Now, we check if point α (confess, confess) is a Pareto solution using Eq. 2.39.

We take point β, $\mathbf{u}_\alpha = (-8,-8)$. Is $\mathbf{u}_\beta = (-8,-8) \geq \mathbf{u}_\alpha(-3,-10)$? The answer is no. Next, we take point ζ. $\mathbf{u}_\zeta = (-10,-3)$. Is $\mathbf{u}_\alpha = (-8,-8) \geq \mathbf{u}_\zeta = (-10,-3)$? The answer is no. Now, we take the final point γ, $\mathbf{u}_\gamma = (-5,-5)$. Is $\mathbf{u}_\alpha = (-8,-8) \geq \mathbf{u}_\gamma(-5,-5)$? The answer is yes. In fact, $\mathbf{u}_\alpha > \mathbf{u}_\gamma$. Therefore, according to Eq. 2.39, $\mathbf{u}$ $\mathbf{u}_\alpha = (-8,-8)$ is not Pareto optimal.

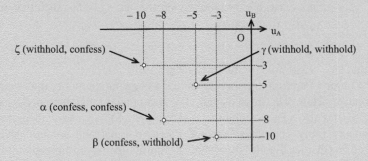

EXAMPLE 2.9

Continue with Example 2.5, except formulating the problem as a cooperative game and looking for the Pareto optimal. From Example 2.5 we have the payoff functions for firms 1 and 2 as, respectively,

$$u_1(q_1,q_2) = -(q_1)^2 + (A - q_2 - c_1)\, q_1, \text{ and} \tag{2.40}$$

$$u_2(q_1,q_2) = -(q_2)^2 + (A - q_1 - c_2)\, q_2 \tag{2.41}$$

We assume $A = 200$, $c_1 = c_2 = 2$. Hence, Eqs 2.40 and 2.41 become

$$u_1(q_1,q_2) = -(q_1)^2 + (198 - q_2)\, q_1, \text{ and} \tag{2.42}$$

$$u_2(q_1,q_2) = -(q_2)^2 + (198 - q_1)\, q_2 \tag{2.43}$$

Solutions

The profit functions u_1 and u_2 are graphed in terms of q_1 and q_2 as shown in the next page (left), indicating that the maxima of u_1 and u_2 are approaching in the opposite direction; u_1 toward the lower right and u_2 toward the upper left. In this case, $q_1 - q_2$ is called the design space. Finding the Pareto optimal in a design space is not straightforward and is not desired. Instead, we graph u_1 and u_2 in the so-called criterion space shown in the next page (right). MATLAB scripts for creating these graphs can be found in Appendix A.

EXAMPLE 2.9—cont'd

Based on the definition of Eq. 2.39, points along the line between A and B are Pareto optimal, also called the Pareto front.

In the noncooperative game shown in Example 2.5, the Nash equilibrium is found at $q_1^* = q_2^* = 66$, and $u_1 = u_2 = 4356$, as shown in the figure below (right), which is not Pareto optimal. Any point on the front is Pareto optimal, in which neither firm is better off without making the other firm worse off.

Also from Example 2.7, the Nash equilibrium of a sequential game is found at $q_1^* = 99$ and $q_2^* = 49.5$, profits $u_1 = 4900.5$, and $u_2 = 2.450.25$, which is not Pareto optimal either.

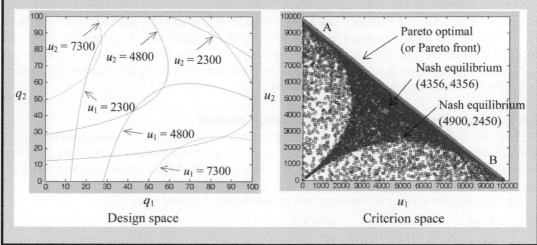

Design space   Criterion space

A cantilever beam of rectangular cross section is loaded by a point force $P = 1000$ N at the tip, as shown in Figure 2.14. The length of the beam is $\ell = 200$ mm, the width is $b = 30$ mm, and height is $h = 40$ mm. The beam is made of aluminum 1060 Alloy, in which the yield strength is $S_y = 27.6$ MPa, modulus of elasticity is $E = 69$ GPa $= 69,000$ MPa, and mass density is $\rho = 2.7 \times 10^{-3}$ g/mm^3.

The weight w, bending stress σ, and vertical displacement z are included for design consideration. They are defined respectively as follows:

$$w = \rho g b h \ell \tag{2.44}$$

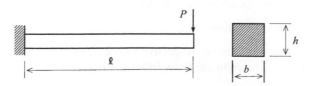

FIGURE 2.14

Cantilever beam example.

where $g = 9806$ mm/s^2 is the gravitational acceleration,

$$\sigma = \frac{6P\ell}{bh^2} \tag{2.45}$$

and

$$z = \frac{4P\ell^3}{Ebh^3} \tag{2.46}$$

For the current design, $b = 30$ mm and $h = 40$ mm; the weight, the maximum bending stress, and maximum vertical displacement are, respectively, $w = 6.35$ N, $\sigma = 25$ MPa, and $z = 0.2415$ mm. Note that weight is very small compared to the external load P; therefore, it is ignored in stress and displacement calculations. Also, the maximum bending stress is less than the material yield strength S_y. Next, we introduce two design problems, unconstrained and constrained, and illustrate the concept of solving the design problems using utility theory.

2.6.1.1 Beam design example 1: Unconstrained problem

The two objective functions included in this design problem are weight $w(b,h)$ and displacement $z(b,h)$, in which the width b and height h of the beam cross section are defined as design variables. Note that the length of the beam is assumed to be fixed. The lower and upper bounds of the design variables are chosen as 10 mm $\leq b \leq$ 60 mm, and 20 mm $\leq h \leq$ 80 mm.

We illustrate the solution in four steps and present results for three cases.

Step 1: Determination of attribute bounds

We determine the bounds—that is, the worst and best designs—of the attributes (in this example, the two objectives) based on the bounds of design variables b and h. For the weight objective, the worst (heaviest) and best designs are, respectively:

$w_{worst} = 25.4$ N, when both design variables reach their respective upper bounds; that is, $b = 60$ mm and $h = 80$ mm; and

$w_{best} = 1.06$ N, when both design variables reach their respective lower bounds; that is, $b = 10$ mm and $h = 20$ mm.

For the displacement objective, the worst (largest displacement) and best designs are, respectively:

$z_{worst} = 5.80$ mm, when both design variables reach their respective lower bounds; that is, $b = 10$ mm and $h = 20$ mm; and

$z_{best} = 0.0151$ mm, when both design variables reach their respective upper bounds; that is, $b = 60$ mm and $h = 80$ mm.

It is obvious that these two objectives are in conflict.

Step 2: Formulation of SAU functions

We define two SAU functions for the respective two attributes following the utility functions defined in Section 2.2.4. For the weight attribute, following Eq. 2.14, we have

$$u_w(s) = \frac{1 - e^{-r_w s_w}}{1 - e^{-r_w}} \tag{2.47}$$

where s_w is a normalized form of the outcome in the weight attribute, as defined in Eq. 2.15; that is,

$$s_w = \frac{w - w_{\text{worse}}}{w_{\text{best}} - w_{\text{worse}}} = \frac{w - 25.4}{1.06 - 25.4} \tag{2.48}$$

For the displacement attribute, we have

$$u_z(s) = \frac{1 - e^{-r_z s_z}}{1 - e^{-r_z}} \tag{2.49}$$

where s_z is a normalized form of the outcome in the displacement attribute, defined as

$$s_z = \frac{z - z_{\text{worse}}}{z_{\text{best}} - z_{\text{worse}}} = \frac{z - 5.80}{0.0151 - 5.80} \tag{2.50}$$

Note that, as discussed in Section 2.4.4, the parameters r_w and r_z determine the shape of the respective utility function curves, which reflect the designer's attitude toward risk. Recall that r_w (or r_z) > 0 implies risk averse, r_w (or r_z) < 0 implies risk prone, and r_w (or r_z) = 0 implies risk neutral. The utility curves of corresponding situations for the two respective attributes are shown in Figure 2.15.

Step 3: Formulation of MAU functions

We define a multiplicative MAU by introducing scaling constants k_w and k_z that represent the designer's preference between attributes w and z. The multiplicative MAU is defined, following Eq. 2.17, as

$$u(w, z) = \frac{1}{K}\left\{ \prod_{i=1}^{2}[1 + Kk_i u_i] - 1 \right\} = \frac{1}{K}\{(1 + Kk_w u_w)(1 + Kk_z u_z) - 1\}$$
$$= k_w u_w + k_z u_z + Kk_w k_z u_w u_z \tag{2.51}$$

in which K satisfies

$$1 + K = \prod_{i=1}^{2}(1 + Kk_i) = (1 + Kk_w)(1 + Kk_z) \tag{2.52}$$

(a)   **(b)**

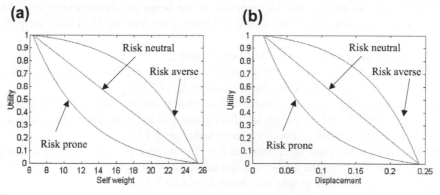

FIGURE 2.15

Utility functions: (a) for attribute: weight w, and (b) for attribute: displacement z.

That is,

$$K = \frac{1 - k_w - k_z}{k_w k_z} \tag{2.53}$$

Our goal is now to maximize the MAU function $u(w,z)$.

Step 4: Search Designs

In this step, we introduce three cases illustrating different aspects of the design scenarios that present concept and essential elements of applying utility theory to engineering design.

Case 1: Base case—Risk neutral ($r_w = r_z = 0$) and no preference ($k_w = 0.5$, $k_z = 0.5$).

As discussed in Section 2.4.4, for $r_w = r_z = 0$, the SAU functions are respectively $u_w = s_w$ and $u_z = s_z$. And for $k_w = 0.5$, $k_z = 0.5$, the multiplicative MAU becomes additive MAU; that is, $u(w,z) = k_w u_w + k_z u_z = 0.5 (s_w + s_z)$:

$$u(w,z) = k_w u_w + k_z u_z = 0.5(s_w + s_z) = 0.5\left(\frac{w - 25.4}{1.06 - 25.4} + \frac{z - 5.80}{0.0151 - 5.80}\right) \tag{2.54}$$

in which w and z are the weight and displacement of the beam defined in Eqs 2.44 and 2.46, respectively. Note that both w and z are functions of design variables b and h.

There are numerous methods to find the maximum of the MAU function defined in Eq. 2.54. Finding maximum or minimum of the objective function in a design problem is called design optimization, which will be discussed in Chapter 3. For this example, we use a brute force approach, the so-called generative method, to find the maximum of the MAU function. In this method, the design variables are divided into number of intervals between their respective upper and lower bounds, from which the maximum is sought by comparing the values of the MAU function at each design to the combination of the design variables in the intervals.

In this example, we define a design interval as $d = 0.1$ mm for both b and h. As a result, design variables b and h are divided into, respectively, $n_b = (b_u - b_\ell)/d + 1 = (60 - 10)/0.1 + 1 = 501$, and $n_h = (h_u - h_\ell)/d + 1 = (80 - 20)/0.1 + 1 = 601$. The MAU function will be evaluated $n_b \times n_h = 501 \times 601$ times. We use MATLAB to find the maximum and create graphs to illustrate the results. The optimum design is found at $b = 10$ mm and $h = 57.7$ mm, in which the MAU function is $u(w,z) = 0.9395$, and the weight and displacement of the beam are, respectively, $w = 3.06$ N and $z = 0.241$ mm. The solution graphs are shown in Figure 2.16, and the MATLAB script is given in Appendix A.

Case 2: Changing preference between attributes w and z.

In this case, we assume risk neutral, but change the preference attributes of w and z. In Case 2a, we assume $k_w = 0.7$, $k_z = 0.3$, and for Case 2b, we assume $k_w = 0.3$, $k_z = 0.7$.

The optimum design for Case 2a is found at $b = 10$ mm and $h = 46.7$ mm, in which the MAU function is $u(w,z) = 0.9365$, and the weight and displacement of the beam are, respectively, $w = 2.47$ N and $z = 0.455$ mm. The solution graphs are shown in Figure 2.17a.

The optimum design for Case 2b is found at $b = 10$ mm and $h = 71.3$ mm, in which the MAU function is $u(w,z) = 0.9529$, and the weight and displacement of the beam are, respectively, $w = 3.76$ N and $z = 0.128$ mm. The solution graphs are shown in Figure 2.17b.

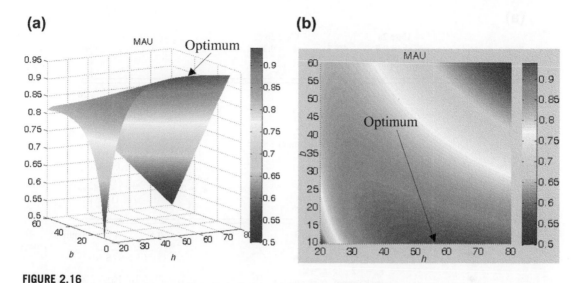

FIGURE 2.16

MAU function and optimum solution of Case 1: (a) iso-view and (b) top view.

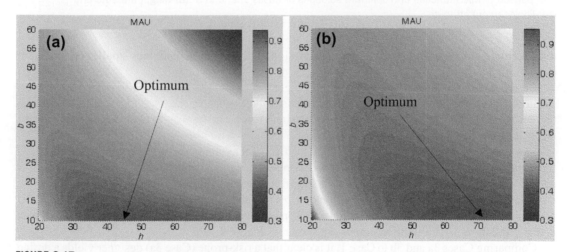

FIGURE 2.17

MAU function and optimum solution of Case 2: (a) Case 2a, $k_w = 0.7$, $k_z = 0.3$. (b) Case 2b, $k_w = 0.3$, $k_z = 0.7$.

Comparing Cases 2a and 2b to Case 1, it is clear that a larger preference factor pushes the solution toward a design that reflects the designer's preference, as illustrated in Figure 2.18. For example, for Case 2a, preference is given to weight attribute ($k_w = 0.7$); the optimal design in weight becomes less compared to Case 1. On the other hand, for Case 2b, preference is given to

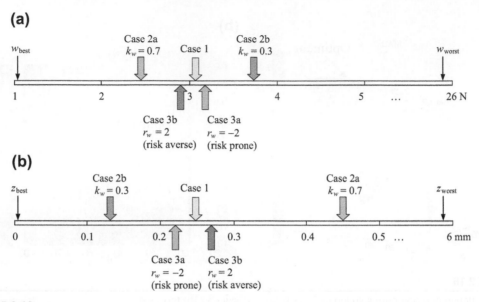

FIGURE 2.18

Comparison of MAU function and optimum solutions of Cases 1, 2, and 3: (a) weight attribute and (b) displacement attribute.

displacement attribute ($k_z = 0.7$); the optimal design in displacement becomes smaller compared to Case 1.

Case 3: Changing attitude toward risk for weight w.

In this case, we assume no preference between attributes w and z (i.e., $k_w = k_z = 0.5$), but change the attitude toward risk for the weight attribute to both risk prone ($r_w = -2$, Case 3a), and risk averse ($r_w = 2$, Case 3b).

The optimum design for Case 3a is found at $b = 10$ mm and $h = 59.1$ mm, in which the MAU function is $u(w,z) = 0.9394$, and the weight and displacement of the beam are, respectively, $w = 3.13$ N and $z = 0.225$ mm. The solution graphs are shown in Figure 2.19a.

The optimum design for Case 3b is found at $b = 10$ mm and $h = 56.6$ mm, in which the MAU function is $u(w,z) = 0.9394$, and the weight and displacement of the beam are respectively, $w = 2.99$ N and $z = 0.256$ mm. The solution graphs are shown in Figure 2.19b.

Comparing Cases 3a and 3b to Case 1, it is clear that a risk-prone (Case 3a) designer yields a design with larger weight, and a risk-averse (Case 3b) designer yields a design with smaller weight, as expected.

Results of the five cases (1, 2a, 2b, 3a, and 3b) are summarized in Table 2.12. In all five cases, at the optimum, the beam width is at its minimum; only the height h is different. Increasing the height h increases weight attribute but decreases the displacement attribute z. Preference or risk attitude that favors the weight attribute, including Cases 2a and 3b, leads to a smaller height, and hence less weight. On the other hand, preference or risk attitude that favors the displacement attribute, including Cases 2b and 3a, leads to a larger height, hence a smaller displacement.

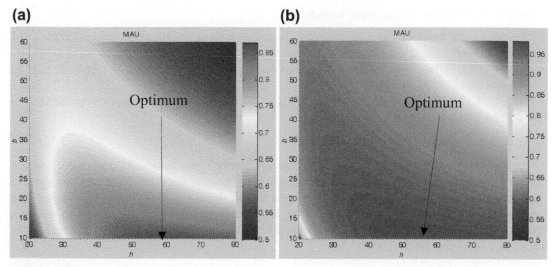

FIGURE 2.19

MAU function and optimum solution of Case 3: (a) Case 3a, $r_w = -2$ (risk prone) and (b) Case 3b (risk averse), $r_w = 2$.

Table 2.12 Result Comparison for Unconstrained Design Problems

Case No.	Problem Setup					Results			
	k_w	k_z	r_w	r_z	b (mm)	h (mm)	w (N)	z (mm)	u
Case 1	0.5	0.5	0	0	10	57.5	3.06	0.241	0.939
Case 2a	0.7	0.3	0	0	10	46.7	2.47	0.455	0.937
Case 2b	0.3	0.7	0	0	10	71.3	3.78	0.128	0.953
Case 3a	0.5	0.5	−2	0	10	59.1	3.13	0.225	0.939
Case 3b	0.5	0.5	2	0	10	56.6	2.99	0.256	0.939

2.6.1.2 Beam design example 2: Constrained problem

This example is identical to that of beam design example 1, except that we add a stress constraint; that is, $\sigma < S_y = 27.6$ MPa. One of the approaches in handling the constrained design problem in the framework of utility theory is to first construct an SAU function for the constraint function, and incorporate the SAU to the MAU function, in which the infeasible design (where constraint is violated) is automatically eliminated from design consideration.

We follow the same four steps as those of the previous example to set up and solve the constrained problem.

Step 1: Determination of attribute bounds

The bounds for the weight and displacement remain unchanged. For the stress constraint, the worst (largest) and best designs are, respectively:

$\sigma_{worst} = 300$ MPa, when both design variables reach their respective lower bounds; that is, $b = 10$ mm and $h = 20$ mm; and

$\sigma_{best} = 3.13$ MPa, when both design variables reach their respective upper bounds; that is, $b = 60$ mm and $h = 80$ mm.

Step 2: Formulation of SAU functions

Again, we only need to construct a utility function for the stress constraint. The utility function must reveal the nature of the constraint; that is, utility is 1 when stress is less than the yield strength (also called feasible design) and 0 when the stress is over the upper limit (infeasible design). One function that satisfies the requirement is a unit step function. However, a step function is discontinuous and is difficult to handle. Therefore, we introduce a function that is continuous and almost reveals the nature of a step function as follows:

$$u(\sigma) = \frac{1}{1 + e^{\frac{\sigma_{0.5} - \sigma}{s}}} \tag{2.55}$$

in which $\sigma_{0.5}$ is the value of the stress at which $u(\sigma) = 0.5$, and s is the slope of the utility function $u(\sigma)$ at utility $u = 0.5$. If we set $\sigma_{0.5} = S_y = 27.6$ MPa, the utility functions for slope $s = -1$ and $s = -0.1$ are graphed in Figures 2.20a and b, respectively. It is clear that when the slope is small, the utility function closely resembles a unit step function as desired. In this example, we set slope $s = -0.01$.

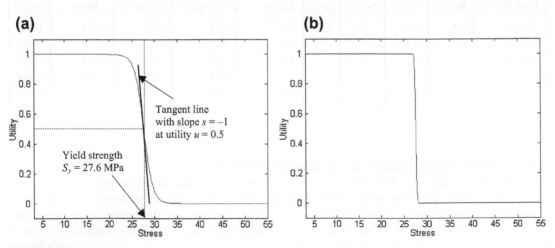

(a)

(b)

Tangent line
with slope $s = -1$
at utility $u = 0.5$

Yield strength
$S_y = 27.6$ MPa

FIGURE 2.20

SAU function for a stress constraint: (a) slope $s = -1$ and (b) slope $s = -0.1$.

Step 3: Formulation of MAU functions

We define a multiplicative MAU by adding the utility function of the stress constraint and introducing an additional parameter k_c as follows (Iyer et al. 1999):

$$u(w, z, \sigma) = [k_c + (1 - k_c)u(w, z)]u(\sigma) \tag{2.56}$$

in which the utility function $u(w,z)$ is defined in Eq. 2.51. Equation 2.56 shows that the utility function $u(w,z,\sigma) = 0$ when $u(\sigma) = 0$, implying that an infeasible design is encountered. For a feasible design, because $u(\sigma) = 1$ as defined in Eq. 2.55, the MAU function $u(w,z,\sigma) = k_c + (1 - k_c)u(w,z)$ with $k_c < 1$. Note that k_c represents the utility of a feasible design with worst possible attribute values, which is introduced to distinguish the situations between a feasible design of zero utility and an infeasible design where utility is always zero, as defined in Eq. 2.56. In this example, we set $k_c = 0.2$.

Step 4: Search Designs

In this step, we introduce three cases illustrating different aspects of the design scenarios.

Case 4: Base case—Risk neutral ($r_w = r_z = 0$) and no preference ($k_w = 0.5$, $k_z = 0.5$)

The optimum design is found at $b = 10$ mm and $h = 66.1$ mm, in which the MAU function is $u(w,z) = 0.9498$, and the weight and displacement of the beam are, respectively, $w = 3.50$ N and $z = 0.161$ mm. The solution graphs are shown in Figure 2.21, and the MATLAB script is given in Appendix A.

Recall in Case 1 that the optimum is found for the unconstrained problem at $b = 10$ mm and $h = 57.7$ mm. Obviously, for the constrained problem, when the stress constraint is taken into consideration, the height of the beam cross section is increased to $h = 66.1$ mm in order to bring the design into the feasible range.

Cases 2 and 3 are repeated with the stress constraint as Cases 5 and 6. That is, in Case 5a, we assume $k_w = 0.7$, $k_z = 0.3$; for Case 5b, we assume $k_w = 0.3$, $k_z = 0.7$. In Case 6a, we change the attitude toward risk for the weight attribute to risk prone ($r_w = -2$); and in Case 6b, to risk averse ($r_w = 2$). Results are listed in Table 2.13. Results of Cases 5a, 6a, and 6b are identical to those of Case 4.

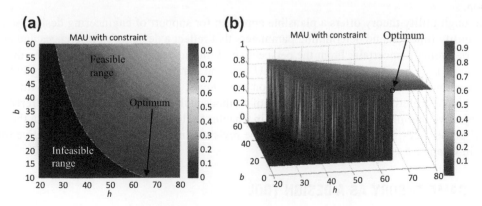

FIGURE 2.21

MAU function and optimum solution of Case 4: (a) top view and (b) iso-view.

Table 2.13 Result Comparison for the Constrained Design Problems

Case No.	Problem Setup				Results					
	k_w	k_z	r_w	r_z	b (mm)	h (mm)	w (N)	z (mm)	σ (MPa)	u
Case 4	0.5	0.5	0	0	10	66.1	3.50	0.161	27.46	0.950
Case 5a	0.7	0.3	0	0	10	66.1	3.50	0.161	27.46	0.938
Case 5b	0.3	0.7	0	0	10	71.3	3.78	0.128	23.60	0.962
Case 6a	0.5	0.5	−2	0	10	66.1	3.50	0.161	27.46	0.906
Case 6b	0.5	0.5	2	0	10	66.1	3.50	0.161	27.46	0.977

This is because that the weight cannot be reduced further without violating the stress constraint, although a preference is given to the weight attribute (Case 5a) and risk prone toward weight attribute is assumed (Case 6a). On the other hand, displacement can be further reduced by increasing the height h without violating the stress constraint. As a result, we observe displacement reduction due to the height increment in Case 5b.

Comparing Cases 1 and 4, with and without the stress constraint, the height h cannot be reduced to be less than 66.1 mm without violating the stress constraint, as shown in the results of Case 4. The same observation is found in the remaining cases, except for Cases 2 and 5b, in which the same results are found: displacement is reduced by increasing height h without violating the stress constraint.

2.6.1.3 Summary of utility theory as a design tool

Employing utility theory to support engineering design offers several advantages. First, SAU utility functions that normalize the respective attributes are constructed so that these attributes can be treated with uniformity. Second, a designer's attitude toward risk is captured and incorporated into design decision making. Third, a designer's preferences are well represented in a MAU function, either in an additive or multiplicative form. Also, constraints can be handled by constructing a continuous function closely resembling a step function, which brings the utility of the constraint function into the MAU function.

Although utility theory offers a plausible approach for support of engineering design, selecting proper utility function and determining parameters that reflect a designer's preferences and attitude toward risk remain uncertain. From the perspective of multiobjective optimization (to be discussed in Chapter 5), by maximizing one single MAU function that is converted by combining individual SAU functions represents one of the solution techniques, referred to as methods with a priori articulation of preferences. The pros and cons of methods with a priori articulation of preferences for multiobjective optimization will be discussed in Chapter 5, along with methods other than those with a priori preferences. The beam examples will be revisited in the context of multiobjective optimization in Chapter 5.

2.6.2 GAME THEORY AS A DESIGN TOOL

A cylindrical pressure vessel example is discussed in this subsection to illustrate the idea of using game theory as a design tool. We formulate the design problem as three different games: a strategy form

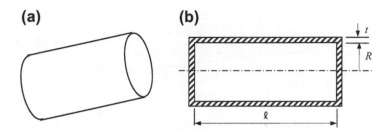

FIGURE 2.22

Cylindrical pressure vessel: (a) iso-view and (b) section view with dimensions.

game, sequential game, and cooperative game. To minimize computation and mathematical equations involved, we use graphs in design space and criterion space to facilitate the concept illustration.

2.6.2.1 The pressure vessel design example

A pressure vessel shown in Figure 2.22 is internally pressurized with pressure $P = 4000$ psi. The length, radius, and thickness of the vessel are, respectively, $\ell = 100$ in., $R = 30$ in., and $t = 1$ in. The weight density and yield strength of the vessel material are, respectively, $\gamma = 0.283$ lb/in. and $S_y = 32{,}000$ psi.

The design objectives are to minimize the weight W and maximize the volume V of the vessel by varying two design variables, radius R and thickness t; that is,

$$\text{Minimize}: \quad W(R,t) = \gamma \left[\pi(R+t)^2(L+2t) - \pi R^2 \ell \right] \tag{2.57}$$

$$\text{Maximize}: \quad V(R,t) = \pi R^2 \ell \tag{2.58}$$

The constraints include the hoop stress and a number of inequality constraints on the geometric dimensions of the design variables, as listed below:

$$\sigma(R,t) = PR/t \leq S_y \tag{2.59}$$

$$5t - R \leq 0 \tag{2.60}$$

$$t + R \leq 40 \tag{2.61}$$

In addition, sides constraints specify the upper and lower bounds for the respective design variables R and t; that is,

$$0.5 \leq t \leq 6 \text{ in.} \tag{2.62}$$

$$0 \leq R \leq 38 \text{ in.} \tag{2.63}$$

The constraints defined in Eqs 2.59 and 2.60 can be combined and simplified into the following constraint:

$$5t \leq R \leq 8t \tag{2.64}$$

Therefore, the feasible region can be identified as a polygon ABCDE by intersecting the four constraint equations (Eqs 2.61 to 2.64), and it is graphed in Figure 2.23. Note that the intersecting points are calculated as A(0.5, 2.5), B(0.5, 4), C(4.44, 35.55), D(6, 34), and E(6, 30).

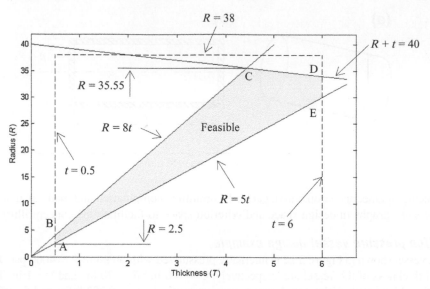

FIGURE 2.23

The feasible range of the pressure vessel design problem.

Note that in Figure 2.23, the upper and lower bounds of the design variables R and t are represented by dashed lines. Apparently, due to constraints other than the side constraints, the original upper and lower bounds of R will never be activated. The feasible bounds for R are now reduced to

$$2.5 \le R \le 35.55 \text{ in.} \tag{2.65}$$

which will be automatically satisfied when all other constraints are not violated. Therefore, the design problem is reduced to Eqs 2.57, 2.58, 2.61, 2.62, 2.64, and 2.65.

We assume two players (or designers) are playing (or providing design decisions) this game (or design problem). Player W wishes to minimize the weight of the vessel W by varying the thickness design variable t. Player V wishes to maximize the volume of the vessel V by varying the radius design variable R.

2.6.2.2 Strategy form game

In this game, we assume that players are noncooperative. They are varying their respective design variable for the best possible objective that can be achieved. The solution to this problem is a Nash equilibrium or Nash solution, as discussed in Section 2.5.2.

First, player V needs to maximize V by changing R for any given strategy of player W; that is,

$$\text{Minimize :} \quad -V = -\pi R^2 \ell \tag{2.66}$$

$$\text{Subject to :} \quad t + R \le 40 \tag{2.67}$$

$$5t \le R \le 8t \tag{2.68}$$

$$2.5 \le R \le 35.55 \tag{2.69}$$

Note that $-V$ decreases monotonically with increasing R, which means player V needs to choose the largest possible R. Therefore, the solution (also called the rational reaction set in literature) is formulated as below, by reviewing constraint equations (Eqs 2.67 and 2.68):

$$R^*(t) = \begin{cases} 40 - t, & \text{if } 0.5 \leq t \leq 4.45 \\ 8t, & \text{if } 4.45 \leq t \leq 6 \end{cases} \tag{2.70}$$

or

$$R^*(t) = \min\{40 - t, \ 8t\} \tag{2.71}$$

which depends on the thickness design variable t, determined by player W.

Note that this solution can be easily obtained using the feasible region diagram (red line BCD in Figure 2.24a), requiring player V to maximize R for each t within the feasible range.

In the meantime, player W needs to minimize weight of the vessel W (by changing t) for any given strategy of player V; that is,

$$\text{Minimize}: \quad W = \rho\left[\pi(R+t)^2(\ell + 2t) - \pi R^2 \ell\right] \tag{2.72}$$

$$\text{Subject to}: \quad t + R \leq 40 \tag{2.73}$$

$$5t \leq R \leq 8t \tag{2.74}$$

$$0.5 \leq t \leq 6 \text{ in.} \tag{2.75}$$

Note that W decreases monotonically with decreasing t, which means player W needs to choose the smallest possible t. Therefore, the solution (rational reaction set) can be obtained as

$$t^*(R) = \begin{cases} R/8, & \text{if } 4 \leq R \leq 35.5 \\ 0.5, & \text{if } 2.5 \leq R \leq 4 \end{cases} \tag{2.76}$$

or

$$t^*(R) = \max\{0.5, \ R/8\} \tag{2.77}$$

which depends on R.

Again, the solution can be easily observed from the feasible space diagram (blue line ABC in Figure 2.24b); that is, player W needs to minimize t for each R within the feasible range.

The Nash solution is where the rational reaction sets of the two players intersect with each other. In this case, the Nash solution is $R = 8t$, and $4 \leq R \leq 35.5$, which is represented by the straight line segment BC (green) in the design space, as shown in Figure 2.25.

2.6.2.3 Sequential game

We present two cases: Case 1, in which player W is the leader, and Case 2, in which player V is the leader.

Case 1: Player W is the leader.

As discussed in the previous section, the design problem for player W is

$$\text{Minimize}: \quad W = \rho\left[\pi(R+t)^2(\ell + 2t) - \pi R^2 \ell\right] \tag{2.78}$$

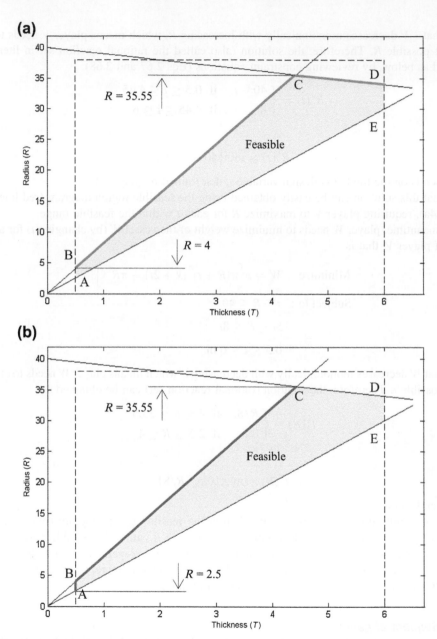

FIGURE 2.24

Exploration of design solutions in the design space: (a) solutions for player *W* and (b) solutions for player *V*.

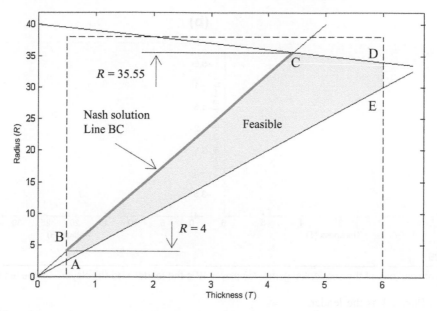

FIGURE 2.25

Nash solution represented on line segment BC.

$$\text{Subject to: } R = R^*$$

$$0.5 \le t \le 6 \text{ in.} \tag{2.79}$$

where R^* is the solution to the follower's (player V) problem, given by

$$\text{Minimize} : - V = -\pi R^2 \ell \tag{2.80}$$

$$\text{Subject to} : t + R \le 40 \tag{2.81}$$

$$5t \le R \le 8t \tag{2.82}$$

$$2.5 \le R \le 35.55 \tag{2.83}$$

Note that the follower's problem has been solved above when calculating the Nash solution, as

$$R^*(t) = \min\{40 - t, \ 8t\} \tag{2.84}$$

Substituting R^* into the leader's design problem, the leader's objective function W then becomes a function of t only, which is plotted in Figure 2.26a.

Because $W(t)$ increases monotonically with respect to t, the solution to the leader's problem (minimize W) is $t = 0.5$ in.; hence, $R = 8t = 4$ in. This solution represents a very long vessel with small thickness and radius.

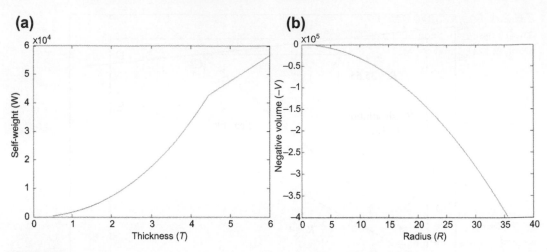

FIGURE 2.26

Illustration of solutions of the sequential game: (a) player *W* is the leader and (b) player *V* is the leader.

Case 2: Player *V* is the leader.

As discussed in the previous section, the design problem for player *V* is

$$\text{Minimize}: \quad -V = -\pi R^2 \ell \tag{2.85}$$

$$\text{Subject to}: \quad t = t^*$$

$$2.5 \leq R \leq 35.55 \tag{2.86}$$

where t^* is the solution to the follower's (player *W*) problem, given by

$$\text{Minimize}: \quad W = \rho\left[\pi(R+t)^2(\ell+2t) - \pi R^2 \ell\right] \tag{2.87}$$

$$\text{Subject to}: \quad t + R \leq 40 \tag{2.88}$$

$$0.5 \leq t \leq 6 \text{ in.} \tag{2.89}$$

$$5t \leq R \leq 8t \tag{2.90}$$

Note that the follower's problem has been solved above when calculating the Nash solution, as $t^*(R) = \max\{0.5, R/8\}$.

Substituting t^* into the leader's design problem, the leader's objective function $-V$ then becomes a function of R only, which is plotted in Figure 2.26b.

Because $-V(R)$ decreases monotonically with respect to R, the solution to the leader's problem (minimize $-V$) is $R = 35.5$ in.; hence, $t = R/8 = 4.44$ in. This solution represents a narrow vessel with large thickness and radius.

Neither a very long vessel with small thickness and radius obtained from the game that player *W* is the leader or a narrow vessel with large thickness and radius (player *V* is the leader) is desirable in the design of the pressure vessel. A compromised design is generally considered to be better than these two extremes.

2.6.2.4 Cooperative game

As discussed in Section 2.5.4, the solution to a cooperative game is Pareto optimal. The design problem is restated as below.

$$\text{Minimize}: \quad W(R,t) = \gamma \left[\pi (R+t)^2 (L+2t) - \pi R^2 \ell \right] \tag{2.91}$$

$$\text{Maximize}: \quad V(R,t) = \pi R^2 \ell \tag{2.92}$$

$$\text{Subject to}: \quad t + R \leq 40 \tag{2.93}$$

$$0.5 \leq t \leq 6 \text{ in.} \tag{2.94}$$

$$5t \leq R \leq 8t \tag{2.95}$$

$$2.5 \leq R \leq 35.55 \text{ in.} \tag{2.96}$$

The solution graphs are shown in Figure 2.27, and the MATLAB script is given in Appendix A. Note that only solutions corresponding to feasible set are plotted. Curve ABC represents the Pareto solution, while the Nash solution in the objective space is curve BC, which is a subset of the Pareto front.

Finally, point B is the solution to the sequential game with player W as the leader; point C is the solution to the sequential game with player V as the leader.

2.6.2.5 Summary on game theory as design tool

Employing game theory to support engineering design broadens the scope of the type and scenarios of the design problems that can be formulated and solved. For a cooperative game, Pareto optimal and relevant solution techniques for solving multiobjective optimization problems have been well developed, as will be discussed in Chapter 5. However, formulating a design problem as a strategy

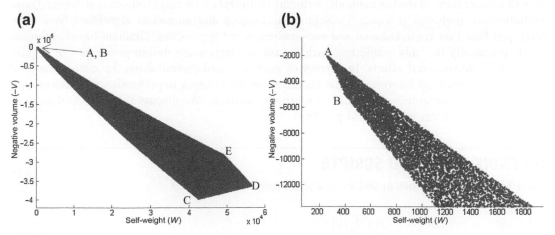

FIGURE 2.27

Pareto optimal of the cooperative game: (a) feasible set enclosed by points A to E and (b) zoomed-in view near points A and B.

form game or a sequential game that supports multiple designers or decentralized design teams offers a plausible approach to solving practical design scenarios that have not been fully explored. Although game theory offers a plausible approach for support of engineering design, more research work needs to be done, such as in developing a systematic approach for formulating and decomposing a design problem into a set of subproblems to be solved by a group of design teams. A practical and viable design tool based on game theory is intriguing and is highly anticipated in the design community.

2.7 SUMMARY

In this chapter, we focused on decision making in support of engineering design. We discussed two conventional methods, the decision matrix and decision tree, which are effective and powerful in support of high-level decision making. Risk and uncertainty are the two major issues that decision makers must confront. We reviewed a few basics of decision theory that support decision making under risk and uncertainty. We also discussed two theories, utility theory and game theory, which were explored to support engineering design. Although not matured yet, design methods based on the theories offer significant advantages over conventional decision making and optimization techniques.

One of our goals in this chapter was to introduce the design theory that led to the development of decision theories and decision-based design. We touched a bit on the topic by presenting two design examples that used utility theory and game theory as design tools, respectively. We hope the discussion provided serves as a gateway to those who are interested in broadening their view through learning new ideas and methods, and for those who are interested in entering the research field of design theory and methods.

We believe this chapter serves its purpose as a prelude to the subject of design theory and methods. We hope this chapter offers adequate breadth and depth that help readers move on to the following chapters. The remaining chapters offer somewhat more mathematical and rigorous discussion on design methods that are well developed and being used in both academia and industry. We will go over some of the key methods, including optimization for single objective (Chapter 3) and multiobjective problems (Chapter 5). Solution techniques and numerical algorithms have been developed based on gradient-based and non-gradient-based approaches. Gradient-based optimization is practically the only viable approach for solving large-scale design problems that require substantial computational efforts. In support of gradient-based optimization, the methods require calculations of gradient information that characterizes the changes in performance measures with respect to design variables—the so-called sensitivity analysis. We discuss this important topic in Chapter 4, with a focus on structural problems.

APPENDIX A. MATLAB SCRIPTS

Script 1: graphing functions u_1 and u_2 in design space in Example 2.9.

```
clear all;
[q1,q2] = meshgrid(0:1:100,0:1:100);
u1 = -q1.^2+(198-q2).*q1;
[C,h] = contour(q1,q2,u1,3);
```

```
set(h,'ShowText','on','TextStep',get(h,'LevelStep')*2)
colormap cool
hold on;
u2 = -q2.^12+(198-q1).*q2;
[C,h] = contour(q1,q2,u2,3);
set(h,'ShowText','on','TextStep',get(h,'LevelStep')*2)

colormap cool.

hold off;
```

Script 2: graphing functions u_1 and u_2 in criterion space in Example 2.9

```
%Cournot example; plot objective domain
clear all; format long
NN = 10,000;     %No. of random points for plotting
%-----------------------------------------------
A = 1; %counting feasible points
B = 1; %counting infeasible points
Smax = 0;
q1o = 0;
q2o = 0;
for ii = 1:NN
        q1 = rand*100;
        q2 = rand*100;
        u1 = -q1^2+(198-q2)*q1;
        u2 = -q2^2+(198-q1)*q2;
        if q1>0 & q2>0;
                feasi(:,A) = [u1,u2];
                if Smax < feasi(1,A);
                        Smax = feasi(1,A);
                        q1o = q1; q2o = q2;
                end
                A = A+1;
        else
                infeasi(:,B) = [u1,u2];
                B=B+1;
        end
end
figure (1)
plot(feasi(1,:),feasi(2,:),'bs','MarkerSize',3)
plot(infeasi(1,:),infeasi(2,:),'rs','MarkerSize',3)
xlabel('u1'); ylabel('u2')
axis equal
```

Script 3: solving the unconstrained problem of the cantilever beam example in Section 2.6.1.1

```
%Unit: mm,N,MPa
clear all
%------------- Parameters that can be adjusted
```

```
k_w = 0.5; k_z = 0.5;              %%%%% Weighting between attributes
r_w = 0.0001;    r_z = -0.0001;    %%%%% Model risk; r > 0 risk adverse;
r < 0 risk adverse; r~0 risk neutral
b_u = 60; b_l = 30;                %%%% Upper and lower bounds of base
h_u = 80; h_l = 40;                %%%% Upper and lower bounds of height
interval = 0.1;                    %%%%% Solution search interval size
%------------ Define properties
l=200;                  P=1000;         E=69,000;        YS=27.6;
rou = 2700*10^(-9);     %Kg/mm3
%----------------------- An example; special case

b=30; h=40; I=b*h3/12; z=*l3/(3*E*I); M=P*l; stress=M*(h/2)/I; w=rou*b*h*l*9.8;

%-----------------------------------
wl=rou*b_u*h_u*l*9.8;              w_u = rou*b_l*h_l*l*9.8;
z_l=P*l^3/(3*E*(b_l*h_l^3/12));    z_u= P*l^3/(3*E*(b_u*h_u^3/12));


x_w = w_l: 0.01:w_u;              x_z = z_l: 0.001:z_u;
s_w=(x_w-w_l)./(w_u-w_l);         s_z=(x_z-z_l)./(z_u-z_l);
u_w=(1-exp(-r_w.*s_w))./(1-exp(-r_w));  u_z=(1-exp(-r_z.*s_z))./(1-exp(-r_z));


figure (1);    % plot utility curves
subplot(1,2,1); plot(x_w,u_w); xlabel('Self weight'); ylabel('Utility')
subplot(1,2,2); plot(x_z,u_z); xlabel('Displacement'); ylabel('Utility')
%--------------------------------------- Build MAU
B = b_l:interval:b_u;
H = h_l:interval:h_u;
W = rou*B.'*H*l*9.8;
Z = P.*l.^3./(3*E*(B.'*H.^3/12));

S_w=(W-w_l)./(w_u-w_l);
S_z=(Z-z_l)./(z_u-z_l);

U_w=(1-exp(-r_w.*S_w))./(1-exp(-r_w));
U_z=(1-exp(-r_w.*S_z))./(1-exp(-r_w));
K=(1-k_w-k_z)/(k_w*k_z);
U_MAU = k_w*U_w + k_z*U_z + K*k_w*k_z*U_w.*U_z;
%------------------------------------------- Plot Result
figure (2); surf(H,B,U_MAU); shading
interp; colormap(jet); colorbar; view(0,90); title('MAU'); xlabel('h'); ylabel('b')
%------------------------------------------- Search for optimum
[Row,Col] = find(U_MAU = = max(max(U_MAU)));
B_H_O = [b_l+(Row-1)*interval h_l+(Col-1)*interval]
W_Z_O = [rou*B_H_O(1)*B_H_O(2)*l*9.8 P*l^3/(3*E*(B_H_O(1)*B_H_O(2)^3/12))]
%end
```

Script 4: solving the constrained problem of the cantilever beam example in Section 2.6.1.2.

```
%Unit: mm,N,MPa
```

```
clear all
%------------ Parameters that can be adjusted
 k_w=0.5; k_z=0.5;                        %%%%% Weighting between attributes
 r_w = 0.0001;      r_z = -0.0001;        %%%%% Model risk; r > 0 risk adverse; r <
0 risk adverse; r~0 risk neutral
 slope = -0.01;                           %%%%% Slope of the constraint utility function
 k_C = 0.2;                               %%%%% parameter for MAU with constraint
 b_u = 60; b_l = 10;                      %%%%% Upper and lower bounds of base
 h_u = 80; h_l = 20;                      %%%%% Upper and lower bounds of height
 interval = 0.1;                          %%%%% Solution search interval size
%------------- Define properties
 l = 200;      P = 1000;          E = 69,000;     YS = 27.6;
 rou = 2700*10^(-9);   %Kg/mm3
%----------------------- An example; special case
b=60; h=80; I=b*h^3/12; z=P*l^3/(3*E*I); M=P*l; stress = M*(h/2)/I; w=rou*b*h*l*9.8;
%-------------------------------- Utility functions for individual attributes
w_l = rou*b_u*h_u*l*9.8;               w_u = rou*b_l*h_l*l*9.8;
z_l = P*l3/(3*E*(b_l*h_l^3/12));       z_u = P*l^3/(3*E*(b_u*h_u^3/12));
stress_l = P*l*(h_u/2)/(b_u*h_u^3/12);   stress_u = P*l*(h_l/2)/(b_l*h_l^3/12);


x_w = w_l: 0.01:w_u;                   x_z = z_l: 0.001:z_u;
s_w=(x_w-w_l)./(w_u-w_l);              s_z=(x_z-z_l)./(z_u-z_l);
u_w=(1-exp(-r_w.*s_w))./(1-exp(-r_w));  u_z=(1-exp(-r_z.*s_z))./(1-exp(-r_z));


figure (1);     % plot utility curves
subplot(1,3,1); plot(x_w,u_w); xlabel('Self weight'); ylabel('Utility')
subplot(1,3,2); plot(x_z,u_z); xlabel('Displacement'); ylabel('Utility')
x_stress = stress_l:0.1:stress_u;
u_stress = 1./(1 + exp((YS-x_stress)./slope));
subplot(1,3,3); plot(x_stress, u_stress);
xlabel('Stress'); ylabel('Utility'); axis([stress_l stress_u -0.1 1.1]);
%-------------------------------- Build MAU
K=(1-k_w-k_z)/(k_w*k_z);


B = b_l:interval:b_u;
H = h_l:interval:h_u;
W = rou*B.'*H*l*9.8;
Z = P.*l.^3./(3*E*(B.'*H.^3/12));


S_w=(W-w_l)./(w_u-w_l);
S_z=(Z-z_l)./(z_u-z_l);


U_w=(1-exp(-r_w.*S_w))./(1-exp(-r_w));
U_z=(1-exp(-r_w.*S_z))./(1-exp(-r_w));
U_MAU = k_w*U_w + k_z*U_z + K*k_w*k_z*U_w.*U_z;     %MAU w/o constraint
%-------------------------------------------- Constraint
STRESS = 6.*P.*l./(B.'*H.^2);
U_C = 1./(1 + exp((YS-STRESS)./slope));
```

```
U_MAU_C=(k_C+(1-k_C)*U_MAU).*U_C;          %MAU with constraint
%-------------------------------------------- Plot Result
figure (2); surf(H,B,STRESS); shading
interp; colormap(jet); colorbar; view(0,90); title('Stress'); xlabel('h'); ylabel('b')
figure (3); surf(H,B,U_C); shading
interp; colormap(jet); colorbar; view(0,90); title('Constraint'); xlabel('h'); ylabel('b')
figure (4); surf(H,B,U_MAU); shading
interp; colormap(jet); colorbar; view(0,90); title('MAU'); xlabel('h'); ylabel('b')
figure (5); surf(H,B,U_MAU_C); shading
interp; colormap(jet); colorbar; view(0,90); title('MAU with constraint'); xlabel('h');
ylabel('b')
%-------------------------------------------- Search for optimum
[Row,Col] = find(U_MAU = = max(max(U_MAU)));
[Row_C,Col_C] = find(U_MAU_C = = max(max(U_MAU_C)));
B_H_O = [b_l+(Row-1)*interval h_l+(Col-1)*interval]
W_Z_O = [rou*B_H_O(1)*B_H_O(2)*l*9.8 P*l^3/(3*E*(B_H_O(1)*B_H_O(2)^3/12))]
B_H_O_C = [b_l+(Row_C-1)*interval h_l+(Col_C-1)*interval]
W_Z_O_C = [rou*B_H_O_C(1)*B_H_O_C(2)*l*9.8 P*l^3/(3*E*(B_H_O_C(1)*B_H_O_C(2)^3/12))]
%end
```

Script 5: graphing the Pareto solution of Figure 2.27 in Section 2.6.2.3.

```
NN = 1,000,000;          % No. of points
%------------------
T_u = 6; T_l = 0.5;                       %%%%% Upper and lower bounds of thickness
R_u = 38; R_l = 0;                        %%%%% Upper and lower bounds of radius
%------------- Define properties
rou=0.283; L=100; St=32; P=4;
%-----------------------
A = 0; B = 0;              % Count number of feasible and infeasible points
for ii = 1:NN
      TT = T_l + rand*(T_u-T_l);
      RR = R_l + rand*(R_u-R_l);
      WGT = rou*pi*((RR + TT)^2*(L+2*TT)-RR^2*L);
      VOL = -pi*RR^2*L;
      stress = P*RR/TT;
      if stress < St && 5*TT ≤ RR && RR ≤ 40-TT
            A = A+1;
            Feasi(:,A) = [WGT,VOL];
      else
            B=B+1;
            Infeasi(:,B) = [WGT,VOL];
      end
end
figure (5)
hold on
plot(Feasi(1,:),Feasi(2,:),'b.','MarkerSize',8)        % Plot feasible solutions
%plot(Infeasi(1,:),Infeasi(2,:),'r.','MarkerSize',8)   % Plot infeasible solutions
%legend('Feasible','Infeasible')
```

```
xlabel('Self-weight (W)'); ylabel('Negative volume (-V)')
%----------------------------------------- Plot Nash solution
T_Nash = 0.5:0.1:4.4;
R_Nash = 8.*T_Nash;
WGT_Nash = rou.*pi.*((R_Nash + T_Nash).^2.*(L+2.*T_Nash)-R_Nash.^2.*L);
VOL_Nash = -pi.*R_Nash.^2.*L;
plot(WGT_Nash, VOL_Nash,'k.','MarkerSize',20)
hold off
```

QUESTIONS AND EXERCISES

1. If the weighting factors listed in Table 2.3 are revised as Strength: 2, Weight: 4, Machinability: 3, Corrosion resistance: 5, and Material cost: 3, which material among the five is desirable to be used?

2. A manufacturing firm is moving to absorb some short-term excess production capacity at one of its plants. The firm is considering a short manufacturing run for either of two new products, a coffee maker or a blender. The market for each product is known if the products can be successfully developed. However, there is some chance that it will not be possible to successfully develop them. Revenue of $1,000,000 would be realized from selling the coffee maker and revenue of $400,000 would be realized from selling the blender. Both of these amounts are net of production cost but do not include development cost. If development is unsuccessful for a product, then there will be no sales, and the development cost will be totally lost. Development cost would be $100,000 for the coffee maker and $10,000 for the blender. Suppose that the probability of development success is 0.5 for the coffee maker and 0.8 for the blender. Sketch a decision tree diagram to describe the problem and calculate the payoffs to help the firm make an adequate decision.

3. The following payoff table was developed. Let $P(S1) = 0.25$, $P(S2) = 0.55$, and $P(S3) = 0.20$. Compute the expected monetary value for each of the alternatives. What decision would you recommend?

	State of Nature		
Alternative	**S1**	**S2**	**S3**
A1	$50	$65	$120
A2	$85	$45	$80
A3	$70	$90	$105

4. An agriculture manufacturing company has seen its business expanded to the point where it needs to increase production beyond its existing capacity. It has narrowed the alternatives to two approaches to increase the maximum production capacity:
   a. Expansion, at a cost of $8 million, or
   b. Modernization at a cost of $5 million

Both approaches would require the same amount of time for implementation. Management believes that over the required payback period, demand will either be high or moderate. Because high demand is considered to be somewhat less likely than moderate demand, the probability of high demand has been estimated at 0.35. If the demand is high, expansion would gross an estimated additional $12 million but modernization only an additional $6 million, due to lower maximum production capability. On the other hand, if the demand is moderate, the comparable figures would be $7 million for expansion and $5 million for modernization. Calculate the EMV for each course of action, (a) and (b).

5. A steak house is contemplating opening a new restaurant on Main Street. It has three different models, each with a different seating capacity. They estimate that the average number of customers per hour will be 80, 120, or 160. The payoff table (profits) for the three models is developed as shown in the table below,

	Average Number of Customers Per Hour		
	S1 = 80	S2 = 120	S3 = 160
Model A	$10,000	$18,000	$24,000
Model B	$6000	$16,000	$12,000
Model C	$3000	$19,000	$28,000

Calculate the expected value for each decision.

6. On a busy highway in an urban area, speeding is common. A speeding ticket costs an amount of $t > 0$. Some people drive aggressively and drive over the speed limit. There is a probability $p > 0$ of being caught, which leads to a fine $f > t$. In order to slow down the drivers on the busy highway, there are two possible concepts: doubling the fine f or doubling the patrols (i.e., the probability p). Assuming that the drivers are risk-averse, risk-neutral, and risk-prone, which is the better concept for these three kinds of drivers respectively—double the fine or double the patrols?

7. Jeff is an e-trader. Suppose that Jeff's utility as a function of the money he has in his trading account, x, is given by $U(x) = \ln x$ (the natural logarithm of x).
   a. Is Jeff risk averse? Explain why or why not.
   b. Jeff now has $10,000 and two possible decisions. For decision 1, he loses $500 for certain (by paying the required fees to keep trading). For decision 2, he loses $0 with probability 0.9 and loses $5000 with probability 0.1. Which decision maximizes the expected utility of his money in the account?

9. Formulate the unconstrained beam design problem, as discussed in Section 2.6.1.1, as a strategy form game. Find the Nash equilibrium of this problem.

10. Solve the beam design problem shown in Section 2.6.1.1. In this case, instead of a cantilever beam, assume that the beam is clamped at both ends. Solve this unconstrained problem for the following cases:
   Case 1: The designer is risk neutral with preferences $k_w = 0.5$, $k_z = 0.5$.
   Case 2: The designer is risk averse for the weight attribute and is risk prone for the displacement attribute with preferences $k_w = 0.5$, $k_z = 0.5$.

Case 3: The designer is risk neutral with preferences $k_w = 0.8$, $k_z = 0.2$.

Compare the results of the three cases and make observations in terms of the influences of attitude toward risk and preferences to the solutions.

REFERENCES

Akao, Y., 2004. QFD: Quality Function Deployment – Integrating Customer Requirements into Product Design. Productivity Press, New York.

Aliprantis, C.D., Chakrabarti, S.K., 2006. Game theory in decision-making. In: Lewis, K.E., Chen, W., Schmidt, L.C. (Eds.), Decision Making in Engineering Design. ASME Press, New York, pp. 245–264 (Chapter 21).

Bernoulli, D., January 1954. Exposition of a new theory on the measurement of risk. Econometrica (The Econometric Society) (Dr. Sommer L, Trans.) 22 (1), 22–36 (Originally published in 1738).

Chang, K.H., 2013. Product Performance Evaluation Using CAD/CAE, the Computer Aided Engineering Design Series. Academic Press, Burlington, MA.

Eatas, A., Jones, J.C., 1996. The Engineering Design Process, second ed. John Wiley and Sons, Inc, New York.

Hazelrigg, G.A., 1996. Systems Engineering: An Approach to Information-Based Design. Prentice Hall, Upper Saddle River, NJ.

Huang, G.Q. (Ed.), 1996. Design for X Concurrent Engineering Imperatives. Springer, Netherlands.

Hyman, B., 2003. Fundamentals of Engineering Design, second ed. Pearson Education, Inc., Upper Saddle River, NJ.

Iyer, H.V., Tang, X., Krishnamurty, S., 1999. Constraint Handling and Iterative Attribute Method Building in Decision-Based Engineering Design, ASME Design Engineering Technical Conference. DETC99/DAC-8582, Las Vegas, NV.

Lewis, K.E., Chen, W., Schmidt, L.C., 2006. Decision Making in Engineering Design. ASME Press, New York.

Lochner, R.H., Matar, J.E., 1990. Designing for Quality: An Introduction to the Best of Taguchi and Western Methods of Statistical Experimental Design. Springer, the Netherlands.

Martin, R., 2004. The st. Petersburg paradox. In: Zalta, E.N. (Ed.), The Stanford Encyclopedia of Philosophy, Fall 2004 ed. Stanford University, Stanford, CA.

National Science Foundation, 1996. Research Opportunities in Engineering Design. NSF Strategic Planning Workshop Final Report. National Science Foundation, Washington, DC.

North, D.W., 1968. A tutorial introduction to decision theory. IEEE Transactions on Systems Science and Cybernetics SSC-4 (3), 200–210.

von Neumann, J., Morgenstern, O., 1947. Theory of Games and Economic Behavior, second ed. Princeton University Press, Princeton, NJ.

Park, S.H., Antony, J., 2008. Robust Design for Quality Engineering and Six Sigma. World Scientific, Singapore; Hackensack, NJ.

Shtub, A., Brad, J.F., Globerson, S., 1994. Project Management: Engineering, Technology, and Implementation. Prentice-Hall, Inc, Englewood Cliffs, NJ.

Suh, N.P., 1990. The Principles of Design. Oxford University Press, Oxford.

Voland, G., 2004. Engineering by Design, second ed. Prentice Hall, Upper Saddle River, NJ.

DESIGN OPTIMIZATION

CHAPTER OUTLINE

Design Theory and Methods using CAD/CAE.
Copyright © 2015 Elsevier Inc. All rights reserved.

In this chapter, we discuss design optimization—one of the mainstream methods in support of engineering design. In design optimization, we minimize (or maximize) an objective function subject to performance constraints by varying a set of design variables, such as part dimensions, material properties, and so on. Usually we deal with one objective function in carrying out the so-called single-objective optimization problem, which is the subject of this chapter. Often, there are more than one objective functions to be minimized simultaneously. Such problems are called multi-objective (or multi-criteria) optimization (MOO) problems, which will be discussed in Chapter 5. The objective function and performance constraints are usually extracted from or defined for a physical problem of a single engineering discipline. For example, if we are designing a load-bearing component, in which we minimize the structural weight and subject to constraints characterizing structural strength performance, we are solving a structural optimization problem that is single disciplinary. In many situations, especially in the context of e-Design, we are dealing with designing a product or a system that involves multiple engineering disciplines, including structural, kinematic and dynamic, manufacturing, product cost, and so forth. Such design problems are called multi-disciplinary optimization (MDO), which will be briefly discussed in Chapter 5, and Projects P5 and S5 through a single-piston engine example.

The existence of optimization methods can be traced to the days of Newton, Lagrange, and Cauchy. The development of differential calculus methods for optimization was possible because of the contributions of Newton and Leibnitz to calculus. The foundations of the calculus of variations, which deals with the minimization of functions, were laid by Bernoulli Euler, Lagrange, and Weistrass. The method of optimization for constrained problems, which involve the addition of unknown multipliers, became known by the name of its inventor, Lagrange. Cauchy made the first application of the steepest descent method to solve unconstrained optimization problems. By the middle of the twentieth century, high-speed digital computers made implementation of the numerical optimization techniques possible and stimulated further research on newer methods. Some major developments include the work of Kuhn and Tucker in the 1950s on the necessary and sufficient conditions for the optimal solution of programming problems, which laid the foundation for later research in nonlinear programming. Early development focused on gradient-based algorithms that employ gradient information of the objective and constraint functions in searching for optimal solutions. In the late twentieth century, non-gradient-based approaches, such as simulated annealing, genetic algorithms, and neural network methods, representing a new class of mathematical programming techniques, came into prominence. Today, optimization techniques are widely employed to support engineering design. The applications for mechanical system or component designs include car suspensions for durability, car bodies for noise/vibration/harshness, pumps and turbines for maximum efficiency, and so forth. There are plenty of applications beyond mechanical designs, such as optimal production planning, controlling, and scheduling; optimum pipeline networks for process industry; controlling the waiting and idle times in production lines to reduce the cost of production; optimum design of control systems; among many others.

This chapter offers an introductory discussion to the single-objective optimization problems. We provide a brief introduction to design optimization for those who are not familiar with the subject. Basic concepts include problem formulation, optimality conditions, and graphical solutions using simple problems for which optimal solutions can be found analytically. In addition, we discuss both linear and nonlinear programming and offer a mathematical basis for design problem formulation and solutions. We include both gradient-based and non-gradient approaches for solving optimization problems. For those who are interested in learning more about optimization or are interested in entering this technical area for research, there are several excellent references in the literature (Arora 2012; Vanderplaats 2005; Rao 2009).

Most of the solution techniques require a large amount of evaluations for the objective and constraint functions. The disciplinary or physical models that are employed for function evaluations are often very complex and must be solved numerically (e.g., using finite element methods). It can take significant amount of computation time for a single function evaluation. As a result, the solution process for an optimization problem can be extremely time-consuming. Therefore, in this chapter, readers should see clearly the limitations of the non-gradient approaches in terms of the computational efforts for large-scale design problems. The gradient-based approaches are more suitable to the typical problems in the context of e-Design.

In addition to offering optimization concept and solution techniques, we use functions provided in MATLAB to solve example problems that are beyond hand calculations. MATLAB scripts developed for numerical examples are provided for reference in Appendix A at the end of this chapter. Moreover, we offer a discussion on the practical aspects of carrying out design optimization for practical engineering problems, followed by a brief review on commercial software tools in hope of ensuring that readers are aware of them and their applicability to engineering applications. We include three case studies to illustrate the technical aspects of performing design optimization using commercial computer-aided design (CAD) and computer-aided engineering (CAE) software. In addition, a simple cantilever beam example, modeled in both Pro/MECHANICA Structure and SolidWorks Simulation, is offered to provide step-by-step details in using the respective software to carry out design optimization. Example files can be found on this book's companion Website (booksite.elsevier.com/). Detailed instructions on using these models and steps for carrying out optimization are given in Projects P5 and S5.

Overall, the objectives of this chapter are (1) to provide basic knowledge in optimization and help readers understand the concept and solution techniques, (2) to familiarize readers with practical applications of the optimization techniques and be able to apply adequate techniques for solving optimization problems using MATLAB, (3) to introduce readers to popular optimization software that is commercially available and typical practical engineering problems in case studies, and (4) to help readers become familiar with the optimization capabilities provided in Pro/MECHANICA Structure and SolidWorks Simulation for basic applications.

We hope this chapter helps readers to learn the basics of design optimization, become acquainted with the optimization subject in general, and be able to move on to the follow-up chapters to gain an in-depth knowledge of the broad topics of design methods.

3.1 INTRODUCTION

Many engineering design problems can be formulated mathematically as single-objective optimization problems, in which one single objective function is to be minimized (or maximized) subject to a set of constraints derived from requirements in, for example, product performance or physical sizes. As a simple example, we design a beer can for a maximum volume with a given amount of surface area, as shown in Figure 3.1a. The geometry of the can is simplified as the cylinder shown in Figure 3.1b with two geometric dimensions, radius r and height h. The volume and surface area of the can are $V = \pi r^2 h$ and $A = 2\pi r(r + h)$, respectively. The can design problem can be formulated mathematically as follows:

$$\text{Maximize :} \quad V(r,h) = \pi r^2 h \tag{3.1a}$$

$$\text{Subject to :} \quad A(r,h) = \pi r(r + 2h) = A_0 \tag{3.1b}$$

$$0 < r, \quad 0 < h \tag{3.1c}$$

(a)   **(b)**

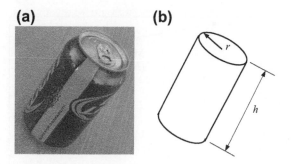

FIGURE 3.1

Beer can design: (a) beer can and (b) beer can simplified as a cylinder.

where A_0 is the given amount of surface area. In this case, $V(r, h)$ is the objective function to be maximized, and $A(r, h) = 2\pi r(r + h) = A_0$ is the constraint, or more precisely, an equality constraint to be satisfied. The radius r and height h are design variables. Certainly, both r and h must be greater than 0.

Solving the beer can design problem is straightforward because both the objective and constraint functions are expressed explicitly in term of the design variables, r and h. For example, from Eq. 3.1b, we have

$$r = -h + \sqrt{h^2 + \frac{A_0}{\pi}}$$

Bringing this equation back to Eq. 3.1a, we have

$$V = \pi h \left(-h + \sqrt{h^2 + \frac{A_0}{\pi}} \right)^2 \tag{3.2}$$

We hence converted a constrained optimization problem of two design variables to an unconstrained problem of single design variable h, which is much easier to solve. How do we solve Eq. 3.2 to find the optimal solution, in this case maximum volume of the beer can? One approach is to graph the volume function in terms of design variable h. For example, if the area is given as $A_0 = \pi$, Eq. 3.2 can be graphed as shown in Figure 3.2, for example, using MATLAB. Readers are referred to Appendix A (Script 1) to find the script that graphs the curve. From the graph, we can easily see that when the height h is about 1.2, volume reaches its maxima around 4.8.

The maximum of Eq. 3.2 can also be found by finding the solution of the derivative of Eq. 3.2, $dV/dh = 0$, and checking if the solution h satisfies $d^2V/dh^2 < 0$ (or if the objective function is concave), as we learned in Calculus. As seen in Figure 3.2, at $h = 1.2$, we have $dV/dh = 0$. Also, the function curve is concave in the neighborhood of $h = 1.2$. Hence, the rate of slope change is negative; that is, $d^2V/dh^2 < 0$.

The beer can design problem represents the simplest kind of optimization problem, in which both the objective and constraint are explicit functions of design variables. In most engineering problems, objective and constraint functions are too complex to be expressed in terms of design variables explicitly. In many cases, they have to be evaluated using numerical methods, for example, finite element methods. When the physical model (created in computer) to be solved for the objective and constraint function evaluations is large, the solution process for an optimization problem can be extremely time-consuming. This is especially true for multidisciplinary problems, in which

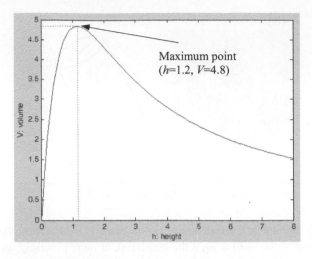

FIGURE 3.2

Graph of volume V in height h for the solution of the beer can example.

performance constraints are evaluated through intensive computations of multiple physical models involved in characterizing the physical behavior of the product. As a result, many methods and algorithms are developed to support design optimization by reducing computation time through minimizing the number of function evaluations.

In this chapter, we use simple and analytical examples to illustrate the optimization concept and some of the most popular solution techniques. When you review the concept and solution methods, please keep in mind that the objective and constraint functions are not necessarily expressed in terms of design variables explicitly. We discuss optimization problem formulation in Section 3.2, and then we introduce optimality conditions in Section 3.3. We include three basic solution approaches: the graphical methods in Section 3.4, and the gradient-based methods for constrained and unconstrained problems in Sections 3.5 and 3.6, respectively. Two popular solution techniques using a non-gradient approach, genetic algorithm and simulated annealing, are briefly discussed in Section 3.7. In Section 3.8, we discuss practical aspects of solving engineering optimization problems, followed by a short review on optimization software in Section 3.9. In Section 3.10, we include three case studies and followed by a tutorial example in Section 3.11.

3.2 OPTIMIZATION PROBLEMS

An optimization problem is a problem in which certain parameters (design variables) need to be determined to achieve the best measurable performance (objective function) under given constraints. In Section 3.1, we introduced the basic idea of design optimization using a very simple beer can example. The design problem is straightforward to formulate, analytical expressions are available for the objective and constraint functions, and the optimal solution is obtained graphically. Although the simple problem illustrates the basic concepts of design optimization, in reality, design problems are much more involved in many aspects, including problem formulation, solutions, and results interpretation.

3.2.1 **PROBLEM FORMULATION**

In general, a single-objective optimization problem can be formulated mathematically as follows:

$$\text{Minimize}: \quad f(\mathbf{x}) \tag{3.3a}$$

$$\text{Subject to}: \quad g_i(\mathbf{x}) \leq 0, \quad i = 1, \, m \tag{3.3b}$$

$$h_j(\mathbf{x}) = 0, \quad j = 1, \, p \tag{3.3c}$$

$$x_k^\ell \leq x_k \leq x_k^u, \quad k = 1, \, n \tag{3.3d}$$

where $f(\mathbf{x})$ is the objective function or goal to be minimized (or maximized); $g_i(\mathbf{x})$ is the ith inequality constraint; m is the total number of inequality constraint functions; $h_j(\mathbf{x})$ is the jth equality constraint; p is the total number of equality constraints; $\mathbf{x}$ is the vector of design variables, $\mathbf{x} = [x_1, x_2, \ldots, x_n]^T$; n is the total number of design variables; and x_k^ℓ and x_k^u are the lower and upper bounds of the kth design variable x_k, respectively. Note that Eq. 3.3d is called side constraints. Equation (3.3) can also be written in a shorthand version as

$$\underset{\mathbf{x} \in S}{\text{Minimize}} \ f(\mathbf{x}) \tag{3.4}$$

in which S is called a feasible set or feasible region, defined as

$$S = \left\{ \mathbf{x} \in R^n \, | \, g_i(\mathbf{x}) \leq 0, \quad i = 1, \, m; \ h_j(\mathbf{x}) = 0, \quad j = 1, \, p; \text{and} \ x_k^\ell \leq x_k \leq x_k^u, \quad k = 1, \, n \right\} \tag{3.5}$$

Note that not all design problems can be formulated mathematically like those in Eq. 3.3 (or Eqs 3.4 and 3.5). However, whenever possible, readers are encouraged to formulate a design problem like the above in order to proceed with solution techniques for solving the design problem.

There are several steps for formulating a design optimization problem. First, the designer or the design team must develop a problem statement. What is the designer trying to accomplish? Is there a clear set of criteria or metrics to determine if the design of the product is successful at the end? Next, the designer or team must collect data and information relevant to the design problem. Is all the information needed to construct and solve the physical models of the design problem available? For example, if the design problem involves structural analysis, is the external load determined accurately?

After the above steps are completed, the designer or team faces two important tasks: design problem formulation and physical modeling.

Design problem formulation transcribes a verbal description (usually qualitative) into a quantitative statement in a mathematical form that defines the optimization problem, like that of Eq. 3.3. This task involves converting qualitative design requirements into quantitative performance measures and identifying objective function (or functions) that determine the performance or outcome of the physical system, such as costs, weight, power output, and so forth. In the meantime, the team identifies the performance constraints that the design must satisfy in accordance with the problem statement or functional and physical requirements identified at the beginning. With the objective and constraint functions identified, the team finds dimensions among other parameters (such as material properties) that largely influence the performance measures and chooses them as design variables with adequate upper or lower bounds. Performance measures are those from which both objective and performance constraints are chosen. Essentially, the goal is to formulate a design problem mathematically, as shown in Eq. 3.3.

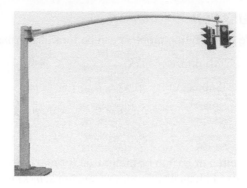

FIGURE 3.3

A traffic light employed for illustration steps of problem formulation.

The physical modeling involves the construction of mathematical models or equations that describe the physical behavior of the system being designed. Note that, in general, a physical problem is too complex to analyze and must be simplified so that it can be solved either analytically or numerically.

Next, we use the cross bar of the traffic light shown in Figure 3.3 to further illustrate the steps in formulating a design optimization problem. In this project, the goal is to design a structurally strong cross bar for the traffic light with a minimum cost. We have collected information needed for the design of the cross bar, including material to be used and its mechanical properties. Additionally, the length of the cross bar has to be 30 ft. to meet a design requirement. The cross bar can be modeled as a cantilever beam with self-weight and a point load due to the light box at the tip of the beam. Without involving detailed geometric modeling and finite element analysis (FEA), we first simplify the problem by assuming the cross bar as a straight beam with constant cross-section. We further assume the weight of the light box is significantly larger than the weight of the cross bar; therefore, the self-weight of the beam can be neglected. We also assume a solid cross-section of rectangular shape with width w and height h. The simplified cross bar is shown schematically in Figure 3.4.

Because we want the beam to be strong and yet as inexpensive as possible, we define the volume of the beam as the objective function to be minimized and we constrain the stress of the beam to be less than its yield strength. The rational is that a beam of lesser volume consumes less material; therefore, it is less expensive. Next, we pick the width and height as design variables. At this point, we need to construct a physical model with equations that govern the behavior of the beam and relate the design variables to the objective and constraint functions. For the objective function, the equation is straightforward; that is, $V(w, h) = wh\ell$. For the stress measure, we use the bending stress equation of

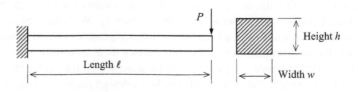

FIGURE 3.4

Cantilever beam example.

the cantilever beam; that is, $\sigma(w, h) = \frac{6P\ell}{wh^2}$. Thereafter, the design problem of the cantilever beam example can be formulated mathematically as

$$\text{Minimize}: \quad V(w, h) = wh\ell \tag{3.6a}$$

$$\text{Subject to}: \quad \sigma(w, h) = \frac{6P\ell}{wh^2} - S_y \le 0 \tag{3.6b}$$

$$w > 0, \quad h > 0 \tag{3.6c}$$

Note that in Eq. 3.6b, S_y is the material yield strength.

In general, physical modeling is not as straightforward as that shown in the cantilever beam example. Numerical simulations, such as finite element methods, are employed for the evaluation of product performance. For topics in physical modeling using numerical simulations, readers are encouraged to refer to Chang (2013a). Function evaluations that support design optimization are carried out by the analysis of the physical models. If cost is involved in the design problem formulation, the readers are referred to Chang (2013b) or other references, such as Ostwald and McLaren (2004), Ostwald (1991), and Clark and Lorenzoni (1996), for more information.

3.2.2 **PROBLEM SOLUTIONS**

Once the problem is formulated, we need to solve for an optimal solution. The solution process involves selecting a most suitable optimization technique or algorithm to find an optimal solution. In general, an optimization problem is solved numerically, in which it is required that designers understand the basic concept and the pros and cons of various optimization techniques. For problems with two or less design variables, the graphical method is an excellent choice. Some simple problems, where objective and constraint functions are written explicitly in terms of design variables, can be solved by using the necessary and sufficient conditions of the optimality. All of these solution techniques will be discussed in this chapter. Most importantly, once an optimal solution is obtained, designers must analyze, interpret, and validate the solutions before presenting the results to others.

For the cantilever beam example, we use the graphical solution technique because there are only two design variables, width w and height h of the beam cross-section. We first graph schematically the stress constraint function σ and side constraints on a plane of two axes w and h, as shown in Figure 3.5a. All designs (w, h) that satisfy the constraints are called feasible designs. The set that collects all feasible designs is called a feasible set or feasible region. For this example, the feasible region S can be written as:

$$S = \{(w, h) \in R^2 \mid \sigma(w, h) - S_y \le 0, \ w > 0, \text{ and } h > 0\} \tag{3.7}$$

Next, we plot the objective function $V(w, h)$ in Figure 3.5b with its iso-lines—in this case, straight lines. The iso-lines of the objective function are decreasing toward the origin of the w–h plane. It is clear that the minimum of the objective function is when the iso-line reaches the origin, in which the objective function V is zero. Certainly, this result is impossible physically. Mathematically, such a solution is called infeasible because the design is not in the feasible region. An optimal solution must be sought in the feasible region. Therefore, we graph the objective function in the feasible region as in Figure 3.5c, in which the objective function intersects the boundary of the feasible region at (w^*, h^*) to reach its minimum $V(w^*, h^*) = w^*h^*\ell$.

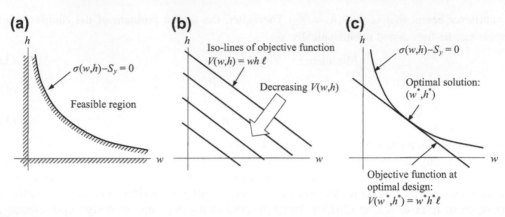

FIGURE 3.5

Cantilever beam design problem solved by using the graphical technique: (a) feasible region, (b) iso-lines of the objective function, and (c) optimal solution identified at (w^*, h^*).

3.2.3 CLASSIFICATION OF OPTIMIZATION PROBLEMS

Both the beer can and beam examples involve constraints, either equality (beer can example) or inequality (beam example). Both are called constrained problems. Occasionally, constrained problems can be converted into unconstrained problems; for example, the constrained problem of the beer can in Eq. 3.1 was converted into the unconstrained problem in Eq. 3.2. Solution techniques for constrained and unconstrained problems and other kinds of problems are different.

Optimization problems can be classified in numerous ways. First, a design problem defined in Eq. 3.3 (and as in the beer can and beam examples) is called a single-objective (or single-criterion) problem because there is one single objective function to be optimized. If a design problem involves multiple objective functions, it is called a multiobjective (or multicriterion) problem. In this case, the design goal is to minimize (or maximize) all objective functions simultaneously. We discuss multiobjective optimization in Chapter 5.

As for the constraint functions, as seen in the examples, there are equality and inequality constraints. An optimization problem may have only equality constraints or inequality constraints, or both. Such problems are called constrained optimization problems. If there are no constraints involved, these problems are called unconstrained problems. Furthermore, based on the nature of expressions for objective and constraint functions, optimization problems can be classified as linear, nonlinear, and quadratic programming problems. That is, if all functions are linear, such problems are called linear optimization problems and they are solved by using linear programming techniques. If one of these functions is nonlinear, they are nonlinear problems, and they are solved by nonlinear programming (NLP) techniques. This classification is extremely useful from a computational point of view because there are special methods or algorithms developed for the efficient solution of a particular class of problems. Thus, the first task a designer needs to investigate is the class of problem encountered or formulated. This will, in many cases, dictate the solution techniques to be adopted in solving the problem.

As to the types of disciplines of the physical models involved in the objective and constraint functions, there are single-disciplinary and MDO problems. Structural optimization, in which only structural performance measures, such as stress, displacement, buckling load, and natural frequency, are involved in the optimization problems, is in general single disciplinary. We will discuss more on the subject of structural design in Chapter 4. MDO usually involves structural, motion, thermal, fluid, manufacturing, and so on. We include a tutorial example in Projects P5 and S5 to illustrate some of the aspects of the topic.

In the beer can and beam examples, all the design variables are permitted to take any real value (in these cases, positive real values), and the optimization problem is called a real-valued programming problem. In many problems, this may not be the case. If one of the design variables is discrete, the problems are called discrete optimization or integer programming problems. Solving discrete optimization problems is a whole lot different than solving problems with continuous design variables of real numbers. In this book, we assume design variables are continuous real numbers. Those who are interested in discrete optimization problems may refer to Kouvelis and Yu (1997) and Syslo et al. (1983) for more in-depth discussions.

Based on the deterministic nature of the variables involved, optimization problems can be classified as deterministic and stochastic programming problems. A stochastic programming problem is an optimization problem in which some or all of the parameters (design variables and/or preassigned parameters) are expressed probabilistically (nondeterministic or stochastic), such as estimates of the life span of structures that have probabilistic inputs of strength and load capacity. If all design variables are deterministic, we have deterministic optimization problems. In this chapter, we focus on deterministic programming problems. In Chapter 5, we briefly discuss stochastic programming problems.

So far, we have assumed that a single designer or single design team is working on the design problems. On some occasions, there are multiple designers or design groups making respective design decisions for the same product, especially for large-scale and complex systems. In these cases, design methods that employ game theory (discussed in Chapter 2) to aid design decision making (Vincent 1983) is still an open topic, which is continuously being explored by the technical community.

3.2.4 **SOLUTION TECHNIQUES**

In general, solution techniques for optimization problems, constrained or unconstrained, can be categorized into three major groups: optimality criteria methods (also called classical methods), graphical methods, and search methods using numerical algorithms, as shown in Figure 3.6.

The classical methods of differential calculus can be used to find the unconstrained maxima and minima of a function of several variables. These methods assume that the function is differentiable twice with respect to the design variables and the derivatives are continuous. For problems with equality constraints, the Lagrange multiplier method can be used. If the problem has inequality constraints, the Karush–Kuhn–Tucker (KKT) conditions can be used to identify the optimum point. However, these methods lead to a set of nonlinear simultaneous equations that may be difficult to solve. The classical methods of optimization are discussed in Section 3.3.

In Section 3.4, we discuss graphical solutions for solving linear and nonlinear examples. Graphical methods provide a clear picture of feasible region and iso-lines of objective functions that are straightforward in identifying optimal solutions. However, they are effective for up to two design

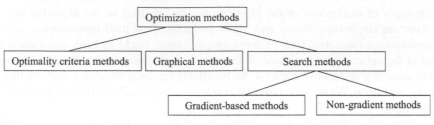

FIGURE 3.6

Classification of solution techniques for optimization problems.

variables, which substantially limit their applications. Note that neither classical methods nor graphical methods require numerical calculations for solutions.

The mainstream solution techniques for optimization problems are search methods involving numerical calculations that search for optimal solution in an iterative process by starting from an initial design. Some techniques rely on gradient information (i.e., derivatives of objective and constraint functions with respect to design variables) to guide the search process. These methods are called gradient-based approaches. Other techniques follow certain rules for search optimal solutions that do not require gradient information. These are called non-gradient-based approaches. We provide a basic discussion on the gradient-based methods in Section 3.5 and narrow into three major algorithms in Section 3.6, including sequential linear programming (SLP), sequential quadratic programming (SQP), and feasible direction method. We include two key algorithms of non-gradient methods in Section 3.7: genetic algorithms and simulated annealing.

3.3 OPTIMALITY CONDITIONS

A basic knowledge of optimality conditions is important for understanding the performance of the various numerical methods discussed later in the chapter. In this section, we introduce the basic concept of optimality, the necessary and sufficient conditions for the relative maxima and minima of a function, as well as the solution methods based on the optimality conditions. Simple examples are used to explain the underlying concepts. The examples will also show the practical limitations of the methods.

3.3.1 BASIC CONCEPT OF OPTIMALITY

We start by recalling a few basic concepts we learned in Calculus regarding maxima and minima, followed by defining local and global optima; thereafter, we illustrate the concepts using functions of one and multiple variables.

3.3.1.1 Functions of a single variable

This section presents a few definitions for basic terms.

Stationary point: For a continuous and differentiable function $f(x)$, a stationary point x^* is a point at which the slope of the function vanishes—that is, $f'(x) = df/dx = 0$ at $x = x^*$, where x^* belongs to its domain of definition. As illustrated in Figure 3.7, a stationary point can be a minimum if $f''(x) > 0$, a maximum if $f''(x) < 0$, or an inflection point if $f''(x) = 0$ in the neighborhood of x^*.

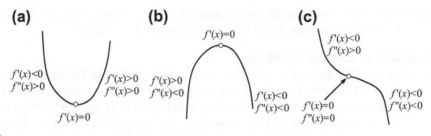

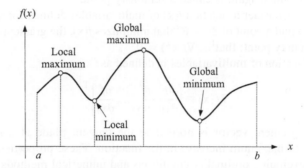

FIGURE 3.7

A stationary point may be (a) a minimum, (b) a maximum, or (c) an inflection point.

FIGURE 3.8

Global and local minimum of a function $f(x)$.

Global and local minimum: A function $f(\mathbf{x})$ is said to have a local (or relative) minimum at $\mathbf{x} = \mathbf{x}^*$ if $f(\mathbf{x}^*) \leq f(\mathbf{x}^* + \delta)$ for all sufficiently small positive and negative values of δ, that is, in the neighborhood of the point $\mathbf{x}^*$. A function $f(\mathbf{x})$ is said to have a global (or absolute) minimum at $\mathbf{x} = \mathbf{x}^*$ if $f(\mathbf{x}^*) \leq f(\mathbf{x})$ for all $\mathbf{x}$ in the domain over which $f(\mathbf{x})$ is defined. Figure 3.8 shows the global and local optimum points of a function $f(x)$ with a single variable x.

Necessary condition: Consider a function $f(x)$ of single variable defined for $a < x < b$. To find a point of $x^* \in (a, b)$ that minimizes $f(x)$, the first derivative of function $f(x)$ with respect to x at $x = x^*$ must be a stationary point; that is, $f'(x^*) = 0$.

Sufficient condition: For the same function $f(x)$ stated above and $f'(x^*) = 0$, then it can be said that $f(x^*)$ is a minimum value of $f(x)$ if $f''(x^*) > 0$, or a maximum value if $f''(x^*) < 0$.

EXAMPLE 3.1

Find a minimum of the function $f(x) = x^2 - 2x$, for $x \in (0, 2)$.

Solutions

The first derivative of $f(x)$ with respect to x is $f'(x) = 2x - 2$. We set $f'(x) = 0$, and solve for $x = 1$, which is a stationary point. This is the necessary condition for $x = 1$ to a minimum of the function $f(x)$.

We take second derivative of $f(x)$ with respect to x, $f''(x) = 2 > 0$, which satisfies the sufficient condition of the function $f(x)$ that has a minimum at $x = 1$, and the minimum value of the function at $x = 1$ is $f(1) = -1$.

The concept illustrated above can be easily extended to functions of multiple variables. We use functions of two variables to provide a graphical illustration on the concepts.

3.3.1.2 Functions of multiple variables

A function of two variables $f(x_1, x_2) = -(\cos^2 x_1 + \cos^2 x_2)^2$ is graphed in Figure 3.9a. Perturbations from point $(x_1, x_2) = (0, 0)$, which is a local minimum, in any direction result in an increase in the function value of $f(\mathbf{x})$; that is, the slopes of the function with respect to x_1 and x_2 are zero at this point of local minimum. Similarly, a function $f(x_1, x_2) = (\cos^2 x_1 + \cos^2 x_2)^2$ graphed in Figure 3.9b has a local maximum at $(x_1, x_2) = (0, 0)$. Perturbations from this point in any direction result in a decrease in the function value of $f(\mathbf{x})$; that is, the slopes of the function with respect to x_1 and x_2 are zero at this point of local maximum. The first derivatives of the function with respect to the variables are zero at the minimum or maximum, which again is called a stationary point.

Necessary condition: Consider a function $f(\mathbf{x})$ of multivariables defined for $\mathbf{x} \in R^n$, where n is the number of variables. To find a point of $\mathbf{x}^* \in R^n$ that minimizes $f(\mathbf{x})$, the gradient of the function $f(\mathbf{x})$ at $\mathbf{x} = \mathbf{x}^*$ must be a stationary point; that is, $\nabla f(\mathbf{x}^*) = 0$.

The gradient of a function of multivariables is defined as

$$\nabla f(\mathbf{x}) \equiv \left[\frac{\partial f}{\partial x_1}, \frac{\partial f}{\partial x_2}, ..., \frac{\partial f}{\partial x_n} \right]^{\mathrm{T}} \tag{3.8}$$

Geometrically, the gradient vector is normal to the tangent plane at a given point $\mathbf{x}$, and it points in the direction of maximum increase in the function. These properties are quite important; they will be used in developing optimality conditions and numerical methods for optimum design. In Example 3.2, the gradient vector for a function of two variables is calculated for illustration purpose.

(a)   **(b)**

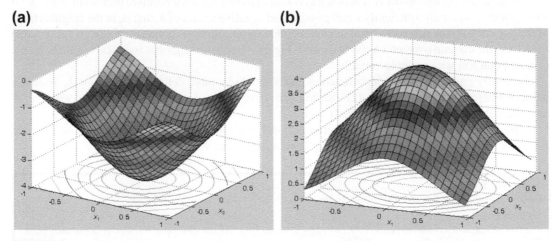

FIGURE 3.9

Functions of two variables (MATLAB Script 2 can be found in Appendix A): (a) $f(x_1, x_2) = -(\cos^2 x_1 + \cos^2 x_2)^2$ with a local minimum at (0, 0) and (b) $f(x_1, x_2) = (\cos^2 x_1 + \cos^2 x_2)^2$ with a local maximum at (0, 0).

EXAMPLE 3.2

A function of two variables is defined as

$$f(x_1, x_2) = x_2 e^{-x_1^2 - x_2^2} \tag{3.9a}$$

which is graphed in MATLAB shown below (left). The MATLAB script for the graph can be found in Appendix A (Script 3). Calculate the gradient vectors of the function at $(x_1, x_2) = (1, 1)$ and $(x_1, x_2) = (1, -1)$.

Solutions

From Eq. 3.8, the gradient vector of the function $f(x_1, x_2)$ is

$$\nabla f(x_1, x_2) = \left[-2x_1 x_2 e^{-x_1^2 - x_2^2}, e^{-x_1^2 - x_2^2} - 2x_2^2 e^{-x_1^2 - x_2^2} \right]^T \tag{3.9b}$$

At $(x_1, x_2) = (1, 1), f(1, 1) = e^{-2} = 0.1353$, and $\nabla f(1, 1) = [-2e^{-2}, -e^{-2}]^T$; and at $(x_1, x_2) = (1, -1), f(1, -1) = -e^{-2} = -0.1353$, and $\nabla f(1, -1) = [2e^{-2}, -e^{-2}]^T$. The iso-lines of $f(1, 1)$ and $f(1, -1)$ as well as the gradient vectors at $(1, 1)$ and $(1, -1)$ are shown in the figure below (right). In this example, gradient vector at a point **x** is perpendicular to the tangent line at **x**, and the vector points in the direction of maximum increment in the function value. The maximum and minimum of the function are shown for clarity.

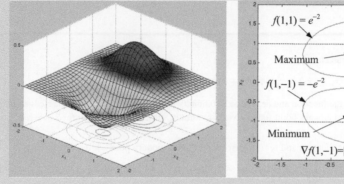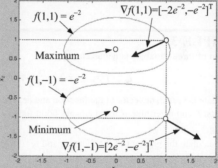

Sufficient condition: For the same function $f(\mathbf{x})$ stated above, let $\nabla f(\mathbf{x}^*) = 0$, then $f(\mathbf{x}^*)$ has a minimum value of $f(\mathbf{x})$ if its Hessian matrix defined in Eq. 3.10 is positive-definite.

$$\mathbf{H} = \nabla^2 f = \left[\frac{\partial^2 f}{\partial x_i \partial x_j} \right] = \begin{bmatrix} \dfrac{\partial^2 f}{\partial x_1^2} & \dfrac{\partial^2 f}{\partial x_1 \partial x_2} & \cdots & \dfrac{\partial^2 f}{\partial x_1 \partial x_n} \\ \dfrac{\partial^2 f}{\partial x_2 \partial x_1} & \dfrac{\partial^2 f}{\partial x_2^2} & \cdots & \dfrac{\partial^2 f}{\partial x_2 \partial x_n} \\ \cdots & \cdots & & \cdots \\ \dfrac{\partial^2 f}{\partial x_n \partial x_1} & \dfrac{\partial^2 f}{\partial x_n \partial x_2} & \cdots & \dfrac{\partial^2 f}{\partial x_n^2} \end{bmatrix}_{n \times n} \tag{3.10}$$

where all derivatives are calculated at the given point $\mathbf{x}^*$. The Hessian matrix is an $n \times n$ matrix, where n is the number of variables. It is important to note that each element of the Hessian is a function in itself that is evaluated at the given point $\mathbf{x}^*$. Also, because $f(\mathbf{x})$ is assumed to be twice continuously differentiable, the cross partial derivatives are equal; that is,

$$\frac{\partial^2 f}{\partial x_i \partial x_j} = \frac{\partial^2 f}{\partial x_j \partial x_i}; \; i, j = 1, \, n \tag{3.11}$$

Therefore, the Hessian is always a symmetric matrix. The Hessian matrix plays a prominent role in exploring the sufficiency conditions for optimality.

Note that a square matrix is positive-definite if (a) the determinant of the Hessian matrix is positive (i.e., $|\mathbf{H}| > 0$) or (b) all its eigenvalues are positive. To calculate the eigenvalues λ of a square matrix, the following equation is solved:

$$|\mathbf{H} - \lambda \mathbf{I}| = 0 \tag{3.12}$$

where $\mathbf{I}$ is an identity matrix of $n \times n$.

EXAMPLE 3.3

A function of three variables is defined as

$$f(x_1, x_2, x_2) = x_1^2 + 2x_1 x_2 + 2x_2^2 + x_3^2 - 2x_1 + x_2 + 8 \tag{3.13a}$$

Calculate the gradient vector of the function and determine a stationary point, if it exists. Calculate a Hessian matrix of the function f, and determine if the stationary point found gives a minimum value of the function f.

Solutions

We first calculate the gradient of the function and set it to zero to find the stationary point(s), if any:

$$\nabla f(x_1, x_2, x_3) = [2x_1 + 2x_2 - 2, \; 2x_1 + 4x_2 + 1, \; 2x_3]^T \tag{3.13b}$$

Setting Eq. 3.13b to zero, we have $\mathbf{x} = [2.5, -1.5, 0]^T$, which is the only stationary point. Now, we calculate the Hessian matrix:

$$\mathbf{H} = \nabla^2 f = \begin{bmatrix} 2 & 2 & 0 \\ 2 & 4 & 0 \\ 0 & 0 & 1 \end{bmatrix} \tag{3.13c}$$

which is positive-definite because

$$|\mathbf{H}| = \begin{vmatrix} 2 & 2 & 0 \\ 2 & 4 & 0 \\ 0 & 0 & 1 \end{vmatrix} = 8 - 4 = 4 > 0 \tag{3.13d}$$

or

$$|\mathbf{H} - \lambda \mathbf{I}| = \begin{vmatrix} 2 - \lambda & 2 & 0 \\ 2 & 4 - \lambda & 0 \\ 0 & 0 & 1 - \lambda \end{vmatrix} = (2 - \lambda)(4 - \lambda)(1 - \lambda) - 4(1 - \lambda) = 0 \tag{3.13e}$$

Solving Eq. 3.13e, we have $\lambda = 1, 0.7639$ and 5.236, which are all positive. Hence, the Hessian matrix is positive-definite; therefore, the stationary point $\mathbf{x}^* = [2.5, -1.5, 0]^T$ is a minimum point, at which the function value is $f(\mathbf{x}^*) = 4.75$.

3.3.2 BASIC CONCEPT OF DESIGN OPTIMIZATION

For an optimization problem defined in Eq. 3.3, we find design variable vector $\mathbf{x}$ to minimize an objective function $f(\mathbf{x})$ subject to the inequality constraints $g_i(\mathbf{x}) \leq 0$, $i = 1$ to m, the equality constraints $h_j(\mathbf{x}) = 0$, $j = 1$ to p, and the side constraints $x_k^\ell \leq x_k \leq x_k^u$, $k = 1, n$. In Eq. 3.5, we define the feasible set S, or feasible region, for a design problem as a collection of feasible designs. For unconstrained problems, the entire design space is feasible because there are no constraints. In general, the optimization problem is to find a point in the feasible region that gives a minimum value to the objective function. From a design perspective, in particular solving Eq. 3.3, we state the followings terms.

Global minimum: A function $f(\mathbf{x})$ of n design variables has a global minimum at $\mathbf{x}^*$ if the value of the function at $\mathbf{x}^* \in S$ is less than or equal to the value of the function at any other point $\mathbf{x}$ in the feasible set S. That is,

$$f(\mathbf{x}^*) \leq f(\mathbf{x}), \ \forall \text{ (for all) } \mathbf{x} \in S \tag{3.14}$$

If strict inequality holds for all $\mathbf{x}$ other than $\mathbf{x}^*$ in Eq. 3.14, then $\mathbf{x}^*$ is called a strong (strict) global minimum; otherwise, it is called a weak global minimum.

Local minimum: A function $f(\mathbf{x})$ of n design variables has a local (or relative) minimum at $\mathbf{x}^* \in S$ if inequality of Eq. 3.14 holds for all $\mathbf{x}$ in a small neighborhood N (vicinity) of $\mathbf{x}^*$. If strict inequality holds, then $\mathbf{x}^*$ is called a strong (strict) local minimum; otherwise, it is called a weak local minimum.

Neighborhood N of point $\mathbf{x}^*$ is defined as the set of points:

$$N = \{\mathbf{x} | \mathbf{x} \in S \text{ with } \|\mathbf{x} - \mathbf{x}^*\| < \delta\} \tag{3.15}$$

for some small $\delta > 0$. Geometrically, it is a small feasible region around point $\mathbf{x}^*$, such as a sphere of radius δ for $n = 3$ (number of design variables $n = 3$).

Next, we illustrate the derivation of the necessary and sufficient conditions using Taylor's series expansion. For the time being, we assume unconstrained problems. In the next subsection, we extend the discussion to constrained problems.

Expanding the objective function $f(\mathbf{x})$ at the inflection point $\mathbf{x}^*$ using Taylor's series, we have

$$f(\mathbf{x}) = f(\mathbf{x}^*) + \nabla f(\mathbf{x}^*)^\mathrm{T} \Delta \mathbf{x} + \frac{1}{2} \Delta \mathbf{x}^\mathrm{T} H(\mathbf{x}^*) \Delta \mathbf{x} + R \tag{3.16}$$

where R is the remainder containing higher-order terms in $\Delta \mathbf{x}$, and $\Delta \mathbf{x} = \mathbf{x} - \mathbf{x}^*$. We define increment $\Delta f(\mathbf{x})$ as

$$\Delta f(\mathbf{x}) = f(\mathbf{x}) - f(\mathbf{x}^*) = \nabla f(\mathbf{x}^*)^\mathrm{T} \Delta \mathbf{x} + \frac{1}{2} \Delta \mathbf{x}^\mathrm{T} H(\mathbf{x}^*) \Delta \mathbf{x} + R \tag{3.17}$$

If we assume a local minimum at $\mathbf{x}^*$, then Δf must be nonnegative due to the definition of a local minimum given in Eq. 3.14; that is, $\Delta f \geq 0$.

Because $\Delta \mathbf{x}$ is small, the first-order term $\nabla f(\mathbf{x}^*)^\mathrm{T} \Delta \mathbf{x}$ dominates other terms, and therefore Δf can be approximated as $\Delta f(\mathbf{x}) \approx \nabla f(\mathbf{x}^*)^\mathrm{T} \Delta \mathbf{x}$. Note that Δf in this equation can be positive or negative depending on the sign of the term $\nabla f(\mathbf{x}^*)^\mathrm{T} \Delta \mathbf{x}$. Because $\Delta \mathbf{x}$ is arbitrary (a small increment in $\mathbf{x}^*$), its components may be positive or negative. Therefore, we observe that Δf can be nonnegative for all possible $\Delta \mathbf{x}$ unless

$$\nabla f(\mathbf{x}^*) = 0 \tag{3.18}$$

In other words, the gradient of the function at **x*** must be zero. In the component form, this necessary condition becomes

$$\frac{\partial f(\mathbf{x}^*)}{\partial x_i} = 0, \quad i = 1, n \tag{3.19}$$

Again, points satisfying Eq. 3.18 or Eq. 3.19 are called stationary points.

Considering the second term on the right-hand side of Eq. 3.17 evaluated at a stationary point **x***, the positivity of Δf is assured if

$$\Delta \mathbf{x}^{\mathrm{T}} \mathbf{H}(\mathbf{x}^*) \, \Delta \mathbf{x} > 0 \tag{3.20}$$

for all $\Delta \mathbf{x} \neq \mathbf{0}$. This is true if the Hessian $\mathbf{H}(\mathbf{x}^*)$ is a positive-definite matrix, which is then the sufficient condition for a local minimum of $f(\mathbf{x})$ at **x***.

3.3.3 LAGRANGE MULTIPLIERS

We begin the discussion of optimality conditions for constrained problems by including only the equality constraints in the formulation in this section; that is, inequalities in Eq. 3.3b are ignored temporarily. More specifically, the optimization problem is restated as

$$\text{Minimize}: \quad f(\mathbf{x}) \tag{3.21a}$$

$$\text{Subject to}: \quad h_j(\mathbf{x}) = 0, \quad j = 1, p \tag{3.21b}$$

$$x_k^{\ell} \leq x_k \leq x_k^u, \quad k = 1, n \tag{3.21c}$$

The reason is that the nature of equality constraints is quite different from that of inequality constraints. Equality constraints are always active for any feasible design, whereas an inequality constraint may not be active at a feasible point. This changes the nature of the necessary conditions for the problem when inequalities are included.

A common approach for dealing with equality constraints is to introduce scalar multipliers associated with each constraint, called Lagrange multipliers. These multipliers play a prominent role in optimization theory as well as in numerical methods, in which a constrained problem is converted into an unconstrained problem that can be solved by using optimality conditions or numerical algorithms specifically developed for them. The values of the multipliers depend on the form of the objective and constraint functions. If these functions change, the values of the Lagrange multipliers also change.

Through Lagrange multipliers, the constrained problem (with equality constraints) shown in Eq. 3.21 is converted into an unconstrained problem as

$$L(\mathbf{x}, \boldsymbol{\lambda}) = f(\mathbf{x}) + \sum_{j=1}^{p} \lambda_j h_j(\mathbf{x}) = f(\mathbf{x}) + \boldsymbol{\lambda}^{\mathrm{T}} \mathbf{h}(\mathbf{x}) \tag{3.22}$$

which is called a Lagrangian function, or simply Lagrangian. If we expand the vector of design variables to include the Lagrange multipliers, then the necessary and sufficient conditions of a local minimum discussed in the previous subsection are applicable to the problem defined in Eq. 3.22.

Before discussing the optimality conditions, we defined an important term called regular point. Consider the constrained optimization problem defined in Eq. 3.21, a point **x*** satisfying the constraint functions $\mathbf{h}(\mathbf{x}^*) = 0$ is said to be a regular point of the feasible set if the objective $f(\mathbf{x}^*)$ is differentiable and gradient vectors of all constraints at the point $\mathbf{x}^*$ are linearly independent. Linear independence means that no two gradients are parallel to each other, and no gradient can be expressed as a linear combination of the others. When inequality constraints are included in the problem definition, then for a point to be regular, gradients of all the active constraints must also be linearly independent.

The necessary condition (or Lagrange multiplier theorem) is stated next.

Consider the optimization problem defined in Eq. 3.21. Let $\mathbf{x}^*$ be a regular point that is a local minimum for the problem. Then, there exist unique Lagrange multipliers $\lambda_j^*, j = 1, p$ such that

$$\nabla L(\mathbf{x}^*, \boldsymbol{\lambda}^*) = \frac{\partial L(\mathbf{x}^*, \boldsymbol{\lambda}^*)}{\partial x_i} = \mathbf{0}; \quad i = 1, n \tag{3.23}$$

Differentiating the Lagrangian $L(\mathbf{x}, \boldsymbol{\lambda})$ with respect to λ_j, we recover the equality constraints as

$$\frac{\partial L(\mathbf{x}^*, \boldsymbol{\lambda}^*)}{\partial \lambda_i} = \mathbf{0}; \quad \Rightarrow h_j(\mathbf{x}^*) = 0; \quad j = 1, p \tag{3.24}$$

The gradient conditions of Eqs 3.23 and 3.24 show that the Lagrangian is stationary with respect to both $\mathbf{x}$ and $\boldsymbol{\lambda}$. Therefore, it may be treated as an unconstrained function in the variables $\mathbf{x}$ and $\boldsymbol{\lambda}$ to determine the stationary points. Note that any point that does not satisfy the conditions cannot be a local minimum point. However, a point satisfying the conditions need not be a minimum point either. It is simply a candidate minimum point, which can actually be an inflection or maximum point.

The second-order necessary and sufficient conditions, similar to that of Eq. 3.20, in which the Hessian matrix includes terms of Lagrange multipliers, can distinguish between the minimum, maximum, and inflection points. More specifically, a sufficient condition for $f(\mathbf{x})$ to have a local minimum at $\mathbf{x}^*$ is that each root of the polynomial in ε, defined by the following determinant equation be positive:

$$\begin{vmatrix} \mathbf{L} - \mathbf{I}\varepsilon & \mathbf{G} \\ \mathbf{G} & \mathbf{0} \end{vmatrix}^{\mathrm{T}} = 0 \tag{3.25}$$

where

$$\mathbf{L} - \mathbf{I}\varepsilon = \begin{bmatrix} L_{11} - \varepsilon & L_{12} & \cdots & L_{1n} \\ L_{21} & L_{22} - \varepsilon & \cdots & L_{2n} \\ \cdots & \cdots & & \cdots \\ L_{n1} & L_{n2} & \cdots & L_{nn} - \varepsilon \end{bmatrix}_{n \times n} \tag{3.26}$$

and

$$\mathbf{G} = \begin{bmatrix} g_{11} & g_{12} & \cdots & g_{1n} \\ g_{21} & g_{22} & \cdots & g_{2n} \\ \cdots & \cdots & & \cdots \\ g_{m1} & g_{m2} & \cdots & g_{mn} \end{bmatrix}_{m \times n} \tag{3.27}$$

Note that L_{ij} is a partial derivative of the Lagrangian L with respect to x_i and λ_j, i.e., $L_{ij} = \frac{\partial^2 L(\mathbf{x}^*, \lambda^*)}{\partial x_i \partial x_j}$, $i, j = 1, n$; and g_{pq} is the partial derivative of g_p with respect to x_q; i.e., $g_{pq} = \frac{\partial g_p(\mathbf{x}^*)}{\partial x_q}$, $p = 1, m$ and $q = 1, n$.

EXAMPLE 3.4

Find the optimal solution for the following problem:

$$\text{Minimize:}\quad f(\mathbf{x}) = 3x_1^2 + 6x_1x_2 + 5x_2^2 + 7x_1 + 5x_2 \tag{3.28a}$$

$$\text{Subject to:}\quad g(\mathbf{x}) = x_1 + x_2 - 5 = 0 \tag{3.28b}$$

Solution

Define the Lagrangian as

$$L(\mathbf{x}, \lambda) = f(\mathbf{x}) + \lambda g(\mathbf{x}) = \left(3x_1^2 + 6x_1x_2 + 5x_2^2 + 7x_1 + 5x_2\right) + \lambda(x_1 + x_2 - 5)$$

Taking derivatives of $L(\mathbf{x}, \lambda)$ with respect to x_1, x_2, and λ, respectively, we have

$$\partial L / \partial x_1 = 6x_1 + 6x_2 + 7 + \lambda = 0$$

From Eq. 3.28b, $6(x_1 + x_2) = 6(5) = -7 - \lambda$. Therefore, $\lambda = -37$.

Also, $\partial L/\partial x_2 = 6x_1 + 10x_2 + 5 + \lambda = 0$. It can be also written as $6(x_1 + x_2) + 4x_2 + 5 + \lambda = 6(5) + 4x_2 + 5 - 37 = 0$. Hence $x_2 = 0.5$ and $x_1 = 4.5$.

We obtain the stationary point $\mathbf{x}^* = [4.5, 0.5]^T$ and $\lambda^* = -37$. Next, we check the sufficient condition of Eq. 3.25; that is, for this example, we have

$$\begin{vmatrix} L_{11} - \varepsilon & L_{12} & g_{11} \\ L_{12} & L_{22} - \varepsilon & g_{21} \\ g_{11} & g_{21} & 0 \end{vmatrix} = 0$$

in which $L_{11} = \left.\frac{\partial^2 L}{\partial x_1^2}\right|_{(\mathbf{x}^*, \lambda^*)} = 6$, $L_{12} = L_{21} = \left.\frac{\partial^2 L}{\partial x_1 \partial x_2}\right|_{(\mathbf{x}^*, \lambda^*)} = 6$, $L_{22} = \left.\frac{\partial^2 L}{\partial x_2^2}\right|_{(\mathbf{x}^*, \lambda^*)} = 10$, $g_{11} = \left.\frac{\partial g}{\partial x_1}\right|_{(\mathbf{x}^*, \lambda^*)} = 1$, and $g_{12} = \left.\frac{\partial g}{\partial x_2}\right|_{(\mathbf{x}^*, \lambda^*)} = 1$. Hence the determinant becomes

$$\begin{vmatrix} 6 - \varepsilon & 6 & 1 \\ 6 & 10 - \varepsilon & 1 \\ 1 & 1 & 0 \end{vmatrix} = 6 + 6 - (10 - \varepsilon) - (6 - \varepsilon) = 0$$

Therefore, $\varepsilon = 2$. Because ε is positive, $\mathbf{x}^*$ and λ^* correspond to a minimum.

3.3.4 KARUSH–KUHN–TUCKER CONDITIONS

Next, we extend the Lagrange multiplier to include inequality constraints and consider the general optimization problem defined in Eq. 3.3.

We first transform an inequality constraint into an equality constraint by adding a new variable to it, called the slack variable. Because the constraint is of the form "$\leq$", its value is either negative or zero at a feasible point. Thus, the slack variable must always be nonnegative (i.e., positive or zero) to make the inequality an equality.

An inequality constraint $g_i(\mathbf{x}) \leq 0$ is equivalent to the equality constraint

$$G_i(\mathbf{x}) = g_i(\mathbf{x}) + s_i^2 = 0 \tag{3.29}$$

where s_i is a slack variable. The variables s_i are treated as unknowns of the design problem along with the design variables. Their values are determined as a part of the solution. When the variable s_i has zero value, the corresponding inequality constraint is satisfied at equality. Such an inequality is called an active (or tight) constraint; that is, there is no "slack" in the constraint. For any $s_i \neq 0$, the corresponding constraint is a strict inequality. It is called an inactive constraint, and has slack given by s_i^2.

Note that once a design point is specified, Eq. 3.29 can be used to calculate the slack variable s_i^2. If the constraint is satisfied at the point (i.e., $g_i(\mathbf{x}) \leq 0$), then $s_i^2 \geq 0$. If it is violated, then s_i^2 is negative, which is not acceptable; that is, the point is not a feasible point and hence is not a candidate minimum point.

Similar to that of Section 3.3.3, through Lagrange multipliers, the constrained problem (with equality and inequality constraints) defined in Eq. 3.3 is converted into an unconstrained problem as

$$L(\mathbf{x}, \boldsymbol{\lambda}, \boldsymbol{\mu}, \mathbf{s}) = f(\mathbf{x}) + \sum_{i=1}^{p} \lambda_i h_i(\mathbf{x}) + \sum_{j=1}^{m} \mu_j \left(g_j(\mathbf{x}) + s_j^2 \right) = f(\mathbf{x}) + \boldsymbol{\lambda}^T \mathbf{h}(\mathbf{x}) + \boldsymbol{\mu}^T \mathbf{G}(\mathbf{x}) \tag{3.30}$$

If we expand the vector of design variables to include the Lagrange multipliers $\boldsymbol{\lambda}$ and $\boldsymbol{\mu}$, and the slack variables $\mathbf{s}$, then the necessary and sufficient conditions of a local minimum discussed in the previous subsection are applicable to the unconstrained problem defined in Eq. 3.30.

Note that derivatives of the Lagrangian L with respect to $\mathbf{x}$ and $\boldsymbol{\lambda}$ lead to Eqs 3.23 and 3.24, respectively. On the other hand, the derivatives of L with respect to $\boldsymbol{\mu}$ yield converted equality constraints of Eq. 3.29. Furthermore, the derivatives of L with respect to $\mathbf{s}$ yield

$$\mu_j s_j = 0, \quad j = 1, m \tag{3.31}$$

which is an additional necessary condition for the Lagrange multipliers of "$\leq$ type" constraints given as

$$\mu_j^* \geq 0; \quad j = 1, m \tag{3.32}$$

where μ_j^* is the Lagrange multiplier for the jth inequality constraint. Equations (3.32) is referred to as the nonnegativity of Lagrange multipliers (Arora 2012).

The necessary conditions for the constrained problem with equality and inequality constraints defined in Eq. 3.3 can be summed up in what are commonly known as the KKT first-order necessary conditions.

Karush–Kuhn–Tucker Necessary Conditions: Let $\mathbf{x}^*$ be a regular point of the feasible set that is a local minimum for $f(\mathbf{x})$, subject to $h_i(\mathbf{x}) = 0$; $i = 1, p$; $g_j(\mathbf{x}) \leq 0$; $j = 1, m$. Then there exist Lagrange multipliers $\boldsymbol{\lambda}^*$ (a p-vector) and μ^* (an m-vector) such that the Lagrangian L is stationary with respect to x_k, λ_i, μ_j, and s_ℓ at the point $\mathbf{x}^*$; that is:

1. Stationarity:

$$\nabla L(x^*, \lambda^*, \mu^*, s^*) = 0; \tag{3.33a}$$

or

$$\frac{\partial L}{\partial x_k} = \frac{\partial f}{\partial x_k} + \sum_{i=1}^{p} \lambda_i \frac{\partial h_i}{\partial x_k} + \sum_{j=1}^{m} \mu_j \frac{\partial g_j}{\partial x_k} = 0; \quad k = 1, n \tag{3.33b}$$

(2) Equality constraints:

$$\frac{\partial L}{\partial \lambda_i} = 0; \quad \Rightarrow h_i(\mathbf{x}^*) = 0; \quad i = 1, p \tag{3.34}$$

(3) Inequality constraints (or complementary slackness condition):

$$\frac{\partial L}{\partial \mu_j} = 0; \quad \Rightarrow g_j(\mathbf{x}^*) + s_j^2 = 0; \quad j = 1, m \tag{3.35}$$

$$\frac{\partial L}{\partial s_j} = 0; \quad \Rightarrow 2\mu_j^* s_j = 0; \quad j = 1, m \tag{3.36}$$

In addition, gradients of the active constraints must be linearly independent. In such a case, the Lagrange multipliers for the constraints are unique.

EXAMPLE 3.5

Solve the following optimization problem using the KKT conditions.

$$\text{Minimize} : f(\mathbf{x}) = -2x_1 - 2x_2 + \frac{1}{2}\left(x_1^2 + x_2^2\right) \tag{3.37a}$$

$$\text{Subject to} : g_1(\mathbf{x}) = 3x_1 + x_2 - 6 \le 0 \tag{3.37b}$$

$$h_1(\mathbf{x}) = x_1 - x_2 = 0 \tag{3.37c}$$

$$0 \le x_1, \quad 0 \le x_2 \tag{3.37d}$$

Solutions

Using Eq. 3.30, we state the Lagrangian of the problem as

$$L = -2x_1 - 2x_2 + \frac{1}{2}\left(x_1^2 + x_2^2\right) + \lambda_1(x_1 - x_2) + \mu_1\left(3x_1 + x_2 - 6 + s_1^2\right) + \mu_2\left(-x_1 + s_2^2\right) + \mu_3\left(-x_2 + s_3^2\right) \tag{3.37e}$$

Taking derivatives of the Lagrangian with respect to $x_1, x_2, \lambda_1, \mu_1, \mu_2, \mu_3, s_1, s_2$, and s_3 and setting them to zero, we have

$$\frac{\partial L}{\partial x_1} = -2 + x_1 + \lambda_1 + 3\mu_1 - \mu_2 = 0 \tag{3.37f}$$

$$\frac{\partial L}{\partial x_2} = -2 + x_2 - \lambda_1 + \mu_1 - \mu_3 = 0 \tag{3.37g}$$

$$\frac{\partial L}{\partial \lambda_1} = x_1 - x_2 = 0 \tag{3.37h}$$

EXAMPLE 3.5—cont'd

$$\frac{\partial L}{\partial \mu_1} = 3x_1 + x_2 - 6 + s_1^2 = 0 \tag{3.37i}$$

$$\frac{\partial L}{\partial \mu_2} = -x_1 + s_2^2 = 0 \tag{3.37j}$$

$$\frac{\partial L}{\partial \mu_3} = -x_2 + s_3^2 = 0 \tag{3.37k}$$

$$\frac{\partial L}{\partial s_1} = 2\mu_1 s_1 = 0 \tag{3.37l}$$

$$\frac{\partial L}{\partial s_2} = 2\mu_2 s_2 = 0 \tag{3.37m}$$

$$\frac{\partial L}{\partial s_3} = 2\mu_3 s_3 = 0 \tag{3.37n}$$

Note that, as expected, Eqs 3.37h–3.37k reduce back to the constraint equations of the original optimization problem in Eqs 3.37b–3.37d. We have total nine equations (Eqs 3.37f–3.37n) and nine unknowns. Not all nine equations are linear. It is not guaranteed that all nine unknowns can be solved uniquely from the nine equations. As discussed before, these equations are solved in different cases. In this example, the equality constraint must be satisfied; hence, from Eq. 3.37h, $x_1 = x_2$. Next, we make assumptions and proceed with different cases.

Case 1: we assume that the inequality constraint, Eq. 3.37i, is active; hence, $s_1 = 0$. From the same equation, we have

$$3x_1 + x_2 - 6 = 4x_1 - 6 = 0. \quad \text{Hence } x_1 = x_2 = 1.5.$$

which implies that the side constraints are not active; hence, from Eqs 3.37j and 3.37k, $s_2 \neq 0$ and $s_3 \neq 0$, implying $\mu_2 = \mu_3 = 0$ from Eqs 3.37m and 3.37n.

Solve μ_1 and λ_1 from Eqs 3.37f and 3.37g, we have $\mu_1 = 0.25$, and $\lambda_1 = -0.25$. Then, from Eq. 3.37a, the objective function is

$$f(1.5, 1.5) = -2(1.5) - 2(1.5) + 1/2(1.5^2 + 1.5^2) = -3.75$$

Case 2: we assume that the side constraints, Eqs 3.37j and 3.37k, are active; hence, $s_2 = s_3 = 0$; and $x_1 = x_2 = 0$. From Eq. 3.37i, $s_1 \neq 0$, hence $\mu_1 = 0$ from Eq. 3.37l. There is no more assumption can be made logically. Equations (3.37f) and (3.37g) consist of three unknowns, λ_1, μ_2, and μ_3; they cannot be solved uniquely. If we further assume $\mu_2 = 0$, then we have $\lambda_1 = 2$ and $\mu_2 = -4$. If we assume $\mu_3 = 0$, then we have $\lambda_1 = -2$ and $\mu_2 = -4$. Then, from Eq. 3.37a, the objective function is

$$f(0, 0) = -2(0) - 2(0) + 1/2(0^2 + 0^2) = 0$$

which is greater than that of Case 1.

Case 3: we assume that the side constraint, Eq. 3.37j, is active; hence, $s_2 = 0$; and $x_1 = 0$. We assume constraints (3.37i) and (3.37k) are not active; hence, $s_1 \neq 0$ and $s_3 \neq 0$; implying $\mu_1 = \mu_3 = 0$. There is no more assumption that can be made logically. Equations (3.37f) and (3.37g) consist of three unknowns, λ_1, μ_2, and x_2; they cannot be solved uniquely. If we further assume $\lambda_1 = 0$, then we have $x_2 = 2$ and $\mu_2 = -2$. Then, from Eq. 3.37a, the objective function is

$$f(0, 2) = -2(0) - 2(2) + 1/2(0^2 + 2^2) = -2$$

which is greater than that of Case 1. We may proceed with a few more cases and find possible solutions. After exhausting all cases, the solution obtained in Case 1 gives a minimum value for the objective function.

As seen in the example above, solving optimization problems using the KKT conditions is not straightforward, even for a simple problem. All possible cases must be identified and carefully examined. In addition, a sufficient condition that involves second derivatives of the Lagrangian is difficult to verify. Furthermore, for practical engineering design problems, objective and constraint functions are not expressed in terms of design variables explicitly, and taking derivatives analytically is not possible. After all, KKT conditions serve well for understanding the concept of optimality.

3.4 GRAPHICAL SOLUTIONS

Graphical methods offer means to seek optimal solutions quickly without involving numerical algorithms. Graphical visualization of the optimization problem and optimal solution in the design space enhances our understanding of the concept and numerical solution techniques to be discussed later. Because graphical methods require designers to plot the objective and constraint functions in the design space, the objective and constraint functions have to be written in terms of design variables explicitly. Moreover, graphical methods are effective only for up to two design variables. We use both linear and nonlinear programming problems to illustrate the use of the methods. In some examples, we use MATLAB plot functions to graph the objective and constraint functions for illustration purposes.

3.4.1 LINEAR PROGRAMMING PROBLEMS

Linear programming (LP) problems involve a linear objective function subject to a set of linear constraint functions. The problems are formulated as

$$\text{Minimize}: \quad f(\mathbf{x}) = a_1 x_1 + a_2 x_2 + \ldots + a_n x_n \tag{3.38a}$$

$$\text{Subject to}: \quad h_i(\mathbf{x}) = b_{i1} x_1 + b_{i2} x_2 + \ldots + b_{in} x_n + b_{i0} = 0, \quad i = 1, m \tag{3.38b}$$

$$g_j(\mathbf{x}) = c_{j1} x_1 + c_{j2} x_2 + \ldots + c_{jn} x_n + c_{j0} \leq 0, \quad j = 1, p \tag{3.38c}$$

$$x_k^\ell \leq x_k \leq x_k^u, \quad k = 1, n \tag{3.38d}$$

In general, solving LP problems using graphical methods involves three steps:

Step 1: Sketch constraint functions (Eqs 3.38b and 3.38c) and side constraints (Eq. 3.38d) on the design plane of two design variables, x_1 and x_2.
Step 2: Identify the feasible region, in which any point in the region satisfies the constraints.
Step 3a: Sketch iso-lines of the objective function on top of the feasible region and identify the optimal point (usually a vertex of the polygon of the feasible region on the x_1–x_2 plane), or
Step 3b: Calculate the values of the vertices of the feasible region, and plug these values into the objective function (Eq. 3.38a) to find the minimum.

Example 3.6 illustrates these steps.

EXAMPLE 3.6

Solve the following LP problem using the graphical method.

$$\text{Minimize}: f(\mathbf{x}) = 2x_1 - 3x_2 \tag{3.39a}$$

$$\text{Subject to}: g_1(\mathbf{x}) = 7 - x_1 - 5x_2 \leq 0 \tag{3.39b}$$

EXAMPLE 3.6—cont'd

$$g_2(\mathbf{x}) = 10 - 4x_1 - x_2 \le 0 \tag{3.39c}$$

$$g_3(\mathbf{x}) = -7x_1 + 6x_2 - 9 \le 0 \tag{3.39d}$$

$$g_4(\mathbf{x}) = -x_1 + 6x_2 - 24 \le 0 \tag{3.39e}$$

Solutions

We follow the steps discussed above to illustrate the graphical method.

Step 1: We first sketch the constraint functions g_1 to g_4 on an x_1-x_2 plane shown below.

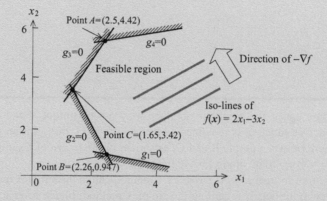

Step 2: We identify the feasible region to the right of a polyline shown above. Note that the region separated by the four straight lines and to the right of line segments ACB shown above is the feasible region.

Step 3a: We sketch iso-lines of the objective function on top of the feasible region. The negative gradient of the iso-lines is $-\nabla f = [-2, 3]^{\mathrm{T}}$, showing the direction of decreasing objective function $f(\mathbf{x})$. The decreasing iso-line intersects the feasible region at Point A before exiting the region. Therefore, the optimal solution to the LP problem is Point A, in which $f(\mathbf{x}) = f(2.5, 4.42) = -8.26$.

Step 3b: We may calculate the values of the vertices of the feasible region, in this case, $A = (2.5, 4.42)$, $B = (2.26, 0.947)$, and $C = (1.65, 3.42)$. Plugging these values into the objective function, we have $f(A) = -8.26$, $f(B) = 1.68$, and $f(C) = -6.96$. Therefore, the optimal solution is at point A.

This LP problem can be also solved by using the MATLAB function linprog. linprog solves an LP problem in the following form:

$$\min \mathbf{f}^{\mathrm{T}}\mathbf{x} \text{ such that } \begin{cases} \mathbf{A} \cdot \mathbf{x} \le \mathbf{b}, \\ \mathbf{Aeq} \cdot \mathbf{x} \le \mathbf{beq}, \\ \mathbf{lb} \le \mathbf{x} \le \mathbf{ub}. \end{cases}$$

where $\mathbf{f}$, $\mathbf{x}$, $\mathbf{beq}$, $\mathbf{lb}$, and $\mathbf{ub}$ are vectors; and $\mathbf{A}$ and $\mathbf{Aeq}$ are matrices. Entities with "eq" imply equality constraints.

The following script solves the LP problem using MATLAB.

```
f=[2;-3];
A=[-1 -5
-4 -1
```

EXAMPLE 3.6—cont'd

```
     -7  6
     -1  6];
b=[-7;-10;9;24];
lb=[];
ub=[];
x0=[]; %initial design
[x,fval, exitflag]=linprog(f,A,b,lb,ub,x0); %returns a value exitflag that
describes the exit condition, x:solution, fval: objective function value
```

The solutions returned by MATLAB are

```
>> x, fval
x =
2.5000
4.4167

fval =
-8.2500
```

which are identical to those obtained by using the graphical method.

Note that, in some cases, the solutions obtained may not be unique. For example, in the above problem, if the objective function is redefined as $f(x) = 4x_1 + x_2$, the solution to this problem is all points in line segment between points B and C, as shown in the figure of Example 3.6. The function value is $f(\mathbf{x}) = 4x_1 + x_2 = 10.02$ along line segment BC.

For an LP problem of more than two variables, a method called the simplex method is powerful and widely accepted. For more details regarding the simplex method, please refer to excellent textbooks, such as Arora (2012).

3.4.2 NONLINEAR PROGRAMMING PROBLEMS

When either the objective or any constraint function is nonlinear in terms of the variables, we have a nonlinear programming (NLP) problem. Steps for solving an NLP problem using graphical method are similar to those of LP problems discussed above, except that graphing these functions may require using software tools, such as MATLAB.

EXAMPLE 3.7

Solve the following NLP optimization problem using the graphical method.

$$\text{Minimize} : f(\mathbf{x}) = (x_1 - 3)^2 + (x_2 - 3)^2 \tag{3.40a}$$

$$\text{Subject to} : g_1(\mathbf{x}) = 3x_1 + x_2 - 6 \le 0 \tag{3.40b}$$

$$0 \le x_1, \text{ and } 0 \le x_2 \tag{3.40c}$$

Solutions

Following the same steps discussed above, we sketch the feasible region bound by constraint $g_1(\mathbf{x})$ and side constraints shown below.

EXAMPLE 3.7—cont'd

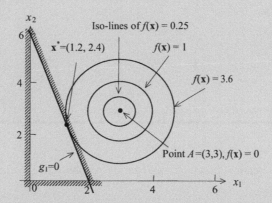

We sketch iso-lines of the objective function, which are concentric circles with center at point $A = (3, 3)$. It is clear in the figure above that the optimal point $\mathbf{x}^*$ is a tangent point of $f(\mathbf{x}) = C$ on the straight line $g_1(\mathbf{x}) = 0$. The tangent point can be calculated by intersecting a straight line of slope 1/3 that passes point A; that is, $x_1 - 3x_2 + 6 = 0$, and the straight line $g_1(\mathbf{x}) = 0$. The point is found as $\mathbf{x}^* = (1.2, 2.4)$, in which $f(\mathbf{x}) = 3.6$.

This NLP problem can also be solved by using MATLAB function `fmincon`, which solves an NLP problem in the following form:

$$\min \mathbf{f}^T\mathbf{x} \text{ such that} \begin{cases} \mathbf{c}(\mathbf{x}) \leq \mathbf{0}, \\ \mathbf{ceq}(\mathbf{x}) = \mathbf{0}, \\ \mathbf{A} \cdot \mathbf{x} \leq \mathbf{b}, \\ \mathbf{Aeq} \cdot \mathbf{x} \leq \mathbf{beq}, \\ \mathbf{lb} \leq \mathbf{x} \leq \mathbf{ub}. \end{cases}$$

The following script solves the NLP problem. We first create the two needed files for objective and constraint functions. We name them, respectively, objfun.m and confun.m. The contents of these two files are shown below. Note that the name of the function needs to match that defined in the main script from the MATLAB window.

```
%contents of the file: objfun.m
function f = objfun(x)
f = (x(1)-3)^2+(x(2)-3)^2;
```

```
%contents of the file: confun.m
function [c, ceq] = confun(x)
% Nonlinear inequality constraints
c = [3*x(1)+x(2)-6;-x(1);-x(2)];
% No nonlinear equality constraints, hence ceq is empty
ceq = [];
```

Enter the following main script:

```
x0 = [0,0]; % Make a starting guess at the solution
Options = optimset('Algorithm','active-set');
% fmincon(objfun,x0,A,b,Aeq,beq,lb,ub,confun, options)
[x,fval] = fmincon(@objfun,x0,[],[],[],[],[],[],@confun, options);
```

Continued

EXAMPLE 3.7—cont'd

The solutions returned by MATLAB are

```
>> x, fval
x =
1.2000 2.4000
fval =
3.6000
```

which are identical to those obtained by using the graphical method.

Next, we use a simple beam, a practical engineering problem, to illustrate the graphical method further.

EXAMPLE 3.8

Formulate optimization problem for the design of a simple support beam of hollow circular cross-section shown below. The beam is loaded with a point force $F = 16{,}000$ N at the mid-section (section A). The yield strength of the material is 1200 MPa and the required safety factor is $n = 3$. Geometric dimensions of the beam are shown in the figure below. We want to design the beam to minimize its volume subject to the maximum bending stress, no greater than the allowable level of 400 MPa (yield strength divided by safety factor), by varying two design variables: the diameter of the hollow section x_1 and length of the left segment of the beam x_2. The minimum thickness of the beam is 7.5 mm. Diameter x_1 must be no less than 30 mm, and length x_2 must be less than half of the beam length. Solve the optimization problem using the graphical method.

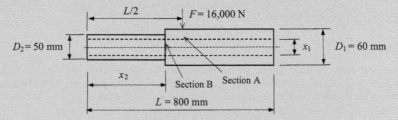

Solutions

We first formulate the optimization problem. The volume of the beam can be calculated as

$$V = \frac{\pi(D_2^2 - x_1^2)}{4}x_2 + \frac{\pi(D_1^2 - x_1^2)}{4}(L - x_2) = \frac{\pi}{4}\left[(2500 - x_1^2)x_2 + (3600 - x_1^2)(800 - x_2)\right]$$

The maximum bending stress is found either at the mid-section (section A) of the beam where the maximum bending moment occurs or at the junction of the two segments (section B) where the bending stiffness of the beam is reduced with a relatively large bending moment of a smaller flexural rigidity. The stress measures at these two locations are calculated as

$$\sigma_A = \frac{Mc}{I} = \frac{\left(\frac{FL}{4}\right)\left(\frac{D_1}{2}\right)}{\frac{\pi}{64}(D_1^4 - x_1^4)} = \frac{\left(\frac{16{,}000 \times 800}{4}\right)\left(\frac{60}{2}\right)}{\frac{\pi}{64}(60^4 - x_1^4)} = \frac{1.956 \times 10^9}{1.296 \times 10^7 - x_1^4}$$

$$\sigma_B = \frac{Mc}{I} = \frac{\left(\frac{Fx_2}{2}\right)\left(\frac{D_2}{2}\right)}{\frac{\pi}{64}(D_2^4 - x_1^4)} = \frac{\left(\frac{16{,}000x_2}{2}\right)\left(\frac{50}{2}\right)}{\frac{\pi}{64}(50^4 - x_1^4)} = \frac{4.074 \times 10^6 x_2}{6.25 \times 10^6 - x_1^4}$$

EXAMPLE 3.8—cont'd

The optimization problem can then be stated as

$$\text{Minimize}: f(x_1,x_2) = \frac{\pi}{4}\left[(2500 - x_1^2)x_2 + (3600 - x_1^2)(800 - x_2)\right] \qquad (3.41a)$$

$$\text{Subject to}: g_1(x_1,x_2) = \frac{1.956 \times 10^9}{1.296 \times 10^7 - x_1^4} - 400 \leq 0 \qquad (3.41b)$$

$$g_2(x_1,x_2) = \frac{4.074 \times 10^6 x_2}{6.25 \times 10^6 - x_1^4} - 400 \leq 0 \qquad (3.41c)$$

$$30 \leq x_1 \leq 45; \quad 0 \leq x_2 \leq 400 \qquad (3.41d)$$

Note that from Eq. 3.41b, we solve for x_1 as $x_1 \leq 53.30$ mm. Because the upper bound of the design variable x_1 is 45 mm, the constraint equation $g_1(\mathbf{x}) \leq 0$ in Eq. 3.41b is never active. Therefore, it is removed from the optimization problem.

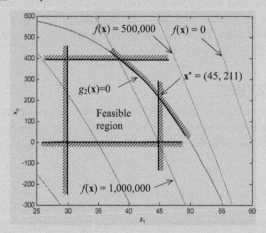

Using the graphical method, we sketch the feasible region bounded by constraint $g_2(\mathbf{x})$ and two side constraints. We also sketch iso-lines of the objective function $f(\mathbf{x})$ shown above. The MATLAB script used for sketching the curves can be found in Appendix A (Script 4). As depicted in the graph, optimal point $\mathbf{x}^*$ is identified as the intersection of $g_2(\mathbf{x}) = 0$ and $x_1 = 45$; that is, $\mathbf{x}^* = (45, 211.0)$, in which $f(\mathbf{x}^*) = f(45, 211.0) = 807{,}300$ mm^3.

As shown in the examples of this section, the concept of the graphical method is straightforward and the method is easy to use. The only limitation is certainly that the number of design variables cannot be greater than 2. On the other hand, as seen in the examples, one key step in using the graphical method is to sketch the feasible region and graph iso-lines of the objective function in the design space. Therefore, familiarity with graphing software, such as MATLAB, becomes important to use the graphical method for solving optimization problems. In the next sections, we introduce more general approaches, including gradient-based and non-gradient approaches for both constrained and unconstrained problems.

3.5 GRADIENT-BASED APPROACH

The gradient-based approach solves optimization problems by searching in the design space based on the gradients of objective and constraint functions that are active using numerical algorithms. This approach solves for both constrained and unconstrained problems of more than two design variables. However, for illustration purposes, we use examples of one or two design variables.

The gradient-based approach starts with an initial design. This approach searches for a local minimum that is closest to the initial design in an iterative manner. Note that for constrained problems, if the initial design is infeasible, the goal of the search is often to first bring the design into the feasible region. In doing so, the objective function may increase if there is a conflict in design between the objective and constraint functions. A convergent criterion needs to be defined to terminate the search when an optimal solution is found. The criterion must stop the search even if a local minimum is not found after certain number of design iterations.

In this section, we discuss unconstrained optimization problems using basic search methods. For illustration purposes, we assume problems with only one design variable. As readers quickly figure the limitation of the simple search methods, we introduce more general gradient-based search methods for more design variables. We offer sample MATLAB scripts for solving example problems. MATLAB scripts can be found in Appendix A.

3.5.1 GENERATIVE METHOD

A generative method is a brute-force approach, in which the design domain is divided into n equal intervals of Δx between its upper bound x^u and lower bound x^ℓ, and then the objective function $f(x)$ is evaluated at individual designs. For example, at the ith design point (or design), the design variable value is determined by

$$x^i = x^\ell + i\Delta x = x^\ell + \frac{x^u - x^\ell}{n} i \tag{3.42}$$

The function values are evaluated at all points and then compared to find a minimum, as illustrated in Figure 3.10. This method finds global minimum in the design domain. However, the closeness of the found solution to the true optimal depends on the size of the interval Δx. A smaller Δx yields a better result, however, requiring more function evaluations. Some commercial software, such as SolidWorks

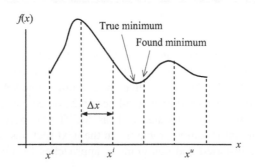

FIGURE 3.10

Illustration of the generative method.

Simulation, uses this method to support design optimization, which is in general inefficient because each function evaluation requires a finite element analysis.

3.5.2 SEARCH METHODS

The simplest search method uses equal intervals in the design variable. As illustrated in Figure 3.11, the minimum of an objective function $f(x)$ of a single variable x is being searched. In general, the explicit expression of the objective function in design variable is not available, in which the objective function is evaluated using a software tool, such as finite element analysis.

In using the equal interval search, the function $f(x)$ is evaluated at points with equal Δx increments. The calculated values of the function at two successive points are then compared. As shown in Figure 3.11, the initial design given is x^0. A search interval Δx is prescribed. If $f(x^0) > f(x^0 + \Delta x)$, then $x^1 = x^0 + \Delta x$ becomes the current design, and the search continues in the positive Δx direction. Otherwise, $x^1 = x^0 - \Delta x$ is the new design, and continue searching in the $-\Delta x$ direction. When an increase in function value at any final point x^f is encountered,

$$f(x^f - \Delta x) < f(x^f)$$

the minimum point has passed. At this point, the search direction is reversed and the increment Δx is usually cut in to half. Now, the search process is restarted by evaluating the function at $x^f - \Delta x/2$. The process repeats until the following convergent criterion is satisfied:

$$|f(x^k) - f(x^{k-1})| < \varepsilon \tag{3.43}$$

in which x^k and x^{k-1} represent two consecutive search points, and ε is a prescribed convergence tolerance. This method finds a local minimum that is close to the start point (or initial design) x^0.

An improved version of the search method is called golden section search. In this method, the size of the search interval is changing by a golden ratio 0.618 from the current to the next search. As shown in Figure 3.12a, the first point is identified as $x^1 = x^\ell + 0.618 (x^u - x^\ell) = x^\ell + 0.618\ell$, then the second point $x^2 = x^u - 0.618\ell$. The method works as follows:

- If $f(x^1) > f(x^2)$, then the minimum is between x^ℓ and x^1, and the region to the right of x^1 is excluded from the search, as illustrated in Figure 3.12b. We assign $x^u = x^1$ to continue the search.

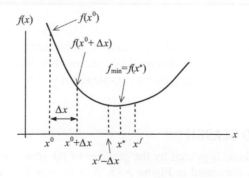

FIGURE 3.11

Illustration of the equal interval search method.

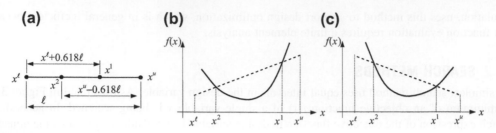

FIGURE 3.12

Illustration of the golden section search method.

- If $f(x^1) < f(x^2)$, then the minimum is between x^2 and x^u, the region to the left of x^2 is excluded from the search, as illustrated in Figure 3.12c. We assign $x^\ell = x^2$ to continue the search.
- If $|f(x^u) - f(x^\ell)| < \varepsilon$, then we found the optimal point at $x = x^u$ or x^ℓ.

This method offers a better convergent rate than that of the equal interval method, implying that less numbers of function evaluations are needed.

EXAMPLE 3.9

Use the golden search method to find the minimum point of the following function in the interval of $1 \leq x \leq 11$. We assume the convergent tolerance is $\varepsilon = 0.0001$.

$$f(x) = 3x^3 + 1500/x$$

Solutions

From the discussion above, the initial two points can be found as $x^1 = x^\ell + 0.618(x^u - x^\ell) = 1 + 6.18 = 7.18$, and $x^2 = x^u - 0.618(x^u - x^\ell) = 11 - 6.18 = 4.82$. Evaluating $f(x)$ at these two points, we have $f(x^1) = 1320$, and $f(x^2) = 647$. Because $f(x^1) > f(x^2)$, we assign $x^u = x^1 = 7.18$ to continue the search.

In the next iteration, we have $x^1 = x^\ell + 0.618(x^u - x^\ell) = 1 + 0.618(7.18 - 1) = 4.82$, and $x^2 = x^u - 0.618(x^u - x^\ell) = 7.18 - 0.618(7.18 - 1) = 3.36$. Evaluating $f(x)$ at these two points, we have $f(x^1) = 647$, and $f(x^2) = 560$. Because $f(x^1) > f(x^2)$, we assign $x^u = x^1 = 4.82$ to continue the search. The process repeats until $|f(x^u) - f(x^\ell)| < \varepsilon = 0.0001$.

The golden search method can be implemented in MATLAB for solving this example, such as the script included in Appendix A (Script 5). The optimal point is found at $x^* = 3.591$, in which $f(x^*) = 556.6$.

These basic search methods are simple and easy to implement. However, they take a lot of function evaluations to find an optimal solution. When the number of design variables increases, the computation time required for performing function evaluations is substantial, making these methods less desirable.

3.5.3 GRADIENT-BASED SEARCH

Another kind of search methods is guided by the gradient of the objective function. The concept of the gradient-based search is illustrated in Figure 3.13. As shown in Figure 3.13, an initial design is

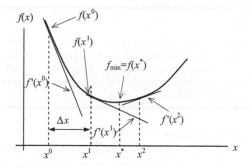

FIGURE 3.13

Illustration of the gradient-based search method.

given as x^0. The derivative (or gradient) of the objective function $f(x)$ is calculated at x^0 as $f'(x^0)$, which is the slope of the objective function curve shown in Figure 3.13. The gradient of the objective function $f'(x^0)$, which is the slope of the objective function, determines the search direction—in this case, in the direction of decreasing the objective function. Once the search direction is determined, the step size along the direction is sought. Usually, a relatively large step size Δx can be assumed to find the variable value of the next iteration; that is, $x^1 = x^0 + \Delta x$. Once x^1 is determined, the objective function value $f(x^1)$ and gradient $f'(x^1)$ are calculated. Usually, the same step size Δx is employed to determine the next design until the search direction is reversed, for example, at x^2, as shown in Figure 3.13. At this point, the step size is reduced to $\Delta x/2$ and the search is resumed. This is called the bisection search. Determining an adequate step size along the search direction is called a line search. There are several prominent algorithms supporting line search in optimization, such as the conjugate gradient method, backtracking line search, or Wolfe conditions.

The process repeats until a convergence criterion is met. Note that the convergence criteria can be defined in different ways. For example, the difference in the objective function values of two consecutive iterations is less than a prescribed convergent tolerance ε_1:

$$\left| f\left(x^k\right) - f\left(x^{k-1}\right) \right| < \varepsilon_1 \tag{3.44a}$$

Or, the magnitude of the gradient of the objective function is less than a prescribed convergent tolerance ε_2:

$$\left| f'\left(x^k\right) \right| < \varepsilon_2 \tag{3.44b}$$

EXAMPLE 3.10

Use the gradient-based method to find the minimum point of the same objective function as that of Example 3.9 in the interval of $1 \leq x \leq 11$. We assume the initial design is given at $x^0 = 1$ and convergent tolerance is $\varepsilon_1 = 0.0001$.

Solutions: First, the derivative of the objective function, $f(x) = 3x^3 + 1500/x$, is

$$f'(x) = 9x^2 - 1500/x^2$$

At the initial design, $f(x^0) = f(1) = 1503$, and $f'(x^0) = f'(1) = -1411$. We assume the initial step size is $\Delta x = (x^u - x^\ell)/2 = (11 - 1)/2 = 5$. Hence, the next design variable is $x^1 = x^0 + \Delta x = 1 + 5 = 6$.

Continued

EXAMPLE 3.10—cont'd

At the new design, we calculate the function value and its gradient as $f(x^1) = f(6) = 898$, and $f'(x^1) = f'(6) = 282$, which is reversed from the previous design point. Hence, the step size is reduced to half, $\Delta x = 5/2 = 2.5$; and the next design point is $x^2 = x^1 - \Delta x = 6 - 2.5 = 3.5$. The process repeats until $|f(x^i) - f(x^{i-1})| < \varepsilon_1 = 0.0001$.

The gradient-based search method can be implemented in MATLAB to solve the example (in Appendix A (Script 6)). The optimal point is found at $x^* = 3.594$, in which $f(x^*) = 556.6$.

Next, we considered objective functions of multiple design variables. In general, the two major factors to determine in gradient-based search methods are search direction and line search. For unconstrained problems, the search direction is determined by the gradient vector of the objective function. Once the search direction is determined, an appropriate step size along the search direction must be determined to locate the next design point, which is called line search. Although several methods can be used to find the optimal solution of multivariable objective functions, the steepest decent method, which is one of the simplest methods, as well as an improvement from the concept of steepest decent called the conjugate gradient method, are discussed. In this subsection, we assume the bisection method discussed just now for line search. More about line search is provided in Section 3.5.4.

3.5.3.1 Steepest decent method

The gradient-based methods rely on the gradient vector to determine the search direction in finding an optimal solution. For an objective function of n variables $f(\mathbf{x})$, the gradient vector is defined as

$$\nabla f(\mathbf{x}) = \begin{bmatrix} \dfrac{\partial f}{\partial x_1} & \dfrac{\partial f}{\partial x_2} & \cdots & \dfrac{\partial f}{\partial x_n} \end{bmatrix}^{\mathrm{T}} \tag{3.45}$$

As discussed in Example 3.2, the gradient vector at a point $\mathbf{x}$ defines the direction of maximum increase in the objective function. Thus, the direction of maximum decrease is opposite to that—that is, negative of the gradient vector $-\nabla f(\mathbf{x})$. Any small move in the negative gradient direction will result in the maximum local rate of decrease in the objective function. The negative gradient vector thus represents a direction of steepest descent for the objective function and is written as

$$\mathbf{n} = -\nabla f(\mathbf{x}) = -\begin{bmatrix} \dfrac{\partial f}{\partial x_1} & \dfrac{\partial f}{\partial x_2} & \cdots & \dfrac{\partial f}{\partial x_n} \end{bmatrix}^{\mathrm{T}} \tag{3.46}$$

or

$$n_i = -\frac{\partial f}{\partial x_i}, \quad i = 1, n \tag{3.47}$$

In general, the method of steepest descent, also called the gradient descent method, starts with an initial point $\mathbf{x}^0$ and, as many times as needed, moves from $\mathbf{x}^k$ to $\mathbf{x}^{k+1}$ by minimizing along the vector $\mathbf{n}^k$ extending from $\mathbf{x}^k$ in the direction of $-\nabla f(\mathbf{x}^k)$, the local downhill gradient. The vector $\mathbf{n}^k$, representing the search direction for minimizing the objective function along the steepest decent direction, is usually normalized as

$$\mathbf{n}^k = \mathbf{n}(\mathbf{x}^k) = -\frac{\nabla f(\mathbf{x}^k)}{\|\nabla f(\mathbf{x}^k)\|} \tag{3.48}$$

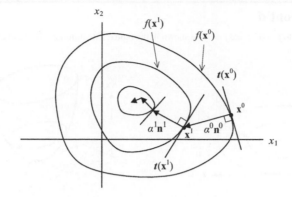

FIGURE 3.14

Illustration of the orthogonal steepest decent path.

Notice the same notation **n** is employed for the normalized search direction.

As mentioned before, the steepest decent direction at a point $\mathbf{x}^k$ determines the direction of maximum decrease in the objective function. Once the search direction is calculated, line search is carried out to find a step size α^k along the search direction. The next design is then determined as

$$\mathbf{x}^{k+1} = \mathbf{x}^k + \alpha^k \mathbf{n}^k \tag{3.49}$$

If we use the bisection search method, the step size α is reduced by half if the function value at the current design is greater than the previous one. The process repeats until the difference in the objective function values of two consecutive iterations is less than a prescribed convergent tolerance ε_1, or the magnitude of the gradient is less than a prescribed tolerance; that is, $\|\nabla f(\mathbf{x}^k)\| < \varepsilon_2$. Note that geometrically, the steepest decent direction of a design $\mathbf{x}^k$ is perpendicular to the tangent line $\mathbf{t}(\mathbf{x}^k)$ of the iso-line of the objective function $f(\mathbf{x}^k)$, as shown in Figure 3.14. Hence, the search path is a zigzag polyline shown in Figure 3.14.

EXAMPLE 3.11

Use the steepest decent method to find the minimum point of the objective function

$$f(x_1, x_2) = \frac{(x_1 - 2)^2}{4} + (x_2 - 1)^2$$

We assume the initial design is given at $\mathbf{x}^0 = (0, 0)$, the step size is $\alpha = 2$, and convergent tolerance is $\varepsilon_1 = 0.00001$.

Solutions

We first plot the function for better illustration. The iso-lines of the objective function are ellipse with center point at (2, 1), and with a ratio of long and short axes 2:1. Two iso-lines of $f = 0.25$ and $f = 1$ are shown in the figure next page (left). The elliptic cone shown in the figure next page (right) depicts the objective function viewed in three-dimensional (3-D) space.

Continued

EXAMPLE 3.11—cont'd

Two ellipses corresponding to $f = 0.25$ and $f = 1$ are shown in the figure. The minimum is found at the center of the ellipses; that is, $(x_1, x_2) = (2, 1)$.

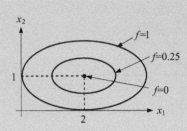

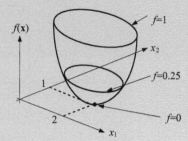

The gradient of the objective function can be calculated as

$$\nabla f(x_1, x_2) = \left[\frac{\partial f}{\partial x_1} \quad \frac{\partial f}{\partial x_2} \right]^{\mathrm{T}} = \left[\frac{1}{2}(x_1 - 2) \quad 2(x_2 - 1) \right]^{\mathrm{T}}$$

As before, we start with an initial design, which is assumed as $\mathbf{x}^0 = (0, 0)$. At this point, the function value and gradient are $f(\mathbf{x}^0) = f(0, 0) = 2$ and $\nabla f(\mathbf{x}^0) = [-1, -2]^{\mathrm{T}}$, respectively. Also, the search direction is calculated using Eq. 3.48 as $\mathbf{n}^0 = \mathbf{n}(\mathbf{x}^0) = [1/\sqrt{5}, 2/\sqrt{5}]^{\mathrm{T}}$.

We assume $\alpha^0 = 2$, and we use the bisection search method as before. Therefore, from Eq. 3.49 we calculate the next point $\mathbf{x}^1$ as

$$\mathbf{x}^1 = \mathbf{x}^0 + \alpha^0 \mathbf{n}^0 = [0, 0]^{\mathrm{T}} + 2\left[1/\sqrt{5}, 2/\sqrt{5}\right]^{\mathrm{T}} = \left[2/\sqrt{5}, 4/\sqrt{5}\right]^{\mathrm{T}} = [0.894, 1.79]^{\mathrm{T}}$$

At the new point, the function value, gradient, and search direction are $f(\mathbf{x}^1) = f(0.894, 1.79) = 0.928$, $\nabla f(\mathbf{x}^1) = [-0.553, 1.58]^{\mathrm{T}}$, and $\mathbf{n}^1 = [0.330, -0.944]^{\mathrm{T}}$, respectively. Then, we calculate the next point $\mathbf{x}^2$ as

$$\mathbf{x}^2 = \mathbf{x}^1 + \alpha^1 \mathbf{n}^1 = [0.894, \ 1.79]^{\mathrm{T}} + 2[0.330, -0.944]^{\mathrm{T}} = [1.56, -0.0986]^{\mathrm{T}}$$

At the new point, the function value is $f(\mathbf{x}^2) = f(1.56, -0.0986) = 1.42 > f(\mathbf{x}^1)$. Hence, step size α is reduced to $\alpha^2 = \alpha^0/2$. The process continues until $|f(\mathbf{x}^k) - f(\mathbf{x}^{k-1})| < \varepsilon_1 = 0.00001$.

The gradient-based search method can be implemented in MATLAB for solving the example (see Script 6 in Appendix A for details). The optimal point is found at $\mathbf{x}^* = (2.0003, 1.0018)$, in which $f(\mathbf{x}^*) = 4.49 \times 10^{-6}$. The first few iterations of the search path are illustrated in the figure below.

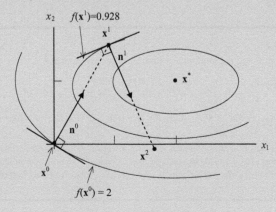

3.5.3.2 Conjugate gradient method

The conjugate gradient method is a simple and effective modification of the steepest descent method. We start with an initial design $\mathbf{x}^0$, set the convergence tolerance ε, and calculate the function value $f(\mathbf{x}^0)$ and gradient vector $\nabla f(\mathbf{x}^0)$. With the gradient vector calculated, we calculate the search direction $\mathbf{n}^0$ using Eq. 3.48. After this first iteration, instead of continuously using Eq. 3.48 to calculate the search direction as in the steepest decent method, the conjugate gradient method searches optimal design along the conjugate direction defined by

$$\mathbf{n}^k = -\frac{\nabla f(\mathbf{x}^k)}{\|\nabla f(\mathbf{x}^k)\|} + \beta^k \mathbf{n}^{k-1} \tag{3.50}$$

in which the parameter β^k is defined in several different ways, for example, by Fletcher and Reeves (1964) as

$$\beta^k = \frac{\|\nabla f(\mathbf{x}^k)\|}{\|\nabla f(\mathbf{x}^{k-1})\|} \tag{3.51}$$

After the first iteration, the only difference between the conjugate gradient and steepest descent methods is in Eq. 3.50. In this equation, the current steepest descent direction is modified by adding a scaled direction that was used in the previous iteration. The scale factor β^k is determined by using lengths of the gradient vector at the two iterations, as shown in Eq. 3.51. Thus, the conjugate direction is nothing but a deflected steepest descent direction. This is a simple modification that requires little additional calculation. In general, this method improves the rate of convergence of the steepest descent method.

EXAMPLE 3.12

Use the conjugate gradient method to find the minimum point of the same function of Example 3.11 with a convergent tolerance $\varepsilon_1 = 0.00001$.

Solutions

The first iteration is identical to the steepest decent method; that is, $\nabla f(\mathbf{x}^0) = [-1, -2]^T$, $\mathbf{n}^0 = \mathbf{n}(\mathbf{x}^0) = [1/\sqrt{5}, 2/\sqrt{5}]^T$, $\mathbf{x}^1 = [0.894, 1.79]^T$, $\nabla f(\mathbf{x}^1) = [-0.553, 1.58]^T$, and $\mathbf{n}^1 = [0.330, -0.944]^T$. From Eq. 3.51:

$$\beta^1 = \frac{\|\nabla f(\mathbf{x}^1)\|}{\|\nabla f(\mathbf{x}^0)\|} = \frac{\|[-0.533, 1.58]^T\|}{\|[-1, -2]^T\|} = \frac{1.67}{2.24} = 0.746$$

Hence, from Eq. 3.50, the search direction becomes

$$\mathbf{n}^1 = -\frac{\nabla f(\mathbf{x}^1)}{\|\nabla f(\mathbf{x}^1)\|} + \beta^1 \mathbf{n}^0 = -\frac{[-0.533, 1.58]^T}{1.67} + 0.764\left[1/\sqrt{5}, 2/\sqrt{5}\right]^T = [0.661, -0.263]^T$$

The vector $\mathbf{n}^1$ is normalized as $\mathbf{n}^1 = [0.929, -0.370]^T$. Then, we calculate the next point $\mathbf{x}^2$ as

$$\mathbf{x}^2 = \mathbf{x}^1 + \alpha^1 \mathbf{n}^1 = [0.894, 1.79]^T + 2[0.929, -0.370]^T = [2.75, 1.05]^T$$

At the new point, the function value is $f(\mathbf{x}^2) = f(2.75, 1.05) = 0.143 < f(\mathbf{x}^1)$. Hence, the process continues with the same step size α. The process continues until $|f(\mathbf{x}^k) - f(\mathbf{x}^{k-1})| < \varepsilon_1 = 0.00001$.

The conjugate gradient method can be implemented in MATLAB to solve the example (see Script 8 in Appendix A for details). The optimal point is found at $\mathbf{x}^* = (1.9680, 1.0133)$, in which $f(\mathbf{x}^*) = 4.34 \times 10^{-4}$.

3.5.3.3 Quasi-Newton method

With the steepest descent method, only first-order derivative information is used to determine the search direction. If second-order derivatives are available, we can use them to represent the objective function more accurately, and a better search direction can be found. With the inclusion of second-order information, we can expect a better rate of convergence as well. For example, Newton's method, which uses the Hessian of the function in calculating the search direction, has a quadratic rate of convergence (meaning that it converges very rapidly when the design point is within certain radius of the minimum point).

The basic idea of the classical Newton's method is to use a second-order Taylor's series expansion of the function around the current design point $\mathbf{x}$, a vector of size of $n \times 1$ (n is the number of design variables):

$$f(\mathbf{x} + \Delta\mathbf{x}) = f(\mathbf{x}) + \nabla f^{\mathrm{T}}\Delta\mathbf{x} + \frac{1}{2}\Delta\mathbf{x}^{\mathrm{T}}\mathbf{H}\Delta\mathbf{x} \tag{3.52}$$

where $\Delta\mathbf{x}$ is a small change in design and $\mathbf{H}_{n\times n}$ is the Hessian of the objective function f at the point $\mathbf{x}$. Employing the optimality condition to Eq. 3.52 by taking derivative of Eq. 3.52 with respect to $\Delta\mathbf{x}$ and setting the derivative to zero—that is, $\partial f/\partial(\Delta\mathbf{x}) = 0$—we have

$$\nabla f + \mathbf{H}\Delta\mathbf{x} = 0 \tag{3.53}$$

Assuming $\mathbf{H}$ to be nonsingular, we get an expression for $\Delta\mathbf{x}$ as

$$\Delta\mathbf{x} = -\mathbf{H}^{-1}\nabla f \tag{3.54}$$

where $\Delta\mathbf{x}$ updates design to the next point; that is, $\mathbf{x}^{k+1} = \mathbf{x}^k + \Delta\mathbf{x}$. Because Eq. 3.52 is just an approximation for f at the current point $\mathbf{x}^k$, the next design point $\mathbf{x}^{k+1}$ updated using $\Delta\mathbf{x}$ will most likely not the precise minimum point of $f(\mathbf{x})$. Therefore, the process will have to be repeated to obtain improved estimates until a minimum is reached.

Newton's method is inefficient because it requires calculation of $n(n + 1)/2$ second-order derivatives to generate the Hessian matrix. For most engineering design problems, calculation of second-order derivatives may be too expensive to perform or entirely infeasible numerically due to poor accuracy. Also, Newton's method runs into difficulties if the Hessian of the function is singular at any iteration.

Several methods were proposed to overcome the drawbacks of Newton's method by generating an approximation for the Hessian matrix or its inverse at each iteration. Only the first derivatives of the function are used to generate these approximations. They are called quasi-Newton methods. We introduce the Broyden–Fletcher–Goldfarb–Shanno (BFGS) method, which updates the Hessian rather than its inverse at every iteration.

3.5.3.4 The BFGS method

The search direction $\mathbf{n}^k$ is given by the solution of the Newton equation in Eq. 3.54:

$$\mathbf{H}^k\mathbf{n}^k = -\nabla f\left(\mathbf{x}^k\right) \tag{3.55}$$

where $\mathbf{H}^k = \mathbf{H}(\mathbf{x}^k)$ is an approximation to the Hessian matrix which is updated iteratively at each design iteration. Note that at the initial design, the Hessian matrix is set to be an identity matrix; that is, $\mathbf{H}^0 = \mathbf{I}$. A line search in the direction $\mathbf{n}^k$ is then carried out to find the next point $\mathbf{x}^{k+1}$, as shown in Eq. 3.49.

The approximate Hessian at the design iteration $k + 1$ is updated by the addition of two matrices:

$$\mathbf{H}^{k+1} = \mathbf{H}^k + \mathbf{D}^k + \mathbf{E}^k \tag{3.56}$$

where the correction matrices $\mathbf{D}^k$ and $\mathbf{E}^k$ are given as

$$\mathbf{D}^k = \frac{\mathbf{y}^k \left(\mathbf{y}^k\right)^{\mathrm{T}}}{\left(\mathbf{y}^k\right)^{\mathrm{T}} \cdot \mathbf{s}^k} \tag{3.57}$$

and

$$\mathbf{E}^k = \frac{\nabla f^k \left(\nabla f^k\right)^{\mathrm{T}}}{\left(\nabla f^k\right)^{\mathrm{T}} \cdot \mathbf{n}^k} \tag{3.58}$$

where $\nabla f^k = \nabla f(\mathbf{x}^k) = [\partial f(\mathbf{x}^k)/\partial x_1, \ldots, \partial f(\mathbf{x}^k)/\partial x_n]^{\mathrm{T}}$, $\mathbf{y}^k = \nabla f^{k+1} - \nabla f^k$, and $\mathbf{s}^k = \alpha^k \mathbf{n}^k = \mathbf{x}^{k+1} - \mathbf{x}^k$.

EXAMPLE 3.13

Use the BFGS method to find the minimum point of the same objective function of Example 3.11 with a convergent tolerance $\varepsilon_2 = 0.001$. In this example, we assume the initial step size to be $\alpha = 1$.

Solutions

The first iteration is identical to the steepest decent method; that is, $\nabla f(\mathbf{x}^0) = [-1, -2]^{\mathrm{T}}$, $\mathbf{n}^0 = [1/\sqrt{5}, 2/\sqrt{5}]^{\mathrm{T}}$, $\mathbf{x}^1 = \mathbf{x}^0 + \alpha \mathbf{n}^0 = [0,0]^{\mathrm{T}} + [1/\sqrt{5}, 2/\sqrt{5}]^{\mathrm{T}} = [0.447, 0.894]^{\mathrm{T}}$, and $\nabla f(\mathbf{x}^1) = [-0.776, -0.211]^{\mathrm{T}}$. Using the BFGS method, we first create a Hessian matrix at the initial design as an identity matrix $\mathbf{H}^0 = \mathbf{I}_{2 \times 2}$. Then, we update the Hessian matrix at $\mathbf{x}^1$ by calculating matrices $\mathbf{D}^0$ and $\mathbf{E}^0$. For the matrix $\mathbf{D}^0$, we first calculate $\mathbf{y}^0$ and $\mathbf{s}^0$ as

$$\mathbf{y}^0 = \nabla f(\mathbf{x}^1) - \nabla f(\mathbf{x}^0) = [-0.776, -0.211]^{\mathrm{T}} - [-1, -2]^{\mathrm{T}} = [0.224, 1.79]^{\mathrm{T}}$$

and

$$\mathbf{s}^0 = \alpha^0 \mathbf{n}^0 = [0.447, 0.894]^{\mathrm{T}} - [0,0]^{\mathrm{T}} = [0.447, 0.894]^{\mathrm{T}}$$

Hence, from Eq. 3.57, we have

$$\mathbf{D}^0 = \frac{\mathbf{y}^0 \left(\mathbf{y}^0\right)^{\mathrm{T}}}{\left(\mathbf{y}^0\right)^{\mathrm{T}} \cdot \mathbf{s}^0} = \frac{\begin{bmatrix} 0.224 \\ 1.79 \end{bmatrix} \begin{bmatrix} 0.224 & 1.79 \end{bmatrix}}{\begin{bmatrix} 0.224 & 1.79 \end{bmatrix} \begin{bmatrix} 0.447 \\ 0.894 \end{bmatrix}} = \frac{\begin{bmatrix} 0.0502 & 0.401 \\ 0.401 & 3.20 \end{bmatrix}}{1.70} = \begin{bmatrix} 0.0295 & 0.236 \\ 0.236 & 1.88 \end{bmatrix}$$

Then, from Eq. 3.58, we have

$$\mathbf{E}^0 = \frac{\nabla f^0 \left(\nabla f^0\right)^{\mathrm{T}}}{\left(\nabla f^0\right)^{\mathrm{T}} \cdot \mathbf{n}^0} = \frac{\begin{bmatrix} -1 \\ -2 \end{bmatrix} \begin{bmatrix} -1 & -2 \end{bmatrix}}{\begin{bmatrix} -1 & -2 \end{bmatrix} \begin{bmatrix} 1/\sqrt{5} \\ 2/\sqrt{5} \end{bmatrix}} = \frac{\begin{bmatrix} 1 & 2 \\ 2 & 4 \end{bmatrix}}{-\sqrt{5}} = \begin{bmatrix} -0.447 & -0.894 \\ -0.894 & -1.79 \end{bmatrix}$$

Continued

EXAMPLE 3.13—cont'd

The Hessian matrix is now updated according to Eq. 3.56 as

$$\mathbf{H}^1 = \mathbf{H}^0 + \mathbf{D}^0 + \mathbf{E}^0$$

$$= \begin{bmatrix} 1 & 0 \\ 0 & 1 \end{bmatrix} + \begin{bmatrix} 0.0295 & 0.236 \\ 0.236 & 1.88 \end{bmatrix} + \begin{bmatrix} -0.447 & -0.894 \\ -0.898 & -1.79 \end{bmatrix} = \begin{bmatrix} 0.583 & -0.658 \\ -0.659 & 1.09 \end{bmatrix}$$

Hence, from Eq. 3.55, we have

$$\mathbf{n}^1 = -(\mathbf{H}^1)^{-1} \nabla f(\mathbf{x}^1) = -\begin{bmatrix} 0.582 & -0.658 \\ -0.658 & 1.09 \end{bmatrix}^{-1} \begin{bmatrix} -0.776 \\ -0.211 \end{bmatrix} = \begin{bmatrix} 4.89 \\ 3.14 \end{bmatrix}$$

The vector $\mathbf{n}^1$ is normalized as $\mathbf{n}^1 = [0.841, 0.540]^T$. Then, we calculate the next point $\mathbf{x}^2$ as

$$\mathbf{x}^2 = \mathbf{x}^1 + \alpha^1 \mathbf{n}^1 = [0.447, 0.894]^T + 1[0.841, 0.540]^T = [1.29, 1.43]^T$$

At the new point, the function value is $f(\mathbf{x}^2) = f(1.29, 1.43) = 0.316 > f(\mathbf{x}^1)$. Hence, step size α is unchanged. The process continues until $||\nabla f(\mathbf{x}^k)|| < \varepsilon_2 = 0.001$.

The gradient-based search method can be implemented in MATLAB for solving the example (see Script 9 in Appendix A for details). The optimal point is found at $\mathbf{x}^* = (2, 1)$, in which $f(\mathbf{x}^*) = 0$.

3.5.4 LINE SEARCH

As mentioned earlier, the purpose of the line search is to determine an appropriate step size α along the search direction $\mathbf{n}$ in searching for an optimal solution. Up to this point in our discussion, we employed interval-reducing methods, such as the bisection method, for line search in our discussion and examples. These methods are simple but can require many function evaluations (many design iterations) to reach an optimum. In engineering design problems, function evaluation requires a significant amount of computational effort. Therefore, these methods may not be desired for practical applications. Because line search is an important step in optimization, we offer a few more details on this subject. We will go over two popular methods: secant method and quadratic curve fitting.

3.5.4.1 Concept of line search

As discussed in Section 3.5.3, when a search direction $\mathbf{n}^k$ is found, the next design point $\mathbf{x}^{k+1}$ is determined by a step size α using Eq. 3.49. At the new design, we expect to have a reduced objective function value:

$$f(\mathbf{x}^k + 1) = f(\mathbf{x}^k + \alpha \mathbf{n}^k) \le f(\mathbf{x}^k) \tag{3.59}$$

A step size α that reduces the objective function the most is desirable. Therefore, a line search problem can be formulated as a subproblem:

$$\underset{\alpha \ge 0}{\text{Minimize}} \, f(\alpha) = f(\mathbf{x}^k + \alpha \mathbf{n}^k) \tag{3.60}$$

Because $\mathbf{x}^k$ and $\mathbf{n}^k$ are known, this problem reduces to a minimization problem of a single variable α. Assuming that $f(\mathbf{x})$ is smooth and continuous, we find its optimum where its first-derivative is set to zero:

$$f'(\alpha) = \left.\frac{df(\mathbf{x}^k + \alpha\mathbf{n}^k)}{d\alpha}\right|_{\alpha=0} = f(\mathbf{x}^k) + \nabla f(\mathbf{x}^k)^{\mathrm{T}}(\alpha\mathbf{n}^k) = 0 \tag{3.61}$$

The optimization problem of Eq. 3.60 is converted into a root-finding problem of Eq. 3.61, in which the step size α is sought. Mathematically, the step size α can be found as

$$\alpha = -\frac{f(\mathbf{x}^k)}{\nabla f(\mathbf{x}^k)^{\mathrm{T}}\mathbf{n}^k} \tag{3.62}$$

in which all the quantities on the right-hand side are known at iteration k. This is nothing but Newton's method.

EXAMPLE 3.14

Use the BFGS method combined with the Newton's method for line search to find the minimum point of the same objective function of Example 3.11 with a convergent tolerance $\varepsilon_2 = 0.001$.

Solutions

The first iteration is identical to the steepest decent method; that is, $\nabla f(\mathbf{x}^0) = [-1, -2]^{\mathrm{T}}$, $\mathbf{n}^0 = [1/\sqrt{5}, 2/\sqrt{5}]^{\mathrm{T}}$. From Eq. 3.62:

$$\alpha^0 = -\frac{f(\mathbf{x}^0)}{\nabla f(\mathbf{x}^0)^{\mathrm{T}}\mathbf{n}^0} = -\frac{2}{[-1 \quad -2]\begin{bmatrix} 1/\sqrt{5} \\ 2/\sqrt{5} \end{bmatrix}} = 0.894$$

Then, $\mathbf{x}^1 = \mathbf{x}^0 + \alpha\mathbf{n}^0 = [0,0]^{\mathrm{T}} + 0.894\,[1/\sqrt{5}, 2/\sqrt{5}]^{\mathrm{T}} = [0.400, 0.800]^{\mathrm{T}}$, and $\nabla f(\mathbf{x}^1) = [-0.800, -0.400]^{\mathrm{T}}$. Using the BFGS method, we first create a Hessian matrix at the initial design as an identity matrix $\mathbf{H}^0 = \mathbf{I}_{2\times2}$. Then we update the Hessian matrix at $\mathbf{x}^1$ by calculating matrices $\mathbf{D}^0$ and $\mathbf{E}^0$. For the matrix $\mathbf{D}^0$ we first calculate $\mathbf{y}^0$ and $\mathbf{s}^0$ as

$$\mathbf{y}^0 = \nabla f(\mathbf{x}^1) - \nabla f(\mathbf{x}^0) = [-0.800, -0.400]^{\mathrm{T}} - [-1, -2]^{\mathrm{T}} = [0.200, 1.60]^{\mathrm{T}}$$

and

$$\mathbf{s}^0 = \alpha^0\mathbf{n}^0 = 0.894[0.447, 0.894]^{\mathrm{T}} - [0, 0]^{\mathrm{T}} = [0.400, 0.800]^{\mathrm{T}}$$

Hence from Eq. 3.57, we have

$$\mathbf{D}^0 = \frac{\mathbf{y}^0(\mathbf{y}^0)^{\mathrm{T}}}{(\mathbf{y}^0)^{\mathrm{T}} \cdot \mathbf{s}^0} = \frac{\begin{bmatrix} 0.200 \\ 1.60 \end{bmatrix}[0.200 \quad 1.60]}{[0.200 \quad 1.60]\begin{bmatrix} 0.400 \\ 0.800 \end{bmatrix}} = \frac{\begin{bmatrix} 0.0400 & 0.320 \\ 0.320 & 2.56 \end{bmatrix}}{1.36} = \begin{bmatrix} 0.0294 & 0.235 \\ 0.235 & 1.88 \end{bmatrix}$$

Then, from Eq. 3.58, we have

$$\mathbf{E}^0 = \frac{\nabla f^0(\nabla f^0)^{\mathrm{T}}}{(\nabla f^0)^{\mathrm{T}} \cdot \mathbf{n}^0} = \frac{\begin{bmatrix} -1 \\ -2 \end{bmatrix}[-1 \quad -2]}{[-1 \quad -2]\begin{bmatrix} 1/\sqrt{5} \\ 2/\sqrt{5} \end{bmatrix}} = \frac{\begin{bmatrix} 1 & 2 \\ 2 & 4 \end{bmatrix}}{-\sqrt{5}} = \begin{bmatrix} -0.447 & -0.894 \\ -0.894 & -1.79 \end{bmatrix}$$

Continued

EXAMPLE 3.14—cont'd

The Hessian matrix is now updated according to Eq. 3.56 as

$$\mathbf{H}^1 = \mathbf{H}^0 + \mathbf{D}^0 + \mathbf{E}^0$$

$$= \begin{bmatrix} 1 & 0 \\ 0 & 1 \end{bmatrix} + \begin{bmatrix} 0.0295 & 0.236 \\ 0.236 & 1.88 \end{bmatrix} + \begin{bmatrix} -0.447 & -0.894 \\ -0.898 & -1.79 \end{bmatrix} = \begin{bmatrix} 0.583 & -0.658 \\ -0.659 & 1.09 \end{bmatrix}$$

Hence, from Eq. 3.55, we have

$$\mathbf{n}^1 = -(\mathbf{H}^1)^{-1} \nabla f(\mathbf{x}^1) = -\begin{bmatrix} 0.582 & -0.658 \\ -0.658 & 1.09 \end{bmatrix}^{-1} \begin{bmatrix} -0.800 \\ -0.400 \end{bmatrix} = \begin{bmatrix} 5.63 \\ 3.76 \end{bmatrix}$$

The vector $\mathbf{n}^1$ is normalized as $\mathbf{n}^1 = [0.832, 0.555]^T$. Then, we calculate the step size:

$$\alpha^1 = -\frac{f(\mathbf{x}^1)}{\nabla f(\mathbf{x}^1)^T \mathbf{n}^1} = \frac{0.680}{[-0.800 \quad -0.400]\begin{bmatrix} 0.832 \\ 0.555 \end{bmatrix}} = 0.766$$

Therefore, the next point $\mathbf{x}^2$ is

$$\mathbf{x}^2 = \mathbf{x}^1 + \alpha^1 \mathbf{n}^1 = [0.400, 0.800]^T + 0.766\,[0.832, 0.555]^T = [1.04, 1.23]^T$$

At the new point, the function value is $f(\mathbf{x}^2) = f(1.04, 1.23) = 0.283$. The process continues until $\|\nabla f(\mathbf{x}^k)\| < \varepsilon_2 = 0.001$. The gradient-based search method can be implemented in MATLAB for solving the example (see Script 10 in Appendix A for details). The optimal point is found at $\mathbf{x}^* = (2, 1)$, in which $f(\mathbf{x}^*) = 0$.

3.5.4.2 Secant method

Because, in practical applications, the gradient vector is expensive to calculate, the secant method is often employed to approximate the gradient vector. To simplify the notation, we assume a function of a single variable for the time being.

We start with two estimates of the design points, x^0 and x^1. The iterative formula, for $n \geq 1$, is

$$x^{k+1} = x^k - \frac{f(x^k)}{q(x^{k-1}, x^k)} \quad \text{or} \quad \Delta x = x^{k+1} - x^k = -\frac{f(x^k)}{q(x^{k-1}, x^k)} \tag{3.63}$$

where

$$q(x^{k-1}, x^k) = \frac{f(x^{k-1}) - f(x^k)}{x^{k-1} - x^k} \tag{3.64}$$

Note that Eq. 3.63 is nothing but Newton's method, except that the denominator in the fraction is replaced by $q(x^{k-1}, x^k)$ defined in Eq. 3.64, which approximates the gradient of the objective function. Note that if x^k is close to x^{k-1}, then $q(x^{k-1}, x^k)$ is close to $f'(x^k)$, and the secant method and Newton's method are virtually identical.

For a function of multivariables, Eqs 3.63 and 3.64 can be written as

$$\mathbf{x}^{k+1} = \mathbf{x}^k + \alpha^k \mathbf{n}^k = \mathbf{x}^k - \frac{f(x^k)}{q(\mathbf{x}^{k-1}, \mathbf{x}^k)^T \mathbf{n}^k} \mathbf{n}^k \tag{3.65a}$$

or

$$\alpha^k = -\frac{f\left(\mathbf{x}^k\right)}{\mathbf{q}(\mathbf{x}^{k-1},\mathbf{x}^k)^{\mathrm{T}}\mathbf{n}^k} \tag{3.65b}$$

Note that in Eq. 3.65b, we have

$$\mathbf{q}\left(\mathbf{x}^{k-1},\mathbf{x}^k\right) = \left[\frac{f\left(\mathbf{x}^{k-1}\right)-f\left(\mathbf{x}_1^k\right)}{x_1^{k-1}-x_1^k} \quad \frac{f\left(\mathbf{x}^{k-1}\right)-f\left(\mathbf{x}_2^k\right)}{x_2^{k-1}-x_2^k} \quad \cdots \quad \frac{f\left(\mathbf{x}^{k-1}\right)-f\left(\mathbf{x}_n^k\right)}{x_n^{k-1}-x_n^k}\right]^{\mathrm{T}} \tag{3.66}$$

where the vector $\mathbf{x}_i^k = [x_1^{k-1},\ldots,x_{i-1}^{k-1}, x_i^k, x_{i+1}^{k-1},\ldots,x_n^{k-1}]^{\mathrm{T}}$ and $f(\mathbf{x}_i^k) = f(x_1^{k-1},\ldots,x_{i-1}^{k-1}, x_i^k, x_{i+1}^{k-1},\ldots,x_n^{k-1})$. That is, only the ith variable is changed from $(k-1)$th to that of the kth iteration.

EXAMPLE 3.15

Continue with Example 3.14, except use the secant method for the line search. Recall that the objective function is

$$f(x_1,x_2) = \frac{(x_1-2)^2}{4} + (x_2-1)^2$$

We assume two points $\mathbf{x}^0 = [0,0]^{\mathrm{T}}$ and $\mathbf{x}^1 = [0.400, 0.800]^{\mathrm{T}}$, and the first search direction $\mathbf{n}^1 = [0.841, 0.540]^{\mathrm{T}}$ is given from Example 3.14. Therefore, we have $f(\mathbf{x}^0) = 2$ and $f(\mathbf{x}^1) = 0.680$.

Solutions

For $k = 1$, we calculate $\mathbf{q}(\mathbf{x}^0, \mathbf{x}^1)$ following Eq. 3.66:

$$\mathbf{q}\left(\mathbf{x}^0,\mathbf{x}^1\right) = \left[\frac{f(\mathbf{x}^0)-f(\mathbf{x}_1^1)}{x_1^0-x_1^1} \quad \frac{f(\mathbf{x}^0)-f(\mathbf{x}_2^1)}{x_2^0-x_2^1}\right]^{\mathrm{T}} = \left[\frac{f(0,0)-f(0.400,0)}{0-0.400} \quad \frac{f(0,0)-f(0,0.800)}{0-0.800}\right]^{\mathrm{T}} = [-0.900 \quad -1.20]^{\mathrm{T}}$$

Note that the analytical gradient vector at $\mathbf{x}^0$ is $\nabla f(\mathbf{x}^0) = [-1, -2]^{\mathrm{T}}$. The approximation given above is not close because $\mathbf{x}^1$ is not close to $\mathbf{x}^0$. Now we calculate α^1 using Eq. 3.65b as

$$\alpha^1 = -\frac{f\left(\mathbf{x}^1\right)}{\mathbf{q}(\mathbf{x}^0,\mathbf{x}^1)^{\mathrm{T}}\mathbf{n}^1} = -\frac{0.680}{[-0.900 \quad -1.20]\begin{bmatrix}0.841\\0.540\end{bmatrix}} = 0.484$$

Hence

$\mathbf{x}^2 = \mathbf{x}^1 + \alpha^1 \mathbf{n}^1 = [0.400, 0.800]^{\mathrm{T}} + 0.484 [0.841, 0.540]^{\mathrm{T}} = [0.407, 0.261]^{\mathrm{T}}$. We then update the Hessian matrix starting from an identity matrix $\mathbf{H}^1 = \mathbf{I}$, and repeat the process until $\|\nabla f(\mathbf{x}^k)\| < \varepsilon_2 = 0.001$.

The search method can be implemented in MATLAB to solve the example (see Script 11 in Appendix A for details). The optimal point is found at $\mathbf{x}^* = (2, 1)$, in which $f(\mathbf{x}^*) = 0$.

3.6 CONSTRAINED PROBLEMS

The major difference between a constrained and unconstrained problem is that for a constrained problem, an optimal solution must be sought in a feasible region; for an unconstrained problem, the feasible region contains the entire design domain. For a constrained problem, bringing an infeasible design into a feasible region is critical, in which gradients of active constraints are taken into consideration when determining the search direction for the next design. In this section, we first outline the nature of the constrained optimization problem and the concept of solution techniques. In Section 3.6.2, we then discuss a widely accepted strategy for dealing with the constraint functions, the so-called ε-active strategy. Thereafter, in Sections 3.6.3–3.6.5 we discuss the mainstream solution techniques for solving constrained optimization problems, including SLP, SQP, and the feasible direction method. These solution techniques are capable of solving general optimization problems with multiple constraints and many design variables. Before closing out this section, we introduce the penalty method, which solves a constrained problem by converting it to an unconstrained problem, and then we solve the unconstrained problem using methods discussed in Section 3.5. For illustration purposes, we use simple examples of one or two design variables. Like Section 3.5, we offer sample MATLAB scripts (see Appendix A) for solving example problems.

3.6.1 BASIC CONCEPT

Recall the mathematical definition of the constrained optimization problem:

$$\text{Minimize}: \quad f(\mathbf{x}) \tag{3.67a}$$

$$\text{Subject to}: \quad g_i(\mathbf{x}) \le 0, \quad i = 1, m \tag{3.67b}$$

$$h_j(\mathbf{x}) = 0, \quad j = 1, p \tag{3.67c}$$

$$x_k^\ell \le x_k \le x_k^u, \quad k = 1, n \tag{3.67d}$$

Similar to solving unconstrained optimization problems, all numerical methods are based on the iterative process, in which the next design point is updated by a search direction $\mathbf{n}$ and a step size α along the direction. The next design point $\mathbf{x}^{k+1}$ is then obtained by evaluating the design at the current design point $\mathbf{x}^k$ (some methods include information from previous design iterations) as

$$\mathbf{x}^{k+1} = \mathbf{x}^k + \Delta\mathbf{x}^k = \mathbf{x}^k + \alpha^k \mathbf{n}^k \tag{3.68}$$

For an unconstrained problem, the search direction $\mathbf{n}$ considers only the gradient of the objective function. For constrained problems, however, optimal solutions must be sought in the feasible region. Therefore, active constraints in addition to objective functions must be considered while determining the search direction as well as the step size. As with the unconstrained problems, all algorithms need an initial design to initiate the iterative process. The difference is for a constrained problem, the starting design can be feasible or infeasible, as illustrated in Figure 3.15a, in which a constrained optimization of two design variables x_1 and x_2 is assumed. The feasible region of the problem is identified on the surface of the objective function as well as projected onto the x_1–x_2 plane.

If an initial design is inside the feasible region, such as points A^0 or B^0, then we minimize the objective function by moving along its descent direction—say, the steepest descent direction—as if we are dealing with an unconstrained problem. We continue such iterations until either a minimum point is reached, such as the search path starting at point A^0, or a constraint is encountered (i.e., the boundary of the feasible region is reached, like the path of initial design at point B^0). Once the constraint boundary is encountered at point B^1, one strategy is to travel along a tangent line to the boundary, such as the direction B^1B^2 illustrated in Figure 3.15b. This leads to an infeasible point from where the constraints are corrected in order to again reach the feasible point B^3. From there, the preceding steps are repeated until the optimum point is reached.

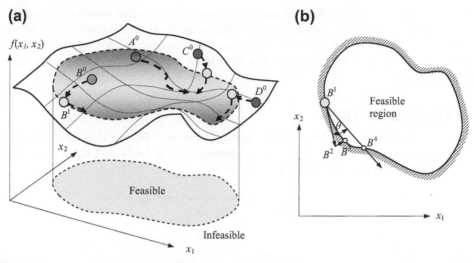

FIGURE 3.15

Concept illustration for solving a constrained optimization problem. (a) Paths illustrating different solution scenarios. (b) Top view of the feasible region, with a design point B residing on the boundary of the feasible region.

Another strategy is to deflect the tangential direction B^1B^2 toward the feasible region by a small angle θ when there are no equality constraints. Then, a line search is performed through the feasible region to reach the boundary point B^4, as shown in Figure 3.15b. The procedure is then repeated from there.

When the starting point is infeasible, like points C^0 or D^0 in Figure 3.15a, one strategy is to correct constraint violations to reach the constraint boundary. From there, the strategies described in the preceding paragraph can be followed to reach the optimum point. For example, for D^0, a similar path to that shown in path $B^1B^2B^3$ or B^1B^4 in Figure 3.15b is followed. The case for point C^0 is easier because the decent direction of objective function also corrects the constraint violations.

A good analogy for finding a minimum of a constrained problem is rolling a ball in a fenced downhill field. The boundary of the feasible region is the fence, and the surface of the downhill field is the objective function. When a ball is released at a location (i.e., the initial design), the ball rolls due to gravity. If the initial point is chosen such that the ball does not encounter the fence, the ball rolls to a local crest (minimum point). If an initial point chosen allows the ball to hit the fence, the ball rolls along the fence to reach a crest. If the initial point is outside the fenced area, the ball has to be thrown into the fenced area before it starts rolling.

Several algorithms based on the strategies described in the foregoing have been developed and evaluated. Some algorithms are better for a certain class of problems than others. In this section, we focus on general algorithms that have no restriction on the form of the objective or the constraint functions. Most of the algorithms that we will describe in this chapter can treat feasible and infeasible initial designs.

In general, numerical algorithms for solving constrained problems start with a linearization of the objective and constraint functions at the current design. The linearized subproblem is solved to determine the search direction **n**. Once the search direction is found, a line search is carried out to find an adequate step size α for the next design iteration. Following the general solution steps, we introduce three general methods: SLP, SQP, and the feasible direction method. Before we discuss the solution techniques, we discuss the ε-active strategy that determines the active constraints to incorporate for design optimization.

3.6.2 ε-ACTIVE STRATEGY

An ε-active constraint strategy (Arora 2012), shown in Figure 3.16, is often employed in solving constrained optimization problems. Inequality constraints in Eq. 3.67b and equality constraints of Eq. 3.67c are first normalized by their respective bounds:

$$b_i = \frac{g_i}{g_i^u} - 1 \le 0, \quad i = 1, m \tag{3.69a}$$

and

$$e_i = \frac{h_j}{h_j^u} - 1, \quad j = 1, p \tag{3.69b}$$

Usually, when b_i (or e_i) is between two parameters CT (usually -0.03) and CTMIN (usually 0.005), g_i is active, as shown in Figure 3.16. When b_i is less than CT, the constraint function is inactive or feasible. When b_i is larger than CTMIN, the constraint function is violated. Note that CTMIN-CT $= \varepsilon$.

3.6.3 THE SEQUENTIAL LINEAR PROGRAMMING ALGORITHM

The original optimization problem stated in Eq. 3.67 is first linearized by writing Taylor's expansions for the objective and constraint functions at the current design $\mathbf{x}^k$ as below.

Minimize the linearized objective function:

$$f\left(\mathbf{x}^{k+1}\right) = f\left(\mathbf{x}^k + \Delta\mathbf{x}^k\right) \approx f\left(\mathbf{x}^k\right) + \nabla f^{\mathrm{T}}\left(\mathbf{x}^k\right)\Delta\mathbf{x}^k \tag{3.70a}$$

subject to the linearized inequality constraints

$$g_i\left(\mathbf{x}^{k+1}\right) = g_i\left(\mathbf{x}^k + \Delta\mathbf{x}^k\right) \approx g_i\left(\mathbf{x}^k\right) + \nabla g_i^{\mathrm{T}}\left(\mathbf{x}^k\right)\Delta\mathbf{x}^k \le 0; \quad i = 1, m \tag{3.70b}$$

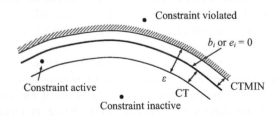

FIGURE 3.16

ε-Active constraint strategy.

and the linearized equality constraints

$$h_j(\mathbf{x}^{k+1}) = h_j(\mathbf{x}^k + \Delta\mathbf{x}^k) \approx h_j(\mathbf{x}^k) + \nabla h_j^\mathrm{T}(\mathbf{x}^k)\Delta\mathbf{x}^k = 0; \quad j = 1, p \tag{3.70c}$$

in which $\nabla f(\Delta\mathbf{x}^k)$, $\nabla g_i(\Delta\mathbf{x}^k)$, and $\nabla h_j(\Delta\mathbf{x}^k)$ are the gradients of the objective function, the ith inequality constraint and the jth equality constraint, respectively, and $\approx$ implies approximate equality.

To simplify the mathematical notations in our discussion, we rewrite the linearized equations in Eq. 3.70 as

$$\text{Minimize}: \quad \bar{f} = \mathbf{c}^\mathrm{T}\mathbf{d} \tag{3.71a}$$

$$\text{Subject to}: \quad \mathbf{A}^\mathrm{T}\mathbf{d} \leq \mathbf{b} \tag{3.71b}$$

$$\mathbf{N}^\mathrm{T}\mathbf{d} = \mathbf{e} \tag{3.71c}$$

$$-\Delta_\ell \leq \mathbf{d} \leq \Delta_u \tag{3.71d}$$

where

$$\mathbf{c}_{n\times 1} = \nabla f(\mathbf{x}^k) = \left[\partial f(\mathbf{x}^k)/\partial x_1, \partial f(\mathbf{x}^k)/\partial x_2, \ldots, \partial f(\mathbf{x}^k)/\partial x_n\right]^\mathrm{T};$$

$$\mathbf{d}_{n\times 1} = \Delta\mathbf{x}^k = \left[\Delta x_1^k, \Delta x_2^k, \ldots, \Delta x_n^k\right]^\mathrm{T};$$

$$\mathbf{A}_{m\times n} = \left[\nabla g_1(\mathbf{x}^k), \nabla g_2(\mathbf{x}^k), \ldots, \nabla g_m(\mathbf{x}^k)\right]_{m\times n},$$

in which

$$\nabla g_i(\mathbf{x}^k) = \left[\partial g_i(\mathbf{x}^k)/\partial x_1, \partial g_i(\mathbf{x}^k)/\partial x_2, \ldots, \partial g_i(\mathbf{x}^k)/\partial x_n\right]_{n\times 1}^\mathrm{T};$$

$$\mathbf{N}_{p\times n} = \left[\nabla h_1(\mathbf{x}^k), \nabla h_2(\mathbf{x}^k), \ldots, \nabla h_p(\mathbf{x}^k)\right]_{p\times n},$$

in which

$$\nabla h_i(\mathbf{x}^k) = \left[\partial h_i(\mathbf{x}^k)/\partial x_1, \partial h_i(\mathbf{x}^k)/\partial x_2, \ldots, \partial h_i(\mathbf{x}^k)/\partial x_n\right]_{n\times 1}^\mathrm{T};$$

$$\mathbf{b}_{m\times 1} = -g_i(\mathbf{x}^k) = \left[-g_1(\mathbf{x}^k), -g_2(\mathbf{x}^k), \ldots, -g_m(\mathbf{x}^k)\right]_{m\times 1};$$

and

$$\mathbf{e}_{m\times 1} = -h_i(\mathbf{x}^k) = \left[-h_1(\mathbf{x}^k), -h_2(\mathbf{x}^k), \ldots, -h_p(\mathbf{x}^k)\right]_{p\times 1}.$$

Note that in Eq. 3.71a, $f(\mathbf{x}^k)$ is dropped. $\Delta_\ell = [\Delta_{1\ell}^k, \Delta_{2\ell}^k, \ldots, \Delta_{n\ell}^k]_{n\times 1}^\mathrm{T}$ and $\Delta_u = [\Delta_{1u}^k, \Delta_{2u}^k, \ldots, \Delta_{nu}^k]_{n\times 1}^\mathrm{T}$ are the move limits—that is, the maximum allowed decrease and increase in the design variables at the kth design iteration. Note that the move limits make the linearized subproblem bounded and give the design changes directly without performing the line search for a step size α. Therefore, no line search is required in SLP. Choosing adequate move limits is critical to the SLP. More about the move limits will be discussed in Example 3.16.

As discussed before, the SLP algorithm starts with an initial design $\mathbf{x}^0$. At the kth design iteration, we evaluate the objective and constraint functions as well as their gradients at the current design $\mathbf{x}^k$. We select move limits $\Delta_{i\ell}^k$ and Δ_{iu}^k to define an LP subproblem of Eq. 3.71. Solve the linearized subproblem for $\mathbf{d}^k$, and update the design for the next iteration as $\mathbf{x}^{k+1} = \mathbf{x}^k + \mathbf{d}^k$.

The process repeats until convergent criteria are met. In general, the convergent criteria for an LP subproblem include

$$g_i(\mathbf{x}^{k+1}) \le \varepsilon_1, \quad i = 1, m; \quad |h_j(\mathbf{x}^{k+1})| \le \varepsilon_1, \quad j = 1, p; \quad \text{and} \quad \|\mathbf{d}^k\| \le \varepsilon_2 \qquad (3.72)$$

EXAMPLE 3.16

Solve Example 3.7 with one additional equality constraint using SLP. The optimization problem is restated as below.

$$\text{Minimize}: \quad f(\mathbf{x}) = (x_1 - 3)^2 + (x_2 - 3)^2 \qquad (3.73a)$$

$$\text{Subject to}: \quad g_1(\mathbf{x}) = 3x_1 + x_2 - 6 \le 0 \qquad (3.73b)$$

$$h_1(\mathbf{x}) = x_1 - x_2 = 0 \qquad (3.73c)$$

$$0 \le x_1, 0 \le x_2 \qquad (3.73d)$$

Solutions

We sketch the feasible region bounded by inequality constraint $g_1(\mathbf{x}) \le 0$, side constraints, and equality constraint $h_1(\mathbf{x}) = 0$, as shown below. As is obvious in the sketch, the optimal solution is found at $\mathbf{x}^* = (1.5, 1.5)$, the intersection of $g_1(\mathbf{x}) = 0$, and $h_1(\mathbf{x}) = 0$, in which $f(\mathbf{x}) = 4.5$.

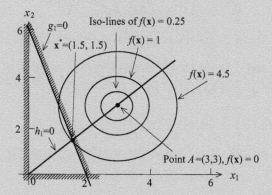

Now we use this example to illustrate the solution steps using SLP. We assume an initial design at $\mathbf{x}^0 = (2, 2)$. At the initial design, we have $f(2, 2) = (x_1 - 3)^2 + (x_2 - 3)^2 = 2$, $g_1(2, 2) = 3x_1 + x_2 - 6 = 2 > 0$, and $h_1(2, 2) = 0$. The inequality constraint g_1 is greater than 0; therefore, this constraint is violated. The initial design is not in the feasible region, as also illustrated in the figure above. The optimization problem defined in Eqs 3.73a–3.73d is linearized as follows:

$$\text{Minimize}: \quad \bar{f} = \mathbf{c}^T \mathbf{d} = [2(x_1 - 3) \quad 2(x_2 - 3)] \begin{bmatrix} \Delta x_1 \\ \Delta x_2 \end{bmatrix} \qquad (3.73e)$$

$$\text{Subject to}: \quad \mathbf{A}^T \mathbf{d} \le \mathbf{b}; \text{ i.e., } [3 \quad 1] \begin{bmatrix} \Delta x_1 \\ \Delta x_2 \end{bmatrix} \le 6, \quad \text{or} \quad \bar{g}_1 = 3\Delta x_1 + \Delta x_2 - 6 \le 0 \qquad (3.73f)$$

$$\mathbf{N}^T \mathbf{d} = \mathbf{e}; \text{ i.e., } [1 \quad -1] \begin{bmatrix} \Delta x_1 \\ \Delta x_2 \end{bmatrix} = 0, \quad \text{or} \quad \bar{h}_1 = \Delta x_1 - \Delta x_2 = 0 \qquad (3.73g)$$

$$-0.2 \le \Delta x_1 \le 0.2, -0.2 \le \Delta x_2 \le 0.2 \qquad (3.73h)$$

EXAMPLE 3.16—cont'd

We have chosen the move limits to be 0.2, which is 10% of the current design variable values, as shown in Eq. 3.73h. At

the initial design, $\mathbf{x}^0 = (2, 2)$, we are minimizing $\bar{f} = \begin{bmatrix} -2 & -2 \end{bmatrix} \begin{bmatrix} \Delta x_1 \\ \Delta x_2 \end{bmatrix} = -2\Delta x_1 - 2\Delta x_2$ subject to constraints

(Eqs 3.73f–3.73h). The subproblem has two variables; it can be solved by referring to the sketch below. Because we chose 0.2 as the move limits, the solution to the LP subproblem must lie in the region of the small dotted square box shown below. It can be seen that there is no feasible solution to this linearized subproblem because the small box does not intersect the line $\bar{g}_1 = 0$. We must enlarge this region by increasing the move limits. Thus, we note that if the move limits are too restrictive, the linearized subproblem may not have a solution.

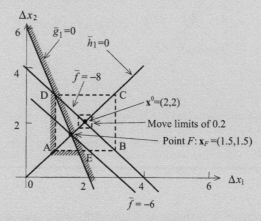

If we choose the move limits to be 1—that is, 50% of the design variable values—then the design must lie within a larger box ABCD of 2×2 as shown above. Hence the feasible region of the LP problem is now the triangle AED intersecting $\bar{h}_1 = 0$ (that is, line segment AF). Therefore, the optimal solution of the LP problem is found at point F: $\mathbf{x}_F = (1.5, 1.5)$, where $\bar{f} = -6$. That is, $\mathbf{d} = [\Delta x_1^0, \Delta x_2^0]^T = [-0.5, -0.5]^T$, and $\mathbf{x}^1 = \mathbf{x}^0 + \mathbf{d} = [2, 2]^T + [-0.5, -0.5]^T = [1.5, 1.5]^T = \mathbf{x}_F$.

In the next design $\mathbf{x}^1$, we evaluate the objective and constraint functions of the original optimization problem as well as their gradients. We have $f(1.5, 1.5) = (x_1 - 3)^2 + (x_2 - 3)^2 = 4.5$, $g_1(1.5, 1.5) = 3x_1 + x_2 - 6 = 0$, and $h_1(1.5, 1.5) = 0$. The design is feasible. Again, at the design iteration $\mathbf{x}^1 = (1.5, 1.5)$, we create the LP problem as

$$\text{Minimizing:} \quad \bar{f} = \begin{bmatrix} -3 & -3 \end{bmatrix} \begin{bmatrix} \Delta x_1 \\ \Delta x_2 \end{bmatrix} = -3\Delta x_1 - 3\Delta x_2$$

$$\text{Subject to:} \quad \bar{g}_1 = 3\Delta x_1 + \Delta x_2 - 6 \leq 0$$

$$\bar{h}_1 = \Delta x_1 - \Delta x_2 = 0$$

$$-1 \leq \Delta x_1 \leq 1, -1\Delta x_2 \leq 1$$

As illustrated in the figure next page, the feasible region of the LP subproblem is now the polygon $A_1E_1F_1D_1$ intersecting $\bar{h}_1 = 0$. Therefore, the optimal solution of the LP problem is found again at $\mathbf{x}^1 = (1.5, 1.5)$, the same point. That is, in this design iteration, $\mathbf{d} = [\Delta x_1^1, \Delta x_2^1]^T = [0, 0]^T$.

Continued

EXAMPLE 3.16—cont'd

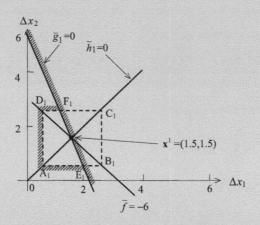

At this point, the convergent criterion stated in Eq. 3.72, for example, $\|\mathbf{d}^1\| = 0 \le \varepsilon_2$, is satisfied; hence, an optimal solution is found at $\mathbf{x}^1 = (1.5, 1.5)$.

In fact, for this particular problem, it takes only one iteration to find the optimal solution. In general, this may not be the case. An iterative process often takes numerous iterations to achieve a convergent solution.

Although the SLP algorithm is a simple and straightforward approach to solving constrained optimization problems, it should not be used as a black-box approach for engineering design problems. The selection of move limits is in essence trial and error and can be best achieved in an interactive mode. Also, the method may not converge to the precise minimum because no descent function is defined, and the line search is not performed along the search direction to compute a step size. Nevertheless, this method may be used to obtain improved designs in practice. It is a good method to include in our toolbox for solving constrained optimization problems.

3.6.4 THE SEQUENTIAL QUADRATIC PROGRAMMING ALGORITHM

The SQP algorithm incorporates second-order information about the problem functions in determining a search direction $\mathbf{n}$ and step size α. A search direction in the design space is calculated by utilizing the values and the gradients of the objective and constraint functions. A quadratic programming subproblem is defined as

$$\text{Minimize}: \quad \bar{f} = \mathbf{c}^T\mathbf{d} + \frac{1}{2}\,\mathbf{d}^T\mathbf{d} \tag{3.74a}$$

$$\text{Subject to}: \quad \mathbf{A}^T\mathbf{d} \le \mathbf{b} \tag{3.74b}$$

$$\mathbf{N}^T\mathbf{d} = \mathbf{e} \tag{3.74c}$$

in which a quadratic term is added to the objective function $\bar{f}$ and the constraint functions (3.74b) and (3.74c) are identical to those of LP subproblem, except that there is no need to define the move limits.

The solution of the QP problem **d** defines the search direction **n** (where $\mathbf{n} = \mathbf{d}/\|\mathbf{d}\|$). Once the search direction is determined, a line search is carried out to find an adequate step size α. The process repeats until the convergent criteria defined in Eq. 3.72 are met.

EXAMPLE 3.17

Solve the same problem of Example 3.16 using SQP.

Solutions

We assume the same initial design at $\mathbf{x}^0 = (2, 2)$. At the initial design, we have $f(2, 2) = (x_1-3)^2 + (x_2-3)^2 = 2$, $g_1(2, 2) = 3x_1 + x_2 - 6 = 2 > 0$, and $h_1(2, 2) = 0$. The initial design is infeasible. The QP subproblem can be written as

$$\text{Minimize}: \quad \bar{f} = \mathbf{c}^T\mathbf{d} + \frac{1}{2}\mathbf{d}^T\mathbf{d} = \begin{bmatrix} 2(x_1 - 3) & 2(x_2 - 3) \end{bmatrix} \begin{bmatrix} \Delta x_1 \\ \Delta x_2 \end{bmatrix} + \frac{1}{2}\begin{bmatrix} \Delta x_1 & \Delta x_2 \end{bmatrix}\begin{bmatrix} \Delta x_1 \\ \Delta x_2 \end{bmatrix} \tag{3.75a}$$

$$\text{Subject to}: \quad \bar{g}_1 = 3\Delta x_1 + \Delta x_2 - 6 \le 0 \tag{3.75b}$$

$$\bar{h}_1 = \Delta x_1 - \Delta x_2 = 0 \tag{3.75c}$$

At the initial design, $\mathbf{x}^0 = (2, 2)$, we are minimizing

$$\bar{f} = \begin{bmatrix} -2 & -2 \end{bmatrix}\begin{bmatrix} \Delta x_1 \\ \Delta x_2 \end{bmatrix} + \frac{1}{2}\begin{bmatrix} \Delta x_1 & \Delta x_2 \end{bmatrix}\begin{bmatrix} \Delta x_1 \\ \Delta x_2 \end{bmatrix} = -2\Delta x_1 - 2\Delta x_2 + \frac{1}{2}\left(\Delta x_1^2 + \Delta x_2^2\right)$$

subject to constraint Eqs 3.75b and 3.75c. The QP subproblem can be solved by either using the KKT condition or graphical method. We use the graphical method for this example.

Referring to the sketch below, the optimal solution to the QP subproblem is found at point F: $\mathbf{x}_F = (1.5, 1.5)$, where $\bar{f} = -3.75$. That is, $\mathbf{d} = [\Delta x_1^0, \Delta x_2^0]^T = [-0.5, -0.5]^T$. Note that the quadratic function $\bar{f}$ is sketched using the MATLAB script shown in Appendix A (Script 12).

For the next iteration, we have $\mathbf{n} = \mathbf{d}/\|\mathbf{d}\| = [-0.707, -0.707]^T$ and assume a step size $\alpha = 1$. Hence, for the next design, $\mathbf{x}^1 = \mathbf{x}^0 + \alpha\mathbf{n}^0 = [2, 2]^T + 1[-0.707, -0.707]^T = [1.293, 1.293]^T$.

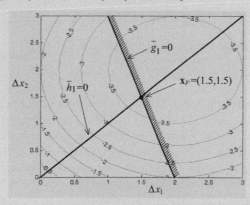

In the next design $\mathbf{x}^1$, we evaluate the objective and constraint functions of the original optimization problem as well as their gradients. We have $f(1.293, 1.293) = (x_1 - 3)^2 + (x_2 - 3)^2 = 5.828$, $g_1(1.293, 1.293) = 3x_1 + x_2 - 6 = -0.828$, and $h_1(1.293, 1.293) = 0$. The design is feasible. Again, at the design iteration $\mathbf{x}^1 = (1.293, 1.293)$, we create the QP problem as

---

Continued

EXAMPLE 3.17—cont'd

Minimizing: $\bar{f} = [-3.414 - 3.414] \begin{bmatrix} \Delta x_1 \\ \Delta x_2 \end{bmatrix} + \frac{1}{2} [\Delta x_1 \; \Delta x_2] \begin{bmatrix} \Delta x_1 \\ \Delta x_2 \end{bmatrix} = -3.414\Delta x_1 - 3.414\Delta x_2 + \frac{1}{2}(\Delta x_1^2 + \Delta x_2^2)$

Subject to: $\bar{g}_1 = 3\Delta x_1 + \Delta x_2 - 6 \leq 0$

$\bar{h}_1 = \Delta x_1 - \Delta x_2 = 0$

The optimal design of the QP problem is found again at $\mathbf{x}^1 = (1.5, 1.5)$, the same point since the constraint functions are unchanged. That is, $\mathbf{d} = [\Delta x_1^1, \Delta x_2^1]^T = [0.207, 0.207]^T$. The process continues until the convergent criterion stated in Eq. 3.72 are satisfied. After several iteration, an optimal solution is found at $\mathbf{x}^* = (1.5, 1.5)$.

3.6.5 FEASIBLE DIRECTION METHOD

The basic idea of the feasible direction method is to determine a search direction that moves from the current design point to an improved feasible point in the design space. Thus, given a design $\mathbf{x}^k$, an improving feasible search direction $\mathbf{n}^k$ is determined such that for a sufficiently small step size $\alpha > 0$, the new design, $\mathbf{x}^{k+1} = \mathbf{x}^k + \alpha^k \mathbf{n}^k$ is feasible, and the new objective function is smaller than the current one; that is, $f(\mathbf{x}^{k+1}) < f(\mathbf{x}^k)$. Note that $\mathbf{n}$ is a normalized vector, defined as $\mathbf{n} = \mathbf{d}/\|\mathbf{d}\|$, where $\mathbf{d}$ is the nonnormalized search direction solved from a subproblem to be discussed.

Because along the search direction $\mathbf{d}^k$ the objective function must decrease without violating the applied constraints, taking in account only inequality constraints, it must result that

$$\nabla f(\mathbf{x}^k) \cdot \mathbf{d}^k < 0 \tag{3.76}$$

and

$$\nabla g_i(\mathbf{x}^k) \cdot \mathbf{d}^k < 0, \quad \text{for } i \in \mathbf{I}_k \tag{3.77}$$

$\mathbf{I}_k$ is the potential constraint set at the current point, defined as

$$\mathbf{I}_k \equiv \left\{ i \middle| g_i(\mathbf{x})^k + \varepsilon \geq 0 \; i = 1, m \right\} \tag{3.78}$$

Note that ε is a small positive number, selected to determine ε-active constraints as discussed in Section 3.6.1. Note that $g_i(\mathbf{x})$ is normalized as in Eq. 3.69a. The inequality constraints enclosed in the set of Eq. 3.78 are either violated or ε-active, meaning they have to be considered in determining a search direction that brings the design into the feasible region. Equations 3.76 and 3.77 are referred to as usability and feasibility requirements, respectively. A geometrical interpretation of the requirements is shown in Figure 3.17 for a two-variable optimization problem, in which the search direction $\mathbf{n}$ points to the usable-feasible region.

This method has been developed and applied mostly to optimization problems with inequality constraints. This is because, in implementation, the search direction $\mathbf{n}$ is determined by defining a linearized subproblem (to be discussed next) at the current feasible point, and the step size α is

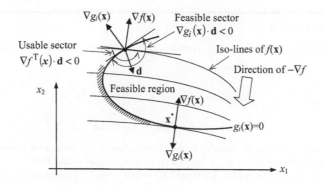

FIGURE 3.17

Geometric description of the feasible direction method.

determined to reduce the objective function as well as maintain feasibility of design. Because linear approximations are used, it is difficult to maintain feasibility with respect to the equality constraints. Although some procedures have been developed to treat equality constraints in these methods, we will describe the method for problems with only inequality constraints.

The desired search direction **d** will meet the requirements of usability and feasibility, and it gives the highest reduction of the objective function along it. Mathematically, it is obtained by solving the following linear subproblem in **d**:

$$\text{Minimize :} \quad \beta \tag{3.79a}$$

$$\text{Subject to :} \quad \nabla f^T(\mathbf{x})\mathbf{d} - \beta \le 0 \tag{3.79b}$$

$$\nabla g_i^T(\mathbf{x})\mathbf{d} - \beta \le 0, \quad \text{for } i \in I_k \tag{3.79c}$$

$$d_j^\ell \le d_j \le d_j^u, \quad \text{for } j = 1, \, n \tag{3.79d}$$

Note that this is a linear programming problem. If $\beta < 0$, then **d** is an improving feasible direction. If $\beta = 0$, then the current design satisfies the KKT necessary conditions and the optimization process is terminated. To compute the improved design in this direction, a step size α is needed.

EXAMPLE 3.18

Solve the optimization problem of Example 3.7 using the feasible direction method. The problem is restated below:

$$\text{Minimize :} \quad f(\mathbf{x}) = (x_1 - 3)^2 + (x_2 - 3)^2 \tag{3.80a}$$

$$\text{Subject to :} \quad g_1(\mathbf{x}) = 3x_1 + x_2 - 6 \le 0 \tag{3.80b}$$

$$0 \le x_1, \quad \text{and} \quad 0 \le x_2 \tag{3.80c}$$

Solutions

Referring to the sketch of Example 3.7, the optimal solution is found as $\mathbf{x}^* = (1.2, 2.4)$, in which $f(\mathbf{x}) = 3.6$. In this example, we present two cases of two respective initial designs, one in the feasible region and the other one in the infeasible region.

Continued

EXAMPLE 3.18–cont'd

Case A: feasible initial design at $x^0 = (1, 1)$. From Eq. 3.79, a subproblem can be written at the initial design as

$$\text{Minimize} : \quad \beta \tag{3.80d}$$

$$\text{Subject to} : \quad q_1 = -4d_1 - 4d_2 - \beta \leq 0 \tag{3.80e}$$

$$-1 \leq d_1 \leq 1, -1 \leq d_2 \leq 1 \tag{3.80f}$$

Note that we do not need to include the linearized constraint equation $g_1 \leq 0$ because the design is feasible. We assume lower and upper bounds of the subproblem as -1 and 1, respectively, as stated in Eq. 3.80f.

We sketch the feasible region defined by Eqs 3.80e and 3.80f with $\beta = 0$ and -8, respectively, as shown below. For $\beta = 0$, the feasible region is the triangle ABC, and for $\beta = -8$, the feasible region reduces to a single point $C = (1, 1)$. As is obvious in the sketches below, the optimal solution of the subproblem is found at $C = (1, 1)$, in which $\beta = -8$.

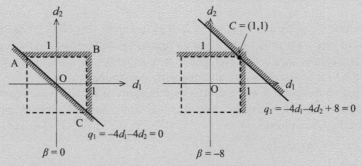

Certainly, if the bounds of d_1 and d_2 are changed, the solution changes as well. However, the search direction defined by $\mathbf{n} = \mathbf{d}/||\mathbf{d}|| = [1, 1]^T/||[1, 1]^T|| = [0.707, 0.707]^T$ remains the same. From Eq. 3.79b, we have

$$\nabla f^T(1, 1)\mathbf{d} = [-4, -4][1, 1]^T = -8(= \beta) < 0$$

Note that $\nabla f^T \mathbf{d}$ is a dot product of ∇f^T and $\mathbf{d}$. Geometrically, $(\nabla f^T \mathbf{d})/||\nabla f^T \mathbf{d}|| = -1$ is the angle between the vectors ∇f^T and $\mathbf{d}$—in this case 180°, as shown below. This is because the initial design is feasible, and the search direction $\mathbf{n}$ (or $\mathbf{d}$) is the negative of the gradient of the objective function ∇f.

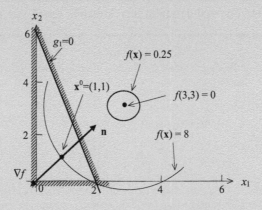

EXAMPLE 3.18—cont'd

Case B: infeasible initial design at $x^0 = (2, 2)$. From Eq. 3.79, a subproblem can be written at the initial design as

$$\text{Minimize}: \quad \beta \tag{3.80g}$$

$$\text{Subject to}: \quad q_1 = -2d_1 - 2d_2 - \beta \le 0 \tag{3.80h}$$

$$q_2 = 3d_1 + d_2 - \beta \le 0 \tag{3.80i}$$

$$-1 \le d_1 \le 1, \quad -1 \le d_2 \le 1 \tag{3.80j}$$

Similar to Case A, we sketch the feasible region defined by of Eqs 3.80h–3.80j with $\beta = 0$ and -0.8, respectively, as shown below. For $\beta = 0$, the feasible region is the triangle ABO, and for $\beta = -0.8$, the feasible region reduces to a single point at $C = (-0.6, 1)$. As is obvious in the sketches, the optimal solution of the subproblem is found at $C = (-0.6, 1)$, in which $\beta = -0.8$.

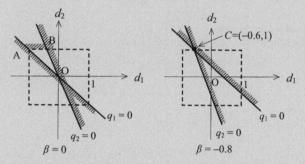

The search direction is found as $\mathbf{n} = \mathbf{d}/\|\mathbf{d}\| = [-0.6, 1]^T/\|[-0.6, 1]\|^T = [-0.441, 0.735]^T$. From Eqs 3.79b and 3.79c, we have

$$\nabla f^T(2, 2)\mathbf{d} = [-2, -2][-0.6, 1]^T = -0.8\Big(= \beta\Big) < 0, \quad \text{and} \quad \nabla g_i^T(2, 2)\mathbf{d} = [3, 1][-0.6, 1]^T = -0.8 < 0$$

Geometrically, they are the respective angles between the vectors ∇f^T and $\mathbf{d}$ and ∇g_i^T and $\mathbf{d}$, as shown below. Because the design is infeasible, the gradient of the active constraint is taken into consideration in calculating the search direction. In fact, because the same parameter β is employed in the constraint equations of the subproblem, the search direction $\mathbf{n}$ points in a direction that splits the angle between $-\nabla f$ and $-\nabla g_1$.

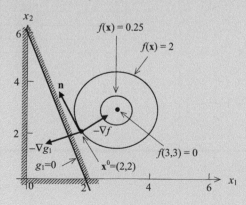

In the constraint equations of the subproblem stated in Eqs 3.79b and 3.79c, the same parameter β is employed. As demonstrated in Case B of Example 3.18, the same β leads to a search direction **n** pointing in a direction that splits the angle between $-\nabla f$ and $-\nabla g_1$, in which g_1 is an active constraint function.

To determine a better feasible direction **d**, the constraints of Eq. 3.79c can be modified as

$$\nabla g_i^T(\mathbf{x})\mathbf{d} - \theta_i\beta \leq 0, \quad \text{for } i \in I_K \tag{3.81}$$

where θ_i is called the push-off factor. The greater the value of θ_i, the more the direction vector **d** is pushed into the feasible region. The reason for introducing θ_i is to prevent the iterations from repeatedly hitting the constraint boundary and slowing down the convergence.

EXAMPLE 3.19

Find the search directions **n** for Case B of Example 3.18, assuming $\theta_1 = 0, 0.5,$ and 1.5.

Solutions

We show the solutions in the following four cases.

Case A: $\theta_1 = 0$. From Eqs 3.79 and 3.81, a subproblem can be written at the initial design as

$$\text{Minimize:} \quad \beta \tag{3.82a}$$

$$\text{Subject to:} \quad q_1 = -2d_1 - 2d_2 - \beta \leq 0 \tag{3.82b}$$

$$q_2 = 3d_1 + d_2 \leq 0 \tag{3.82c}$$

$$-1 \leq d_1 \leq 1, \quad -1 \leq d_2 \leq 1 \tag{3.82d}$$

Following the similar approach as in Example 3.18, the solution to the subproblem defined in Eqs 3.82a–3.82d is found at $\mathbf{d} = (-1/3, 1)$ with $\beta = -4/3$. In fact, the search direction points in a direction that is parallel to the active constraint g_1 at the current design $\mathbf{x}^0$; i.e., the search direction is perpendicular to the gradient of the constraint function ∇g_1, as shown below in the vector $\mathbf{d}(0)$.

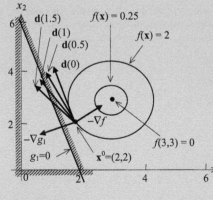

Case	θ_1	**d**	β
A	0	$\mathbf{d}(0) = (-1/3, 1)$	$-4/3$
B	0.5	$\mathbf{d}(0.5) = (-1/2, 1)$	-1
C	1	$\mathbf{d}(1) = (-0.6, 1)$	-0.8
D	1.5	$\mathbf{d}(1.5) = (-2/3, 1)$	$-2/3$

EXAMPLE 3.19–cont'd

Case B: $\theta_1 = 0.5$. The constraint equation q_2 of the subproblem becomes

$$q_2 = 3d_1 + d_2 - 0.5\beta \leq 0 \tag{3.82e}$$

The solution to the subproblem is found at $\mathbf{d} = (-1/2, 1)$ with $\beta = -1$. The search direction $\mathbf{n}$ leans to $-\nabla g_1$, as shown in the figure above in the vector $\mathbf{d}(0.5)$. That is, the design is pushed more into the feasible region compared to the case where $\theta_1 = 0$.

Case C: $\theta_1 = 1$. The constraint equation q_2 of the subproblem becomes

$$q_2 = 3d_1 + d_2 - \beta \leq 0 \tag{3.82f}$$

The solution to the subproblem is found at $\mathbf{d} = (-0.6, 1)$ with $\beta = -0.8$. The search direction $\mathbf{d}$ leans more to $-\nabla g_1$, as shown in the figure above in the vector $\mathbf{d}(1)$.

Case D: $\theta_1 = 1.5$. The constraint equation q_2 of the subproblem becomes

$$q_2 = 3d_1 + d_2 - 1.5\beta \leq 0 \tag{3.82g}$$

The solution to the subproblem is found at $\mathbf{d} = (-2/3, 1)$ with $\beta = -2/3$. The search direction $\mathbf{d}$ leans more to $-\nabla g_1$, as shown in the figure above in the vector $\mathbf{d}(1.5)$.

3.6.6 PENALTY METHOD

A penalty method replaces a constrained optimization problem by a series of unconstrained problems whose solutions ideally converge to the solution of the original constrained problem. The unconstrained problems are formed by adding a term, called a penalty function, to the objective function that consists of a penalty parameter multiplied by a measure of violation of the constraints. The measure of violation is nonzero when the constraints are violated and is zero in the region where constraints are not violated.

Recall that a constrained optimization problem considered is defined as

$$\underset{\mathbf{x} \in S}{\text{Minimize}} \quad f(\mathbf{x}) \tag{3.83}$$

where S is the set of feasible designs defined by equality and inequality constraints. Using the penalty method, Eq. 3.83 is first converted to an unconstrained problem as

$$\text{Minimize } \Phi(\mathbf{x}, r_p) = f(\mathbf{x}) + r_p p(\mathbf{x}) \tag{3.84}$$

where $f(\mathbf{x})$ is the original objective function, $p(\mathbf{x})$ is an imposed penalty function, and r_p is a multiplier that determines the magnitude of the penalty. The function $\Phi(\mathbf{x}, r_p)$ is called pseudo-objective function.

There are numerous ways to create a penalty function. One of the easiest is called exterior penalty (Vanderplaats 2007), in which a penalty function is defined as

$$p(\mathbf{x}) = \sum_{i=1}^{m} \{\max[0, g_i(\mathbf{x})]\}^2 + \sum_{j=1}^{p} [h_j(\mathbf{x})]^2 \tag{3.85}$$

From Eq. 3.85, we see that no penalty is imposed if all constraints are satisfied. However, whenever one or more constraints are violated, the square of these constraints is included in the penalty function.

If we choose a small value for the multiplier r_p, the pseudo-objective function $\Phi(\mathbf{x}, r_p)$ may be solved easily, but may converge to a solution with large constraint violations. On the other hand, a large value of r_p ensures near satisfaction of all constraints but may create a poorly conditioned optimization problem that is unstable and difficult to solve numerically. Therefore, a better strategy is to start with a small r_p and minimize $\Phi(\mathbf{x}, r_p)$. Then, we increase r_p by a factor of γ (say $\gamma = 10$), and proceed with minimizing $\Phi(\mathbf{x}, r_p)$ again. Each time, we take the solution from the previous optimization problem as the initial design to speed up the optimization process. We repeat the steps until a satisfactory result is obtained. In general, solutions of the successive unconstrained problems will eventually converge to the solution of the original constrained problem.

EXAMPLE 3.20

Solve the following optimization problem using the penalty method.

$$\text{Minimize}: \quad f(x) = x \tag{3.86a}$$

$$\text{Subject to}: \quad g_1(x) = 1 - x \leq 0 \tag{3.86b}$$

$$g_2(x) = \frac{1}{2}x - 1 \leq 0 \tag{3.86c}$$

$$0 \leq x \leq 3 \tag{3.86d}$$

Solutions

We show the objective and constraint functions in the sketch below. It is obvious that the feasible region is $[1, 2]$, and the optimal solution is at $x = 1, f(1) = 1$.

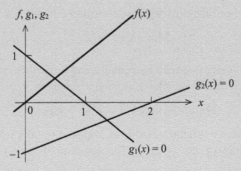

Now we solve the problem using the penalty method. We convert the constrained problem to an unconstrained problem using Eq. 3.84 as

$$\text{Minimize } \Phi(x, r_p) = x + r_p\left\{[\max(0, 1 - x)]^2 + [\max(0, 0.5x - 1)]^2\right\} \tag{3.86e}$$

We start with $r_p = 1$ and use the golden search to find the solution. The MATLAB script for finding the solution to Eq. 3.86e can be found in Appendix A (Script 13).

> **EXAMPLE 3.20—cont'd**
>
> For $r_p = 1$, the golden search found a solution of $x = 0.5$, $\Phi(0.5, 1) = 0.75$, with constraint functions $g_1(0.5) = 0.5$ (violated) and $g_2(0.5) = -0.75$ (satisfied).
>
> We increase $r_p = 10$. The golden search found a solution of $x = 0.950$, $\Phi(0.950, 10) = 0.975$, with constraint functions $g_1(0.950) = 0.05$ (violated) and $g_2(0.950) = -0.525$ (satisfied).
>
> We increase $r_p = 100$. The golden search found a solution of $x = 0.995$, $\Phi(0.995, 100) = 0.9975$, with constraint functions $g_1(0.995) = 0.005$ (violated) and $g_2(0.995) = -0.5025$ (satisfied).
>
> When we increase $r_p = 10,000$, we have $x = 0.9999$, $\Phi(0.9999, 100) = 1.000$, with constraint functions $g_1(0.9999) = 0.00001$ (violated) and $g_2(0.9999) = -0.500$ (satisfied). At this point, the objective function is $f(0.9999) = 0.9999$. The convergent trend is clear from results of increasing the r_p value.

3.7 NON-GRADIENT APPROACH

Unlike the gradient-based approach, the non-gradient approach uses only the function values in the search process, without the need for gradient information. The algorithms developed in this approach are very general and can be applied to all kinds of problems—discrete, continuous, or nondifferentiable functions. In addition, the methods determine global optimal solutions as opposed to the local optimal determined by a gradient-based approach. Although the non-gradient approach does not require the use of gradients of objective or constraint functions, the solution techniques require a large amount of function evaluations. For large-scale problems that require significant computing time for function evaluations, the non-gradient approach is too expensive to use. Furthermore, there is no guarantee that a global optimum can be obtained. The computation issue may be overcome to some extent by the use of parallel computing or supercomputers. The issue of global optimal solution may be overcome to some extent by allowing the algorithm to run longer.

In this section, we discuss two popular and representative algorithms of the non-gradient approach: the genetic algorithm (GA) and simulated annealing (SA). We introduce concepts and solution process for the algorithms to provide readers with a basic understanding of these methods.

3.7.1 GENETIC ALGORITHMS

Recall that the optimization problem considered is defined as

$$\underset{\mathbf{x} \in S}{\text{Minimize}} \quad f(\mathbf{x}) \tag{3.87}$$

where S is the set of feasible designs defined by equality and inequality constraints. For unconstrained problems, S is the entire design space. Note that to use a genetic algorithm, the constrained problem is often converted to an unconstrained problem using, for example, the penalty method discussed in Section 3.6.6.

3.7.1.1 Basic concepts

A genetic algorithm starts with a set of designs or candidate solutions to a given optimization problem and moves toward the optimal solution by applying the mechanism mimicking evolution principle in nature—that is, survival of the fittest. The set of candidate solutions is called a population. A candidate solution in a population is called individual, creature, or phenotype. Each candidate solution has a set of properties represented in its chromosomes (or genotype) that can be mutated and altered. Traditionally, candidate solutions are represented in binary strings of 0 and 1, but other encodings are also possible. The evolution usually starts from a population of randomly generated individuals. The solution process is iterative, with the population in each iteration called a generation. The size of the population in each generation is unchanged. In each generation, the fitness of every individual in the population is evaluated; the fitness is usually the value of the objective (or pseudo-objective converted from a constrained problem) function in the optimization problem being solved. The better fit individuals are stochastically selected from the current population, and each individual's genome is modified (recombined and possibly randomly mutated) to form a new generation. Because better fit members of the set are used to create new designs, the successive generations have a higher probability of having designs with better fitness values. The new generation of candidate solutions is then used in the next iteration of the algorithm. Commonly, the algorithm terminates when either a maximum number of generations has been reached or a satisfactory fitness level has been achieved for the population.

3.7.1.2 Design representation

In a genetic algorithm, an individual (or a design point) consists of a chromosome and a fitness function.

The chromosome represents a design point, which contains values for all the design variables of the design problem. The gene represents the value of a particular design variable. The simplest algorithm represents each chromosome as a bit string. Typically, an encoding scheme is prescribed, in which numeric parameters can be represented by integers, although it is possible to use floating point representations as well. The basic algorithm performs crossover and mutation at the bit level. For example, we assume an optimization problem of three design variables: $\mathbf{x} = [x_1, x_2, x_3]^T$. We use a string length of 4 (or 4 bits) for each design variable, in which $2^4 = 16$ discrete values can be represented. If, at the current design, the design variables are $x_1 = 4$, $x_2 = 7$, and $x_3 = 1$, the chromosome length is $C = 12$:

$$\underbrace{0\,1\,0\,0}_{x_1=4}\ \underbrace{0\,1\,1\,1}_{x_2=7}\ \underbrace{0\,0\,0\,1}_{x_3=1}$$

With a method to represent a design point defined, the first population consisting of N_p design points (or candidate solutions) needs to be created. N_p is the size of the population. This means that N_p strings need to be created. In some cases, the designer already knows some good usable designs for the system. These can be used as seed designs to generate the required number of designs for the population using some random process. Otherwise, the initial population can be generated randomly via the use of a random number generator. Back to the example of three design variables $\mathbf{x} = [x_1, x_2, x_3]^T$, N_p number of random variables of $C = 12$ digits can be generated. Let us say one of them is 803590043721. A rule can be set to convert the random number to a string of 0 and 1. The rule can be, for example, converting any value between zero and four to "0," and between five and nine to "1." Following this rule, the random number is converted to a string 100110000100, representing a design point of $x_1 = 9$ (decoded from string: 1001), $x_2 = 8$ (string: 1000), and $x_3 = 4$ (string: 0100).

Once a design point is created, its fitness function is evaluated. The fitness function defines the relative importance of a design. A higher fitness value implies a better design. The fitness function may be defined in several different ways. One commonly employed fitness function is

$$F_i = f_{max} - f_i \tag{3.88}$$

where f_{max} is the maximum objective function value obtained by evaluating at each design point of the current population, and f_i and F_i are the objective function value and fitness function value of the ith design point, respectively.

3.7.1.3 Selection

The basic idea of a genetic algorithm is to generate a new set of designs (population) from the current set such that the average fitness of the population is improved, which is called a reproduction or selection process. Parents are selected according to their fitness; for example, each individual is selected with a probability proportional to its fitness value, say 50%, meaning that only half of the population with the best fitness functions is selected as parents to breed the next generation by undergoing genetic operations, to be discussed next. By doing so, weak solutions are eliminated and better solutions survive to form the next generation. The process is continued until a stopping criterion is satisfied or the number of iterations exceeds a specified limit.

3.7.1.4 Reproduction process and genetic operations

There are many different strategies to implement the reproduction process; usually a new population (children) is created by applying recombination and mutation to the selected individuals (parents). Recombination creates one or two new individuals by swapping (crossing over) the genome of a parent with another. A recombined individual is then mutated by changing a single element (genome) to create a new individual. Crossover and mutation are the two major genetic operations commonly employed in the genetic algorithms.

Crossover is the process of combining or mixing two different designs (chromosomes) into the population. Although there are many methods for performing crossover, the most common ones are the one-cut-point and two-cut-point methods. A cut point is a position on the genetic string. In the one-cut-point method, a position on the string is randomly selected that marks the point at which two parent design points (chromosomes) split. The resulting four halves are then exchanged to produce new designs (children). For example, string $A = 0100\ 0111\ 0001$, representing design $\mathbf{x}^A = (4, 7, 1)$ and string $B = 1001\ 1000\ 0100$ representing $\mathbf{x}^B = (9, 8, 4)$ are two design points of the current generation. If the cut point is randomly chosen at the seventh digit, as shown in Figure 3.18a, the new strings become $A' = 0100\ 0110\ 0100$, in which the last five digits were replaced by those of string B, and $B' = 0100\ 0110\ 0100$, in which the first seven digits were replaced by those of string A. As a result, two new designs are created for the next generation, i.e., $\mathbf{x}^{A'} = (4, 6, 4)$ and $\mathbf{x}^{B'} = (4, 6, 4)$. These two new design points (children) $\mathbf{x}^{A'}$ and $\mathbf{x}^{B'}$ are identical, which is less desirable.

Similarly, the two-cut-point method is illustrated in Figure 3.18b, in which the two cut points are chosen as 3 and 7, respectively. The two new strings become $A'' = 0101\ 1001\ 0001$, in which the third

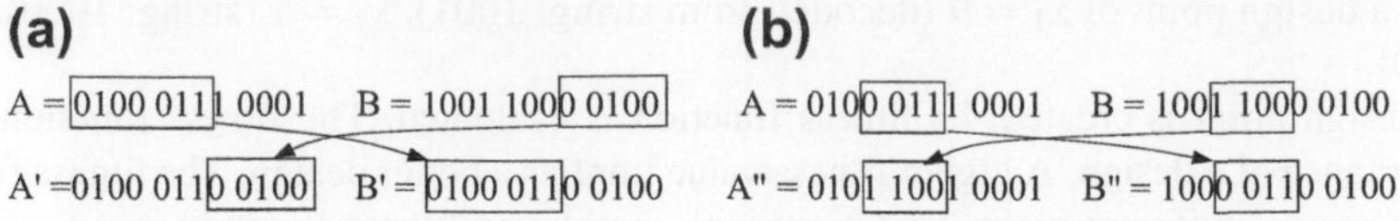

(a)   **(b)**

A = 0100 011 1 0001   B = 1001 1000 0100   A = 0100 011 1 0001   B = 1001 1000 0100

A' = 0100 011 0 0100   B' = 0100 011 0 0100   A" = 010 1 100 1 0001   B" = 1000 0110 0100

FIGURE 3.18

Crossover operations: (a) with one cut point and (b) with two cut points.

to sixth digits were replaced by those of string B, and similarly $B'' = 1000\ 0110\ 0100$. As a result, two new designs are created for the next generation: $\mathbf{x}^{A''} = (5, 9, 1)$ and $\mathbf{x}^{B'} = (8, 3, 4)$.

Selecting how many or what percentage of chromosomes to crossover, and at what points the crossover operation occurs, is part of the heuristic nature of genetic algorithms. There are many different approaches, and most are based on random selections.

Mutation is another important operator that safeguards the process from a complete premature loss of valuable genetic material during crossover. In terms of a binary string, this step corresponds to the selection of a few members of the population, determining a location on the strings at random, and switching the 0 to 1 or vice versa.

Similar to the crossover operator, the number of members selected for mutation is based on heuristics, and the selection of location on the string for mutation is based on a random process.

Let us select a design point as "1000 1110 1001" and select the 7th digit to mutate. The mutation operation involves replacing the current value of 1 at the seventh location with 0 as "1000 1100 1001."

For example, string $C = 1110\ 0111\ 0101$ represents design $\mathbf{x}^C = (14, 7, 5)$. If we choose the seventh digit to mutate, the new strings become $C' = 1110\ 0101\ 0101$. As a result, the new design is created for the next generation as $\mathbf{x}^C = (14, 5, 5)$.

In numerical implementation, not all individuals in a population go through crossovers and mutations. Too many crossovers can result in a poorer performance of the algorithm because it may produce designs that are far away from the mating designs (designs of higher fitness value). The mutation, on the other hand, changes designs in the neighborhood of the current design; therefore, a larger amount of mutation may be allowed. Note also that the population size needs to be set to a reasonable number for each problem. It may be heuristically related to the number of design variables and the number of all possible designs determined by the number of allowable discrete values for each variable. Key parameters, such as the number of crossovers and mutations, can be adjusted to fine-tune the performance of the algorithm. In general, the probability of crossover being applied is typically less than 0.1, and probability of mutation is between 0.6 and 0.9.

3.7.1.5 Solution process

A typical solution process of a genetic algorithm is illustrated in Figure 3.19. The initial population is usually generated randomly in accordance with the encoding scheme. Once a population is created, fitness function is assigned or evaluated to individuals in the population, for example, using Eq. 3.88. Parents are then selected according to their fitness. Genetic operations, including crossover and mutation, are performed to create children. The fitness of the new population is evaluated and the process is repeated until stopping criteria are met. The stopping criteria include (a) the improvement for the best objective (or pseudo-objective) function value is less than a small tolerance ε_0 for the last

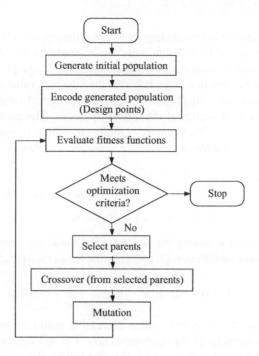

FIGURE 3.19

A typical solution process of a genetic algorithm.

n consecutive iterations (n is chosen by the users), or (b) if the number of iterations exceeds a specified value.

3.7.2 SIMULATED ANNEALING

Simulated annealing is a stochastic approach that locates a good approximation to the global minimum of a function. The main advantage of SA is that it can be applied to broad range of problems regardless of the conditions of differentiability, continuity, and convexity that are normally required in conventional optimization methods. Given a long enough time to run, an algorithm based on this concept finds global minima for an optimization problem that consists of continuous, discrete, or integer variables with linear or nonlinear functions that may not be differentiable.

The name of the approach comes from the annealing process in metallurgy. This process involves heating and controlled cooling of a material to increase the size of its crystals and reduce their defects. Simulated annealing emulates the physical process of annealing and was originally proposed in the domain of statistical mechanics as a means of modeling the natural process of solidification and formation of crystals. During the cooling process, it is assumed that thermal equilibrium (or quasi-equilibrium) conditions are maintained. The cooling process ends when the material reaches a state of minimum energy, which, in principle, corresponds with a perfect crystal.

3.7.2.1 Basic concept

The basic idea for implementation of this analogy to the annealing process is to generate random points in the neighborhood of the current best design point and evaluate the problem functions there. If the objective function value is smaller than its current best value, the design point is accepted and the best function value is updated. If the function value is higher than the best value known thus far, the point is sometimes accepted and sometimes rejected. Nonimproving (inferior) solutions are accepted in hope of escaping an local optimum in search of the global optimum. The probability of accepting non-improving solutions depends on a "temperature" parameter, which is typically nonincreasing with each iteration. One commonly employed acceptance criterion is stated as

$$
P(\mathbf{x}') = \begin{cases} e^{-\dfrac{f(\mathbf{x}') - f(\mathbf{x})}{T^k}}, & \text{if } f(\mathbf{x}') - f(\mathbf{x}) \geq 0 \\ 1, & \text{if } f(\mathbf{x}') - f(\mathbf{x}) < 0 \end{cases} \tag{3.89}
$$

where $P(\mathbf{x}')$ is the probability of accepting the new design $\mathbf{x}'$ randomly generated in the neighborhood of $\mathbf{x}$, $\mathbf{x}$ is the current best point (of iteration k), and T^k is the temperature parameter at the kth iteration, such that

$$
T^k > 0, \quad \text{for all } k; \quad \text{and} \quad \lim_{k \to \infty} T^k = 0 \tag{3.90}
$$

where temperature parameter T^k is positive, and is decreasing gradually to zero when the number of iterations k reaches a prescribed level. In implementation, k is not approaching infinite but a sufficiently large number. Also, as shown in Eq. 3.89, the probability of accepting an inferior design depends on the temperature parameter T^k. At earlier iterations, T^k is relative large; hence, the probability of accepting an inferior design is higher, providing a means to escape the local minimum by allowing so-called hill-climbing moves. As the temperature parameter T^k is decreased in later iterations, hill-climbing moves occur less frequently, and the algorithm offers a better chance to converge to a global optimal. Hill-climbing is one of the key features of the simulated annealing method.

As shown in Eq. 3.90, the algorithm demands a gradual reduction of the temperature as the simulation proceeds. The algorithm starts initially with T^k set to a high value, and then it is decreased at each step following some annealing schedule—which may be specified by the user, or set by a parameter, such as

$$
T^{k+1} = r\,T^k, r < 1 \tag{3.91}
$$

where is T^{k+1} the temperature for the next iteration $k + 1$.

3.7.2.2 Solution process

We start the solution process by choosing an initial temperature T^0 and a feasible trial point $\mathbf{x}^0$. Compute objective function $f(\mathbf{x}^0)$. Select an integer L (which sets a limit on the number of iterations), and a parameter $r < 1$.

At the kth iteration, we then generate a new point $\mathbf{x}'$ randomly in a neighborhood of the current point $\mathbf{x}^k$. If the point is infeasible, generate another random point until feasibility is satisfied. Check if the new point $\mathbf{x}'$ is acceptable. If $f(\mathbf{x}') < f(\mathbf{x}^k)$, then accept $\mathbf{x}'$ as the new best point, set $\mathbf{x}^{k+1} = \mathbf{x}'$, and continue the iteration. If $f(\mathbf{x}') \geq f(\mathbf{x}^k)$, then calculate $P(\mathbf{x}')$ using Eq. 3.89. In the meantime, we generate a random number z uniformly distributed in [0, 1]. If $P(\mathbf{x}') > z$, then accept $\mathbf{x}'$ as the new best point and set $\mathbf{x}^{k+1} = \mathbf{x}'$ to continue the iteration. If not, we generate another new point $\mathbf{x}''$ randomly, and repeat the same steps until a new best point is found. We repeat the iteration, until k reaches the prescribed maximum number of iterations L.

3.8 PRACTICAL ENGINEERING PROBLEMS

We have introduced numerous optimization problems and their solution techniques. So far, all example problems we presented assume explicit expressions of objective and constraint functions in design variables. In practice, there are only a very limited number of engineering problems, such as a cantilever beam or two-bar truss systems, that are simple enough to allow for explicit mathematical expressions in relating functions with design variables. In general, explicit expressions of functions in design variables are not available. In these cases, software tools such as MATLAB are not applicable.

Solving design optimization problems that involve function evaluations of physical problems that do not have explicit function expressions in terms of design variables deserve our attention because they are very common to design engineers in practice. Some of these problems require substantial computation time for function evaluations. In either case, we must rely on commercial software to solve these problems.

Commercial CAD or CAE tools with optimization capabilities offer viable capabilities for solving practical engineering problems. They should be the first options that designers seek for solutions. One key factor to consider in using CAD tools for optimization is design parameterization using geometric dimensions. The part or assembly must be fully parameterized so that the solid models can be updated in accordance with design changes during optimization iterations. We briefly discuss design parameterization in Section 3.8.1 from optimization perspective. For more details, readers are referred to Chang (2014). We also provide a brief overview on some of the popular commercial software in Section 3.9 and offer tutorial examples for further illustrations.

In some situations, commercial software may not offer sufficient or adequate capabilities that support design needs. For one, commercial tools may not offer function evaluations for the specific physical problems at hand. For example, if fatigue life is to be maximized for a load-carrying component, the commercial tool employed must provide adequate fatigue life computation capability. In addition, the tool must allow users to include fatigue life as an objective or constraint function for defining the optimization problem, and underline optimization algorithm must be fully integrated with the fatigue life computation capability. Another situation could be that the optimization capabilities offer in the commercial tool employs solution techniques that require many function evaluations, such as genetic algorithm. For large-scale problems, carrying out many analyses for function evaluations may not be feasible computation-wise. One may face a situation that multiple commercial codes need to be integrated to solve the problems at hand. We discuss tool integration for optimization in Section 3.8.2, which provides readers with the basic ideas and a sample case for such a need.

Finally in many cases, a true minimum is not necessarily sought; instead, an improved design is sufficient, especially for large-scale problems that require days for function evaluations. In this situation, batch mode optimization that takes several design iterations to converge may not be feasible computation wise. We present an interactive design process in Section 3.8.3 that significantly reduces the number of function evaluations and design iterations. This interactive process supports solving large-scale problems in just one of two design iterations.

3.8.1 DESIGN PARAMETERIZATION

Before solving a design optimization problem, the design problem must be formulated mathematically as defined in Eq. 3.3, in which the objective and constraint functions as well as design variables must be identified. From a design perspective, the part or assembly must be fully parameterized. In general, CAD solid models serve well as a design model for optimization. In this case, designers must define design variables by relating dimensions of the part features and creating assembly mating constraints between parts to parameterize the product model through the parametric modeling technique offered by CAD systems (Chang 2014). With the parameterized model, the designer can make a design change simply by changing geometric dimension values and asking the CAD software to automatically regenerate the parts that are affected by the change—hence, the entire assembly. This is essential for updating the design model when a new set of design variables are available at the beginning of a new design iteration.

For example, the bore diameter of an engine case is defined as the design variable, as shown in Figure 3.20a. When the diameter is changed from 1.2 to 1.6 in., the engine case is regenerated first by properly updating its solid features that are affected by the change. As shown in Figure 3.20b, the engine case becomes wider and the distance between the two exhaust manifolds is larger, just to name a few. At the same time, the change propagates to other parts in the assembly, including the piston, piston pin, cylinder head, cylinder sleeve, cylinder fins, and crankshaft, as illustrated in Figure 3.20b. More important, the parts stay intact, maintaining adequate assembly mating constraints, and the change does not induce interference nor leave excessive gaps between parts. With such parametric models, design optimization can be carried out without the interruption of a failed model regenerating during the design iterations.

In addition to geometric dimensions, material parameters (e.g., modulus of elasticity) and physical parameters (e.g., spring constant) can also be included as design variables for design optimization. On the other hand, for design problems involving idealized structural models that are analyzed by using finite element analysis, such as a thin shell modeled in surface and beams as line models, thickness and

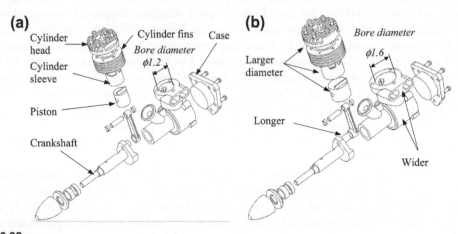

FIGURE 3.20

A single-piston engine in the exploded view: (a) bore diameter 1.2 in and (b) bore diameter 1.6 in.

cross-section properties (e.g., area or moment of inertia) defined at individual finite elements can be defined as design variables. Although these design variables are less intuitive physically, they are commonly employed for design optimization in FEA-based optimization tools, such as ANSYS. More about FEA-based optimization tools are discussed in Section 3.9.2.

3.8.2 TOOL INTEGRATION FOR DESIGN OPTIMIZATION

Three major tools are critical for support of design optimization, modeling, analysis, and optimization. Modeling tool provides designers the capability of creating product design model, either part (in CAD or FEA tools) or assembly. Analysis tools support function evaluations, depending on the kinds of physical problems being solved. The physical problems can be single disciplinary, which involve a single engineering discipline, such as structural analysis with function evaluations for stress, displacement, buckling load factor, and so forth. On the other hand, the physical problems can be multidisciplinary, in which two or more engineering disciplines are involved, such as motion analysis of a mechanism (for reaction force calculation), structural analysis for load carrying components of the mechanism, machining simulation (for machining time calculation and manufacturing cost estimate), and so forth. Finally, optimization tools that offer desired optimization algorithms for searching for optimal design are essential. In many situations, gradient calculations (also called design sensitivity analysis) play an important role in design optimization. This is because, in general, gradient-based optimization techniques that require much less function evaluations compared to the non-gradient approaches are the only viable approach to support design problems involved large-scale physical models. Moreover, accurate gradient information facilitates the search for optimal solutions, and often requires fewer design iterations. More about gradient calculations are discussed in Chapter 4.

In the following, we present a case of tool integration for CAD-based mechanism optimization, in which kinematic and dynamic analysis is required for function evaluations. This integrated system has been applied to support the suspension design of a high-mobility multipurpose wheeled vehicle (HMMWV) (Chang 2013a).

In this integrated system, commercial codes are first sought to support modeling, analysis, and optimization. In addition to the commercial codes, a number of software modules need to be implemented to support the tool integration, such as interface modules for data retrieval and model update modules for updating CAD and simulation models in accordance with design changes. The overall flowchart of the software system that supports CAD-based mechanism optimization is illustrated in Figure 3.21. The system consists of Pro/ENGINEER and SolidWorks for product model representation, Dynamic Analysis and Design System (DADS; Haug and Smith 1990) for kinematic and dynamic analysis of mechanical systems including ground vehicles, and Design Optimization Tool (DOT; www.vrand.com/dot.html) for a gradient-based design optimization. In this case, the overall finite difference method has been adopted to support gradient calculations.

In this system, engineers will create parts and assemblies of the product in a CAD tool. The solid model will be parameterized by properly generating part features and assembly constraints, as well as relating geometric dimensions to capture the design intents, as discussed in Section 3.8.1. Independent geometric dimensions of significant influence on the motion characteristics of the mechanical system are chosen as design variables. Consequently these variables help engineers achieve the design objectives more effectively.

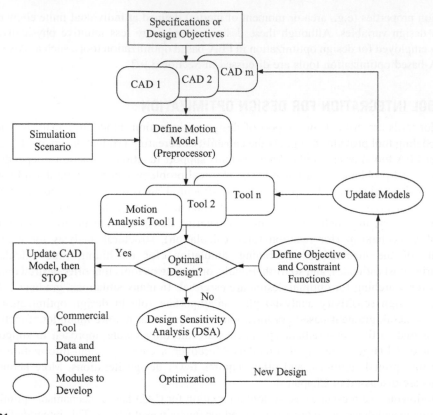

FIGURE 3.21

Overall flow of the CAD-based mechanism optimization.

The preprocessor supports design engineers in defining a complete motion model derived from the CAD solid model. Key steps include assigning a body local coordinate system (usually the default coordinate system in CAD), defining connections (or joints) between bodies, specifying initial conditions, and creating loads and drivers for dynamic and kinematic analyses. More about motion model creation and simulation can be found in Chang 2013a.

Design sensitivity analysis (DSA) calculates gradients of motion performance measures of the mechanical system with respect to dimension design variables in CAD. This is critical for optimization. The gradient information provides engineers with valuable information for making design decisions interactively. At the same time, it supports gradient-based optimization algorithms in searching for an optimal design. The gradient information and performance measure values are provided to the optimization algorithms in order to find improved designs during optimization iterations. In general, an analytical DSA method for gradient calculations is desirable, in which the derivative of a motion performance ψ with respect to CAD design variables $\mathbf{x}$ can be expressed as follows:

$$\frac{\partial \psi(\mathbf{d}, \mathbf{c})}{\partial \mathbf{x}} = \frac{\partial \psi(\mathbf{d}, \mathbf{c})}{\partial \mathbf{m}} \frac{\partial \mathbf{m}}{\partial \mathbf{d}} + \frac{\partial \psi(\mathbf{d}, \mathbf{c})}{\partial \mathbf{p}} \frac{\partial \mathbf{p}}{\partial \mathbf{d}} + \frac{\partial \psi(\mathbf{d}, \mathbf{c})}{\partial \mathbf{c}} \tag{3.92}$$

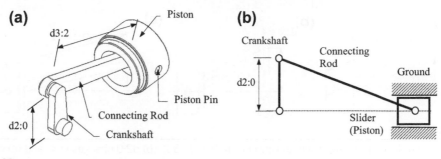

FIGURE 3.22

A slider-crank mechanism: (a) CAD solid model and (b) schematic view.

where $\mathbf{x} = [\mathbf{d}, \mathbf{c}]^T$ is the vector of design variables; $\mathbf{d}$ is the vector of dimension design variables captured in CAD solid models; $\mathbf{c}$ is a vector of physical parameters included in a load or a driver, such as the spring constant of a spring force created in the analysis model; $\mathbf{m}$ is the vector of mass property design variables, including mass center, total mass, and moment of inertia; and $\mathbf{p}$ is the vector of joint position design variables. In many situations, analytical methods for gradient calculations are not available; the overall finite difference method is an acceptable alternative, in which the derivative of a motion performance ψ with respect to CAD design variables $\mathbf{x}$ can be expressed as follows:

$$\frac{\partial \psi(\mathbf{x})}{\partial x_i} \approx \frac{\Delta \psi(\mathbf{x})}{\Delta x_i} = \frac{\psi(\mathbf{x} + \Delta x_i) - \psi(\mathbf{x})}{\Delta x_i} \tag{3.93}$$

where $\psi(\mathbf{x})$ is a dynamic performance of the mechanical system at the current design, and $\psi(\mathbf{x} + \Delta x_i)$ is the performance at the perturbed design with a design perturbation Δx_i for the ith design variable. Note that the design perturbation Δx_i is usually very small.

The motion model must be updated after a new design is determined in the optimization iterations. Mass properties and joint locations of the new design must be recalculated according to the new design variable values. The new properties and locations will replace the existing values in the input data file or binary database of the motion model for motion analysis in the next design iteration. Note that the definition of the motion model is assumed to be unchanged in design iterations; for example, no new body or joint can be added, driving conditions cannot be altered, and the same road condition must be kept during design iterations. Mass properties, joint locations, and physical parameters that define forces or torque (e.g., spring constant) are allowed to change during design iterations.

A simple slider crank example shown in Figure 3.22a is presented to demonstrate the feasibility of the integrated system. Schematically, the mechanism is a standard 4-bar linkage, as illustrated in Figure 3.22b. Moreover, geometric features in the crankshaft and connecting rod have been created with proper dimensions and references such that when their lengths are changed, the entire parts vary accordingly. At the assembly level, when either of the two length dimensions d2:0 or d3:2 is changed, the change is propagated to the affected parts. The remaining parts are kept unchanged, and the entire assembly is kept intact, as illustrated in Figure 3.23.

The CAD and motion models are created in SolidWorks and SolidWorks Motion, respectively. The mechanism is driven by a constant torque of 10 in-lb. applied to the crankshaft for 5 s.

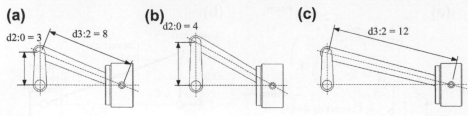

FIGURE 3.23

Change of length dimensions: (a) length dimensions d2:0 and d3:2, (b) d2:0 changes to 4, and (c) d3:2 changes to 12.

Note that friction is assumed to be nonexisting in any joint. The optimization problem is formulated as follows:

$$\text{Minimize}: \quad \phi(\mathbf{x}) \tag{3.94a}$$

$$\text{Subject to}: \quad \psi_1(\mathbf{x}) \leq 9000 \text{ lb} \tag{3.94b}$$

$$\psi_1(\mathbf{x}) \leq 9000 \text{ lb} \tag{3.94c}$$

$$1.0 \leq x_1 \leq 4.0 \text{ in} \tag{3.94d}$$

$$0.2 \leq x_2 \leq 1.0 \text{ in} \tag{3.94e}$$

$$5.0 \leq x_3 \leq 8.0 \text{ in} \tag{3.94f}$$

where the objective function $\phi(\mathbf{x})$ is the total volume of the mechanism; $\psi_1(\mathbf{x})$ and $\psi_2(\mathbf{x})$ are the maximum reaction forces in magnitude at the joints between crank and bearing, and between crank and rod, respectively, during a 5-s simulation period; and x_1, x_2, and x_3 are design variables specifying the crankshaft length (d2:0), crankshaft width, and rod length (d3:2), respectively. At the initial design, we have $x_1 = 3$ in, $x_2 = 0.5$ in, and $x_3 = 8$ in, as shown in Figure 3.24a. The maximum reaction forces are

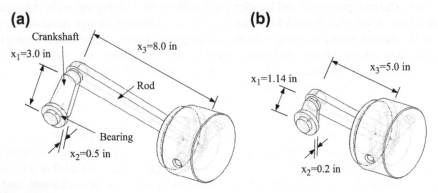

FIGURE 3.24

Design optimization of the slider-crank mechanism: (a) initial design with design variables and (b) optimal design (Chang and Joo 2006).

Table 3.1 Design Optimization of the Slider-Crank Mechanism (Chang and Joo 2006)

Measure	Initial Design	Optimal Design	% Change
$\phi(\mathbf{x})$	28.42 in^3	25.21 in^3	−11.3
$\psi_1(\mathbf{x})$	10,130 lb	9015 lb	−9.9
$\psi_2(\mathbf{x})$	9670 lb	8854 lb	−8.4
x_1	3 in	1.14 in	−62.0
x_2	0.5 in	0.20 in	−60.0
x_3	8.0 in	5.0 in	−37.5

10,130 and 9670 lb, respectively; therefore, the initial design is infeasible. The optimization took 15 iterations to converge using the modified feasible direction (MFD) algorithm (Vanderplaats 2005) in DOT. At the optimum, the overall volume of the mechanism is reduced by 11%, both performance constraints are satisfied, and two out of the three design variables reached their respective lower bounds, as listed in Table 3.1. The optimized mechanism is shown in Figure 3.24b. Note that all three design variables are reduced to achieve an optimal design because the design directions that minimize the total volume and reduce reaction forces between joints (due to mass inertia) are consistent.

3.8.3 INTERACTIVE DESIGN PROCESS

For practical design problems, a true optimal is not necessarily sought. Instead, an improved design that eliminates known problems or deficiencies is often sufficient. Also, as mentioned earlier, function evaluations require substantial computation time for large-scale problems. Therefore, even using the gradient-based solution techniques, the computation may take too long to converge to an optimal solution.

In addition to batch-mode design optimization, an interactive approach that reduces number of function evaluations and design iterations is desired, especially for large-scale problems. In this subsection, we discuss one such approach called three-step design process, involving sensitivity display, what-if study, and trade-off analysis. The interactive design approach mainly supports the designer in better understanding the behavior of the design problem at the current design and suggests design changes that effectively improve the product performance in one or two design iterations.

3.8.3.1 Sensitivity display

The derivatives (or gradients) of functions with respect to design variables are called design sensitivity coefficients or gradients. These coefficients can be calculated, for example, using the overall finite difference method mentioned in Section 3.8.2. Once calculated, the coefficients can be shown in a matrix form, such as in a spreadsheet, as shown in Figure 3.25. Individual rows in the matrix display numerical numbers of gradients for a specific function (e.g., von Mises stress in Row 3 in Figure 3.25) with respect to all design variables; that is, $\partial\psi_i/\partial x_j, j = 1$, number of design variables (which is three in the matrix of Figure 3.25). Each row of the matrix shows the influence of design variables to the specific function (either objective or constraint). On the other hand, individual columns in the matrix display gradients of all functions with respect to a single design variable; that is, $\partial\psi_i/\partial x_j, i = 1$, number of functions. Each column shows the influence of the design variable to all functions.

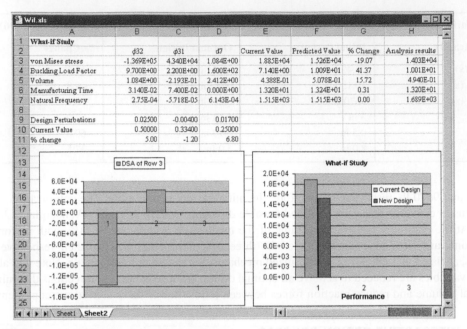

FIGURE 3.25

Spreadsheet showing sensitivity matrix and bar charts for better visualization.

The design sensitivity matrix contains valuable information for the designer to understand structural behavior so as to make appropriate design changes. For example, if the von Mises stress is greater than the allowable stress, then the data in Row 3 show a plausible design direction that reduces the magnitude of the stress measure effectively. In the bar chart of the Row 3 data shown in the lower left of Figure 3.25, increasing the first design variable reduces stress significantly because the gradient is negative with a largest magnitude. On the other hand, decreasing the second design variable reduces the stress because the gradient of the design variable is positive. Changing the third design variable does not impact the stress because its gradient is very small. In fact, data in each row show the steepest ascent direction for the respective function (in increasing its function value). Reversing the direction gives the steepest descent direction for design improvement. By reviewing row or column data of the sensitivity matrix in bar charts, designers gain knowledge in the behavior of the design problem and possibly come up with a design that improves respective functions. Once a design change is determined, the designer can carry out what-if study that calculates the function values at the new design using first-order approximations without going over analyses for function evaluations.

3.8.3.2 What-if study

A what-if study, also called a parametric study, offers a quick way for the designer to find out "What will happen if I change a design variable this small amount?" In what-if study, the function values at the new design are approximated by the first-order Taylor's series expansion:

$$\psi_i(\mathbf{x} + \delta x_j) \approx \psi_i(\mathbf{x}) + \frac{\partial \psi_i}{\partial x_j} \delta x_j \tag{3.95}$$

where ψ_i is the ith function of the design problem, $\partial\psi_i/\partial x_j$ is the design sensitivity coefficient of the ith function with respect to the jth design variable, and δx_j is the design perturbation of the jth design variable. The what-if study gives quick first-order approximation for product performance measures at the perturbed design without going through a new analysis, for example, by using a finite element code.

In general, a design perturbation $\delta\mathbf{x}$ is provided by the designer, calculated along the steepest descent direction of a function with a given step size, or computed in the direction found by the trade-off determination to be discussed next, with a step size α; that is, $\delta\mathbf{x} = \alpha\mathbf{n}$. The what-if study results and the current function values can be displayed in a spreadsheet or bar charts, such as that shown in Figure 3.25 (lower right). These kinds of displays allow the designer to instantly identify the predicted performance values at the perturbed design and compare them with those at the current design.

Once a better design is obtained from the what-if study (in approximation), the designer may commit to the changes by updating the CAD and analysis models, then carrying out analyses for the updated model to confirm that the design is indeed better. This completes one design iteration in an interactive fashion. Note that to ensure reasonably accurate function predictions using Eq. 3.95, the step size α must be small so that the perturbation $\frac{\partial\psi_i}{\partial\mathbf{x}}\delta\mathbf{x}$ is, as a rule of thumb, less than 10% of the function value $\psi_i(\mathbf{x})$.

3.8.3.3 *Trade-off determination*

In many situations, functions are in conflict due to a design change. A design may bring an infeasible design into a feasible region by reducing constraint violations; in the meantime, the same change could increase the objective function value that is undesirable. Therefore, very often in the design process, the designer must carry out design trade-offs between objective and constraint functions.

The design trade-off analysis method presented in this section assists the designer in finding the most appropriate design search direction of the optimization problem formulated in Eq. 3.3, using four possible options: (1) reduce cost (objective function value), (2) correct constraint neglecting cost, (3) correct constraint with a constant cost, and (4) correct constraint with a cost increment. As a general rule of thumb, the first algorithm, reduce cost, can be chosen when the design is feasible. When the current design is infeasible, among the other three algorithms, generally one may start with the third option, correct constraint with a constant cost. If the design remains infeasible, the fourth option, correct constraint with a cost increment of, say, 10%, may be chosen. If a feasible design is still not found, the second algorithm, correct constraint neglecting cost, can be selected. A QP subproblem, discussed in Section 3.6.4, is formulated to find the search direction numerically corresponding to the option selected.

The QP subproblem for the first option (cost reduction) can be formulated as

$$\text{Minimize}: \quad \bar{f} = \mathbf{c}^T\mathbf{d} + \frac{1}{2}\mathbf{d}^T\mathbf{d} \tag{3.96a}$$

$$\text{Subject to}: \quad \mathbf{A}^T\mathbf{d} \leq \mathbf{b} \tag{3.96b}$$

$$\mathbf{N}^T\mathbf{d} = \mathbf{e} \tag{3.96c}$$

which is identical to those of Eq. 3.74 in Section 3.6.4.

For the second option (constraint correction neglecting cost), the QP subproblem is formulated as

$$\text{Minimize}: \quad \bar{f} = \frac{1}{2}\mathbf{d}^T\mathbf{d} \tag{3.97a}$$

$$\text{Subject to}: \quad \mathbf{A}^T\mathbf{d} \leq \mathbf{b} \tag{3.97b}$$

$$\mathbf{N}^T\mathbf{d} = \mathbf{e} \tag{3.97c}$$

Note that Eq. 3.97 is similar to 3.96, except that the first term of the objective function in Eq. 3.96a is deleted in Eq. 3.97a because the cost (objective function) is neglected.

For the third option (constraint correction with a constant cost), the QP subproblem is the same as Eq. 3.97, with an additional constraint $\mathbf{c}^T\mathbf{d} \leq 0$ (implying no cost increment). For the last option (constraint correction with a specified cost), the QP subproblem is the same as Eq. 3.97, with an additional constraint $\mathbf{c}^T\mathbf{d} \leq \Delta$, where Δ is the specified cost increment (implying the cost cannot increase more than Δ). The QP subproblems can be solved using a QP solver, such as MATLAB.

After a search direction $\mathbf{d}$ is found by solving the QP subproblem for the respective option, it is normalized to the vector $\mathbf{n} = \mathbf{d}/\|\mathbf{d}\|$. Thereafter, a number of step sizes α_i can be used to perturb the design. Objective and constraint function values, represented as ψ_i, at a perturbed design $\mathbf{x} + \alpha_i\mathbf{n}$ can be approximated by carrying out a what-if study. Once a satisfactory design is identified after trying out different step sizes in an approximation sense, the design model can be updated to the new design for the next design iteration.

A particular advantage of the interactive design approach is that the designer can choose proper option to perform design trade-offs and carry out what-if studies efficiently, instead of depending on design optimization algorithms to find a proper design by carrying out line searches using several function evaluations, which is expensive for large-scale problems. The result of the interactive design process can be a near-optimal design that is reliable. A case study is presented in Section 3.10.1 using a tracked-vehicle roadwheel example to further illustrate the interactive design process.

3.9 OPTIMIZATION SOFTWARE

There are numerous commercial optimization software tools that support engineering design. We have seen throughout this chapter the use of MATLAB to carry out optimization for both mathematical and engineering problems. In using MATLAB, the objective and constraint functions must be explicitly expressed in terms of design variables. Although easy to use, there are only limited number of engineering problems, such as a cantilever beam or two-bar truss, that are simple enough to allow for explicit mathematical expressions in relating functions with design variables.

In this section, we provide a brief review on commercial software tools by which engineering optimization problems beyond simple beams and trusses can be solved. We categorize them into optimization in CAD and optimization in FEA that offer general optimization capabilities. We also include special-purpose codes that are tailored for solving specific types of optimization problems.

3.9.1 **OPTIMIZATION IN CAD**

In most cases, optimization capability is implemented as a software module in commercial software that offers more engineering capabilities than just optimization. Some of them are embedded in CAD software, such as Pro/ENGINEER and SolidWorks, and more are available in CAE software, such ANSYS (www.ansys.com), and MSC/NASTRAN (www.mscsoftware.com/product/msc-nastran), and LS-OPT (www.lstc.com/products/ls-opt).

Pro/MECHANICA Structure (or Mechanica) embedded and fully integrated in Pro/ENGINEER supports optimization for structural problems, in which geometric dimensions and physical parameters, such as material properties, of a part or assembly can be included as design variables. Functions, such as stress, displacement, buckling load factors, and weight, can be incorporated as objective and constraint functions. Mechanica employs standard gradient-based optimization technique in searching for optimal solutions. If the part or assembly in Pro/ENGINEER is fully parameterized, optimization can be carried out fully automatically.

SolidWorks Simulation also supports engineering optimization. Similar to Mechanica, Simulation supports optimization for structural problems. Instead of using the gradient-based solution technique, Simulation employs generative method for search a solution, as discussed in Section 3.5.1. As a result, more FEAs are needed, yielding a near-optimal solution.

In addition to Pro/ENGINEER and SolidWorks, commercial CAD software tools, such as CATIA V5 and NX CAE 8.5, support design optimization for structural problems.

The key advantages of solving optimization problems using CAD are twofold. First, CAD supports parametric modeling. Hence, design variables are defined by choosing part dimensions. As long as the part or assembly is fully parameterized, optimization can be carried out fully automatically in most cases. Second, the FEA and optimization modules are fully integrated with CAD. Therefore, the design capabilities, including choosing objective and constraint functions, as well as selecting design variables are straightforward and easy to use. Tutorial examples are provided in Section 3.11 and Projects P5 and S5 to illustrate details in using Mechanica and Simulation for structural optimization.

3.9.2 **OPTIMIZATION IN FEA**

Another group of optimization codes are embedded in commercial FEA software, including MSC/NASTRAN, ANSYS DesignXplore (www.ansys.com), GENESIS (www.vrand.com/Genesis.html), and OptiStruct (www.altairhyperworks.com).

MSC/NASTRAN, developed by MSC Software Corporation in Newport Beach, California, is one of the most popular FEA software in industry. The software offers a set of most complete CAE capabilities, including structural, thermal, fluid, fatigue, dynamics, and more. Its solution 200: Design Optimization and Sensitivity Analysis supports sensitivity analysis and gradient-based optimization for three major types of structural design problems: sizing, shape, and topology optimization. With response functions and constraints supported across multiple disciplines, users do not have to perform multiple optimization runs for each discipline. It is possible to combine all these disciplines into a single run, so that users gain efficiency and obtain better designs.

ANSYS is another powerful and popular CAE software. DesignXplorer of ANSYS offers capabilities for designers to explore design alternatives, including optimization. DesignXplore employs a generative approach similar to that of SolidWorks Simulation, in which a design space is subdivided to create a series of simulation experiments for exploring better designs. With the simulation results obtained from simulation experiments, DesignXplore employs so-called response surface technologies that interpolate between the data points in multidimensional design space. The interpolated results can be visualized as a 2-D or 3-D description of the relationships between design variables and performance functions. Optimal design is then searched on the response surface.

GENESIS, developed by Vanderplaats Research & Development, is a fully integrated finite element analysis and design optimization software package. Analysis is based on the finite element method for static, normal modes, direct and modal frequency analysis, random response analysis, heat transfer, and system buckling calculations. Design is based on the gradient-based solution techniques—more specifically the SLP, SQP, and feasible direction methods. These approximate problems generated using analysis and sensitivity information, are used for the optimization, which is performed by DOT (or BIGDOT) optimizers. When the optimum of the approximate problem has been found, a new finite element analysis is performed and the process is repeated until the solution has converged to an optimum. Many design options are available for users, including shape, sizing, and topology.

OptiStruct is one of the software modules of HyperWorks, which is the principal product offered by Altair Engineering—a product design and development, engineering software, and cloud computing software company in Detroit, Michigan. OptiStruct supports topology, sizing, and shape optimization to create better and more alternative design proposals leading to structurally sound and lightweight design. OptiStruct has been widely accepted by the automotive industry.

3.9.3 SPECIAL-PURPOSE CODES

In addition to optimization capability embedded in CAD and FEA software, there are optimization software tools that integrate commercial FEA and provide capabilities that support solving general design optimization problems. This software includes Tosca Structure (www.fe-design.com/en/products/tosca-structure), LS-OPT, and so forth.

Tosca Structure offers structural optimization by integrating with industry-standard FEA packages, including ABAQUS, ANSYS, and MSC/NASTRAN. It allows for rapid and reliable design of lightweight, stiff, and durable components and systems. Tosca offers topology and sizing design capabilities through two modules: Structure.topology and Structure.sizing, respectively.

LS-OPT is a graphical optimization tool that interfaces with LS-DYNA and allows the designer to structure the design process, explore the design space, and compute optimal designs according to specified constraints and objectives. The optimization capability in LS-OPT is based on response surface and design of experiments (similar to that of DesignXplore). The software allows the combination of multiple disciplines and/or cases for the improvement of a unique design.

3.10 CASE STUDIES

We present three case studies, all involving FEA for structural analysis. The first case, sizing optimization of roadwheel, aims to minimize volume and constrain deformation of a tracked-vehicle roadwheel by varying its thicknesses. The second case study optimizes the geometric shape of an engine connecting rod using a p-version FEA code. The third case study focuses on optimizing the geometric shape of an engine turbine blade parameterized in Pro/ENGINEER.

3.10.1 SIZING OPTIMIZATION OF ROADWHEEL

In this case study, we present a sizing optimization for a tracked-vehicle roadwheel, in which thicknesses of shell finite elements are varying for better designs. Function evaluations are carried out using FEA, in which shell elements (instead of solid elements) are created from the surface geometric model created in a geometric modeling tool, MSC/PATRAN (www.mscsoftware.com/product/patran).

The roadwheels shown in Figures 3.26a and b are heavy load carrying components of the tracked vehicle suspension system. There are seven wheels on each side of the vehicle. The geometric model of the roadwheel is created in MSC/PATRAN as quadrilateral surface patches, as shown in Figure 3.26c. The objective of this design problem is to minimize its volume with prescribed allowable deformation at the contact area, by varying its thicknesses.

3.10.1.1 Geometric modeling and design parameterization

Due to symmetry, only half of the wheel is modeled for design and analysis. The outer diameter of the wheel is 25 in., with two cross-section thicknesses, 1.25 in. at rim section (dv5, dv6, and dv7) and 0.58 in. at hub section (dv1 to dv4), as shown in Figure 3.27a. To model the wheel, 216 Coons patches and 432 triangular finite elements are created in the geometric and finite element models, respectively, using PATRAN.

Thicknesses of the wheel are defined as design variables, which are linked with surface patches along the circumferential direction of the wheel to maintain a symmetric design, as illustrated in Figures 3.27b and c. Figure 3.27d shows patches along the inner edge of the wheel, in which patch thicknesses are linked together as design variable dv1. In a similar fashion, addition six design variables are defined for the wheel, as shown in Figures 3.28b and c. In the current design, we have dv1 = dv2 = dv3 = dv4 = 0.58 in., and dv5 = dv6 = dv7 = 1.25 in.

3.10.1.2 Analysis model

ANSYS plate elements STIF63 are employed for finite element analysis. There are 432 triangular plate elements and 1650 degrees of freedom defined in the model, as shown in Figure 3.28a. This wheel is made up of aluminum with modulus of elasticity, $E = 10.5 \times 10^6$ psi, shear modulus, $G = 3.947 \times 10^6$ psi, and Poisson's ratio, $v = 0.33$.

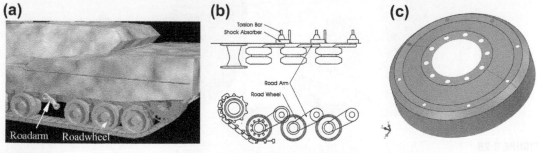

(a)

Roadarm   Roadwheel

(b)

Torsion Bar
Shock Absorber

Road Arm
Road Wheel

(c)

FIGURE 3.26

Tracked vehicle roadwheel. (a) Suspension showing roadarm and roadwheel. (b) Schematic view of the suspension, front and top views. (c) Geometric model of roadwheel in MSC/PATRAN (recreated in SolidWorks).

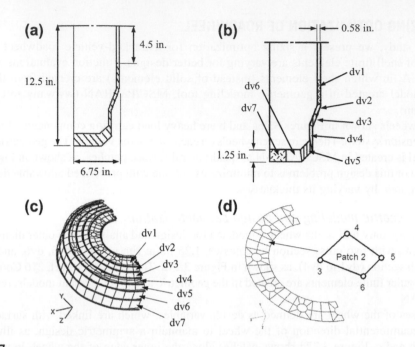

FIGURE 3.27

Roadwheel geometric model and design parameterization. (a) Half wheel section view with key dimensions. (b) The seven thickness design variables in section view. (c) Thicknesses of patches along the circumferential direction linked as design variables. (d) A closer view of the patches in the inner edge of the wheel.

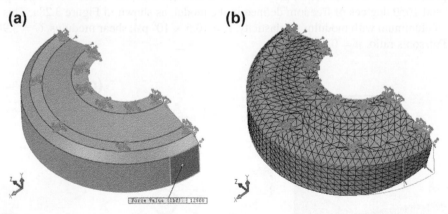

FIGURE 3.28

Finite element model of the roadwheel. (a) Triangular mesh, boundary conditions, and loads applied. (b) Deformed shape obtained from FEA using ANSYS (Chang et al. 1992) (recreated in SolidWorks Simulation).

Table 3.2 Design Sensitivity Matrix

Performance	dv1	dv2	dv3	dv4	dv5	dv6	dv7
Displacement (in.)	−0.045865	−0.01553	−0.011015	−0.019515	−0.019420	−0.056306	−0.099343
Volume (in.^{3})	35.1301	26.2068	29.8408	34.8080	47.7519	93.8567	92.4915

The circumference of the six small holes in the hub area is fixed. Symmetric conditions are imposed at the cutoff section, and a distributed load of total 12,800 lb is applied to the six elements in the area where the wheel contacts the track. A deformed wheel shape, obtained from ANSYS analysis results, is displayed in PATRAN for result evaluation, as shown in Figure 3.26b. From the analysis results, it is found that the maximum displacement occurs at contact area, more specifically node 266, in the y-direction with magnitude 0.108 in. Volume of the wheel is 361.9 in^3.

3.10.1.3 Performance measures
The maximum displacement (found at 266 in the y-direction) and wheel volume are defined as performance measures for the design problem.

3.10.1.4 Design sensitivity results and display
In this case, gradients of the displacement and volume are calculated using the sensitivity analysis method based on the continuum formulation to be discussed in Chapter 4. Design sensitivity coefficients computed are listed in Table 3.2.

To display design sensitivity coefficients, both a PATRAN fringe plot and bar chart are utilized. Figure 3.29 shows the design sensitivity coefficients of the displacement performance measure. The design sensitivity plots clearly indicate that increasing the thickness at outer edge of the wheel; that is,

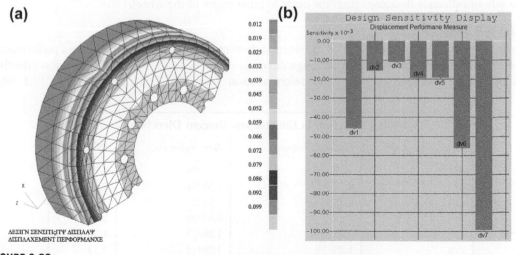

FIGURE 3.29

Design sensitivity of displacement performance measure: (a) fringe plot in PATRAN and (b) bar chart.

(a)

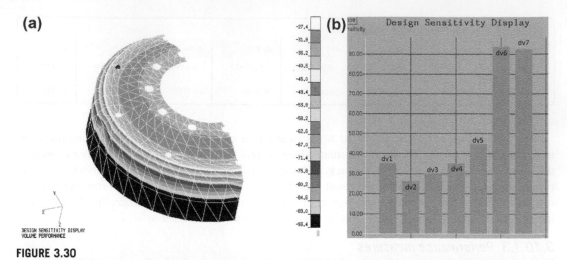

FIGURE 3.30

Design sensitivity of volume performance measure: (a) fringe plot in PATRAN and (b) bar chart.

design variable dv7 reduces the displacement most significantly. As shown in the bar chart, a 1-in. increment of thickness at outer edge of the wheel yields a 0.0993 in. reduction in the displacement. This influence decreases from the outer to inner edges of the wheel. At the inner edge of the wheel, the influence is increasing to about 40% of the maximum value.

Figure 3.30 shows the design sensitivity coefficients of the volume performance measure. The result shows that increasing thickness around outer edge of the wheel, dv6 and dv7, increases the volume performance measure most significantly. As shown in the bar chart, a 1-in increment of thickness around the outer edge of the wheel yields 93.4 in.^{3} and 92.5 in.^{3} gain in wheel volume. The rate of influence decreases from the outer to inner edges of the wheel.

3.10.1.5 *What-if study*

A what-if study is carried out based on the steepest descent direction of the displacement performance measure with a step size of 0.1 in. The design direction shown in Table 3.3 suggests the most effective design change to reduce the maximum deformation at node 266. Table 3.4 shows the first-order

Table 3.3 Design Direction in the Steepest Descent Direction			
Design Variable	**Current Value (in.)**	**New Value (in.)**	**% Change**
dv1	0.58	0.61596	6.20
dv2	0.58	0.59218	2.10
dv3	0.58	0.58864	1.49
dv4	0.58	0.59530	2.64
dv5	1.25	1.26523	1.22
dv6	1.25	1.29415	3.53
dv7	1.25	1.32790	6.23

Table 3.4 What-if Results and Verification

Performance Measure	Current Value	Predicted Value	FEA Results	Accuracy
Volume	361.941 in.^{3}	376.345 in.^{3}	376.339 in.^{3}	100.0
Displacement	0.108173 in.	0.095420 in.	0.096425 in.	101.0

prediction of displacement and volume performance values using the design sensitivity coefficients and design perturbation. It is shown that the displacement performance is reduced from 0.1082 to 0.0954 in., following such a design change. However, volume increases from 361.94 to 376.35 in.^{3}. A finite element analysis is carried out at the perturbed design to verify if the predictions are accurate. The finite element analysis results given in Table 3.4 show that the predicted performance values are very close to the results from finite element analysis because the sensitivity coefficients are accurate and the design perturbation is within a small range.

3.10.1.6 Trade-off determination

From the design sensitivity displays and what-if study, a conflict is found in the design in reducing structural volume and maximum deformation. To find the best design direction, a trade-off study is carried out. To support the trade-off study, the volume performance measure is selected as the objective function, and the displacement performance measure is defined as constraint function, with an upper bound of 0.1 in. Notice that, in the current design, the displacement performance measure is 0.1082 in., which is greater than the bound. Therefore, the current design is infeasible. The side constraints are defined for all the design variables with bounds 0.1 and 10.0 in.

With an infeasible design, the second option, constraint correction, is selected for trade-off study. A QP subproblem is formed and solved to determine a design direction. Table 3.5 shows the design direction obtained from solving the QP subproblem.

A what-if study is carried out again following the design direction suggested by the trade-off determination, using a step size 0.1 in. The results of the what-if study are listed in Table 3.6, which show the approximation of objective and constraint function values using the design sensitivity coefficients and design perturbation. In this case, constraint violation is completely corrected with a small increment in the objective function (volume). A finite element analysis is carried out at the perturbed design to verify the approximations are accurate, as given in Table 3.6.

Table 3.5 Design Direction for Trade-off Determination

Design Variables	Current Value (in.)	Direction (in.)	Perturbation
dv1	0.58	0.2305	0.0231
dv2	0.58	0.07805	0.0078
dv3	0.58	0.05536	0.0055
dv4	0.58	0.09807	0.0098
dv5	1.25	0.09759	0.0097
dv6	1.25	0.02830	0.0028
dv7	1.25	0.4992	0.0499

Table 3.6 What-if Study Results and Verification

Objective and Constraint	Current Value	Predicted Value	FEA Results	Accuracy
Objective	361.941 in.^{3}	371.172 in.^{3}	371.169 in.^{3}	100.0
Constraint	0.1082 in.	0.1000 in.	0.1004 in.	100.4

3.10.1.7 Design optimization

To perform design optimization, the same objective, constraint, and side constraints defined in trade-off determination are used. DOT, ANSYS, and sensitivity computation and model update programs similar to those discussed in Section 3.8.2 are integrated to perform design optimization. After four iterations, a local minimum is achieved. The optimization histories for objective, constraint, and design variables are shown in Figures 3.31a, b, and c, respectively.

From Figure 3.31a, the objective function starts around 362 in.^{3} and jumps to 382 in.^{3} immediately to correct constraint violation. Then, the objective function is reduced further until a minimum point, 354 in.^{3} is reached. Also, the constraint function history graph shows that 80% violation is found at the initial design, and the violation is reduced significantly to 65% below the bound at the first iteration. Then, the constraint function is stabilized and stays feasible for the rest of iterations. At optimum, the constraint is 4% below the bound, the maximum displacement becomes 0.09950 in., and the design is feasible. The most interesting observation is that, from Figure 3.31c, all design variables are decreasing in the design iterations, except for dv1 and dv7. Design variable dv1 increases from 0.58 to 0.65 in. at the optimum. However, the most significant design change is dv7 from 1.25 to 1.44 in., which contributes largely to the reduction in the deformation, as listed in Table 3.7. Decrement of the rest of the design variables contributes to the volume reduction.

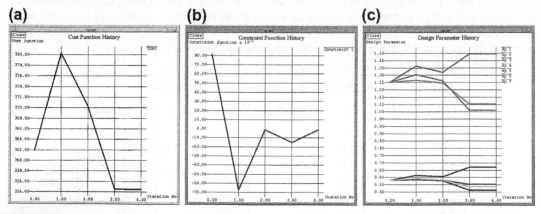

FIGURE 3.31

Design optimization history: (a) objective function, (b) constraint function, and (c) design variables.

Table 3.7 Design Variable Values at Optimum

Design Variables	Initial Design (in.)	Optimum Design (in.)
dv1	0.58	0.650
dv2	0.58	0.556
dv3	0.58	0.512
dv4	0.58	0.540
dv5	1.25	1.113
dv6	1.25	1.053
dv7	1.25	1.442

3.10.1.8 Postoptimum study

At optimum, the designer can still acquire significant information to assist in the design or manufacturing process. The design sensitivity plot for a displacement performance measure at optimum is shown in Figure 3.32. The sensitivity plot shows that thickness at the outer edge has a significant effect on the maximum displacement at the contact area. In other areas, sensitivity coefficients are relatively small. This plot suggests that in the manufacturing process, restrictive tolerance needs to be applied to the thickness at the outer edge because small errors made in the outer rim will impact the displacement.

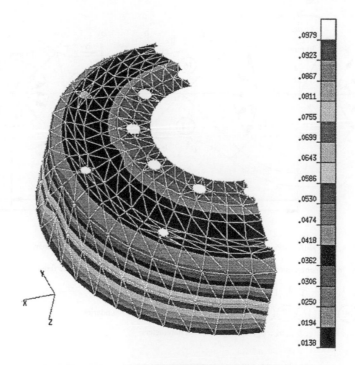

FIGURE 3.32

Design sensitivity of displacement performance measure at the optimum displayed in fringe plot in PATRAN.

3.10.2 SHAPE OPTIMIZATION OF THE ENGINE CONNECTING ROD

This case study involves shape optimization of an engine connecting rod, in which the geometric shape of the rod is parameterized to achieve an optimal design considering essential structural performance measures, such as stresses. In this case, geometric and finite element models are created in a *p*-version FEA code, called STRESSCHECK (www.esrd.com). FEA is performed also using STRESSCHECK. Shape design sensitivity computation is carried out using the continuum-based method to be discussed in Chapter 4, and optimization is carried out using DOT. STRESSCHECK and DOT are integrated with the sensitivity analysis code to support a batch mode optimization for this example.

3.10.2.1 Geometric and finite element models

The connecting rod is modeled as plane stress problem using 25 quadrilateral p-elements. The shape of the connecting rod is to be determined to minimize its weight subject to a set of stress constraints. The geometric model, finite element mesh, physical dimensions, and design variables are shown in Figure 3.33. The material properties are modulus of elasticity $E = 2.07 \times 10^5$ MPa and Poisson's ratio $v = 0.298$.

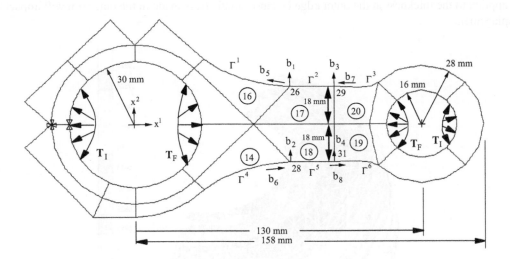

$E = 2.07 \times 10^5$ MPa, $v = 0.298$, $\rho = 7.81 \times 10^{-6}$ kg/mm^3, Plane Stress Problem

⊻ : movement x^2-direction is fixed,   ✳ : movements x^1-and x^2-direction are fixed

Γ^1, Γ^2, Γ^3, Γ^4, Γ^5, Γ^6 : design boundaries

⑭ ⑯ ⑰ ⑱ ⑲ ⑳ : finite elements with design boundaries

b_1, b_2, b_3, b_4, b_5, b_6, b_7, b_8 : design parameters

FIGURE 3.33

The engine connecting rod model (Hwang et al. 1997).

In this problem, two loadings are considered: the firing load T_F, which occurs during the combustion cycle, and the inertia load T_I, which occurs during the suction cycle of the exhaust stroke. These loads are defined as follows (Hwang et al. 1997):

$$T_F = \begin{cases} 357.589\theta^2 - 0.0263131\theta - 175.390, \text{ at left inner circle}, -40° \leq \theta \leq 40° \\ 518.622\theta^2 - 3258.60\theta + 4812.67, \text{ at right inner circle}, 140° \leq \theta \leq 220° \end{cases} \quad (3.98)$$

$$T_I = \begin{cases} 21.7327\theta^4 - 282.180\theta^3 + 1335.71\theta^2 - 2723.33\theta + 1998.7, \\ \quad \text{at left inner circle}, 105° \leq \theta \leq 225° \\ 49.6133\theta^4 1.22975\theta^3 - 76.8156\theta^2 - 0.823978\theta 12.4547, \\ \quad \text{at right inner circle}, -75° \leq \theta \leq 75° \end{cases} \quad (3.99)$$

where θ is the angle measured counterclockwise from the positive x^1-axis.

3.10.2.2 Design parameterization and problem definition

A design boundary that consists of boundary segments Γ^1 to Γ^6 is parameterized using Hermit cubic curves. Eight design variables are shown in Figure 3.33 and listed in Table 3.8. For the firing load, the upper bound of the allowable principal stress is $\sigma_{UF} = 37$ MPa, and the lower bound is $\sigma_{UF} = -279$ MPa. For the inertia load, the upper bound of the allowable principal stress is $\sigma_{UI} = 136$ MPa, and the lower bound is $\sigma_{UI} = -80$ MPa. There are 488 stress constraints imposed along boundaries Γ^1 to Γ^6 and at the interior of elements 14, 16, 17, 18, 19, and 20, as shown in Figure 3.34.

Table 3.8 Design Variables of Engine Connecting Rod	
Design Variable	**Definition**
b_1	Position of node 26 in x^2-direction
b_2	Position of node 28 in x^2-direction
b_3	Position of node 29 in x^2-direction
b_4	Position of node 31 in x^2-direction
b_5	Slope at node 26
b_6	Slope at node 28
b_7	Slope at node 29
b_8	Slope at node 31

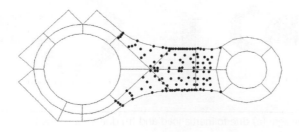

FIGURE 3.34

Locations of stress constraint points of the connection rod.

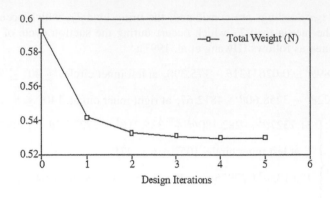

FIGURE 3.35

Objective function history of connecting rod.

3.10.2.3 Design optimization

At the initial design, the total weight of the rod is 0.5932 N and no stress constraints are violated. For design optimization, the modified feasible direction method of DOT is used. After five design iterations, an optimum design is obtained. The total weight at the optimum design is 0.5298 N. The objective function history is shown in Figure 3.35. The total weight has been reduced by 10.7%. Because the firing load is larger than the inertia load, a number of stress constraints are active under the firing load at the optimum design. Figures 3.36 and 3.37 are the stress contours at the initial and optimum designs, respectively. As shown in Figure 3.37, all active stress constraints are due to the firing load.

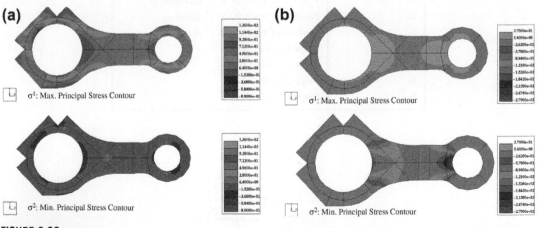

FIGURE 3.36

Stress contour at initial design: (a) due to inertia load and (b) due to firing load.

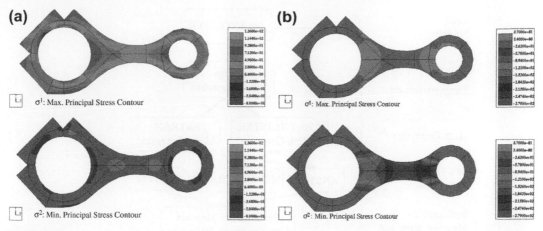

FIGURE 3.37

Stress contour at optimal design: (a) due to inertia load and (b) due to firing load.

Figure 3.38 shows the design and finite element models at the initial and optimum designs. A reduced optimum weight is obtained by distributing the high stress points along the design boundary Γ^1 to Γ^6 and making stress constraints active. At the optimum design, the neck region become thinner while stress constraints are not violated.

3.10.3 SHAPE OPTIMIZATION OF ENGINE TURBINE BLADE

In this case study, we briefly present a shape optimization of a 3-D engine turbine blade using the integrated tool environment shown in Figure 3.39. In this tool set, Pro/ENGINEER is employed to support solid modeling, in which CAD dimensions are defined as design variables for shape optimization. Both Pro/MESH of Pro/ENGINEER and MSC/PATRAN are used to support finite element mesh generation. Commercial FEA codes, including ANSYS, MAS/NASTRA, and PolyFEM (PolyFEM 1994) are integrated to support FEA. Optimization is carried out using DOT.

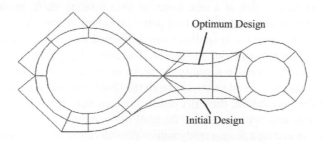

FIGURE 3.38

Geometry model of the connecting rod at initial and optimum designs.

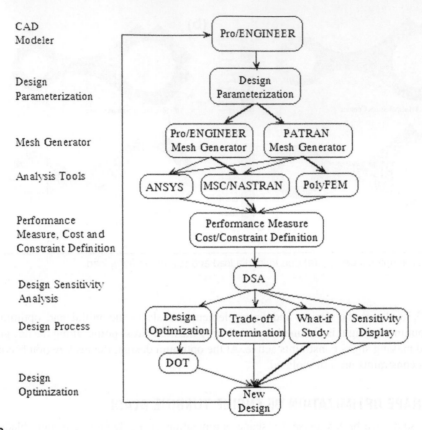

CAD
Modeler

Design
Parameterization

Mesh Generator

Analysis Tools

Performance
Measure, Cost and
Constraint Definition

Design Sensitivity
Analysis

Design Process

Design
Optimization

FIGURE 3.39

Integrated tool environment for the shape optimization of the engine turbine blade (Hardee et al. 1999).

3.10.3.1 The turbine blade example

A turbine blade is inserted in a slot of a disc mounted on a rotating shaft, as shown in Figure 3.40 schematically. The blade can be divided into four parts: airfoil, platform, shank, and dovetail. When the turbine is operating, fluid pressure is applied to the surface of the air foil, and an inertia body force is applied to the whole blade structure due to rotation. The inertia body force contributes significantly to the blade structural deformation. To simplify the problem, fluid pressure is, neglected in the finite element and sensitivity analyses. In addition, although the shank sustains a major stress flow from the dovetail to the airfoil due to rotation, the FEA results confirm that the platform does not contribute significantly to blade structural behavior. Thus, the platform is removed in the modeling process. The profile of the airfoil is determined from aerodynamics consideration and is not considered for shape design changes. However, the shape of the shank and the position of the dovetail can be modified to improve structural performance of the blade.

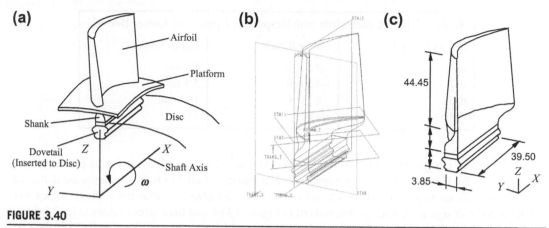

FIGURE 3.40

The turbine blade example: (a) schematic view of the physical model, (b) CAD solid model created in Pro/ENGINEER, and (c) model showing key dimensions (in mm).

3.10.3.2 CAD and finite element models

The CAD model of the turbine blade shown in Figure 3.40b is created using Pro/ENGINEER. The CAD model is imported into PATRAN for mesh generation. The finite element model shown in Figures 3.34a and b contains 315 3-D 20-node elements (ANSYS STIF95) and 2388 nodes. The material properties are modulus of elasticity $E = 2.1 \times 10^5$ MPa, Poisson's ratio $v = 0.29$, and mass density $\rho = 7.8 \times 10^{-6}$ kg/mm^3. For the boundary conditions, displacements of the nodes at two sides of the dovetail are fixed in all directions and a constant angular velocity $\omega = 1570.8$ rad/s (15,000 rpm) is applied to create the inertia body force in the entire blade.

3.10.3.3 Design parameterization for the blade

Four design variables, dp1 to dp4 shown in Figure 3.42, are defined in the model. They are respectively the offset of the dovetail in the X-, Y-, and Z-directions, and rotation of the dovetail about an axis that is parallel to the Z-axis. The four design variables characterize the design changes by repositioning the dovetail. Thus, the dovetail translates with constant cross-sectional shape and straight boundaries, although the shape of the shank changes. Both the shape changes of the shank and translation of the dovetail affect the structural behavior of the blade because the mass is redistributed due to these changes.

3.10.3.4 Design problem definition

The design optimization problem is defined to minimize the volume of the blade and keep stress below a prescribed limit. An optimal design can be achieved by repositioning the dovetail, thereby changing the shape of the shank, which is characterized by the four design variables discussed before. Note that side constraints are defined for the four design variables, as listed in Table 3.9, where design variable dp4 is restricted in the range of $-15°$ to $15°$ to restrict the dovetail rotation angle outside a $\pm15°$ range. The bounds defined for the first design variable ensure the proper alignment of the airfoil in the blade.

Table 3.9 Side Constraints with Respective Upper and Lower Bounds

Design Variable	Lower Bound	Upper Bound
dp1	−5.0 mm	5.0 mm
dp2	−5.0 mm	5.0 mm
dp3	−8.7 mm	16.7 mm
dp4	−15.0°	15.0°

At the initial design, structural volume is 15,300 mm^3. The von Mises stress at element Gauss points (points for Gauss integration in numerical calculations) of all 315 finite elements are defined as constraints with an upper bound of 4.5 MPa. The 10 finite elements that have stress values over 4.1 MPa in the design iterations are pointed out in Figure 3.41a, and their stress values at initial design are listed in Table 3.10. Also, Z-displacement of nodes at the top surface of the airfoil (77 nodes) and X-displacement of nodes at the front and rear corners of the airfoil (8 nodes) must be less than 5 mm to maintain clearance between the blade and engine casing. These areas are pointed out in Figure 3.41a. Thus, in total, 400 design constraints are defined. As shown in Table 3.10, stress in element 171 is larger than 4.5 MPa; therefore, the initial design is infeasible.

3.10.3.5 Design optimization

An optimal design is found in three design iterations. The optimization history graphs for objective, selected constraints, and design variables are shown in Figures 3.43a, b and c, respectively. The history graphs show that the objective function starts from 15,300 mm^3 and reduces to 14,500 mm^3 at the first design iteration. During last two iterations, the objective function converges to an optimum criterion defined in the DOT. Note that at the initial design (infeasible), objective function reduction is possible because the design search direction that corrects the constraint violation (stress at element 171) also

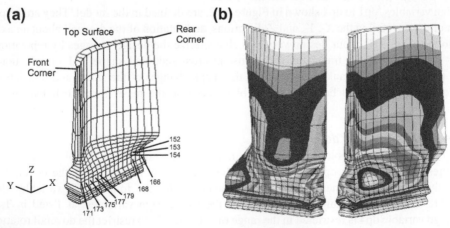

FIGURE 3.41

Turbine blade model: (a) finite element mesh with elements of high stresses and (b) stress fringe plots.

Table 3.10 Selected Gauss Point Stress Measures for the Initial Design

Element	Stress (MPa)	Upper Bound (MPa)	Status
171	4.819	4.5	Violated
173	4.226	4.5	Inactive
175	4.191	4.5	Inactive
177	4.159	4.5	Inactive
179	4.102	4.5	Inactive
168	2.748	4.5	Inactive
166	2.158	4.5	Inactive
154	2.712	4.5	Inactive
153	2.335	4.5	Inactive
152	2.227	4.5	Inactive

reduces the objective function value. The objective function value is reduced further until the design reaches a minimum of blade volume 14,150 mm^3. Note that the displacement constraints are inactive throughout the design iterations.

At the optimum, all stresses are below the bound. As shown in Figures 3.44a and b, stress in element 171 is initially greater than the upper bound and is reduced to 3.32 MPa at the optimum design. The highest stress at the optimum is 3.58 MPa, found in element 179. Comparison of stress at element 179 at the optimum to stress at element 171 at initial design shows that stress has been reduced by 25.7%, and the stress in the shank is distributed more evenly at the optimum design, as shown in Figure 3.44b.

At the optimum, the dovetail rotates 2.1°, which is within the ±15° limit, to adjust stress distribution in the shank. The blade finite element models at the initial and optimum designs are shown in Figure 3.45 for comparison.

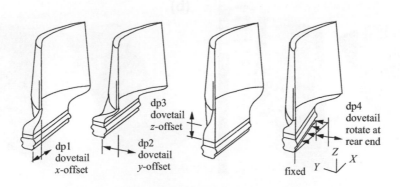

FIGURE 3.42

Design parameterization for the turbine blade (Chang and Choi 1993).

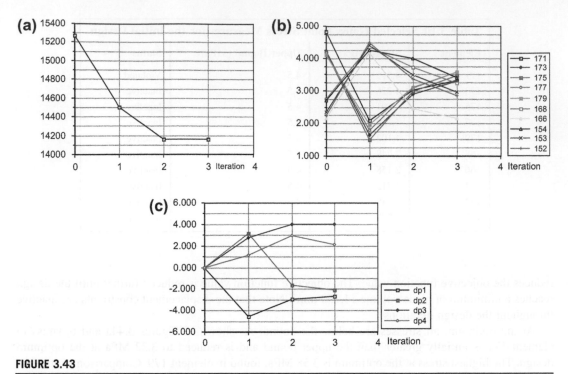

FIGURE 3.43

Optimization history graphs: (a) objective function, (b) selected constraint functions, and (c) design variables.

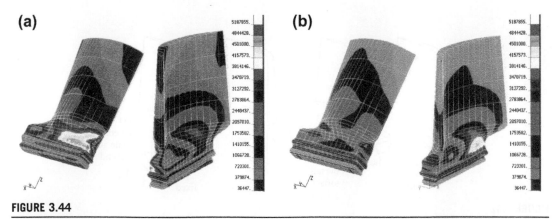

FIGURE 3.44

Stress distribution: (a) initial design and (b) optimal design.

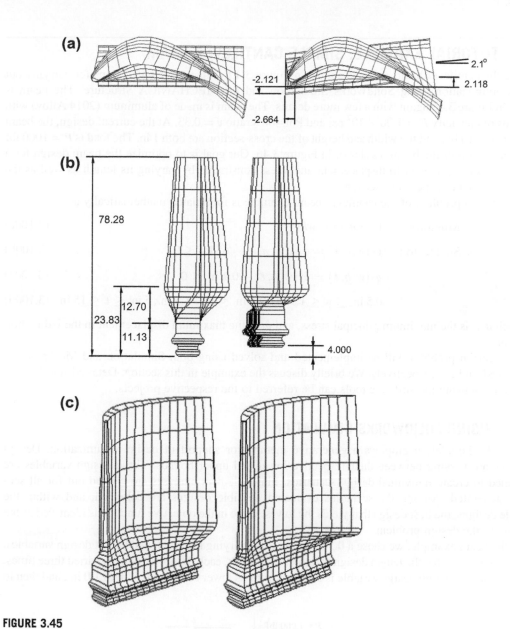

FIGURE 3.45

Finite element models of the blade at initial (left) and optimum (right) designs: (a) top view, (b) front view, and (c) iso-view.

3.11 TUTORIAL EXAMPLE: SIMPLE CANTILEVER BEAM

We use the cantilever beam example shown in Figure 3.4 to illustrate detailed steps for carrying out design optimization using SolidWorks Simulation and Pro/MECHANICA Structure. The beam is shown in Figure 3.46 again with a few more details. The beam is made of aluminum (2014 Alloy) with modulus of elasticity $E = 1.06 \times 10^7$ psi and Poisson's ratio $v = 0.33$. At the current design, the beam length is $\ell = 10$ in., and the width and height of the cross-section are both 1 in. The load is $P = 1000$ lbf acting at the tip of the beam, as shown in Figure 3.46. Our goal is to optimize the beam design for a minimum volume subject to displacement and stress constraints by varying its length as well as the width and height of the cross-section.

The design problem of the cantilever beam example is formulated mathematically as

$$\text{Minimize}: \quad V(w, h, \ell) = wh\ell \tag{3.100a}$$

$$\text{Subject to}: \quad g_2(w, h, \ell) = \sigma_{\text{mp}}(w, h, \ell) - 65 \text{ ksi} \leq 0 \tag{3.100b}$$

$$g_2(w, h, \ell) = \delta_y(w, h, \ell) - 0.1 \text{ in.} \leq 0 \tag{3.100c}$$

$$0.5 \text{ in.} \leq w \leq 1.5 \text{ in.}, 0.5 \text{ in.} \leq h \leq 1.5 \text{ in.}, 5 \text{ in.} \leq \ell \leq 15 \text{ in.} \tag{3.100d}$$

in which σ_{mp} is the maximum principal stress, and δ_y is the maximum displacement in the y-direction (vertical).

This design problem will be implemented and solved using both Simulation and Mechanica in Projects S1 and P1, respectively. We briefly discuss the example in this section. Detailed step-by-step operations in using the software tools can be referred to the respective projects.

3.11.1 USING SOLIDWORKS SIMULATION

SolidWorks Simulation employs a generative method for support of design optimization. Design variables are varying between their respective lower and upper bounds. These design variables are combined to create individual design scenarios. Finite element analyses are carried out for all scenarios generated. Among the scenarios evaluated, feasible designs are collected; and within the feasible designs, the best design that yields the lowest value in the objective function is identified as the solution to the design problem.

In the beam example, we chose a 0.5 in. interval for varying the width and height design variables, and a 5 in. interval for the length design variable. As a result, each design variable is varied three times. For example, the width design variable is changed from its lower bound of 0.5 in. to 1.0 in., and then to

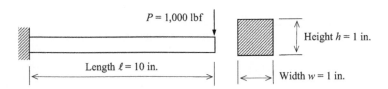

FIGURE 3.46

Cantilever beam example for tutorial projects P1 and S1.

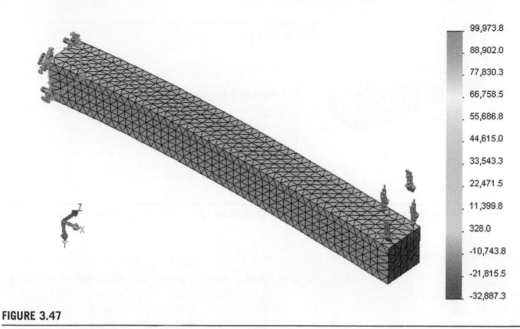

	99,973.8
	88,902.0
	77,830.3
	66,758.5
	55,686.8
	44,615.0
	33,543.3
	22,471.5
	11,399.8
	328.0
	-10,743.8
	-21,815.5
	-32,887.3

FIGURE 3.47

Finite element model of the cantilever beam in SolidWorks Simulation.

its upper bound 1.5 in. Similarly, three changes take place for the height and length design variables, respectively. These changes in design variables are combined to create 27 ($3 \times 3 \times 3$) scenarios to search for an optimal solution.

A static design study is created for the beam with boundary and load conditions shown in Figure 3.47. Also shown in Figure 3.47 are the finite element mesh (default median mesh) and the fringe plot of the maximum principal stress (or first principal stress).

A total of 29 static analyses (27 plus initial and current designs) will be carried out for the study. In the graphics window, the designs of all scenarios will appear in the beam with changing dimensions, as shown in Figure 3.48.

At the end, an optimal solution is found for Scenario 19 (see Figure 3.49), in which width, height, and length become 0.5, 5, and 1.5 in., respectively. Displacement is 0.0296 in. (<0.1 in.), stress is 42,076 psi (<65 ksi), and the total mass is 0.379 lb (reduced from 1.01 lb from initial design).

3.11.2 USING PRO/MECHANICA STRUCTURE

Unlike Simulation, Mechanica employs a gradient-based solution technique for solving design problems. Accessing the optimization capability in Mechanica is similar to that of defining and solving an FEA, which is straightforward.

A static design study is created for the beam with boundary and load conditions shown in Figure 3.50a. Figure 3.50a also shows the finite element mesh (13 tetrahedron solid elements). The maximum bending stress fringe plot is shown in Figure 3.50b, which shows maximum tensile and compressive stresses, respectively, at the top and bottom fibers of the beam close to the root end.

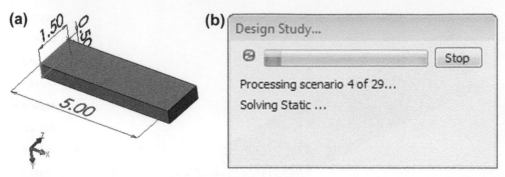

FIGURE 3.48

Design optimization underway: (a) beam with varying dimensions and (b) status dialog box showing progress.

29 of 29 scenarios ran successfully. Design Study Quality: High

		Current	Initial	Optimal (19)	Scenario 16	Scenario 17	Scenario 18	Scenario 19	Scenario 20	Scenario 21
Length		10in	10in	5in	5in	10in	15in	5in	10in	15in
Width		1in	1in	0.5in	1.5in	1.5in	1.5in	0.5in	0.5in	0.5in
Height		1in	1in	1.5in	1in	1in	1in	1.5in	1.5in	1.5in
Stress1	< 65000 psi	99974 psi	99974 psi	42076 psi	28487 psi	67068 psi	83643 psi	42076 psi	89354 psi	1.1789e+005 psi
Displacement1	< 0.1in	0.37721in	0.37721in	0.0296in	0.03171in	0.25065in	0.84573in	0.0296in	0.22621in	0.75782in
Mass1	Minimize	1.01156 lb	1.01156 lb	0.379336 lb	0.758673 lb	1.51735 lb	2.27602 lb	0.379336 lb	0.758673 lb	1.13801 lb

FIGURE 3.49

Optimal solution reported by Simulation.

FEA results are also given in the status dialog box shown in Figure 3.50c, in which maximum displacement and principal stress are 0.373 in. and 71.92 ksi, respectively.

An optimal solution is obtained in seven design iterations. At the optimum, the three design variables are length $\ell = 5$ in., width $w = 0.5$ in., and height $h = 1.08$ in. Constraint functions are stress $\sigma_{mp} = 64.99$ ksi, and displacement $\delta_x = 0.0766$ in.; both are feasible. The total mass is 0.0007056 lbf s^2/in. (or 0.272 lbm), reduced from 0.002669 lbf s^2/in. (or 1.03 lbm) from the initial design. The optimization history graphs for objective and constraint functions are shown in Figures 3.51a, b, and c, respectively.

3.12 SUMMARY

In this chapter, we discussed a broad range of topics in design optimization. We started with a simple beer can design problem, then moved into basic optimization concepts—specifically, the optimality conditions. We introduced three major solution approaches: optimality criteria, the graphical method, and the search method, for both constrained and unconstrained problems of linear and nonlinear functions. We pointed out that the search methods using numerical algorithms, including gradient-based and non-gradient-based approaches, are most general. We also mentioned that the

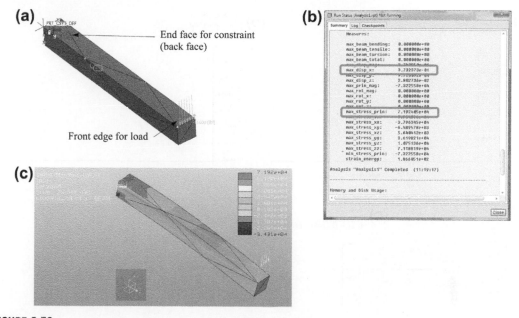

FIGURE 3.50

Finite element model of the cantilever beam in Mechanica: (a) boundary and load condition with mesh, (b) status dialog box with measures, and (c) maximum bending stress fringe plot.

gradient-based approach requires far fewer function evaluations; therefore, it is better suited to large-scale problems. We discussed several classical and widely accepted algorithms of the gradient-based search methods for both constrained and unconstrained problems. We also provided simple examples for explaining the concept and illustrating details of these algorithms. We included two representative algorithms for the non-gradient-based approach: genetic algorithm and simulated annealing. With a basic understanding of the concept and solution techniques of design optimization, we also discussed practical aspects in solving practical design problems using commercial CAD and CAE codes. We reviewed commercial CAD, CAE, and optimization tools, which offer plausible capabilities for solving general design optimization problems. We also presented three case studies that examined more in-depth in the kind of practical problems that the optimization techniques are able to support and how they are solved using commercial tools and in-house software modules.

We hope this chapter provided readers with adequate depth and breadth on this important and widely employed topic in engineering design. As we pointed out numerous times, the gradient-based solution techniques are most suitable to solving general engineering problems, especially those requiring substantial computation time. The key ingredient of the gradient-based techniques is the gradient calculation, also called design sensitivity analysis.

With a basic understanding of the subject of design optimization, we narrow our focus into structural design in Chapter 4, in which we discuss structural design problems, including sizing, material, shape, and topology. We introduce design sensitivity analysis methods and integration of sensitivity analysis with FEA and optimization algorithms for solving structural optimization

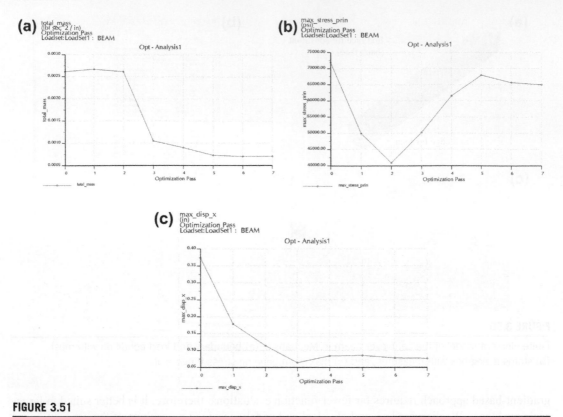

FIGURE 3.51

The optimization history graphs: (a) objective function, (b) stress constraint function, and (c) displacement constraint function.

problems. Looking ahead to Chapter 5, we discuss MOO encountered in practices, especially when we solve design problems that involve multiple engineering disciplines. Important topics to be discussed in Chapters 4 and 5 should help us become more capable and competent in solving engineering design problems.

APPENDIX A. MATLAB SCRIPTS

Script 1: Graphing volume function for the solution of the beer can example.

```
%Beer can example; plot volume as function of h
clear all;
h=0:0.1:8;
A0=pi;
V=pi*h.*(-h+(h.^2+4*A0/pi).^0.5).^2;
plot(h,V);
xlabel('h: height'); ylabel('V: volume');
```

Script 2: Graphing surface function $f(x_1, x_2)$ of Figures 3.9a and b.

```
%Figure 3.9a
clear all;
[x1,x2]=meshgrid([-1:0.075:1]);
f=-(cos(x1).^2+cos(x2).^2).^2;
surfc(x1,x2,f)
xlabel('x1');
ylabel('x2');
colormap (hsv)

%Figure 3.9b
clear all;
[x1,x2]=meshgrid([-1:0.075:1]);
f=(cos(x1).^2+cos(x2).^2).^2;
surfc(x1,x2,f)
xlabel('x1');
ylabel('x2');
colormap (hsv)
```

Script 3: Graphing surface function $f(x_1, x_2)$ and iso-lines of $f(x_1, x_2)$ at $(x_1, x_2) = (1, 1)$ and $(x_1, x_2) = (1, -1)$ of Example 3.2.

```
%Surface figure of Example 3.2
clear all;
[x1,x2] = meshgrid([-2:0.125:2]);
f = x2.*exp(-x1.^2-x2.^2);
surfc(x1,x2,f,'FaceColor','interp')
xlabel('x1');
ylabel('x2');
colormap (hsv)

%Iso-lines of objective function of Example 3.2.
clear all;
[x1,x2] = meshgrid([-2:0.125:2]);
f=x2.*exp(-x1.^2-x2.^2);
contour(x1,x2,f,0.1353);
xlabel('x1');
ylabel('x2');
hold on;
contour(x1,x2,f,-0.1353);
hold off;
```

Script 4: Graphing the feasible region and iso-lines of Example 3.8.

```
clear all;
y1=25:1:55;
y2=400/(4.074*10^6)*(6.26*10^6-y1.^4);
plot(y1,y2);
xlabel('x1');
```

```
ylabel('x2');
hold on;
[x1,x2]=meshgrid(25:1:60,-300:1:600);
F=pi/4*((2500-x1.^2).*x2+(3600-x1.^2).*(800-x2));
[C,h] = contour(x1,x2,F);
set(h,'ShowText','on','TextStep',get(h,'LevelStep')*1)
colormap cool
hold off;
```

Script 5: Solving Example 3.9.

```
clear all;
x_l=1;x_u=11;eps=0.0001;
f_l=3*x_l^3+1500/x_l;
f_u=3*x_u^3+1500/x_u;
while (abs(f_u-f_l) > eps)
    x1=x_l+0.618*(x_u-x_l);
    x2=x_u-0.618*(x_u-x_l);
    f1=3*x1^3+1500/x1;
    f2=3*x2^3+1500/x2;
    if (f1>f2)x_u=x1;
    elseif (f2>f1)x_l=x2;
    end
    f_l=3*x_l^3+1500/x_l;
    f_u=3*x_u^3+1500/x_u;
end
```

Script 6: Solving Example 3.10.

```
clear all;
x=1;dx=5;eps=0.0001;
fc=3*x3+1500/x;
fpc=9*x^2-1500/x^2;
f=0;
fp=0;
while abs(f-fc) > eps
    if fpc<0;
        x=x+dx;
    else
        x=x-dx;
    end
    f=fc;
    fp=fpc;
    fc=3*x^3+1500/x;
    fpc=9*x^2-1500/x^2;
    if sign(fp)~=sign(fpc);
        dx=dx/2;
    end
end
```

Script 7: Solving Example 3.11.

```
clear all;
x=[0 0];alp=2;eps=0.00001;
fc=(x(1)-2)^2/4+(x(2)-1)^2;
gfc=[(x(1)-2)/2 2*(x(2)-1)];
nc=-gfc./norm(gfc);
f=0;
while abs(f-fc) > eps
    f=fc;
    x=x+alp*nc;
    fc=(x(1)-2)^2/4+(x(2)-1)^2;
    gfc=[(x(1)-2)/2 2*(x(2)-1)];
    nc=-gfc./norm(gfc);
    if fc > f;
        alp=alp/2;
    end
end
```

Script 8: Solving Example 3.12.

```
% Conjugate gradient method
clear all;
alp=2;eps=0.00001;
% Inital design x0
x=[0 0];
f=(x(1)-2)^2/4+(x(2)-1)^2;
gf=[(x(1)-2)/2 2*(x(2)-1)];
n=-gf./norm(gf);
% First design iteration x1
x=x+alp*n;
gfc=[(x(1)-2)/2 2*(x(2)-1)];
fc=(x(1)-2)^2/4+(x(2)-1)^2;
beta=norm(gfc)/norm(gf);
nc=-gfc./norm(gfc)+beta*n;
nc=nc./norm(nc);
f=0;
% After 1st iteration
while abs(f-fc) > eps
    x=x+alp*nc;
    f=fc; gf=gfc; n=nc;
    fc=(x(1)-2)^2/4+(x(2)-1)^2;
    gfc=[(x(1)-2)/2 2*(x(2)-1)];
    beta=norm(gfc)/norm(gf);
    nc=-gfc./norm(gfc)+beta*n;
    nc=nc./norm(nc);
    if fc > f;
        alp=alp/2;
    end
end
```

Script 9: Solving Example 3.13.

```
% BFGS method
clear all;
alp=1;eps=0.001;
% Inital design x0
x=[0 0]';
H=[1 0
   0 1];
f=(x(1)-2)^2/4+(x(2)-1)^2;
gf=[(x(1)-2)/2 2*(x(2)-1)]';
n=-gf./norm(gf);
% First design iteration x1
x=x+alp*n;
gfc=[(x(1)-2)/2 2*(x(2)-1)]';
fc=(x(1)-2)^2/4+(x(2)-1)^2;
D=(gfc-gf)*(gfc-gf)'./((gfc-gf)'*(alp*n));
E=gf*gf'./(gf'*n);
Hc=H+D+E;
nc=-inv(Hc)*gfc;
nc=nc./norm(nc);
f=0;
% After 1st iteration
while norm(nc) > eps
    x=x+alp*nc;
    f=fc; gf=gfc; n=nc; H=Hc;
    fc=(x(1)-2)^2/4+(x(2)-1)^2;
    gfc=[(x(1)-2)/2 2*(x(2)-1)]';
    D=(gfc-gf)*(gfc-gf)'./((gfc-gf)'*(alp*n));
    E=gf*gf'./(gf'*n);
    Hc=H+D+E;
    nc=-inv(Hc)*gfc;
    nc=nc./norm(nc);
    if fc > f;
        alp=alp/2;
    end
end
```

Script 10: Solving Example 3.14.

```
% BFGS method with Newton's Method for line search
clear all;
eps=0.001;
% Inital design x0
x=[0 0]';
H=[1 0
   0 1];
f=(x(1)-2)^2/4+(x(2)-1)^2;
gf=[(x(1)-2)/2 2*(x(2)-1)]';
```

```
n=-gf./norm(gf);
alp=-f/(gf'*n);
% First design iteration x1
x=x+alp*n;
gfc=[(x(1)-2)/2 2*(x(2)-1)]';
fc=(x(1)-2)^2/4+(x(2)-1)^2;
D=(gfc-gf)*(gfc-gf)'./((gfc-gf)'*(alp*n));
E=gf*gf'./(gf'*n);
Hc=H+D+E;
nc=-inv(Hc)*gfc;
nc=nc./norm(nc);
alp=-fc/(gfc'*nc);
f=0;i=0;
% After 1st iteration
while norm(nc) > eps
     i=i+1;
     x=x+alp*nc;
     f=fc; gf=gfc; n=nc; H=Hc;
     fc=(x(1)-2)^2/4+(x(2)-1)^2;
     gfc=[(x(1)-2)/2 2*(x(2)-1)]';
     D=(gfc-gf)*(gfc-gf)'./((gfc-gf)'*(alp*n));
     E=gf*gf'./(gf'*n);
     Hc=H+D+E;
     nc=-inv(Hc)*gfc;
     nc=nc./norm(nc);
     alp=-fc/(gfc'*nc);
end
```

Script 11: Solving Example 3.15.

```
% BFGS method with Newton's Method for line search
clear all;
eps=0.01;
% Inital two points x0 and x1
x=[0 0]';
xc=[0.4 0.8]';
f=(x(1)-2)^2/4+(x(2)-1)^2;
gfc=[(f-((xc(1)-2)^2/4+(x(2)-1)^2))/(x(1)-xc(1)) (f-((x(1)-2)^2/4+(xc(2)-1)^2))/
(x(2)-xc(2))]';
fc=(xc(1)-2)^2/4+(xc(2)-1)^2;
nc=[0.841 0.540]';
alp=-fc/(gfc'*nc);
Hc=[1 0.
   0 1];
f=0;i=0;
% After 1st iteration
while norm(nc) > eps
     i=i+1;
```

```
        f=fc; n=nc; H=Hc; gf=gfc; x=xc;
        xc=x+alp*nc;
        fc=(xc(1)-2)^2/4+(xc(2)-1)^2;
        gfc=[(f-((xc(1)-2)^2/4+(x(2)-1)^2))/(x(1)-xc(1)) (f-((x(1)-2)^2/4+
(xc(2)-1)^2))/(x(2)-xc(2))]';
        D=(gfc-gf)*(gfc-gf)'./((gfc-gf)'*(alp*n));
        E=gf*gf'./(gf'*n);
        Hc=H+D+E;
        nc=-inv(Hc)*gfc;
        nc=nc./norm(nc);
        alp=-fc/(gfc'*nc);
end
```

Script 12: Solving Example 3.17.

```
%Iso-lines of objective function of Example 3.17.
clear all;
[x1,x2] = meshgrid([0:0.25:3]);
f=-2*x1-2*x2+0.5*(x1.^2+x2.^2);
[C,h] = contour(x1,x2,f);
set(h,'ShowText','on','TextStep',get(h,'LevelStep')*1)
colormap cool
hold off;
```

Script 13: Solving Example 3.20.

```
clear all;
x_l=0;x_u=3;eps=0.0001;rp=10000;%increase rp value to approach feasible solution
f_l=x_l+rp*((max(0,1-x_l))^2+(max(0,0.5*x_l-1))^2);
f_u=x_u+rp*((max(0,1-x_u))^2+(max(0,0.5*x_u-1))^2);
while (abs(f_u-f_l) > eps)
     x1=x_l+0.618*(x_u-x_l);
     x2=x_u-0.618*(x_u-x_l);
     f1=x1+rp*((max(0,1-x1))^2+(max(0,0.5*x1-1))^2);
     f2=x2+rp*((max(0,1-x2))^2+(max(0,0.5*x2-1))^2);
     if (f1>f2)x_u=x1;
     elseif (f2>f1)x_l=x2;
     end
     f_l=x_l+rp*((max(0,1-x_l))^2+(max(0,0.5*x_l-1))^2);
     f_u=x_u+rp*((max(0,1-x_u))^2+(max(0,0.5*x_u-1))^2);
end
```

REFERENCES

Arora, J.S., 2012. Introduction to Optimum Design. Academic Press.

Chang, K.H., 2013a. Product Performance Evaluation Using CAD/CAE, the Computer Aided Engineering Design Series. Academic Press, Burlington, MA.

Chang, K.H., 2013b. Product Manufacturing and Cost Estimate Using CAD/CAE, the Computer Aided Engineering Design Series. Academic Press, Burlington, MA.

Chang, K.H., 2014. Product Design Modeling Using CAD/CAE, the Computer Aided Engineering Design Series. Academic Press, Burlington, MA.

Chang, K.H., Joo, S.H., 2006. Design parameterization and tool integration for CAD-based mechanism optimization. Adv. Eng. Softw. 37, 779–796.

Chang, K.H., Choi, K.K., Perng, J.H., 1992. Design sensitivity analysis and optimization tool for sizing design applications. In: Fourth AIAA/AIR Force/NASA/OAI Symposium on Multidisciplinary Analysis and Optimization, Paper No. 92-4798, pp. 867–877. Cleveland, Ohio.

Chang, K.H., Choi, K.K., 1993. Shape design sensitivity analysis and optimization for spatially rotating objects. J. Struct. Opt. 6 (No. 4), 216–226.

Clark, F., Lorenzoni, A.B., 1996. Applied Cost Engineering, third ed. CRC.

Fletcher, R., Reeves, R.M., 1964. Function minimization by conjugate gradients. Comput. J. 7, 149–160.

Hardee, E., Chang, K.H., Tu, J., Choi, K.K., Grindeanu, I., Yu, X., 1999. CAD-based shape design sensitivity analysis and optimization. Adv. Eng. Softw. 30 (No. 3), 153–175.

Haug, E.J., Smith, R.C., 1990. DADS-Dynamic Analysis and Design System, Multibody Systems Handbook, pp. 161–179.

Hwang, H.Y., Choi, K.K., Chang, K.H., 1997. Shape design sensitivity analysis and optimization using p-version finite element analysis. Mech. Struct. Mach. 25 (No. 1), 103–137.

Kouvelis, P., Yu, G., 1997. Robust Discrete Optimization and Its Applications. Kluwer Academic Publishers, P.O. Box 17, 3300 AA Dordrecht, The Netherlands.

Ostwald, P.F., 1991. Engineering Cost Estimating, third ed. Prentice Hall.

Ostwald, P.F., McLaren, T.S., 2004. Cost Analysis and Estimating for Engineering and Management. Pearson/Prentice Hall.

PolyFEM Development Group, 1994. PolyFEM-Pro/ENGINEER User Guide and Reference. Release 2.0. IBM Almaden Research Center, San Jose, CA.

Rao, S.S., 2009. Engineering Optimization: Theory and Practice, fourth ed. Wiley.

Syslo, M.M., Deo, N., Kowalik, J.S., 1983. Discrete Optimization Algorithms: With Pascal Programs. Dover Publications, Inc., Mineola, NY.

Vanderplaats, G.N., 2005. Numerical Optimization Techniques for Engineering Design, fourth ed. Vanderplaats Research & Development, Inc.

Vanderplaats, G.N., 2007. Multidiscipline Design Optimization. Vanderplaats Research & Development, Inc., Colorado Springs, CO.

Vincent, T.L., 1983. Game theory as a design tool. J. Mech. Transm. 105, 165–170.

QUESTIONS AND EXERCISES

1. A function of two variables is defined as

$$f(x_1, x_2) = \left(\cos^2 x_1 + \cos^2 x_2\right)^2$$

Sketch by hand or use MATLAB to graph the function, and calculate the gradient vectors of the function at $(x_1, x_2) = (0, 0)$, $(x_1, x_2) = (1, -1)$, and $(x_1, x_2) = (-1, 1)$. Sketch the gradient vectors on the x_1–x_2 plane together with their respective iso-lines.

2. Design a rectangular box with a square cross-sectional area of width x and height h. The goal of the design problem is to maximize the volume of the box subject to the sum of the length of the

12 edges to a fixed number, say, $L = 192$. Formulate the design problem as a constrained optimization problem. Convert the constrained problem to an unconstrained problem and solve the unconstrained problem by finding stationary point(s) and checking the second derivatives of the objective function.

3. A function of three variables is defined as

$$f(x_1, x_2, x_2) = 3x_1^2 + 2x_1 x_2 + 8x_2^2 + x_3^2 - 2x_1 + 5x_2 - 9$$

Calculate the gradient vector of the function, and determine if a stationary point exists. If it does, calculate a Hessian matrix of the function f, and determine if the stationary point found gives a minimum value of the function f.

4. Continue with Problem 2. First, convert the constrained problem into an unconstrained problem using Lagrange multiplier and solve the optimization problem by using the KKT conditions.

5. Solve the following LP problems using the graphical method.
   Problem (a)

$$\text{Minimize} : \ f(\mathbf{x}) = 4x_1 - 5x_2$$

$$\text{Subject to} : \ g_1(\mathbf{x}) = -4 - x_1 + x_2 \le 0$$

$$g_2(\mathbf{x}) = -7 + x_1 + x_2 \le 0$$

Problem (b)

$$\text{Minimize} : \ f(\mathbf{x}) = 2x_1 - x_2$$

$$\text{Subject to} : \ g_1(\mathbf{x}) = -12 + 4x_1 + 3x_2 \le 0$$

$$g_2(\mathbf{x}) = -4 + 2x_1 + x_2 \le 0$$

$$g_3(\mathbf{x}) = -4 + x_1 + 2x_2 \le 0$$

$$0 \le x_1, \ 0 \le x_2$$

6. Solve the following NLP optimization problems using the graphical method.
   Problem (a)

$$\text{Minimize} : \ f(\mathbf{x}) = x_1^2 - 3x_1 x_2 + x_2^2$$

$$\text{Subject to} : g_1(\mathbf{x}) = x_1^2 + x_2^2 - 6 \le 0$$

$$0 \le x_1, \quad \text{and} \quad 0 \le x_2$$

Problem (b)

$$\text{Minimize} : \ f(\mathbf{x}) = 2x_1^3 - 8x_1 x_2 + 15x_2^2 - 4x_1$$

$$\text{Subject to} : h_1(\mathbf{x}) = x_1^2 + x_1 x_2 + 1 = 0$$

$$g_1(\mathbf{x}) = 4x_1 + x_2^2 - 4 \le 0$$

7. Use the golden search method to find the minimum point of the following functions in the interval of $1 \le x \le 10$. We assume the convergent tolerance is $\varepsilon_1 = 0.0001$.

Function (a): $f(x) = x^2 + 500/x^3$

Function (b): $f(x) = x^2 - e^{x^2+1}$

8. Use the gradient-based method to find the minimum point of the same objective functions (a) and (b) as Problem 7. We assume the initial design is given at $x^0 = 1$ and convergent tolerance is $\varepsilon_1 = 0.0001$. Use a proper method to determine the step size.

9. Use the steepest decent method to find the minimum point of the objective function

$$f(x_1, x_2, x_3) = x_1^2 + 2x_2^2 + 3x_3^2 + 2x_1x_2 + 7x_2x_3$$

We assume the initial design is given at $\mathbf{x}^0 = (2, 4, 5)$, the initial step size $\alpha = 1$, and convergent tolerance is $\varepsilon_1 = 0.00001$.

10. Use the conjugate gradient method to find the minimum point of the same function of Problem 9.

11. Use the BFGS method to find the minimum point of the same objective function of Problem 9.

12. Use the BFGS method combined with the Newton's method for a line search to find the minimum point of the same objective function of Problem 9.

13. Continue with Problem 12, except using the secant method for a line search.

14. Solve the Problems 6(a) and 6(b) using SLP.

15. Solve Problem 14, except using SQP.

16. Solve Problem 14, except using the feasible direction method.

17. Solve Problem 14, except using the penalty method.

18. A simple two-bar truss structure shown below supports a vertical force F without structural failure. The cross-sectional areas of the bars are A_1 and A_2, and the lengths of the bars are ℓ_1 and ℓ_2. Cross-sectional areas of the bars are allowed to change.

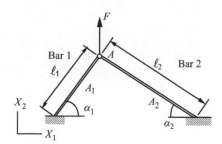

To simplify the physical model, the structure is assumed to be homogeneous with material properties: modulus of elasticity E and mass density ρ. Also, the bars are assumed to be straight before and after deformation; that is, no bending effect is considered in this problem. Because the problem is simple and can be solved analytically, there is no need to use finite element method.

A design problem is formulated as:

$$\text{Minimize} : \quad M = \rho_1 A_1 \ell_1 + \rho_2 A_2 \ell_2 \tag{a}$$

$$\text{Subject to} : \quad \sigma_1 \leq \bar{\sigma} \tag{b}$$

$$\sigma_2 \leq \bar{\sigma} \tag{c}$$

$$0 < A_1 \tag{d}$$

$$0 < A_2 \tag{e}$$

where M is mass of the structure; σ_1 and σ_2 are the stresses in bar 1 and bar 2, respectively, which must be less than the stress failure limit $\bar{\sigma}$; and the areas must be positive.

a. Sketch a free body diagram of force acting at point A, as shown in the figure in previous page, formulate equilibrium equations, and solve for the stresses σ_1 and σ_2.

b. Calculate the derivatives of the objective and stress constraint functions with respect to design variables A_1 and A_2, respectively.

c. Sketch the feasible region and iso-lines of the objective function on the A_1–A_2 plane and find the optimal solution to the optimization problem using the graphical method. The numerical numbers of the properties are given as follows: modulus of elasticity $E = 2.1 \times 10^5$ MPa, Poisson's ratio $v = 0.29$, mass density $\rho_1 = \rho_2 = 7.8 \times 10^{-6}$ kg/mm^3, $\ell_1 = 300$ mm, $\alpha_1 = 45°$, $\alpha_2 = 30°$, $A_1 = A_2 = 10$ mm^2, $\bar{\sigma} = 45$ MPa, and $F = 1000$ N.

STRUCTURAL DESIGN SENSITIVITY ANALYSIS

CHAPTER OUTLINE

In this chapter, we provide a broad but brief discussion on the topic of design sensitivity analysis (DSA), which calculates gradients (or derivatives) of structural performance measures with respect to design variables. Sensitivity analysis is essential for gradient-based optimization algorithms, in which accurate and efficient computation methods are desirable. In this chapter, we narrow our focus on structural problems, more specifically, linear elastic structures under static loads. We introduce basic concepts and solution methods in DSA with enough depth so that readers become familiar with this topic and are able to implement some of the techniques for support in engineering design. We include in this chapter basic topics that are relevant to structural problems, such as sizing, shape, and topology designs. We introduce a broad range of methods in sensitivity analysis, including the brute-force finite difference, popular semianalytical methods, and mathematically rigorous and accurate continuum-based approaches. Some of the methods discussed can be easily extended to support engineering applications beyond structural problems; for example, the overall finite difference methods are general and can be implemented for problems such as mechanism design optimization

(Chang and Joo 2006). We also offer case studies to illustrate practical applications of the methods for structural problems, including a coupled problem of continuum with molecular dynamics (MD).

Some of the discussions are quite mathematically involved. We assume readers are familiar with basic finite element analysis (FEA) theory and the relevant energy principles associated with it. We intend to use simple examples to illustrate basic concepts and solution techniques so as to avoid complex mathematical derivations. Readers who are proficient in FEA theory and desire to learn more about this topic, especially the continuum-based approaches, are referred to Choi and Kim (2006a,b).

The objectives of this chapter include (1) providing readers with the basic concepts and solution methods for structural sensitivity analysis and using simple examples to help readers grasp a basic understanding on the topic, (2) familiarizing readers with practical applications of the sensitivity analysis through case studies, and (3) providing a brief discussion on the implementation aspect of the continuum-discrete method to assist readers in future endeavors on this and other relevant topics.

Although only simple examples are employed for illustration, readers should be mindful that the concept and solution techniques introduced in this chapter are aimed at solving general and large-scale structural problems.

4.1 INTRODUCTION

As discussed in Chapter 3, numerical optimization techniques can be categorized as gradient-based and nongradient algorithms. Gradient-based algorithms often lead to a local optimum. Nongradient algorithms usually converge to a global optimum, but they require a substantial amount of function evaluations. For large-scale problems, as are often encountered in engineering design, nongradient algorithms are less desirable. Gradient-based algorithms require gradient or sensitivity information, in addition to function evaluations, to determine adequate search directions for better designs during optimization iterations.

In optimization problems, the objective and constraint functions are often called performance measures. Sensitivity, sensitivity coefficient, and gradient are terms used interchangeably to define the rate of change in a performance measure with respect to the change in design variable. They support not only gradient-based optimization but reveal critical design information that could guide designers to achieve improved designs in just one or two design iterations—for example, through the interactive design steps discussed in Chapter 3.

As shown in Figure 4.1, a change in the height h or width w of a cantilever beam with the rectangular cross-section affects the maximum bending stress inside the beam. The bending stress can be formulated as

$$\sigma(w, h) = \frac{6P\ell}{wh^2} \tag{4.1}$$

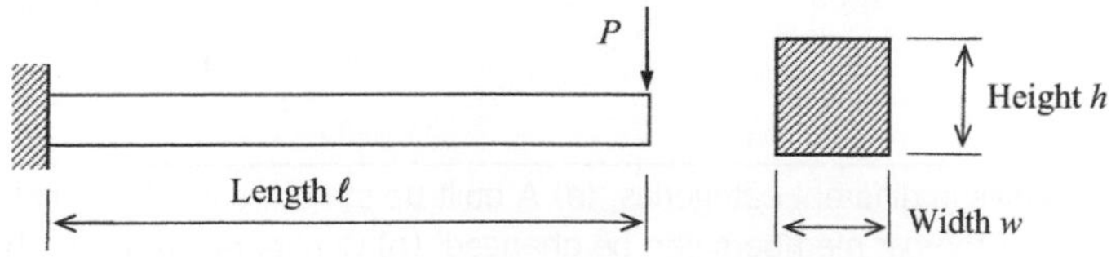

FIGURE 4.1

Cantilever beam with a rectangular cross-section.

where P is a point force applied at the tip of the beam. The gradient of the bending stress can be calculated by taking derivatives of the stress in Eq. 4.1 with respect to height and width of the beam cross-section, respectively, as

$$\frac{\partial \sigma}{\partial w} = -\frac{6P\ell}{w^2 h^2} \tag{4.1a}$$

and

$$\frac{\partial \sigma}{\partial h} = -\frac{12P\ell}{wh^3} \tag{4.1b}$$

From Eqs 4.1a and 4.1b, both gradients are negative, implying that increasing either height or width reduces the bending stress. Physically, it is obvious that increasing either height or width increases the moment of inertia I of the beam cross-section. In this case, $I = wh^3/12$, therefore reducing the bending stress. Between the two design variables, increasing the height dimension h is two times more effective than the width w in reducing the bending stress if $w = h$ (a square cross-section at the current design). Equations, such as Eqs 4.1a and 4.1b, provide desired information that supports designer to achieve improved design effectively.

The formulation of DSA can vary significantly depending on what type of design variables are being considered. In general, a structure consists of bars (or trusses), beams, membranes, shells, and/or elastic solid structural components. Depending on the constituents of the structure being designed, there are five different types of design variables—material, sizing, configuration, shape, and topology, as illustrated in Figure 4.2. For example, for a bar or beam structure shown in Figure 4.2a, material design variables can be the mass density ρ or modulus of elasticity E, while sizing design variables can

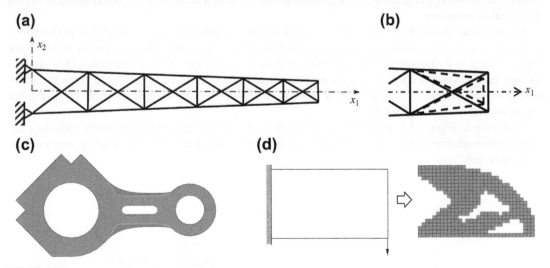

FIGURE 4.2

Illustration of design variables in different categories. (a) A built-up structure in which material parameters and cross-sectional areas of the bar members can be changed. (b) Configuration design by adjusting the orientations and lengths of truss members (Twu and Choi 1993). (c) Shape design for a 2-D engine connecting rod (Edke and Chang 2011). (d) Topology optimization of a solid beam (Tang and Chang 2001).

be the cross-sectional areas A of individual bar members or the area moment of inertias I of the beam members. Configuration design variables are related to the orientations of components in built-up structures, such as the change shown in Figure 4.2b, in which orientation angles of individual members are altered in addition to length changes. Shape design variables describe the change in the length of one-dimensional (1-D) structures or the geometric shape of two-dimensional (2-D) and three-dimensional (3-D) structures, as illustrated in Figure 4.2c. Topology optimization determines the layout of the structure, as depicted in Figure 4.2d.

To carry out sensitivity analysis, it is obvious that the performance measure is presumed to be a differentiable function of the design, at least in the neighborhood of the current design. In general for structural designs, essential structural responses, such as displacement, stress, buckling load, frequency, and so forth, are differentiable functions with respect to design.

Sensitivity analysis for performance measures expressed explicitly in design variables, such as in Eq. 4.1, is straightforward. One derivative leads to sensitivity equations such as those in Eqs 4.1a and 4.1b. No expensive computation is required.

However, in most cases, performance measures cannot be expressed explicitly in design variables. In fact, in many cases, function evaluations must be carried out numerically using FEA. In these cases, there is no equation to take derivatives for. So, how do we calculate gradients to explore viable design alternatives and support batch mode design optimization?

In this chapter, we discuss sensitivity analysis for sizing and shape designs, as well as topology optimization. We use simple and analytical examples to illustrate the sensitivity analysis concept and detailed solution techniques. We start by introducing the basic formulation of a simple bar structure in Section 4.2, which serves as an example problem to facilitate our follow-up discussion. We then provide an overview on sensitivity analysis methods in Section 4.3. In Section 4.4, we discuss sensitivity analysis for sizing and material design variables, especially using the continuum formulation. In Section 4.5, we discuss shape DSA, in which design parameterization and design velocity field computations are first discussed. Shape DSA methods are followed. In Section 4.6, we discuss topology optimization, which is effective for generating structural layouts in support of concept design. We offer two case studies in Section 4.7 to showcase some of the typical design applications in practice. We assume that the performance measures cannot be expressed explicitly in design. Therefore, although simple problems are employed for illustration, readers should be mindful that in practice function evaluations are carried out numerically, such as using FEA.

4.2 SIMPLE BAR EXAMPLE

In this section, we introduce basic concept in structural analysis that is essential for DSA. We use a simple bar example to illustrate the concept for the purpose of avoiding complex mathematical formulations that would dilute our focus. The bar example and formulation discussed in this section will be used in the next section to illustrate some of the solution techniques for sensitivity analysis.

We start by deriving a governing equation in the differential form with which we are familiar. Then, in Section 4.2.2, we introduce the energy equation that is equivalent to the differential governing equation but is more relaxed in the smoothness requirement on the solution function. The energy form is more general and is suitable to serve as the starting point when formulating equations for both FEA and sensitivity analysis. We use the simple bar example in Section 4.2.3 to illustrate the finite element formulation.

4.2.1 DIFFERENTIAL EQUATION

A 1-D bar under uniformly distributed load f (force per length) is shown in Figure 4.3a. The bar has a uniform cross-sectional area A and length ℓ. The modulus of elasticity is E. We first formulate the differential equation that governs the structural response of the bar, and then we present the energy-based formulation for the same structure. Thereafter, we introduce the finite element equations of the problem derived from the energy formulation.

The governing differential equation of the bar being stretched by the distributed force f can be obtained from the free-body diagram in Figure 4.3b, in which an arbitrary small element is cut off with forces acting on it. Note that $F(x + \Delta x)$ and $F(x)$ are internal forces acting on the small element and $f\Delta x$ is the distributed load acting on the small element. The internal force $F(x + \Delta x)$ can be written in a Taylor series expansion as

$$F(x + \Delta x) = F(x) + F(x)_{,1}\Delta x + \text{higher order term} \tag{4.2}$$

in which the shorthand notation $F(x)_{,1} = \partial F(x)/\partial x$ is employed. The equilibrium equation of the small element can be formulated as

$$F(x + \Delta x) - F(x) + f\Delta x = 0 \tag{4.3}$$

Hence,

$$-F(x)_{,1}\Delta x + \text{higher order term} = f\Delta x \tag{4.4}$$

Dividing both sides of Eq. 4.4 by Δx and setting Δx to zero, we have

$$-F(x)_{,1} = f \tag{4.5}$$

Note that force $F(x)$ is stress σ times area A, and stress is the modulus of elasticity E times strain $\varepsilon = z_{,1}$, in which the shorthand notation $z_{,1} = \partial z/\partial x$ is employed again. Here, z is the axial displacement of the bar due to the distributed load f. Therefore, stress is $\sigma = Ez_{,1}$. Hence, Eq. 4.5 becomes

$$-\left(EAz_{,1}\right)_{,1} = f \tag{4.6}$$

Boundary conditions of the bar can be written as

$$z(0) = 0, \text{ and} \tag{4.6a}$$

$$EAz_{,1}(\ell) = 0 \tag{4.6b}$$

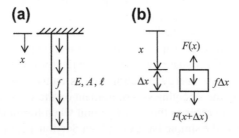

FIGURE 4.3

A one-dimensional bar example under static load. (a) Schematic view. (b) The forces acting on an arbitrary small element Δx.

Displacement at the top ($x = 0$) is fixed and there is no traction force applied at the bottom end of the bar ($x = \ell$); hence, stress at the end times area is equal to zero. Note that Eq. 4.6a is called the essential boundary condition, and Eq. 4.6b is the natural boundary condition.

If the distributed load f is assumed a constant, then Eq. 4.6 implies that the derivative of E, A, and $z_{,1}$ with respect to x must exist and be at least a constant because the distributed force f is a constant. Mathematically, we write $E \in C^1[0, \ell]$ and $A \in C^1[0, \ell]$, in which $C^1[0, \ell]$ stands for the first-order derivative continuous in domain $[0, \ell]$. Hence, the second derivative of the solution z must exist and be at least a constant (or continuous). Therefore, the solution space of the differential equation of Eq. 4.6 and boundary conditions in Eq. 4.6a can be defined as

$$\tilde{S} = \left\{ z \in C^2[0, \ell] \,\big|\, z(0) = 0 \right\} \tag{4.7}$$

in which we include only the essential boundary condition $z(0) = 0$, and $C^2[0, \ell]$ requires functions with the second derivative continuous in $[0, \ell]$. The governing equation in the differential form stated in Eqs 4.6, 4.6a, and 4.6b is often called the "strong form" due to the high smoothness requirement of the solution function z—in this case, C^2.

Solving the differential equation gives the solution to the structural problem as

$$z^{\text{Exact}}(x) = -\frac{f}{2EA}x^2 + \frac{f\ell}{EA}x \tag{4.8}$$

If we are interested in designing the bar to reduce the displacement at the bottom end $z(\ell)$, a displacement performance measure can be defined as

$$\psi = z(\ell) = \frac{f\ell^2}{2EA} \tag{4.9}$$

in which the displacement can be altered by changing material design variable E, sizing design variable A (cross-sectional area), or shape design variable ℓ (length of the bar).

4.2.2 ENERGY EQUATION

In practice, only a handful of structural problems can be formulated and solved analytically. In general, structural problems are solved by using the finite element method (FEM). The formulation of FEM usually starts from an energy equation (or variational equation), as will be illustrated next.

We derive the energy equation of the structural governing equation by multiplying both sides of Eq. 4.6 with an arbitrary virtual displacement $\bar{z}$ and then integrating over the structural domain $[0, \ell]$:

$$\int_0^\ell \left(-EAz_{,1} \right)_{,1} \bar{z}\, dx = \int_0^\ell f\, \bar{z}\, dx \tag{4.10}$$

Carrying out integration in part on Eq. 4.10 one time, we have

$$-\left(EAz_{,1}\bar{z} \right)\Big|_0^\ell + \int_0^\ell EAz_{,1}\bar{z}_{,1}\, dx = \int_0^\ell f\, \bar{z}\, dx \tag{4.11}$$

Both z and $\bar{z}$ belong to the space of kinematically admissible virtual displacement,

$$S = \{z \in H^1[0, \ell] | z(0) = 0\} \tag{4.12}$$

where $z(0) = 0$ is the essential (or kinematic) boundary condition and $H^1[0, \ell]$ is the first-order Sobolev space. Because the boundary terms in Eq. 4.7 can be eliminated by considering $\bar{z}(0) = 0$ (essential boundary condition) and applying the natural boundary condition $z_{,1}(\ell) = 0$, the following energy equation is obtained for the bar problem:

$$a(z, \bar{z}) = \int_0^\ell EAz_{,1}\bar{z}_{,1}dx = \int_0^\ell f \bar{z}dx = \ell(\bar{z}) \tag{4.13}$$

which holds for all $\bar{z} \in S$ (or $\forall \bar{z} \in S$). Note that $a(z, \bar{z})$ and $\ell(\bar{z})$ are called the energy bilinear form and load linear form, respectively, thus implying that $a(z, \bar{z})$ is linear in both z and $\bar{z}$ while $\ell(\bar{z})$ is linear in $\bar{z}$.

If we replace the virtual displacement $\bar{z}$ with the physical displacement z for the energy bilinear form in Eq. 4.13, then

$$a(z, z) = \int_0^\ell A(Ez_{,1})z_{,1}dx = \int_0^\ell A\sigma\varepsilon dx = 2U \tag{4.14}$$

U is the strain energy defined as

$$U = \int_\Omega u dV = \int_0^\ell \left(\frac{1}{2}\sigma\varepsilon\right)A dx = \frac{1}{2}a(z, z) \tag{4.15}$$

where u is the strain energy per volume V. If we do the same for the load linear form, then we have

$$\ell(z) = \int_0^\ell f z dx \tag{4.16}$$

which is the work done by the external force f. Therefore, Eq. 4.13 is nothing but the equilibrium of work and energy in virtual displacement $\bar{z}$, which is formally called the principle of virtual work. The principle of virtual work states that, at equilibrium, the strain energy change due to a small virtual displacement is equal to the work done by the forces in moving through the virtual displacement. More specifically, the virtual displacement $\bar{z}$ is a small imaginary change in configuration that is also an admissible displacement and satisfies the essential boundary condition.

From structural analysis perspective, the principle of virtual work states that the solution z is a function in the trial function space, which satisfies Eq. 4.13 for all virtual displacement $\bar{z}$ in the test function space. Note that, in general, the trial function space and test function space are identical for problems that involves only homogeneous essential boundary conditions; both are the kinematically admissible virtual displacement space S defined in Eq. 4.12. Homogeneous boundary conditions are either zero displacements or zero slope at the boundary. Note that the solution z is called the trial function, and the virtual displacement $\bar{z}$ is called the test function.

Note that solutions z in $H^1[0, \ell]$ imply that the virtual energy terms in Eq. 4.13—both $a(z, \bar{z})$ and $\ell(\bar{z})$—are finite; that is, $a(z, \bar{z}) < \infty$ and $\ell(\bar{z}) < \infty$. It is obvious that as long as $z_{,1}$ and $\bar{z}_{,1}$ exist and are integrable, not even being continuous, the integrals of $a(z, \bar{z})$ and $\ell(\bar{z})$ are finite. The requirements for the smoothness of the solution functions are significantly relaxed from that of space $\tilde{S}$ defined in Eq. 4.7. For structural problems, work and energy are finite for a given external load that is finite, as seen in practice. The energy equation of Eq. 4.13 is also called the generalized formulation or "weak form" because the requirements for the solution functions are not as strong as that of differential equations. Without going through mathematic arguments, we state that the solution z of Eq. 4.13 exists and is unique. Rigourous arguments and proofs can be found in (Haug et al. 1986).

4.2.3 FINITE ELEMENT FORMULATION

Consider discretizing the bar using two truss elements, each with length $\ell/2$, as shown in Figure 4.4. In this example, linear interpolation is used to describe the displacement field between nodal points, as follows:

$$
\mathbf{z}(x) = \begin{cases} \left[1 - \dfrac{2x}{\ell} \quad \dfrac{2x}{\ell} \right] \cdot \begin{bmatrix} Z_1 \\ Z_2 \end{bmatrix}, & 0 \le x \le \ell/2 \\[3ex] \left[1 - \dfrac{2(x - \ell/2)}{\ell} \quad \dfrac{2(x - \ell/2)}{\ell} \right] \cdot \begin{bmatrix} Z_2 \\ Z_3 \end{bmatrix}, & \ell/2 < x \le \ell \end{cases} \tag{4.17}
$$

where the row vectors contain the shape functions of individual elements, and Z_1, Z_2, and Z_3 represent nodal displacements at nodes 1, 2, and 3, respectively. Note that the virtual displacement $\bar{z}$ can be interpolated in the same way as the actual displacement z shown in Eq. 4.17. For clarification, lower-case z represents displacement function, and capital Z represents the nodal displacement obtained from FEA.

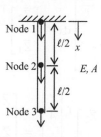

FIGURE 4.4

Finite element model of the 1-D bar structure. Two truss finite elements are used to discretize the structural domain.

Discretizing the left- and right-hand sides of the energy equation (Eq. 4.13) using shape functions gives

$$\int_0^{\ell} EA z_{,1} \bar{z}_{,1} dx = \int_0^{\ell/2} EA z_{,1} \bar{z}_{,1} dx + \int_{\ell/2}^{\ell} EA z_{,1} \bar{z}_{,1} dx$$

$$\stackrel{\text{discretize}}{=} \begin{bmatrix} \bar{Z}_1 \\ \bar{Z}_2 \end{bmatrix}^T \int_0^{\ell/2} EA \begin{bmatrix} -\dfrac{2}{\ell} \\ \dfrac{2}{\ell} \end{bmatrix} \begin{bmatrix} -\dfrac{2}{\ell} & \dfrac{2}{\ell} \end{bmatrix} dx \begin{bmatrix} Z_1 \\ Z_2 \end{bmatrix} + \begin{bmatrix} \bar{Z}_2 \\ \bar{Z}_3 \end{bmatrix}^T \int_0^{\ell/2} EA \begin{bmatrix} -\dfrac{2}{\ell} \\ \dfrac{2}{\ell} \end{bmatrix} \begin{bmatrix} -\dfrac{2}{\ell} & \dfrac{2}{\ell} \end{bmatrix} dx \begin{bmatrix} Z_2 \\ Z_3 \end{bmatrix}$$

$$= \begin{bmatrix} \bar{Z}_1 \\ \bar{Z}_2 \end{bmatrix}^T \frac{2EA}{\ell} \begin{bmatrix} 1 & -1 \\ -1 & 1 \end{bmatrix} \begin{bmatrix} Z_1 \\ Z_2 \end{bmatrix} + \begin{bmatrix} \bar{Z}_2 \\ \bar{Z}_3 \end{bmatrix}^T \frac{2EA}{\ell} \begin{bmatrix} 1 & -1 \\ -1 & 1 \end{bmatrix} \begin{bmatrix} Z_2 \\ Z_3 \end{bmatrix}$$

$$= \begin{bmatrix} \bar{Z}_1 \\ \bar{Z}_2 \\ \bar{Z}_3 \end{bmatrix}^T \cdot \frac{2EA}{\ell} \begin{bmatrix} 1 & -1 & 0 \\ -1 & 2 & -1 \\ 0 & -1 & 1 \end{bmatrix} \cdot \begin{bmatrix} Z_1 \\ Z_2 \\ Z_3 \end{bmatrix} = \bar{Z}_g K_g Z_g$$

(4.18)

and

$$\int_0^{\ell} f \bar{z} dx = \int_0^{\ell/2} f \bar{z} dx + \int_{\ell/2}^{\ell} f \bar{z} dx$$

$$\stackrel{\text{discretize}}{=} \begin{bmatrix} \bar{Z}_1 \\ \bar{Z}_2 \end{bmatrix}^T \int_0^{\ell/2} f \begin{bmatrix} 1 - \dfrac{2x}{\ell} \\ \dfrac{2x}{\ell} \end{bmatrix} dx + \begin{bmatrix} \bar{Z}_2 \\ \bar{Z}_3 \end{bmatrix}^T \int_{\ell/2}^{\ell} f \begin{bmatrix} 1 - \dfrac{2(x - \ell/2)}{\ell} \\ \dfrac{2(x - \ell/2)}{\ell} \end{bmatrix} dx \qquad (4.19)$$

$$= \begin{bmatrix} \bar{Z}_1 \\ \bar{Z}_2 \end{bmatrix}^T \begin{bmatrix} \dfrac{\ell}{4} \\ \dfrac{\ell}{4} \end{bmatrix} dx + \begin{bmatrix} \bar{Z}_2 \\ \bar{Z}_3 \end{bmatrix}^T \begin{bmatrix} \dfrac{\ell}{4} \\ \dfrac{\ell}{4} \end{bmatrix} = \begin{bmatrix} \bar{Z}_1 \\ \bar{Z}_2 \\ \bar{Z}_3 \end{bmatrix}^T \cdot \begin{bmatrix} f\ell/4 \\ f\ell/2 \\ f\ell/4 \end{bmatrix} = \bar{Z}_g F_g$$

where K_g is the generalized global stiffness matrix. Z_g, $\bar{Z}_g$, and F_g are the global displacement, virtual displacement, and force vectors, respectively. Note that the distributed load f has been converted into point loads acting upon the nodal points due to discretization.

Because the left-hand sides of Eqs 4.18 and 4.19 are equal to each other, the global finite element matrix equation can be obtained by eliminating the arbitrary virtual displacement $\overline{\mathbf{Z}}_g$, as follows:

$$\mathbf{K}_g \mathbf{Z}_g = \frac{2EA}{\ell} \begin{bmatrix} 1 & -1 & 0 \\ -1 & 2 & -1 \\ 0 & -1 & 1 \end{bmatrix} \cdot \begin{bmatrix} Z_1 \\ Z_2 \\ Z_3 \end{bmatrix} = \begin{bmatrix} f\ell/4 \\ f\ell/2 \\ f\ell/4 \end{bmatrix} = \mathbf{F}_g \tag{4.20}$$

This equation cannot be solved due to the singularity of $\mathbf{K}_g$. By applying the essential boundary condition $z(0) = 0$, we can remove Z_1 from Eq. 4.20, giving

$$\mathbf{KZ} = \frac{2EA}{\ell} \begin{bmatrix} 2 & -1 \\ -1 & 1 \end{bmatrix} \cdot \begin{bmatrix} Z_2 \\ Z_3 \end{bmatrix} = \begin{bmatrix} f\ell/2 \\ f\ell/4 \end{bmatrix} = \mathbf{F} \tag{4.21}$$

where

$$\mathbf{K} = \frac{2EA}{\ell} \begin{bmatrix} 2 & -1 \\ -1 & 1 \end{bmatrix} \tag{4.22}$$

This is called the reduced global stiffness matrix, which is nonsingular (or more precisely, positive definite). Solving the reduced matrix equation (Eq. 4.22) by inversing the stiffness matrix gives the nodal displacement solution:

$$\mathbf{Z} = \mathbf{K}^{-1}\mathbf{F} = \frac{\ell}{2EA} \begin{bmatrix} 1 & 1 \\ 1 & 2 \end{bmatrix} \begin{bmatrix} f\ell/2 \\ f\ell/4 \end{bmatrix} = \begin{bmatrix} \dfrac{3f\ell^2}{8EA} \\ \dfrac{f\ell^2}{2EA} \end{bmatrix} \tag{4.23}$$

Based on the nodal point solutions given above, the displacement at an arbitrary location in the domain $[0, \ell]$ can be interpolated using the shape functions defined in Eq. 4.17, as follows:

$$\mathbf{z}(x) = \begin{cases} \begin{bmatrix} 1 - \dfrac{2x}{\ell} & \dfrac{2x}{\ell} \end{bmatrix} \cdot \begin{bmatrix} 0 \\ \dfrac{3f\ell^2}{8EA} \end{bmatrix} = \dfrac{2x}{\ell} \dfrac{3f\ell^2}{8EA}, & 0 \leq x \leq \ell/2 \\[3em] \begin{bmatrix} 1 - \dfrac{2(x-\ell/2)}{\ell} & \dfrac{2(x-\ell/2)}{\ell} \end{bmatrix} \cdot \begin{bmatrix} \dfrac{3f\ell^2}{8EA} \\ \dfrac{f\ell^2}{2EA} \end{bmatrix} = \left(1 - \dfrac{2(x-\ell/2)}{\ell}\right) \dfrac{3f\ell^2}{8EA} + \dfrac{2(x-\ell/2)}{\ell} \dfrac{f\ell^2}{2EA}, & \ell/2 < x \leq \ell \end{cases}$$

$$\tag{4.24}$$

Note that in computer implementation, the integration of Eqs 4.18 and 4.19 is carried out numerically (e.g., using Gauss integration), and the stiffness matrix $\mathbf{K}$ is decomposed or factorized numerically (e.g., using Lower Upper (LU) decomposition; Atkinson 1989) for solving the displacement vector $\mathbf{Z}$, instead of fully inversed as stated in Eq. 4.23.

It is of note that the finite element displacement solution $\mathbf{z}(x)$ is piecewise linear (linear in individual elements); however, the exact solution obtained by solving the differential governing equation is a quadratic function of x (Eq. 4.8). They are not identical. This is because linear shape functions are used for interpolation finite element solution in Eq. 4.17, whereas the analytical solution to the problem is a quadratic function of x. However, the finite element solution matches the exact solution only at nodal points; that is, $Z_2 = z(\ell/2)$ and $Z_3 = z(\ell)$.

The stress in the bar can be calculated using the exact solution (Eq. 4.8) as

$$\sigma^{\text{Exact}} = E\left[z^{\text{Exact}}(x)\right]_{,1} = -\frac{f}{A}x + \frac{f\ell}{A} = \frac{f}{A}(\ell - x) \tag{4.25}$$

which is always greater or equal to zero, implying that the bar is stretched due to the distributed load f.

Using the finite element solution of Eq. 4.24, stress in the bar is calculated as

$$\sigma(x) = Ez_{,1}(x) = \begin{cases} E\left[-\dfrac{2}{\ell} \quad \dfrac{2}{\ell}\right] \cdot \begin{bmatrix} 0 \\ \dfrac{3f\ell^2}{8EA} \end{bmatrix} = \dfrac{3f\ell}{4A}, & 0 \le x \le \ell/2 \\[4ex] E\left[-\dfrac{2}{\ell} \quad \dfrac{2}{\ell}\right] \cdot \begin{bmatrix} \dfrac{3f\ell^2}{8EA} \\ \dfrac{f\ell^2}{2EA} \end{bmatrix} = \dfrac{f\ell}{4A}, & \ell/2 < x \le \ell \end{cases} \tag{4.26}$$

which is a piecewise constant function. It is obvious that the stress solved by FEA is not continuous across finite elements. In this bar example, stress at node 2 is not continuous, which is referred to as stress jump, whereas the analytical stress is a linear function of x, as shown in Eq. 4.25. Stress jump is physically impossible. This piecewise constant stress is due to the fact that we use linear shape functions to interpolate displacement results. An important point to note is that, as the finite element mesh is refined (element size reduced), a better solution revealing less stress jump is expected.

4.3 SENSITIVITY ANALYSIS METHODS

In general, structural performance measures are categorized as global and local measures. Global measures, such as volume or weight, vibration frequency, compliance, and buckling load, are measured for the entire structure. On the other hand, measures such as stresses or displacements are defined at specific points or small areas (e.g., average stress in a finite element). Design variables in general include shape and nonshape. Nonshape design variables do not alter the geometric shape of the structure, such as sizing and material. Design variables that alter the geometric shape of the structure include domain shape and configuration of frame structures. In addition, topology optimization considers element density as design variables in general (further discussed in Section 4.6).

Four approaches can be employed for DSA: the analytical derivative, overall finite difference, discrete approach, and continuum approach. As stated at the beginning of Section 4.2, sensitivity analysis calculates the gradient or sensitivity of the performance measure(s) ψ with respect to design variable b. For simplicity, we assume a single performance measure and a single design variable for the time being.

The analytical derivative method (or analytical method) calculates the sensitivity coefficients by taking the derivative of the measure with respect to the design variable analytically. This method requires explicit expression of the measure in terms of the design variable. In practice, only a handful of measures can be expressed in the design variable explicitly. Therefore, the analytical method is very limited in practical applications.

Using the overall finite difference method, sensitivity coefficients are obtained by rerunning structural analysis at a perturbed design and calculating the difference of the measure at the current and perturbed designs. The sensitivity coefficient is then approximated by dividing the difference in performance measure values by the design perturbation.

In the discrete approach, the sensitivity formulation is obtained by taking derivatives of the finite element matrix equations. If the derivatives of the finite element equations are obtained analytically, it is called the discrete-analytical method, in which discrete refers to its formulation, and analytical refers to the nature of the derivatives. If the derivatives are obtained using finite differences, then the method is called a discrete–discrete or semianalytical method. The semianalytical method is probably the most employed approach other than the overall finite difference in practice. Although both the overall finite difference and semianalytical methods are very general, they require perturbing the design variable(s) by a small amount. An adequate design perturbation depends on the characteristics of the function and is sometimes difficult to acquire.

Another approach for sensitivity analysis is called the continuum approach, in which the sensitivity formulation is obtained by taking derivatives of the energy equations in the integral or continuum form, instead of the finite element equations that are discretized—hence the name of the approach. The derivatives of the energy equations are then discretized in finite element formulation and solved by FEA; such a method is called continuum-discrete. In some rare cases, the solutions of the structural problems are obtained analytically without an FEA. These analytical solutions can then be plugged into the continuum sensitivity expression without discretization; this is called the continuum-analytical method. However, because the analytical solution of structural problems is rare, the applications of the continuum-analytical method are limited. One important feature of the continuum approach is that neither the continuum-discrete or continuum-analytical method requires a design perturbation; they are superior to the overall finite difference and semianalytical methods. Also, in many structural design problems, the continuum-discrete method is equivalent to the discrete-analytical method.

The sensitivity analysis methods mentioned above are listed in Figure 4.5, in which we assume a linear elastic problem under static load. Problems other than static may see a slightly different classification than that in Figure 4.5.

The sensitivity analysis methods listed above will be further discussed next. Note that, in practice, the continuum-discrete approach is considered the best because it is general, accurate, and efficient. More importantly, this method can be implemented outside commercial FEA codes that are employed for analysis. In the following, we discuss individual methods and point out their pros and cons. We use the simple bar example discussed in Section 4.2 to illustrate the detailed derivations.

4.3.1 ANALYTICAL DERIVATIVE METHOD

The analytical derivative method (or analytical method) is straightforward. For the simple bar example, the displacement is obtained by solving the differential equation in Section 4.2.1 as

$$z(x) = -\frac{f}{2EA}x^2 + \frac{f\ell}{EA}x \tag{4.27}$$

If we are interested in designing the bar with a reduced displacement at its bottom end, a displacement performance measure can be defined as

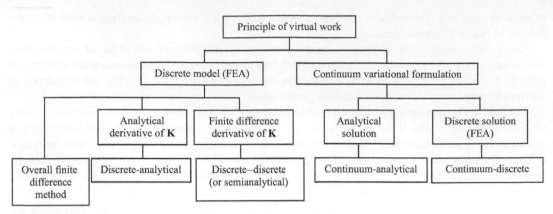

FIGURE 4.5

Major approaches for DSA.

$$\psi = z(\ell) = \frac{f\ell^2}{2EA} \tag{4.28}$$

If the cross-section area A is the design variable, the sensitivity can be calculated by taking derivative of the measure with respect to the design variable analytically:

$$\frac{d\psi}{dA} = -\frac{f\ell^2}{2EA^2} \tag{4.29}$$

The negative sensitivity coefficient implies that increasing the cross-sectional area A decreases the displacement at the bottom end of the bar.

Next, we consider the stress in the bar as the performance measure, which can be calculated, as in Eq. 4.25, as

$$\sigma^{\text{Exact}} = E\left(z^{\text{Exact}}\right)_{,1}(x) = -\frac{f}{A}x + \frac{f\ell}{A} \tag{4.30}$$

We define a stress measure at the root of the bar ($x = 0$):

$$\psi = \frac{f\ell}{A} \tag{4.31}$$

The sensitivity of stress with the area design variable is

$$\frac{d\psi}{dA} = -\frac{f\ell}{A^2} \tag{4.32}$$

This equation shows that increasing the cross-sectional area A decreases the stress at the root of the bar. Another frequently used stress measure is the average stress, which is defined as

$$\psi = \sigma_{\text{ave}} = \frac{\int_0^\ell \sigma dx}{\ell} = \frac{f\ell}{2A} \tag{4.33}$$

The sensitivity of the average stress can be calculated as

$$\frac{d\psi}{dA} = -\frac{f\ell}{2A^2} \tag{4.34}$$

The discussion above assumes analytical expression of measure ψ in design variable b. In general, in rare cases such an expression exists.

4.3.2 OVERALL FINITE DIFFERENCE

The easiest way to compute the sensitivity information of a performance measure without analytical solutions is by using the overall finite difference method. The overall finite difference method computes the sensitivity by evaluating performance measures at different values of design variables, or perturbed designs. Although the given design problem may have many design variables, a single design variable is perturbed at a time. We assume one design variable in the following explanation. If b is the current design, then the analysis results provide the value of performance measure $\psi(b)$—for example, using FEA.

In addition, if the design is perturbed to $b + \Delta b$, where Δb represents a small change in the design, then the sensitivity of $\psi(b)$ can be approximated as

$$\frac{\partial \psi}{\partial b} \approx \frac{\psi(b + \Delta b) - \psi(b)}{\Delta b} \tag{4.35}$$

This is called the forward difference method because the design is perturbed by $+\Delta b$. If $-\Delta b$ is substituted in Eq. 4.35 for $+\Delta b$, then the equation is defined as the backward difference method. Additionally, if the design is perturbed in both directions, such that the design sensitivity is approximated by

$$\frac{\partial \psi}{\partial b} \approx \frac{\psi(b + \Delta b) - \psi(b - \Delta b)}{2\Delta b} \tag{4.36}$$

then the equation is defined as the central difference method. The advantage of the finite difference method is obvious. If structural analysis can be performed and the performance measure can be obtained as a result of structural analysis, then the expressions in Eqs 4.35 and 4.36 become virtually independent of the problem types considered. Consequently, this method is very general and has been widely employed in engineering designs beyond structural problems.

However, sensitivity computation costs become expensive. If n represents the number of design variables, then $n + 1$ number of analyses (including the current design b) have to be carried out for either forward or backward differences, and $2n$ analyses are required for the central differences. This method, although easy to implement, may not be feasible for large-scale problems with many design variables due to extensive computational efforts.

Another major disadvantage of the finite difference method is the degree of accuracy of its sensitivity results. In Eqs 4.35 and 4.36, accurate results can be expected when Δb approaches zero. Figure 4.6 shows some sensitivity results using the finite difference method. The tangential slope of the curve at b_0—that is, $\frac{\partial \psi}{\partial b}\Big|_{b=b_0}$ —is the exact sensitivity value, which is also the true slope of the tangent line at b_0. Depending on the perturbation size, the sensitivity results are quite different; for example,

$$\left(\frac{\Delta \psi}{\Delta b}\right)_1 = \frac{\psi(b_1) - \psi(b_0)}{b_1 - b_0}, \quad \left(\frac{\Delta \psi}{\Delta b}\right)_2 = \frac{\psi(b_2) - \psi(b_0)}{b_2 - b_0}, \quad \text{and} \quad \left(\frac{\Delta \psi}{\Delta b}\right)_3 = \frac{\psi(b_3) - \psi(b_0)}{b_3 - b_0}$$

For a mildly nonlinear performance measure, a relatively large perturbation can still provide a reasonable estimation of sensitivity results. However, for a highly nonlinear performance measure, even a small perturbation yields inaccurate results. Thus, the determination of perturbation size Δb greatly affects the accuracy of the sensitivity coefficients.

On the other hand, although it may be necessary to choose a very small perturbation, numerical noise becomes dominant for a perturbation size that is too small. With a too small perturbation, no reliable difference can be found in the numerical analysis results. Because the behavior of a performance measure in design space is unknown a priori, it is difficult to determine design perturbation sizes that work for all problems.

Usually, a convergence study in design perturbation size Δb is carried out first for a given problem, in which a relatively large perturbation size is assumed at the beginning. By reducing the perturbation size, if the finite difference values shown in Eqs 4.35 or 4.36 approach or converge to a specific value, then the true sensitivity, or the true slope of the tangent line at current design b_0, is accurately approximated. Theoretically, when the perturbation size Δb approaches zero, the finite difference value equals the true slope; that is,

$$\frac{\partial \psi}{\partial b} = \lim_{\Delta b \to 0} \frac{\psi(b + \Delta b) - \psi(b)}{\Delta b} \tag{4.37}$$

However, as discussed, the perturbation size cannot be so small as to induce numerical noise.

EXAMPLE 4.1

For a sinusoidal function $f(b) = \sin b$, calculate its sensitivity at $b = 0$. Use the forward finite difference method to approximate the sensitivity for $\Delta b = 1, 0.5, 0.1, 0.01$.

Solutions

The derivative of $f(b) = \sin b$ with respect to b is

$$\frac{df}{db} = \cos b$$

which gives the analytical sensitivity coefficient at $b_0 = 0$, $df/db = 1$. Now, we use Eq. 4.35 to approximate the sensitivity of the function $f(b)$ for $\Delta b = 1, 0.5, 0.1, 0.01$.

$$\left(\frac{\Delta f}{\Delta b}\right)_{\Delta b = 1} = \frac{\sin(1) - \sin(0)}{1} = \frac{0.8415 - 0}{1} = 0.8415$$

$$\left(\frac{\Delta f}{\Delta b}\right)_{\Delta b = 0.5} = \frac{\sin(0.5) - \sin(0)}{0.5} = \frac{0.4794 - 0}{0.5} = 0.9589$$

$$\left(\frac{\Delta f}{\Delta b}\right)_{\Delta b = 0.1} = \frac{\sin(0.1) - \sin(0)}{0.1} = \frac{0.09983 - 0}{0.1} = 0.9983$$

$$\left(\frac{\Delta f}{\Delta b}\right)_{\Delta b = 0.01} = \frac{\sin(0.01) - \sin(0)}{0.01} = \frac{0.0099998 - 0}{0.01} = 0.99998$$

As shown above, reducing Δb leads to more accurate approximations of the sensitivity of $f(b)$. In fact, for the sine function, when Δb is 0.1, the error in the approximation using the forward finite difference method is less than 1% because the function is not highly nonlinear.

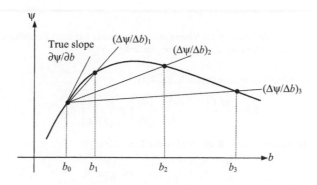

FIGURE 4.6

Influence of step size in the forward finite difference method.

4.3.3 DISCRETE APPROACH

The sensitivity formulation of the discrete approach is obtained by taking derivatives of the discretized finite element equations. For a static problem, as discussed in the simple bar example in Section 4.2, we have

$$KZ = F \tag{4.38}$$

where K is the stiffness matrix, F is the vector of the external load, and Z is the nodal point displacement of the structure obtained by solving the matrix equations. Taking the derivatives of the finite element equations with respect to the design variable b yields

$$K\frac{\partial Z}{\partial b} + \frac{\partial K}{\partial b}Z = \frac{\partial F}{\partial b} \tag{4.39}$$

Moving $\frac{\partial K}{\partial b}Z$ to the right-hand side, we have

$$K\frac{\partial Z}{\partial b} = \frac{\partial F}{\partial b} - \frac{\partial K}{\partial b}Z \tag{4.40}$$

in which the terms on the right-hand side are a vector of so-called fictitious load. The derivative of the nodal displacements can be solved by treating the right-hand side of Eq. 4.40 as an additional load case:

$$\frac{\partial Z}{\partial b} = K^{-1}\left(\frac{\partial F}{\partial b} - \frac{\partial K}{\partial b}Z\right) \tag{4.41}$$

in which the decomposed stiffness matrix used for solving Z in Eq. 4.38 can be used again to solve for $\partial Z/\partial b$, which makes the approach more efficient than the overall finite difference method.

EXAMPLE 4.2

Calculate the sensitivity of the displacement at the bottom end of the simple bar example using the discrete-analytical approach, assuming the modulus of elasticity E as the design variable.

Solutions

Recall the finite element equations of the bar example in Eq. 4.21:

$$\mathbf{KZ} = \frac{2EA}{\ell} \begin{bmatrix} 2 & -1 \\ -1 & 1 \end{bmatrix} \cdot \begin{bmatrix} Z_2 \\ Z_3 \end{bmatrix} = \begin{bmatrix} f\ell/2 \\ f\ell/4 \end{bmatrix} = \mathbf{F} \tag{4.42}$$

Taking derivatives of the stiffness matrix $\mathbf{K}$ and load vector $\mathbf{F}$, we have

$$\frac{\partial \mathbf{K}}{\partial E} = \frac{2A}{\ell} \begin{bmatrix} 2 & -1 \\ -1 & 1 \end{bmatrix} \text{ and } \frac{\partial \mathbf{F}}{\partial E} = \mathbf{0} \tag{4.43}$$

Therefore, from Eq. 4.41, we have

$$\frac{\partial \mathbf{Z}}{\partial E} = \begin{bmatrix} \dfrac{\partial Z_2}{\partial E} \\[2mm] \dfrac{\partial Z_3}{\partial E} \end{bmatrix} = \mathbf{K}^{-1}\left(\frac{\partial \mathbf{F}}{\partial E} - \frac{\partial \mathbf{K}}{\partial E}\mathbf{Z} \right) = \left(\frac{\ell}{2EA} \begin{bmatrix} 1 & 1 \\ 1 & 2 \end{bmatrix} \right)\left(\mathbf{0} - \frac{2A}{\ell}\begin{bmatrix} 2 & -1 \\ -1 & 1 \end{bmatrix} \begin{bmatrix} \dfrac{3f\ell^2}{8EA} \\[2mm] \dfrac{f\ell^2}{2EA} \end{bmatrix} \right) = \begin{bmatrix} -\dfrac{3f\ell^2}{8E^2 A} \\[2mm] -\dfrac{f\ell^2}{2E^2 A} \end{bmatrix} \tag{4.44}$$

Note that the second term in the result vector is $\frac{\partial Z_3}{\partial E} = \frac{\partial z(\ell)}{\partial E} = -\frac{f\ell^2}{2E^2 A}$, which can be also obtained by directly taking the derivative of $z(\ell) = \frac{f\ell^2}{2EA}$ with respect to E.

Equation (4.41) represents a direct differentiation method of the discrete-analytical approach, in which the gradient is obtained analytically by taking the derivative of the finite element equations that are discretized.

For this simple example, the stiffness matrix and load vector are expressed in terms of the material design variable E explicitly. As a result, $\partial \mathbf{K}/\partial E$ and $\partial \mathbf{F}/\partial E$ can be obtained analytically. However, in general, such explicit expressions are not available. They are either too complex to formulate or entirely impossible. Especially when we use commercial FEA software, the stiffness matrix and force vector can only be retrieved as numerical data. Differentiating the matrix and vector in numerical data is out of the question. So, how do we calculate the gradients when we use commercial FEA software for structural analysis? How do we take derivatives of the stiffness matrix and force vector numerically? One possibility is to use finite differences, which can be stated mathematically as

$$\frac{\partial \mathbf{Z}}{\partial b} \approx \mathbf{K}^{-1}\left(\frac{\Delta \mathbf{F}}{\Delta b} - \frac{\Delta \mathbf{K}}{\Delta b}\mathbf{Z} \right) \tag{4.45}$$

in which

$$\frac{\Delta \mathbf{K}}{\Delta b} = \frac{\mathbf{K}(b + \Delta b) - \mathbf{K}(b)}{\Delta b} \tag{4.46}$$

and

$$\frac{\Delta \mathbf{F}}{\Delta b} = \frac{\mathbf{F}(b + \Delta b) - \mathbf{F}(b)}{\Delta b} \tag{4.47}$$

Note that in Eq. 4.46, $\mathbf{K}(b + \Delta b)$ can be obtained by perturbing the design variable b (e.g., the modulus of elasticity E) by a small amount Δb, and then asking the FEA code to generate the stiffness matrix at the perturbed design $b + \Delta b$. $\mathbf{F}(b + \Delta b)$ can be obtained in a similar way, although in Example 4.2, we know $\partial \mathbf{F}/\partial E = 0$; therefore, $\mathbf{F}(E + \Delta E)$ is not needed in this case.

Equations (4.45) to (4.47) represent the so-called discrete–discrete or semianalytical approach because the formulations are based on the discretized formulation in FEA, and the derivatives of the matrix and vector are obtained numerically using finite differences. Although this method is general, determining an adequate design perturbation Δb for accurate derivatives is not straightforward, as discussed in Section 4.3.2. Moreover, retrieving stiffness matrices and load vectors for the current and perturbed designs from commercial FEA codes and approximating their derivatives are nontrivial and sometimes inefficient for large-scale problems in implementation.

4.3.3.1 Direct differentiation method

Recall that Eq. 4.41 represents the direct differentiation method of the discrete-analytical approach. Equation (4.41) is good for displacement sensitivity. For structural design, performance measures are more than just displacements. In general, a performance measure ψ depends on the design explicitly and implicitly. That is, the performance measure ψ is in general a function of the design variable b and displacement (also called the state variable in the literature) $\mathbf{z}(b)$, as follows:

$$\psi = \psi(\mathbf{z}(b), b) \tag{4.48}$$

In general $\mathbf{z} = [z_1, z_2, z_3]^{\mathrm{T}}$, and $\mathbf{z} = z_1$ for one-dimensional problems. The sensitivity of function ψ can thus be expressed using the chain rule. Example 4.3 illustrates such a chain rule for differentiation.

EXAMPLE 4.3

A performance measure is defined as

$$\psi = \psi(z(b), b) = bz(b)^2 \tag{4.49}$$

Calculate the sensitivity of ψ with respect to the design variable b using the chain rule.

Solutions

Taking the derivative of ψ with respect to b, we have

$$\frac{d\psi}{db} = \frac{\partial \psi}{\partial b} + \frac{\partial \psi}{\partial z}\frac{\partial z}{\partial b} = z^2 + 2bz\frac{\partial z}{\partial b} \tag{4.50}$$

Note that the first term on the right-hand side of Eq. 4.50 shows the explicit dependence of ψ on b, which is the partial derivative of ψ with respect to b; that is, $\frac{\partial \psi}{\partial b} = z^2$. The second term consists of $\frac{\partial \psi}{\partial z} = \frac{\partial(bz^2)}{\partial z} = 2bz$, and $\frac{\partial z}{\partial b}$. Note that the second term quantifies the implicit dependence of ψ on b through $\frac{\partial z}{\partial b}$.

If $z = b^2$, then the performance measure becomes $\psi = bz(b)^2 = b^5$. Using Eq. 4.50, the sensitivity is

$$\frac{d\psi}{db} = z^2 + 2bz\frac{\partial z}{\partial b} = b^4 + 2b(b^2)(2b) = 5b^4 \tag{4.51}$$

The same result can be obtained by taking the derivative directly for $\psi = b^5$. In a structural problem, the derivative of displacement $\mathbf{z}$ on the design variable b cannot be obtained by taking the derivative explicitly because an expression such as $z = b^2$ does not exist.

With the understanding of the chain rule shown in Example 4.3, we take the derivative of the performance measure stated in Eq. 4.48 with respect to b as

$$\frac{d\psi}{db} = \frac{\partial\psi}{\partial b} + \frac{\partial\psi}{\partial z}\frac{\partial z}{\partial b} \tag{4.52a}$$

in which z is a continuous function of location x and design b, or

$$\frac{d\psi}{db} = \frac{\partial\psi}{\partial b} + \frac{\partial\psi}{\partial \mathbf{Z}}\frac{\partial \mathbf{Z}}{\partial b} \tag{4.52b}$$

Here, $\mathbf{Z}$ represents the nodal point displacements obtained from FEA. Note that Eq. 4.52b is employed more often. The first term $\frac{\partial\psi}{\partial b}$ represents the explicit dependence of ψ on design variable b, and the second term shows the implicit dependence of ψ on design variable b through $\frac{\partial\mathbf{Z}}{\partial b}$. In structural problems, the calculation of $\frac{\partial\psi}{\partial b}$ is straightforward because the explicit dependence of ψ on design variable b is usually available. The second term in Eq. 4.52 consists of $\frac{\partial\psi}{\partial\mathbf{Z}}$ and $\frac{\partial\mathbf{Z}}{\partial b}$. $\frac{\partial\psi}{\partial\mathbf{Z}}$ can be calculated easily because measure ψ is usually explicitly defined in terms of displacement $\mathbf{Z}$. The term that shows the actual implicit dependence of the performance measure ψ on design variable b is $\frac{\partial\mathbf{Z}}{\partial b}$, which can be calculated in two different ways; both involve FEA for additional load cases using the decomposed stiffness matrix $\mathbf{K}$. One approach is the direct differentiation method, in which Eqs 4.41 or 4.45 is used to solve $\frac{\partial\mathbf{Z}}{\partial b}$.

Once $\partial\mathbf{Z}/\partial b$ is calculated, the sensitivity of a performance measure ψ, defined as a function of (discretized) state variable $\mathbf{Z}$, can be obtained using the direct differentiation method as

$$\frac{d\psi}{db} = \frac{\partial\psi}{\partial b} + \frac{\partial\psi}{\partial\mathbf{Z}}\mathbf{K}^{-1}\left(\frac{\partial\mathbf{F}}{\partial b} - \frac{\partial\mathbf{K}}{\partial b}\mathbf{Z}\right) \tag{4.53}$$

EXAMPLE 4.4

We define the stress of the second element (with ends of node 2 and node 3) of the simple bar shown in Figure 4.4 as a performance measure. As shown in Eq. 4.26, the element stress is defined in terms of nodal point displacement $\mathbf{Z} = [Z_2, Z_3]^T$ as

$$\psi = \sigma = Ez_{,1} = E\left[1 - \frac{2(x-\ell/2)}{\ell} \quad \frac{2(x-\ell/2)}{\ell}\right]_{,1}\cdot\begin{bmatrix}Z_2\\Z_3\end{bmatrix} = E\begin{bmatrix}-\frac{2}{\ell} & \frac{2}{\ell}\end{bmatrix}\cdot\begin{bmatrix}Z_2\\Z_3\end{bmatrix} \tag{4.54}$$

in which the dependence of ψ in design E is both explicit and implicit (through $\mathbf{Z}$). Assuming modulus E and area A as the design variables, calculate the sensitivity coefficients of the stress measure ψ using the direct differentiation method of the discrete-analytical approach.

EXAMPLE 4.4—cont'd

Solutions

For modulus design variable E, the sensitivity of the stress measure ψ can be calculated using Eq. 4.53, as follows:

$$\frac{d\psi}{dE} = \frac{\partial \psi}{\partial E} + \frac{\partial \psi}{\partial \mathbf{Z}}\mathbf{K}^{-1}\left(\frac{\partial \mathbf{F}}{\partial E} - \frac{\partial \mathbf{K}}{\partial E}\mathbf{Z}\right) = \begin{bmatrix} -\dfrac{2}{\ell} & \dfrac{2}{\ell} \end{bmatrix} \cdot \begin{bmatrix} Z_2 \\ Z_3 \end{bmatrix} + \begin{bmatrix} \dfrac{\partial \psi}{\partial Z_2} & \dfrac{\partial \psi}{\partial Z_3} \end{bmatrix}\mathbf{K}^{-1}\left(\frac{\partial \mathbf{F}}{\partial E} - \frac{\partial \mathbf{K}}{\partial E}\mathbf{Z}\right)$$

$$= \begin{bmatrix} -\dfrac{2}{\ell} & \dfrac{2}{\ell} \end{bmatrix} \cdot \begin{bmatrix} \dfrac{3f\ell^2}{8EA} \\ \dfrac{f\ell^2}{2EA} \end{bmatrix} + \begin{bmatrix} -\dfrac{2E}{\ell} & \dfrac{2E}{\ell} \end{bmatrix}\left(\frac{\ell}{2EA}\begin{bmatrix} 1 & 1 \\ 1 & 2 \end{bmatrix}\right)\left(0 - \frac{2A}{\ell}\begin{bmatrix} 2 & -1 \\ -1 & 1 \end{bmatrix}\begin{bmatrix} \dfrac{3f\ell^2}{8EA} \\ \dfrac{f\ell^2}{2EA} \end{bmatrix}\right) \tag{4.55}$$

$$= \begin{bmatrix} -\dfrac{2}{\ell} & \dfrac{2}{\ell} \end{bmatrix} \cdot \begin{bmatrix} \dfrac{3f\ell^2}{8EA} \\ \dfrac{f\ell^2}{2EA} \end{bmatrix} + \begin{bmatrix} -\dfrac{2E}{\ell} & \dfrac{2E}{\ell} \end{bmatrix}\begin{bmatrix} -\dfrac{3f\ell^2}{8E^2A} \\ -\dfrac{f\ell^2}{2E^2A} \end{bmatrix} = \frac{f\ell}{4EA} - \frac{f\ell}{4EA} = 0$$

in which Eq. 4.54 is used to calculate the explicit dependent term $\frac{\partial \psi}{\partial E}$, which can be obtained easily once the solution $\mathbf{Z}$ is available. In this example, the explicit dependence is $\frac{\partial \psi}{\partial E} = \frac{f\ell}{4EA}$. The implicit term is $\frac{\partial \psi}{\partial \mathbf{Z}}\mathbf{K}^{-1}\left(\frac{\partial \mathbf{F}}{\partial E} - \frac{\partial \mathbf{K}}{\partial E}\mathbf{Z}\right) = -\frac{f\ell}{4EA}$, which requires the derivatives of the stiffness matrix $\mathbf{K}$ (and in some cases, the derivative of the load vector $\mathbf{F}$ may not be zero), and solving the finite element equations using the decomposed stiffness matrix $\mathbf{K}$.

As shown in Eq. 4.55, the sensitivity of the stress measure with respect to modulus E is zero, implying that stress is insensitive to the modulus E in this case. In fact, recall Eq. 4.26, the stress of the second element is obtained using FEA:

$$\sigma = E z_{,1} = E\begin{bmatrix} -\dfrac{2}{\ell} & \dfrac{2}{\ell} \end{bmatrix} \cdot \begin{bmatrix} \dfrac{3f\ell^2}{8EA} \\ \dfrac{f\ell^2}{2EA} \end{bmatrix} = \frac{f\ell}{4A} \tag{4.56}$$

which does not depend on the modulus E.

For area design variable A, the sensitivity of the stress measure ψ can be calculated using Eq. 4.53:

$$\frac{d\psi}{dA} = \frac{\partial \psi}{\partial A} + \frac{\partial \psi}{\partial \mathbf{Z}}\mathbf{K}^{-1}\left(\frac{\partial \mathbf{F}}{\partial A} - \frac{\partial \mathbf{K}}{\partial A}\mathbf{Z}\right)$$

$$= 0 + \begin{bmatrix} \dfrac{\partial \psi}{\partial Z_2} & \dfrac{\partial \psi}{\partial Z_3} \end{bmatrix}\left(\frac{\ell}{2EA}\begin{bmatrix} 1 & 1 \\ 1 & 2 \end{bmatrix}\right)\left(\mathbf{0} - \frac{2E}{\ell}\begin{bmatrix} 2 & -1 \\ -1 & 1 \end{bmatrix}\begin{bmatrix} \dfrac{3f\ell^2}{8EA} \\ \dfrac{f\ell^2}{2EA} \end{bmatrix}\right) \tag{4.57}$$

$$= 0 + \begin{bmatrix} -\dfrac{2E}{\ell} & \dfrac{2E}{\ell} \end{bmatrix}\begin{bmatrix} -\dfrac{3f\ell^2}{8EA^2} \\ -\dfrac{f\ell^2}{2EA^2} \end{bmatrix} = -\frac{f\ell}{4A^2}$$

Equation (4.57) implies that increasing the area reduces stress. Note that, in this case, the explicit dependence term is $\frac{\partial \psi}{\partial A} = 0$ because stress is defined as $\sigma = E z_{,1}$, which does not depend on A explicitly.

4.3.3.2 Adjoint variable method

Another way of calculating the implicit dependence $\frac{\partial Z}{\partial b}$ in Eq. 4.52b is the adjoint variable method. Note that the term $(\partial \psi / \partial Z) K^{-1}$ in Eq. 4.53 is independent of design. We set the result of $(\partial \psi / \partial Z) K^{-1}$ as λ^T:

$$\frac{\partial \psi}{\partial Z} K^{-1} = \lambda^T \tag{4.58}$$

in which λ is called the adjoint response or adjoint solution. Transpose the row vectors in Eq. 4.58 to column vectors:

$$K^{-T} \left(\frac{\partial \psi}{\partial Z} \right)^T = \lambda \tag{4.59}$$

By multiplying both sides of Eq. 4.59 by K^T, we have the following adjoint equation:

$$K\lambda = \left(\frac{\partial \psi}{\partial Z} \right)^T \tag{4.60}$$

where the symmetric property of the stiffness matrix, $K = K^T$, is used. Note that the adjoint response λ of the above adjoint equation can be solved by using the same stiffness matrix K of the structure. The only difference is the term on the right-hand side $\partial \psi / \partial Z$, which is called the adjoint load. The adjoint solution λ is not dependent on design but on the performance measure ψ. Thus, the adjoint solution is required per performance measure. Once the adjoint solution is available, the sensitivity can be calculated from Eq. 4.53 as

$$\frac{d\psi}{db} = \frac{\partial \psi}{\partial b} + \lambda^T \left(\frac{\partial F}{\partial b} - \frac{\partial K}{\partial b} Z \right) \tag{4.61}$$

Calculating the sensitivity using Eq. 4.61 is called the adjoint variable method—or more specifically, the adjoint variable method of the discrete-analytical approach.

We consider a general design problem of linear static with number of design variables (NDV) and number of performance measures (NPF) defined in the number of load cases (NL). Note that for a static problem of multiple load cases, performance measures such as stresses or displacements can be defined for respective load cases. If we use the direct differentiation method shown in Eq. 4.53, we need to calculate $\partial Z/\partial b_i$, $i = 1$, NDV for NL × NDV times by solving Eq. 4.41. If we use the adjoint variable method, we need to solve Eq. 4.60 NPF times. Solving the matrix equations for additional fictitious loads or adjoint loads constitutes the major computation time in calculating sensitivity coefficients. Therefore, when the number of performance measures is greater than that of the number of design variables times the number of load cases—that is, NPF > NL × NDV—the direct differential method is more efficient. Otherwise, the adjoint variable method is more desirable computation-wise.

EXAMPLE 4.5

Recall the displacement performance measure defined at the bottom end of the bar shown in Figure 4.4, written as

$$\psi = z(\ell) = \frac{f\ell^2}{2EA} = Z_3 \tag{4.62}$$

Calculate the adjoint load for the displacement performance measure, and calculate its sensitivity for the modulus design variable E using the adjoint variable method.

Solutions

In finite element formulation, the displacement of bar is written as in Eq. 4.24. Because the performance measure is the displacement at the bottom end, which involves only the second finite element, the adjoint load of the displacement measure ψ can be obtained using Eq. 4.60:

$$\frac{\partial\psi}{\partial\mathbf{Z}} = \begin{bmatrix} \dfrac{\partial\psi}{\partial Z_2} & \dfrac{\partial\psi}{\partial Z_3} \end{bmatrix} = \begin{bmatrix} \dfrac{\partial Z_3}{\partial Z_2} & \dfrac{\partial Z_3}{\partial Z_3} \end{bmatrix} = \begin{bmatrix} 0 & 1 \end{bmatrix} \tag{4.63}$$

Hence, the adjoint load for the displacement is nothing but a unit force acting downward (in the same direction as the displacement measure) at node 3 (the location where the displacement measure is defined).

The adjoint responses can be calculated as

$$\boldsymbol{\lambda} = \mathbf{K}^{-1}\left(\frac{\partial\psi}{\partial\mathbf{Z}}\right)^{\mathrm{T}} = \left(\frac{\ell}{2EA}\begin{bmatrix} 1 & 1 \\ 1 & 2 \end{bmatrix}\right)\begin{bmatrix} 0 \\ 1 \end{bmatrix} = \begin{bmatrix} \dfrac{\ell}{2EA} \\ \dfrac{\ell}{EA} \end{bmatrix} \tag{4.64}$$

Hence, the sensitivity of the stress measure can be calculated using Eq. 4.61:

$$\frac{d\psi}{dE} = \frac{\partial\psi}{\partial E} + \boldsymbol{\lambda}^{\mathrm{T}}\left(\frac{\partial\mathbf{F}}{\partial E} - \frac{\partial\mathbf{K}}{\partial E}\mathbf{Z}\right)$$

$$= 0 + \begin{bmatrix} \dfrac{\ell}{2EA} & \dfrac{\ell}{EA} \end{bmatrix}\left(-\frac{2A}{\ell}\begin{bmatrix} 2 & -1 \\ -1 & 1 \end{bmatrix}\begin{bmatrix} \dfrac{3f\ell^2}{8EA} \\ \dfrac{f\ell^2}{2EA} \end{bmatrix}\right) = -\frac{f\ell^2}{2E^2A} \tag{4.65}$$

This can be also obtained by taking derivative of ψ ($\psi = Z_3$) with respect to E directly using Eq. 4.62.

EXAMPLE 4.6

Repeat Example 4.4 using the adjoint variable method, only for modulus design variable E.

Solutions

The stress of the second element of the simple bar is obtained in Example 4.4 as

$$\psi = \sigma = E\begin{bmatrix} -\dfrac{2}{\ell} & \dfrac{2}{\ell} \end{bmatrix}\cdot\begin{bmatrix} Z_2 \\ Z_3 \end{bmatrix} \tag{4.66}$$

Hence, the adjoint load of the stress measure can be obtained using Eq. 4.60:

$$\frac{\partial\psi}{\partial\mathbf{Z}} = \begin{bmatrix} \dfrac{\partial\psi}{\partial Z_2} & \dfrac{\partial\psi}{\partial Z_3} \end{bmatrix} = \begin{bmatrix} -\dfrac{2E}{\ell} & \dfrac{2E}{\ell} \end{bmatrix} \tag{4.67}$$

Continued

EXAMPLE 4.6–cont'd

This consists of a point force $-2E/\ell$ (acting in the upward direction) at node 2, and another point force $2E/\ell$ (acting in the downward direction at node 3). The adjoint responses are

$$\boldsymbol{\lambda} = \mathbf{K}^{-1}\left(\frac{\partial \psi}{\partial \mathbf{Z}}\right)^{\mathrm{T}} = \left(\frac{\ell}{2EA}\begin{bmatrix} 1 & 1 \\ 1 & 2 \end{bmatrix}\right)\begin{bmatrix} -\dfrac{2E}{\ell} \\ \dfrac{2E}{\ell} \end{bmatrix} = \begin{bmatrix} 0 \\ \dfrac{1}{A} \end{bmatrix} \tag{4.68}$$

Hence, the sensitivity of the stress measure can be calculated using Eq. 4.53, for example, assuming modulus E as the design variable:

$$\frac{\mathrm{d}\psi}{\mathrm{d}E} = \frac{\partial \psi}{\partial E} + \boldsymbol{\lambda}^{\mathrm{T}}\left(\frac{\partial \mathbf{F}}{\partial E} - \frac{\partial \mathbf{K}}{\partial E}\mathbf{Z}\right)$$

$$= \begin{bmatrix} -\dfrac{2}{\ell} & \dfrac{2}{\ell} \end{bmatrix} \cdot \begin{bmatrix} \dfrac{3f\ell^2}{8EA} \\ \dfrac{f\ell^2}{2EA} \end{bmatrix} + \begin{bmatrix} 0 & \dfrac{1}{A} \end{bmatrix}\left(-\frac{2A}{\ell}\begin{bmatrix} 2 & -1 \\ -1 & 1 \end{bmatrix}\begin{bmatrix} \dfrac{3f\ell^2}{8EA} \\ \dfrac{f\ell^2}{2EA} \end{bmatrix}\right) = 0 \tag{4.69}$$

4.3.4 CONTINUUM APPROACH

In the continuum approach, the sensitivity is obtained by taking the variation of the energy equation that governs the structural behavior. The energy equation of a special case—the simple bar example—can be found in Eq. 4.13:

$$a_b(z, \bar{z}) = \int_0^\ell EAz_{,1}\bar{z}_{,1}\mathrm{d}x = \int_0^\ell f\,\bar{z}\mathrm{d}x = \ell_b(\bar{z}), \quad \forall \bar{z}\in S \tag{4.70}$$

in which the subscript b emphasizes the dependence of energy forms on design b.

A variation on the energy equation leads to a set of sensitivity equations in integral form that are then solved numerically, usually (but not necessarily) with the same discretization in a finite element formulation as was used for the original structural response. This is the so-called continuum-discrete method. For some simple but rare cases, state variable $\mathbf{z}$ can be obtained analytically, which can then be plugged into the sensitivity equations to calculate the sensitivity coefficients in integral form without discretization. This is called the continuum-analytical method. The continuum-discrete method is general and will be the focus of our discussion.

Before formally introducing the variation operator, we simply state that the variation of a function $f(b)$ is written as

$$f' = \frac{\partial f}{\partial b}\delta b = f'_{\delta b} \tag{4.71}$$

in which δb is the variation in the variable b. More about variation will be discussed in Section 4.4.2.

4.3.4.1 Direction differentiation method

We take variation of the energy bilinear form of the bar example, considering, for example, the area A as the design variable:

$$[a_b(z,\bar{z})]' = \left[\int_0^\ell EAz_{,1}\bar{z}_{,1}\mathrm{d}x \right]' = \int_0^\ell Ez_{,1}\bar{z}_{,1}\mathrm{d}x\delta A + \int_0^\ell EAz'_{,1}\bar{z}_{,1}\mathrm{d}x \tag{4.72}$$

$$= a'_{\delta b}(z,\bar{z}) + a_b(z',\bar{z})$$

where $a'_{\delta b}(z,\bar{z}) = \int_0^\ell E\delta Az_{,1}\bar{z}_{,1}\mathrm{d}x$ represents the explicit dependence of the energy bilinear form on design, and $a_b(z',\bar{z}) = \int_0^\ell EAz'_{,1}\bar{z}_{,1}\mathrm{d}x$ shows the implicit dependence through $z'_{,1}$. According to Eq. 4.71, $z'_{,1} = \frac{\partial z_{,1}}{\partial A}\delta A$. Now, the variation of the load linear form is written as

$$[\ell_b(\bar{z})]' = \ell'_{\delta b}(\bar{z}) \tag{4.73}$$

Hence, the sensitivity equation of the continuum approach can be obtained as

$$a_b(z',\bar{z}) = -a'_{\delta b}(z,\bar{z}) + \ell'_{\delta b}(\bar{z}), \quad \forall \bar{z} \in S \tag{4.74}$$

The term on the left-hand side of Eq. 4.74 is similar to that of Eq. 4.70, except that displacement z in Eq. 4.70 is replaced by z' in Eq. 4.74. The right-hand side of Eq. 4.74 defines a fictitious load, which is written explicitly in terms of the design variable. Thus, solving the sensitivity equation (Eq. 4.74) is the same as solving the original structural equilibrium equation (Eq. 4.70) with fictitious load(s).

For the simple bar example, $[\ell_b(\bar{z})]' = [\int_0^\ell f\, \bar{z}\mathrm{d}x]' = \int_0^\ell f'\, \bar{z}\mathrm{d}x\delta A = 0 = \ell'_{\delta b}(\bar{z})$ because f is the distributed load inside the bar, which was assumed to be constant; that is, $f' = 0$. Hence, the sensitivity equation for the bar example can be written as

$$\int_0^\ell EAz'_{,1}\bar{z}_{,1}\mathrm{d}x = - \int_0^\ell E\delta Az_{,1}\bar{z}_{,1}\mathrm{d}x \tag{4.75}$$

EXAMPLE 4.7

Calculate the sensitivity of the displacement at the bottom end of the simple bar example with respect to area design variable A using the direct differential method of the continuum-discrete approach.

Solutions

We discretize the sensitivity equation of Eq. 4.75 using the same shape function defined in Eq. 4.17:

$$\mathbf{z}(x) = \begin{cases} \left[1 - \dfrac{2x}{\ell} \quad \dfrac{2x}{\ell} \right] \cdot \begin{bmatrix} Z_1 \\ Z_2 \end{bmatrix}, & 0 \leq x \leq \ell/2 \\[2ex] \left[1 - \dfrac{2(x-\ell/2)}{\ell} \quad \dfrac{2(x-\ell/2)}{\ell} \right] \cdot \begin{bmatrix} Z_2 \\ Z_3 \end{bmatrix}, & \ell/2 < x \leq \ell \end{cases} \tag{4.76}$$

Continued

EXAMPLE 4.7–cont'd

We have the finite element matrix equations on the left-hand side (similar to Eq. 4.18), as follows:

$$\int_0^\ell EA z'_{,1} \bar{z}_{,1} \mathrm{d}x \stackrel{\text{discretize}}{=} \begin{bmatrix} \bar{Z}_1 \\ \bar{Z}_2 \\ \bar{Z}_3 \end{bmatrix}^{\mathrm{T}} \cdot \frac{2EA}{\ell} \begin{bmatrix} 1 & -1 & 0 \\ -1 & 2 & -1 \\ 0 & -1 & 1 \end{bmatrix} \cdot \begin{bmatrix} Z'_1 \\ Z'_2 \\ Z'_3 \end{bmatrix} = \bar{\mathbf{Z}}_{\mathrm{g}} \mathbf{K}_{\mathrm{g}} \mathbf{Z}'_{\mathrm{g}} \tag{4.77}$$

where

$$\mathbf{Z}'_{\mathrm{g}} = \begin{bmatrix} Z'_1 \\ Z'_2 \\ Z'_3 \end{bmatrix} = \begin{bmatrix} \dfrac{\partial Z_1}{\partial A} \\[4pt] \dfrac{\partial Z_2}{\partial A} \\[4pt] \dfrac{\partial Z_3}{\partial A} \end{bmatrix} \delta A \tag{4.78}$$

Then, the right-hand side of Eq. 4.75 is discretized with the same shape function:

$$-\int_0^\ell E z_{,1} \bar{z}_{,1} \mathrm{d}x \delta A = -\int_0^{\ell/2} E z_{,1} \bar{z}_{,1} \mathrm{d}x \delta A - \int_{\ell/2}^\ell E z_{,1} \bar{z}_{,1} \mathrm{d}x \delta A$$

$$= -\int_0^{\ell/2} E \left\{ \begin{bmatrix} \bar{Z}_1 & \bar{Z}_2 \end{bmatrix} \begin{bmatrix} -\dfrac{2}{\ell} \\[4pt] \dfrac{2}{\ell} \end{bmatrix} \right\} \frac{3f\ell}{4EA} \mathrm{d}x \delta A - \int_{\ell/2}^\ell E \left\{ \begin{bmatrix} \bar{Z}_2 & \bar{Z}_3 \end{bmatrix} \begin{bmatrix} -\dfrac{2}{\ell} \\[4pt] \dfrac{2}{\ell} \end{bmatrix} \right\} \frac{f\ell}{4EA} \mathrm{d}x \delta A \tag{4.79}$$

$$= -\begin{bmatrix} \bar{Z}_1 & \bar{Z}_2 \end{bmatrix} \begin{bmatrix} -\dfrac{3f\ell}{4A} \\[4pt] \dfrac{3f\ell}{4A} \end{bmatrix} \delta A - \begin{bmatrix} \bar{Z}_2 & \bar{Z}_3 \end{bmatrix} \begin{bmatrix} \dfrac{f\ell}{4A} \\[4pt] \dfrac{f\ell}{4A} \end{bmatrix} \delta A = \begin{bmatrix} \bar{Z}_1 & \bar{Z}_2 & \bar{Z}_3 \end{bmatrix} \begin{bmatrix} \dfrac{3f\ell}{4A} \\[4pt] -\dfrac{f\ell}{2A} \\[4pt] -\dfrac{f\ell}{4A} \end{bmatrix} \delta A = \bar{\mathbf{Z}}_{\mathrm{g}} \mathbf{F}_{\mathrm{g}} \delta A$$

By equating Eqs 4.79 with 4.77 and applying boundary condition, we have $\mathbf{KZ}' = \mathbf{F}_{\mathrm{fic}} \delta A$:

$$\mathbf{KZ}' = \frac{2EA}{\ell} \begin{bmatrix} 2 & -1 \\ -1 & 1 \end{bmatrix} \cdot \begin{bmatrix} \dfrac{\partial Z_2}{\partial A} \\[4pt] \dfrac{\partial Z_3}{\partial A} \end{bmatrix} \delta A = \begin{bmatrix} -\dfrac{f\ell}{2A} \\[4pt] -\dfrac{f\ell}{4A} \end{bmatrix} \delta A = \mathbf{F}_{\mathrm{fic}} \delta A \tag{4.80}$$

Solving Eq. 4.80 and removing δA, we have

$$\begin{bmatrix} \dfrac{\partial Z_2}{\partial A} \\[4pt] \dfrac{\partial Z_3}{\partial A} \end{bmatrix} = \mathbf{K}^{-1} \mathbf{F}_{\mathrm{fic}} = \frac{\ell}{2EA} \begin{bmatrix} 1 & 1 \\ 1 & 2 \end{bmatrix} \begin{bmatrix} -\dfrac{f\ell}{2A} \\[4pt] -\dfrac{f\ell}{4A} \end{bmatrix} = \begin{bmatrix} -\dfrac{3f\ell^2}{8EA^2} \\[4pt] -\dfrac{f\ell^2}{2EA^2} \end{bmatrix} \tag{4.81}$$

The major advantage of the continuum approach is that the sensitivity formulation is independent of the discrete model and numerical schemes. Once the continuum sensitivity equation is obtained, it can be discretized in the same manner as the original analysis. In addition, there is no need to take partial derivatives of the matrix or vector, as required, in the discrete approach, and no need to find design perturbation sizes, as is needed in the semianalytical and overall finite difference methods. More importantly, for practical implementation, the continuum-discrete method can be implemented outside of commercial FEA codes by only using the FEA results. More details are provided in Section 4.4.4.

4.3.4.2 Adjoint variable method

In the continuum approach, the performance measure is expressed in an integral form. For example, the displacement at the bottom end of the bar is defined as

$$\psi = z(\ell) = \int_0^\ell [z(x)\delta(x - \ell)]dx \tag{4.82}$$

Here, $\delta(x - \ell)$ is a direct delta function, defined as

$$\delta(x - \ell) \equiv \begin{cases} \infty, & x = \ell \\ 0, & x \neq \ell \end{cases} \tag{4.83}$$

Taking the variation of Eq. 4.82, we have

$$\psi' = \int_0^\ell [z'(x)\delta(x - \ell)]dx \tag{4.84}$$

The basic idea of the adjoint variable method is to avoid directly calculating z' in Eq. 4.84 by introducing an adjoint structure that is identical to the original structure but with a different set of load—that is, the adjoint load. The adjoint structure for the simple bar example is defined as

$$a_b(\lambda, \bar{\lambda}) = \int_0^\ell [\delta(x - \ell)\bar{\lambda}(x)]dx, \quad \forall \bar{\lambda} \in S \tag{4.85}$$

whose right-hand side is identical to that of Eq. 4.84 by replacing z' with the virtual adjoint response $\bar{\lambda}$. Note that the left-hand side of Eq. 4.85 is the energy bilinear form in terms of adjoint response λ and virtual adjoint response $\bar{\lambda}$. The right-hand side defines the adjoint load, which depends on the type of performance measure. The adjoint response can be obtained in the same way as the original structural response z, for example, using finite element equations, as long as the adjoint load vector is available.

We first evaluate Eq. 4.85 at $\bar{\lambda} = z'$:

$$a_b(\lambda, z') = \int_0^\ell [\delta(x - \ell)z'(x)]dx, \quad \forall z' \in S \tag{4.86}$$

in which the right-hand side equals that of Eq. 4.84.

Recall the sensitivity equation of the direct differentiation method in Eq. 4.74:

$$a_b(z', \bar{z}) = -a'_{\delta b}(z, \bar{z}) + \ell'_{\delta b}(\bar{z}), \quad \forall \bar{z} \in S \qquad (4.87)$$

We evaluate Eq. 4.74 at $\bar{z} = \lambda$:

$$a_b(z', \lambda) = -a'_{\delta b}(z, \lambda) + \ell'_{\delta b}(\lambda) \qquad (4.88)$$

Because the energy bilinear form is symmetric—that is, $a_b(\lambda, z') = a_b(z', \lambda)$—Eq. 4.88 equals Eq. 4.86. Therefore,

$$\psi' = \int_0^\ell [\delta(x - \ell)z'(x)]\mathrm{d}x = -a'_{\delta b}(z, \lambda) + \ell'_{\delta b}(\lambda) \qquad (4.89)$$

This can be calculated once the original structural response z and adjoint response λ are available. Note that additional term(s) may exist in Eq. 4.89 for performance measures other than displacement, which will be illustrated in Section 4.4.3.

EXAMPLE 4.8

Solve for the same problem as Example 4.7 using the adjoint variable method. We defined displacement at node 3 as the performance measure, which is written in an integral form as

$$\psi = z(\ell) = \int_0^\ell [z(x)\delta(x - \ell)]\mathrm{d}x \qquad (4.90)$$

Solutions

We first create an adjoint structure by determining the adjoint load applied to the simple bar. From Eq. 4.85, we have

$$a_b(\lambda, \bar{\lambda}) = \int_0^\ell [\delta(x - \ell)\bar{\lambda}(x)]\mathrm{d}x \qquad (4.91)$$

Discretizing Eq. 4.91 using finite element shape function, we have

$$a_b(\lambda, \bar{\lambda}) = \int_0^\ell [\delta(x - \ell)\bar{\lambda}(x)]\mathrm{d}x = \int_0^{\ell/2} [\delta(x - \ell)\bar{\lambda}(x)]\mathrm{d}x + \int_{\ell/2}^\ell [\delta(x - \ell)\bar{\lambda}(x)]\mathrm{d}x$$

$$= 0 + \int_{\ell/2}^\ell [\delta(x - \ell)\bar{\lambda}(x)]\mathrm{d}x = [\bar{\lambda}_2 \quad \bar{\lambda}_3] \int_{\ell/2}^\ell \delta(x - \ell) \begin{bmatrix} 1 - \dfrac{2(x - \ell/2)}{\ell} \\ \dfrac{2(x - \ell/2)}{\ell} \end{bmatrix} \mathrm{d}x \qquad (4.92)$$

$$= [\bar{\lambda}_2 \quad \bar{\lambda}_3] \begin{bmatrix} 1 - \dfrac{2(\ell - \ell/2)}{\ell} \\ \dfrac{2(\ell - \ell/2)}{\ell} \end{bmatrix} = [\bar{\lambda}_2 \quad \bar{\lambda}_3] \begin{bmatrix} 0 \\ 1 \end{bmatrix}$$

which shows that the adjoint load is a point load applied at the bottom end of the bar (at node 3) in the downward direction, which is identical to those obtained using the discrete approach.

EXAMPLE 4.8—cont'd

Solve the adjoint response by using the finite element equation:

$$\boldsymbol{\lambda} = \mathbf{K}^{-1}\mathbf{F} = \frac{\ell}{2EA}\begin{bmatrix} 1 & 1 \\ 1 & 2 \end{bmatrix}\begin{bmatrix} 0 \\ 1 \end{bmatrix} = \begin{bmatrix} \dfrac{\ell}{2EA} \\ \dfrac{\ell}{EA} \end{bmatrix} \tag{4.93}$$

Bring the original responses z and adjoint responses λ to the sensitivity equation (Eq. 4.89):

$$\psi' = -a'_{\delta b}(z,\lambda) + \ell'_{\delta b}(\lambda) = -\int_0^\ell Ez_{,1}\lambda_{,1}\mathrm{d}x\delta A + 0$$

$$= -\int_0^{\ell/2} E\frac{3f\ell}{4EA}\begin{bmatrix} -\dfrac{2}{\ell} & \dfrac{2}{\ell} \end{bmatrix}\cdot\begin{bmatrix} 0 \\ \dfrac{\ell}{2EA} \end{bmatrix}\mathrm{d}x\delta A - \int_{\ell/2}^\ell E\frac{f\ell}{4EA}\begin{bmatrix} -\dfrac{2}{\ell} & \dfrac{2}{\ell} \end{bmatrix}\cdot\begin{bmatrix} \dfrac{\ell}{2EA} \\ \dfrac{\ell}{EA} \end{bmatrix}\mathrm{d}x\delta A \tag{4.94}$$

$$= -\frac{f\ell^2}{2E^2A}$$

4.4 SIZING AND MATERIAL DESIGNS

We briefly introduced three general approaches finite difference, discrete, and continuum, for calculating the sensitivity of structural performance measures. In these approaches, the overall finite difference is the simplest and easiest to implement with commercial finite element codes. However, the finite difference is inefficient and requires adequate design perturbations. The gradient-based optimization method implemented in Pro/MECHANICA Structure (www.ptc.com) uses the overall finite difference approach for sensitivity analysis. The discrete approach formulates the sensitivity expressions by taking the derivatives of the discretized finite element equations. The derivatives of the finite element equations can be obtained analytically if the explicit expressions of the equations in terms of design variables are available, which is called the discrete-analytical approach. If not, the finite difference method is employed to approximate the derivatives of the equations. This approach is called discrete–discrete or semianalytical method. The solution 200: Design Optimization and Sensitivity Analysis offered by MSC/NASTRAN (www.mscsoftware.com) employs the semianalytical approach for sensitivity analysis. The discrete approach is more efficient than the overall finite difference. However, the semianalytical method still requires adequate design perturbations for accurate sensitivity information. The continuum approach formulates the sensitivity expressions by taking a variation of the energy equations of the structure in an integral form. The expressions can then be discretized using finite element shape functions for numerical evaluations, which is called the continuum-discrete approach. This approach is efficient and does not require any design perturbations.

Considering accuracy and efficiency, the best method is probably the continuum-discrete. In this section, we focus on this method. We include sensitivity analysis for sizing and material design variables that do not affect the geometric shape of the structures; therefore, the sensitivity analysis methods and formulations are similar. In Section 4.5, we introduce shape sensitivity analysis, in which design variables alter the geometric shape of the structure. The shape sensitivity analysis method and formulations are quite different from those of the sizing and material design variables.

We start by introducing energy equations for a general structure in Section 4.4.1. We formally introduce variation in Section 4.4.2, and then focus on static problems in Section 4.4.3, in which we use beam examples for illustration. We discuss implementation of the continuum-discrete approach with commercial FEA codes in Section 4.4.4.

4.4.1 PRINCIPLE OF VIRTUAL WORK

We discussed the energy equation and principle of virtual work for the simple bar example in Section 4.3. The energy equation or the principle of virtual work is the basis of FEA and sensitivity analysis, either the discrete or continuum approach. In this subsection, we present the energy equation of a general two-dimensional structure. Equations for three-dimensional structures can also be obtained with more complex formulations. Similar to the simple bar example, we start by stating the governing equation in differential form, and then carrying out integration by parts to obtain the energy equations. Note that for most structural problems, such as the simple bar example, energy equations can be obtained by starting with the differential equations and carrying out integration by parts. The energy equations for a specific problem and type of structure can also be obtained by using the energy equation of the general structure, with terms and state variables specialized to the problem and type of the structure. We use a beam example to illustrate this point in Section 4.4.3.

4.4.1.1 Differential equation

A general two-dimensional or planar structure shown in Figure 4.7a is loaded with a traction force (force per length) $\mathbf{T} = [T_1, T_2]^T$ at the boundary Γ_1 and an in-plane distributed force (or body force) per area $f = [f_1, f_2]^T$, and fixed at the boundary Γ_0. Note that T_1 and T_2 are the traction forces in the x_1 and x_2 directions, respectively. The same is true for f_1 and f_2. The stress element A inside the structure domain Ω is shown in Figure 4.7b, in which three stress components σ_{11}, σ_{22}, and σ_{12} are present. Note that σ_{12} is the shear stress τ_{12}, and $\sigma_{12} = \sigma_{21}$. A stress element B at the boundary Γ_1 is shown in Figure 4.7c, in which $\mathbf{n} = [n_1, n_2]^T$ is the normal vector of a given point at boundary Γ_1. For static problems, forces f and T are independent of time; hence, the displacement solution $\mathbf{z} = [z_1, z_2]^T$ is not function of time. For dynamic problems, either $\mathbf{f}$ or $\mathbf{T}$ or both forces are time functions, in which the structure is involved

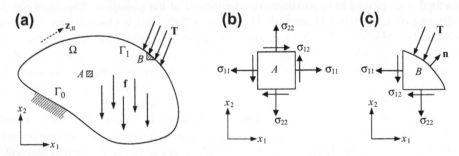

FIGURE 4.7

A general structure with (a) load and boundary conditions, (b) an internal stress element, and (c) a stress element at the traction boundary Γ_1.

with inertia force caused by acceleration $\mathbf{z}_{,tt}$. Note that the shorthand notation, tt in subscript denotes the second derivative of the function with respect to time t. The displacement $\mathbf{z}$ is function of spatial variable $\mathbf{x}$ and time t—that is, $\mathbf{z} = \mathbf{z}(\mathbf{x}, t)$, in this case.

The governing equations in differential form are stated as follows:

$$\begin{cases} \sigma_{11,1} + \sigma_{12,2} + f_1 = \rho z_{1,tt} \\ \sigma_{12,1} + \sigma_{22,2} + f_2 = \rho z_{2,tt} \end{cases}, \quad \text{in } \Omega \qquad (4.95a)$$

The boundary conditions are

$$T_1 = \sigma_{11} n_1 + \sigma_{12} n_2 \text{ and } T_2 = \sigma_{21} n_1 + \sigma_{22} n_2, \text{ on } \Gamma_1, \text{ and} \qquad (4.95b)$$

$$z_1(x, t) = 0 \text{ and } z_2(x, t) = 0, \text{ on } \Gamma_0 \qquad (4.95c)$$

The initial displacement and initial velocity conditions, respectively, are

$$z_1(\mathbf{x}, 0) = z_1^0 \text{ and } z_2(\mathbf{x}, 0) = z_2^0, \text{ in } \Omega \qquad (4.95d)$$

$$z_{1,t}(\mathbf{x}, 0) = z_{1,t}^0 \text{ and } z_{2,t}(\mathbf{x}, 0) = z_{2,t}^0, \text{ in } \Omega \qquad (4.95e)$$

Note that in Eq. 4.95a, the shorthand notations of subscript, 1 and, 2 represent the derivatives of the function with respect to x_1 and x_2, respectively.

Equations (4.96a) to (4.96e) can be further written in a compact form as

$$\sigma_{ij,j} + f_i = \rho z_{i,tt}, \text{ in } \Omega \qquad (4.96a)$$

The boundary and initial conditions are

$$T_i = \sigma_{ij} n_j, \text{ on } \Gamma_1, \text{ and} \qquad (4.96b)$$

$$z_i(\mathbf{x}, t) = 0, \text{ on } \Gamma_0 \qquad (4.96c)$$

$$z_i(\mathbf{x}, 0) = z_i^0, \text{ in } \Omega \qquad (4.96d)$$

$$z_{i,t}(\mathbf{x}, 0) = z_{i,t}^0, \text{ in } \Omega \qquad (4.96e)$$

For two-dimensional problems, $i, j = 1, 2$; for three-dimensional problems, $i, j = 1, 3$.

If we assume that the external loads do not depend on time t, the differential form of the governing equations of the structure shown in Figure 4.7a is reduced to

$$\sigma_{ij,j} + f_i = 0, \text{ in } \Omega \qquad (4.97a)$$

with boundary conditions

$$T_i = \sigma_{ij} n_j, \text{ on } \Gamma_1, \text{ and} \qquad (4.97b)$$

$$z_i(x) = 0, \text{ on } \Gamma_0 \qquad (4.97c)$$

Similar to the simple bar example, we derive the energy equation of the structural governing equation by multiplying both sides of Eq. 4.97a with an arbitrary virtual displacement $\bar{z}$ and then integrating over the structural domain Ω:

$$\int_\Omega \sigma_{ij,j} \bar{z}_i d\Omega + \int_\Omega f_i \bar{z}_i d\Omega = 0 \qquad (4.98)$$

Carrying out integration by part for the terms on the left-hand side of Eq. 4.98, we have

$$\int_{\Gamma_1} \sigma_{ij}\bar{z}_i n_j d\Gamma_1 - \int_{\Omega} \sigma_{ij}\bar{z}_{i,j} d\Omega + \int_{\Omega} f_i\bar{z}_i d\Omega = 0 \qquad (4.99)$$

in which the first term on the left-hand side can be evaluated, using the boundary condition of Eq. 4.97b, as

$$\int_{\Gamma_1} \sigma_{ij}\bar{z}_i n_j d\Gamma_1 = \int_{\Gamma_1} T_i\bar{z}_i d\Gamma_1 \qquad (4.100)$$

The second term on the left-hand side can be written as

$$\int_{\Omega} \sigma_{ij}\bar{z}_{i,j} d\Omega = \int_{\Omega} \sigma_{ij}\bar{\varepsilon}_{ij} d\Omega \qquad (4.101)$$

in which $\bar{\varepsilon}_{ij}$ is the virtual strain in virtual displacement $\bar{z}_i$. Note that the relationship $\sigma_{ij}\bar{z}_{i,j} = \sigma_{ij}\bar{\varepsilon}_{ij}$ is employed in Eq. 4.101, for the following two reasons:

$$\sigma_{ij}\bar{z}_{i,j} = \sigma_{ji}\bar{z}_{j,i} = \sigma_{ij}\bar{z}_{j,i} \qquad (4.102)$$

and

$$\sigma_{ij}\bar{z}_{i,j} = \sigma_{ji}\left[\frac{1}{2}\left(\bar{z}_{i,j} + \bar{z}_{j,i}\right)\right] = \sigma_{ij}\bar{\varepsilon}_{ij} \qquad (4.103)$$

Note that in Eq. 4.102, $\sigma_{ij}\bar{z}_{i,j} = \sigma_{ji}\bar{z}_{j,i}$ is simply obtained by changing the index, and $\sigma_{ji}\bar{z}_{j,i} = \sigma_{ij}\bar{z}_{j,i}$ is due to the symmetric of the stress tensor σ_{ij} (e.g., $\sigma_{12} = \sigma_{21}$). In Eq. 4.103, the following relationship, called the Cauchy strain tensor, is employed:

$$\varepsilon_{ij} = \frac{1}{2}\left(z_{i,j} + z_{j,i}\right) \qquad (4.104)$$

Bringing Eqs 4.100 and 4.101 into Eq. 4.99, we have

$$\int_{\Omega} \sigma_{ij}\bar{\varepsilon}_{ij} d\Omega = \int_{\Omega} f_i\bar{z}_i d\Omega + \int_{\Gamma_1} T_i\bar{z}_i d\Gamma_1 \qquad (4.105)$$

Equation (4.105) can be rewritten as

$$a(\mathbf{z}, \bar{\mathbf{z}}) = \ell(\bar{\mathbf{z}}) \qquad (4.106)$$

Here, $a(\mathbf{z}, \bar{\mathbf{z}})$ and $\ell(\bar{\mathbf{z}})$ are the energy bilinear form (or virtual strain energy) and load linear form (virtual work), defined respectively as

$$a(\mathbf{z}, \bar{\mathbf{z}}) = \int_{\Omega} \sigma_{ij}\bar{\varepsilon}_{ij} d\Omega \qquad (4.107)$$

and

$$\ell(\bar{\mathbf{z}}) = \int_\Omega f_i \bar{z}_i \mathrm{d}\Omega + \int_{\Gamma_1} T_i \bar{z}_i \mathrm{d}\Gamma_1 \tag{4.108}$$

The principle of virtual work states, from a structural analysis perspective, that the solution $\mathbf{z}$ is sought in a trial function space S. That is, $\mathbf{z} \in S$, which is the space of kinematically admissible virtual displacement, defined as

$$S = \left\{ \mathbf{z} \in [H^m(\Omega)]^2 \, \middle| \, \mathbf{z} = 0 \text{ at } \Gamma_0 \right\} \tag{4.109}$$

such that Eq. 4.106 is satisfied for all virtual displacements $\bar{\mathbf{z}} \in S$.

In Eq. 4.109, $\mathbf{z} = 0$ at Γ_0 is the essential (or kinematic) boundary condition and $[H^m(\Omega)]^2$ is the Sobolev space of order m, where m is a positive integer, determined by the type of the structural components—for example $m = 1$ for bars and $m = 2$ for beams. The superscript 2 implies a two-dimensional problem. Similar to the discussion in Section 4.2 for the bar example, the solution $\mathbf{z}$ in $[H^m(\Omega)]^2$ implies that the virtual energy terms $a(\mathbf{z}, \bar{\mathbf{z}})$ and $\ell(\bar{\mathbf{z}})$ in Eq. 4.106 are finite. Without going through rigorous mathematical arguments, we state that the solutions $\mathbf{z}$ of Eq. 4.106 exist and are unique.

4.4.2 VARIATIONS

In Section 4.3.4, we simply stated that the variation of a function $f(b)$ is written as

$$f' = \frac{\partial f}{\partial b} \delta b = f'_{\delta b} \tag{4.110}$$

In fact, the variation we stated is the first variation in the calculus of variations.

4.4.2.1 Definition

In the continuum approach, sensitivity can be understood as variation of a function. Let us consider that a vector of design variable $\mathbf{b}$ is perturbed to $\mathbf{b} + \tau \delta \mathbf{b}$, in which τ is a scalar that measures the perturbation size and $\delta \mathbf{b}$ is the direction of design change. For simplicity, we assume for the time being material or sizing design variables that are not affecting the geometric shape of the structure. Shape sensitivity analysis will be discussed in Section 4.5. We also assume a single design variable b. The variation of the state variable (assuming a scalar for simplicity), $z(x; b)$, which is a function of both spatial variable x (location) and design variable b, is defined as

$$z' \equiv \left. \frac{\mathrm{d}}{\mathrm{d}\tau} z(x; b + \tau \delta b) \right|_{\tau=0} = \lim_{\tau \to 0} \frac{z(x; b + \tau \delta b) - z(x; b)}{\tau} \tag{4.111}$$

Taking Taylor's series expansion for $z(x; b + \tau \delta b)$ at b, we have

$$z(x; b + \tau \delta b) = z(x; b) + \frac{\partial z}{\partial b} \tau \delta b + \frac{1}{2} \frac{\partial^2 z}{\partial b^2} (\tau \delta b)^2 + \text{higher order terms} \tag{4.112}$$

Plugging Eq. 4.112 into Eq. 4.111, we have

$$z' = \frac{\partial z(x; b)}{\partial b} \delta b \qquad (4.113)$$

which is called the variation of z with respect to b. Note that z' in Eq. 4.113 is also written as $z' = z'_{\delta b}$ to emphasize that the variation is taken specifically with respect to variable b.

Note that the coefficient of δb in Eq. 4.113 $\frac{\partial z}{\partial b}$ is the sensitivity of the state variable z—or more specifically, first-order sensitivity—which is equivalent to the derivative discussed in the discrete approach.

Because the direction of design change δb can be arbitrary, Eq. 4.113 must be linear with respect to δb. Therefore, geometrically, $\frac{\partial z}{\partial b}$ is slope of the curve $z(b)$, as illustrated in Figure 4.8. Again, for a simpler illustration, we assume a single design variable b. Note that, mathematically, linearity of a function must have the following two important properties:

$$z'_{\delta b_1 + \delta b_2} = \frac{\partial z}{\partial b} (\delta b_1 + \delta b_2) = \frac{\partial z}{\partial b} \delta b_1 + \frac{\partial z}{\partial b} \delta b_2 = z'_{\delta b_1} + z'_{\delta b_2} \qquad (4.114)$$

and

$$z'_{\alpha \delta b} = \frac{\partial z}{\partial b} (\alpha \delta b) = \alpha \frac{\partial z}{\partial b} \delta b = \alpha z'_{\delta b} \qquad (4.115)$$

where α is a nonzero real number. Note that in Eq. 4.114, δb_1 and δb_2 represent two design perturbations of a single design variable b. For multiple design variables, δb_1 and δb_2 represent design perturbations of two different design variables or two different sets of design variables.

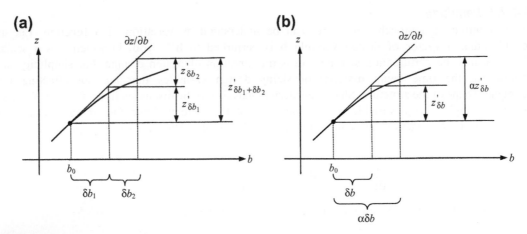

FIGURE 4.8

Illustration of the linearity properties of variations of the state variable z on design b: (a) $z'_{\delta b_1 + \delta b_2} = z'_{\delta b_1} + z'_{\delta b_2}$ and (b) $z'_{\alpha \delta b} = \alpha z'_{\delta b}$.

Geometrically, $z'_{\delta b_1 + \delta b_2}$ is the vertical distance of the slope $\partial z/\partial b$ times the design perturbation $\delta b_1 + \delta b_2$, which is equal to the sum of the slope multiplied by the individual design perturbations δb_1 and δb_2; that is, $z'_{\delta b_1} + z'_{\delta b_2}$, as illustrated in Figure 4.8a. Similarly, the vertical distance of the slope $\partial z/\partial b$ times the design perturbation $\alpha \delta b$ is equal to the slope multiplied by the design perturbation δb times the constant α, as illustrated in Figure 4.8b.

Note also that the variation and partial derivative are commutative, implying the order of the operations is interchangeable:

$$(z_{,i})' = (z')_{,i} \quad \text{or} \quad \left(\frac{\partial z}{\partial x_i}\right)' = \frac{\partial(z')}{\partial x_i} \tag{4.116}$$

and

$$(z_{,ij})' = \left[(z_{,i})'\right]_{,j} = (z')_{,ij} \tag{4.117}$$

if the second derivative is involved.

EXAMPLE 4.9

A function f is defined as

$$f = f(b,z) = b + z^2 \tag{4.118}$$

We assume z is not a function b. Calculate $\frac{\partial f}{\partial b}$, $\frac{\partial f}{\partial z}$, f', $f'_{\delta b}$, and $f'_{\delta z}$.

Solutions

Taking derivatives of f with respect to b and z, respectively we have

$$\frac{\partial f}{\partial b} = 1, \quad \text{and} \quad \frac{\partial f}{\partial z} = 2z \tag{4.119}$$

Now we use definition of variation (Eq. 4.111) to calculate the variations f', $f'_{\delta b}$, and $f'_{\delta z}$.

$$f'_{\delta b}(b,z) \equiv \frac{\mathrm{d}}{\mathrm{d}\tau} f(b + \tau \delta b, z)\Big|_{\tau = 0} = \frac{\mathrm{d}}{\mathrm{d}\tau}\left(b + \tau \delta b + z^2\right)\Big|_{\tau=0} = \delta b = f_{,b}\delta b \tag{4.120}$$

$$f'_{\delta z}(b,z) \equiv \frac{\mathrm{d}}{\mathrm{d}\tau} f(b, z + \tau \delta z)\Big|_{\tau = 0} = \frac{\mathrm{d}}{\mathrm{d}\tau}\left(b + (z + \tau \delta z)^2\right)\Big|_{\tau=0}$$
$$= \frac{\mathrm{d}}{\mathrm{d}\tau}\left(b + z^2 + 2z\tau\delta z + \tau^2 \delta z^2\right)\Big|_{\tau=0} = \left(2z\delta z + 2\tau \delta z^2\right)\big|_{\tau=0} = 2z\delta z = f_{,z}\delta z \tag{4.121}$$

and

$$f' \equiv \frac{\mathrm{d}}{\mathrm{d}\tau} f(b + \tau \delta b, z + \tau \delta z)\Big|_{\tau = 0} = \frac{\mathrm{d}}{\mathrm{d}\tau}\left(b + \tau \delta b + (z + \tau \delta z)^2\right)\Big|_{\tau=0}$$
$$= \frac{\mathrm{d}}{\mathrm{d}\tau}\left(b + \tau \delta b + z^2 + 2z\tau\delta z + \tau^2 \delta z^2\right)\Big|_{\tau=0} = \left(\delta b + 2z\delta z + 2\tau\delta z^2\right)\big|_{\tau=0} \tag{4.122}$$
$$= \delta b + 2z\delta z = f_{,b}\delta b + f_{,z}\delta z = f'_{\delta b} + f'_{\delta z}$$

EXAMPLE 4.10

A function f is defined as

$$f = f(b, z(b)) = b^2 + z(b)^2 \tag{4.123}$$

We assume z is a function b. Calculate the variation of function f.

Solutions

We use Eq. 4.111 to calculate the variation of f:

$$f'(b, z(b)) \equiv \frac{\mathrm{d}}{\mathrm{d}\tau} f(b + \tau\delta b, z(b + \tau\delta b))\Big|_{\tau=0} = \frac{\mathrm{d}}{\mathrm{d}\tau}\left((b + \tau\delta b)^2 + z(b + \tau\delta b)^2\right)\Big|_{\tau=0}$$

$$= \frac{\mathrm{d}}{\mathrm{d}\tau}\left((b + \tau\delta b)^2\right)\Big|_{\tau=0} + \frac{\mathrm{d}}{\mathrm{d}\tau}\left(z(b + \tau\delta b)^2\right)\Big|_{\tau=0} \tag{4.124}$$

$$= 2b\delta b + 2z\frac{\mathrm{d}z(b + \tau\delta b)}{\mathrm{d}\tau}\Big|_{\tau=0} = f'_{\delta b} + 2z\frac{\partial z}{\partial b}\delta b$$

where $f'_{\delta b} = 2b\delta b = f_{,b}\delta b = [f(b, \tilde{z})]'$, and $\frac{\mathrm{d}z(b + \tau\delta b)}{\mathrm{d}\tau}\Big|_{\tau=0} = \frac{\partial z}{\partial b}\delta b$. Note that $\tilde{z}$ implies that the dependence of z on b is suppressed in taking the variation. The first term of Eq. 4.124 $f'_{\delta b}$ represents the explicit dependence of f on variable b, and the second term represents the implicit dependence of f on variable b through $z(b)$.

EXAMPLE 4.11

A function f is defined as

$$f = f(b, z, z_{,b}) = b + z^2 + (z_{,b})^2 \tag{4.125}$$

We assume z is a function of b. Calculate the variation of function f.

Solutions

We use Eq. 4.111 to calculate the variation f':

$$f'(b, z, z_{,b}) \equiv \frac{\mathrm{d}}{\mathrm{d}\tau} f(b + \tau\delta b, z(b + \tau\delta b), z_{,b}(b + \tau\delta b))\Big|_{\tau=0} = \frac{\mathrm{d}}{\mathrm{d}\tau}\left(b + \tau\delta b + z(b + \tau\delta b)^2 + z_{,b}(b + \tau\delta b)^2\right)\Big|_{\tau=0}$$

$$= \frac{\mathrm{d}}{\mathrm{d}\tau}(b + \tau\delta b)\Big|_{\tau=0} + \frac{\mathrm{d}}{\mathrm{d}\tau}\left(z(b + \tau\delta b)^2\right)\Big|_{\tau=0} + \frac{\mathrm{d}}{\mathrm{d}\tau}\left(z_{,b}(b + \tau\delta b)^2\right)\Big|_{\tau=0}$$

$$= \delta b + 2z\frac{\mathrm{d}z(b + \tau\delta b)}{\mathrm{d}\tau}\Big|_{\tau=0} + 2z_{,b}\frac{\mathrm{d}z_{,b}(b + \tau\delta b)}{\mathrm{d}\tau}\Big|_{\tau=0} = f_{,b}\delta b + 2z\frac{\partial z}{\partial b}\delta b + 2z_{,b}\frac{\partial z_{,b}}{\partial b}\delta b = f_{,b}\delta b + f_{,z}z' + f_{,z_{,b}}z'_{,b} \tag{4.126}$$

4.4.2.2 Variations of the energy bilinear and load linear forms

Now, we employ the definition of variation in Eq. 4.111 to the governing equation of general structure in variational form stated in Eq. 4.106:

$$a_b(\mathbf{z}, \bar{\mathbf{z}}) = \ell_b(\bar{\mathbf{z}}), \quad \forall \bar{\mathbf{z}} \in S \tag{4.127}$$

in which the subscript b is added to the energy bilinear and load linear forms to emphasize their dependence on design.

We assume a material or sizing design variable b. Employing the definition of variation in Eq. 4.111 to the energy bilinear form $a_b(\mathbf{z}, \bar{\mathbf{z}})$ on the left-hand side of Eq. 4.127, we have

$$[a_b(\mathbf{z}, \bar{\mathbf{z}})]' \equiv \frac{\mathrm{d}}{\mathrm{d}\tau}[a_b(\mathbf{z}(b + \delta b), \bar{\mathbf{z}})]\bigg|_{\tau=0} = a'_{\delta b}(\mathbf{z}, \bar{\mathbf{z}}) + a_b(\mathbf{z}', \bar{\mathbf{z}}) \qquad (4.128)$$

where

$$a'_{\delta b}(\mathbf{z}, \bar{\mathbf{z}}) = \frac{\mathrm{d}}{\mathrm{d}\tau}[a_b(\tilde{\mathbf{z}}, \bar{\mathbf{z}})]\bigg|_{\tau=0} = [a_b(\tilde{\mathbf{z}}, \bar{\mathbf{z}})]' \qquad (4.129)$$

which shows the explicit dependence of the energy bilinear form $a_b(\mathbf{z}, \bar{\mathbf{z}})$ on design variable b, which is similar to the first term on the right-hand side of Eq. 4.124 in Example 4.10—that is, $f'_{\delta b}$. The second term on the right-hand side, $a_b(\mathbf{z}', \bar{\mathbf{z}})$, represents the implicit dependence of $a_b(\mathbf{z}, \bar{\mathbf{z}})$ on design variables b through $\mathbf{z}'$, which is similar to the second term on the right-hand side of Eq. 4.124 in Example 4.10 that includes $\frac{\partial \mathbf{z}}{\partial b} \delta b$.

By applying the definition of variation in Eq. 4.111 to the load linear form, on the right-hand side of Eq. 4.127, we have

$$[\ell_b(\bar{\mathbf{z}})]' \equiv \frac{\mathrm{d}}{\mathrm{d}\tau}[\ell_b(\bar{\mathbf{z}})]\bigg|_{\tau=0} = \ell'_{\delta b}(\bar{\mathbf{z}}) \qquad (4.130)$$

Hence, the sensitivity equation of the continuum approach of a static problem can be obtained as

$$a_b(\mathbf{z}', \bar{\mathbf{z}}) = -a'_{\delta b}(\mathbf{z}, \bar{\mathbf{z}}) + \ell'_{\delta b}(\bar{\mathbf{z}}), \qquad \forall \bar{\mathbf{z}} \in S \qquad (4.131)$$

The principle of virtual work states that the solution $\mathbf{z}'$ is sought in the function space S, such that Eq. 4.131 is satisfied for all $\bar{\mathbf{z}} \in S$. Solving $\mathbf{z}'$ using Eq. 4.130 is nothing but the direct differentiation method of the continuum approach. Similar to Eq. 4.106, the solutions $\mathbf{z}'$ of Eq. 4.131 exist and are unique.

Several important notes are stated below:

1. The terms on the left-hand sides of Eqs 4.131 and 4.127 are similar—that is, $a_b(\mathbf{z}', \bar{\mathbf{z}})$ and $a_b(\mathbf{z}, \bar{\mathbf{z}})$—except that displacement $\mathbf{z}$ in Eq. 4.127 is replaced by $\mathbf{z}'$ in Eq. 4.131.
2. $a_b(\mathbf{z}', \bar{\mathbf{z}})$ is bilinear because $a_b(\mathbf{z}, \bar{\mathbf{z}})$ is bilinear; that is, $a_b(\mathbf{z}', \bar{\mathbf{z}})$ is linear in $\mathbf{z}'$ and $\bar{\mathbf{z}}$, and $\ell'_{\delta b}(\bar{\mathbf{z}})$ is linear in $\bar{\mathbf{z}}$.
3. The right-hand side of Eq. 4.131 defines a pseudo-load (or fictitious load), which can be expressed explicitly in terms of the design variables. Thus, solving the sensitivity equation is the same as solving the original structural equation with a fictitious load per design variable.
4. $\mathbf{z}'$ is as smooth as $\mathbf{z}$; that is, both are in the Sobolev space $H^m(\Omega)$ and satisfy the essential boundary conditions of $\mathbf{z}' = 0$ on Γ_0.
5. For a static problem, $-a'_{\delta b}(\mathbf{z}, \bar{\mathbf{z}})$ is equivalent to $\frac{\partial \mathbf{K}}{\partial b} \mathbf{Z}$ in the discrete approach, and $\ell'_{\delta b}(\bar{\mathbf{z}})$ is equivalent to $\frac{\partial \mathbf{F}}{\partial b}$.

4.4.3 STATIC PROBLEMS

In this subsection, we discuss sensitivity analysis for static problems using the continuum approach. We use a beam example for illustration. We first derive the energy equation for the beam structure by specializing the general energy equations derived in Section 4.4.1 (instead of starting with a

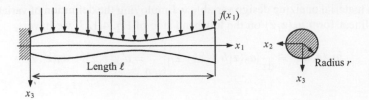

FIGURE 4.9

Cantilever beam example.

differential equation and carrying out integration by parts), and then introducing shape functions to formulate finite element matrix equations. We then discuss sensitivity analysis using both direct differentiation and adjoint variable methods.

Consider a cantilever beam with a circular cross-section under a distributed load $f(x_1)$, as shown in Figure 4.9. The differential governing equation of the beam can be written as

$$\left(EIz_{3,11}\right)_{,11} = f(x_1) \tag{4.132}$$

with essential boundary conditions

$$z_3(0) = z_{3,1}(0) = 0 \tag{4.133}$$

4.4.3.1 Energy equation

We derive the energy equation for the cantilever beam shown in Figure 4.9 by specializing the general energy equation of Eq. 4.105.

For a cantilever beam under bending, the displacement of a deformed beam shown in Figure 4.10a in the longitudinal (x_1-) direction is, as illustrated in Figure 4.10b,

$$z_1 = -x_3 z_{3,1} \tag{4.134}$$

We assume a long beam (or so-called classical beam) where the transverse shear effect is neglected. As a result, the neutral axis is perpendicular to the section of the beam before and after deformation, as depicted in Figure 4.10b. Note that the deformed configuration shown in Figure 4.10 shows $z_{3,1} > 0$ (following the sign convention we assume); therefore, a negative sign in front of z_1 is added in Eq. 4.134.

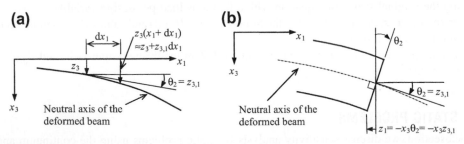

FIGURE 4.10

Deformed cantilever beam. (a) Neutral axis in a deformed configuration with major physical parameters. (b) Zoomed-in view at a given cross-section.

Therefore, the strain along the longitudinal direction, as illustrated in Figure 4.10b, is

$$\varepsilon_{11} = z_{1,1} = \left(-x_3 z_{3,1}\right)_{,1} = -x_3 z_{3,11} \tag{4.135}$$

which is the only nonzero strain. Similarly, stress components consist of only bending stress σ_{11}, obtained as

$$\sigma_{11} = -E x_3 z_{3,11} \tag{4.136}$$

Using Eq. 4.105, we have

$$\int_0^\ell \int_A \sigma_{11}\bar{\varepsilon}_{11} dA dx = \int_0^\ell \int_A \left(-E x_3 z_{3,11}\right)\left(-x_3 \bar{z}_{3,11}\right) dA dx_1$$

$$= \int_0^\ell \left(\int_A x_3^2 dA\right) E z_{3,11} \bar{z}_{3,11} dx_1 = \int_0^\ell EI z_{3,11} \bar{z}_{3,11} dx_1 = \int_0^\ell f(x_1) \bar{z}_3 dx_1 \tag{4.137}$$

Note that the traction force T is zero for the beam example. The principle of virtual work states that the solution z_3 is sought in the space of kinematically admissible virtual displacement S such that Eq. 4.137 is satisfied for all $\bar{z}_3 \in S$. The kinematically admissible virtual displacement S for the cantilever beam is defined as

$$S = \left\{ z_3 \in H^2[0, \ell] \,\middle|\, z_3(0) = z_{3,1}(0) = 0 \right\} \tag{4.138}$$

where H^2 is the Sobolev space of order 2, implying the solution z_3 whose second-order derivative must be integrable.

EXAMPLE 4.12

Derive the energy equation for the cantilever beam shown in Figure 4.9 using integration by parts from the differential equations of Eqs 4.132 and 4.133.

Solution

We introduce the virtual displacement $\bar{z}_3$ that satisfies the essential boundary conditions of Eq. 4.133, and multiply it to both sizes of Eq. 4.132:

$$\int_0^\ell \left(EI z_{3,11}\right)_{,11} \bar{z}_3 dx_1 = \int_0^\ell f(x_1) \bar{z}_3 dx_1 \tag{4.139}$$

We carry out integration by parts twice for Eq. 4.139 to obtain

$$\left(EI z_{3,11}\right)_{,1} \bar{z}_3 \Big|_0^\ell - EI z_{3,11} \bar{z}_{3,1} \Big|_0^\ell + \int_0^\ell EI z_{3,11} \bar{z}_{3,11} dx_1 = \int_0^\ell f(x_1) \bar{z}_3 dx_1 \tag{4.140}$$

Equation (4.140) is called the variational identity and is true for any boundary condition.

Because $\bar{z}_3(0) = \bar{z}_{3,1}(0) = 0$ and the bending moment and transverse shear force at the tip are zero—that is, $EI z_{3,11}(\ell) = 0$, $(EI z_{3,11})_{,1}(\ell) = 0$—the first two terms on the left-hand side of Eq. 4.140 vanish. Then, Eq. 4.140 becomes

$$\int_0^\ell EI z_{3,11} \bar{z}_{3,11} dx_1 = \int_0^\ell f(x_1) \bar{z}_3 dx_1, \quad \forall \bar{z}_3 \in S \tag{4.141}$$

where S is the space of kinematically admissible virtual displacement defined in Eq. 4.138.

4.4.3.2 Finite element discretization

For a beam element under bending, as shown in Figure 4.11b, there are two degrees of freedom at each node: the displacement z_3 and rotation angle θ. For simplicity, we assume constant area A and constant modulus E for the cantilever beam (see Figure 4.11a). Also, the distributed load is assumed to be constant—that is, $f(x_1) = q$. In FEA, a cubic shape function is employed for a (classical) beam element. For a cantilever beam with a point load at the tip, one beam element is sufficient to provide the exact solution because the exact solution is a cubic function of location x_1. However, for a beam with a uniformly distributed load q, the exact displacement solution is a fourth-order function of x_1. A finite element with cubic shape functions will not give an exact solution, except at nodes.

Using cubic shape functions, the displacement function $z_3(x)$ in the beam element can be interpolated as

$$z_3(x) = \mathbf{N}^T\mathbf{Z} = [N_1\ N_2\ N_3\ N_4] \begin{bmatrix} z_{3i} \\ \theta_i \\ z_{3j} \\ \theta_j \end{bmatrix} \tag{4.142}$$

where

$$N_1 = 1 - \frac{3x^2}{\ell^2} + \frac{2x^3}{\ell^3}, \quad N_2 = x - \frac{2x^2}{\ell} + \frac{x^3}{\ell^2},$$

$$N_3 = \frac{3x^2}{\ell^2} - \frac{2x^3}{\ell^3}, \quad N_4 = -\frac{x^2}{\ell} + \frac{x^3}{\ell^2} \tag{4.143}$$

Hence, the second derivative of the displacement can be interpolated as

$$z_{3,11} = \left(\mathbf{N}^T\mathbf{Z}\right)_{,11} = \mathbf{N}_{,11}^T\mathbf{Z} = \left[-\frac{6}{\ell^2} + \frac{12x}{\ell^3} \quad -\frac{4}{\ell} + \frac{6x}{\ell^2} \quad \frac{6}{\ell^2} - \frac{12x}{\ell^3} \quad -\frac{2}{\ell} + \frac{6x}{\ell^2} \right] \begin{bmatrix} z_{3i} \\ \theta_i \\ z_{3j} \\ \theta_j \end{bmatrix}$$

Similarly, $\bar{z}_{3,11} = \left(\mathbf{N}^T\bar{\mathbf{Z}}\right)_{,11} = \mathbf{N}_{,11}^T\bar{\mathbf{Z}}$.

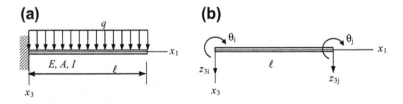

FIGURE 4.11

A simple cantilever beam example. (a) A beam with a uniformly distributed load q. (b) A beam element with displacement and rotation degrees of freedom.

Note that the shape functions of Eq. 4.143 are derived in accordance with the sign conventions shown in Figure 4.11b. Also, to simplify the notation, we omitted the subscript 1 of x_1 in the above equations and those that follow.

Plug $z_{3,11}$ and $\bar{z}_{3,11}$ into the energy bilinear form of Eq. 4.137 and carry out the integrations, which yields

$$\int_0^\ell EI z_{3,11} \bar{z}_{3,11} \, dx = \bar{\mathbf{Z}}^{\mathrm{T}} \mathbf{K}_g \mathbf{Z} \tag{4.144}$$

Here, $\mathbf{K}_g$ is the global stiffness matrix:

$$\mathbf{K}_g = \frac{EI}{\ell^3} \begin{bmatrix} 12 & 6\ell & -12 & 6\ell \\ 6\ell & 4\ell^2 & -6\ell & 2\ell^2 \\ -12 & -6\ell & 12 & -6\ell \\ 6\ell & 2\ell^2 & -6\ell & 4\ell^2 \end{bmatrix}$$

and

$$\int_0^\ell f(x)\bar{z}\,dx = \bar{\mathbf{Z}}^{\mathrm{T}} \int_0^\ell q\mathbf{N}dx = \bar{\mathbf{Z}}^{\mathrm{T}} \begin{bmatrix} \dfrac{1}{2}q\ell \\[4pt] \dfrac{1}{12}q\ell^2 \\[4pt] \dfrac{1}{2}q\ell \\[4pt] -\dfrac{1}{12}q\ell^2 \end{bmatrix} = \bar{\mathbf{Z}}^{\mathrm{T}} \mathbf{F}_g \tag{4.145}$$

Here, $\mathbf{F}_g$ is the global force vector. Equation (4.137) becomes $\bar{\mathbf{Z}}^{\mathrm{T}} \mathbf{K}_g \mathbf{Z} = \bar{\mathbf{Z}}^{\mathrm{T}} \mathbf{F}_g$. Because $\bar{\mathbf{Z}}$ is a virtual displacement that is arbitrary in space S, it can be removed. Hence, we have

$$\mathbf{K}_g \mathbf{Z} = \mathbf{F}_g$$

or

$$\frac{EI}{\ell^3} \begin{bmatrix} 12 & 6\ell & -12 & 6\ell \\ 6\ell & 4\ell^2 & -6\ell & 2\ell^2 \\ -12 & -6\ell & 12 & -6\ell \\ 6\ell & 2\ell^2 & -6\ell & 4\ell^2 \end{bmatrix} \begin{bmatrix} z_{3_i} \\ \theta_i \\ z_{3_j} \\ \theta_j \end{bmatrix} = \begin{bmatrix} \dfrac{1}{2}q\ell \\[4pt] \dfrac{1}{12}q\ell^2 \\[4pt] \dfrac{1}{2}q\ell \\[4pt] -\dfrac{1}{12}q\ell^2 \end{bmatrix} \tag{4.146}$$

Impose the boundary conditions. In this case, $z_{3i} = \theta_i = 0$. The reduced matrix equations become $\mathbf{KZ} = \mathbf{F}$, or

$$\frac{EI}{\ell^3} \begin{bmatrix} 12 & -6\ell \\ -6\ell & 4\ell^2 \end{bmatrix} \begin{bmatrix} z_{3_j} \\ \theta_j \end{bmatrix} = \begin{bmatrix} \dfrac{1}{2}q\ell \\ -\dfrac{1}{12}q\ell^2 \end{bmatrix} \tag{4.147}$$

Inverse the matrix and solve for the nodal point displacements, yielding

$$\begin{bmatrix} z_{3_j} \\ \theta_j \end{bmatrix} = \frac{\ell^3}{EI} \begin{bmatrix} 12 & -6\ell \\ -6\ell & 4\ell^2 \end{bmatrix}^{-1} \begin{bmatrix} \dfrac{1}{2}q\ell \\ -\dfrac{1}{12}q\ell^2 \end{bmatrix} = \frac{\ell}{12EI} \begin{bmatrix} 4\ell^2 & 6\ell \\ 6\ell & 12 \end{bmatrix} \begin{bmatrix} \dfrac{1}{2}q\ell \\ -\dfrac{1}{12}q\ell^2 \end{bmatrix} = \begin{bmatrix} \dfrac{q\ell^4}{8EI} \\ \dfrac{q\ell^3}{6EI} \end{bmatrix} \tag{4.148}$$

Note that from the strength of materials, the exact solution for the cantilever beam with a uniformly distributed load q is

$$z_3^{\text{Exact}}(x) = \frac{q}{24EI} \left(6x^2\ell^2 - 4x^3\ell + x^4 \right) \tag{4.149}$$

The finite element solutions in Eq. 4.148 match those of the exact solutions at the ends, both the displacement and rotation angles.

The finite element solution for displacement $z_3(x)$ for the entire element can then be obtained from Eq. 4.142 as

$$z_3(x) = \mathbf{N}^{\mathsf{T}}\mathbf{Z} = [N_3 N_4] \begin{bmatrix} z_{3_j} \\ \theta_j \end{bmatrix} = \begin{bmatrix} \dfrac{3x^2}{\ell^2} - \dfrac{2x^3}{\ell^3}, & -\dfrac{x^2}{\ell} + \dfrac{x^3}{\ell^2} \end{bmatrix} \begin{bmatrix} \dfrac{q\ell^4}{8EI} \\ \dfrac{q\ell^3}{6EI} \end{bmatrix} = \frac{q\ell}{24EI} \left(5x^2\ell - 2x^3 \right) \tag{4.150}$$

where $z_3(x)$ represents the beam displacements due to the distributed load q, which is a cubic function because cubic shape functions were employed for the finite element formulation.

Apparently, the FEM does not give the exact solution because the element shape function is a cubic function, and the exact solution is a fourth-order polynomial function. When taking the second derivatives of $z_3(x)$ and $z_3^{\text{Exact}}(x)$ from Eqs 4.150 and 4.149, respectively, we have $z_{3,11}(x) = \frac{q\ell}{12EI}(5\ell - 6x)$ and $z_{3,11}^{\text{Exact}}(x) = \frac{q}{2EI}(\ell^2 - 2x\ell + x^2)$. At $x = \ell$, we have $z_{3,11}(\ell) = -\frac{q\ell^2}{12EI} \neq 0$ and $z_{3,11}^{\text{Exact}}(\ell) = 0$. Note that $EIz_{3,11}(x)$ is the bending moment in the beam. At the tip $(x = \ell)$, the bending moment of the cantilever beam should be zero. It is apparent that the exact solution $z_3^{\text{Exact}}(x)$ captures this important behavior; however, the approximate solution obtained from FEA does not. In general, when more elements are employed for the cantilever beam (i.e., by dividing the beam into more elements of smaller size), the finite element solution eventually approaches the exact solution.

4.4.3.3 Direct differentiation method of the continuum-discrete approach

For linear static problems, we focus on displacement, stress, and compliance performance measures. We have discussed bars in Section 4.3. In this section, we use beams for illustration.

We assume a rectangular cross-section for the cantilever beam shown in Figure 4.11, with width w and height h. We use the direct differentiation method of the continuum-discrete approach to calculate the sensitivity of the bending stress σ with the height h as the design variable.

Taking the variation of the energy equation (Eq. 4.137) with respect to the area moment of inertia I, we have

$$\int_0^\ell EIz'_{3,11}\bar{z}_{3,11}\,dx = -\int_0^\ell E\delta Iz_{3,11}\bar{z}_{3,11}\,dx, \quad \forall \bar{z} \in S \tag{4.151}$$

in which $I = \frac{wh^3}{12}$, $I' = \delta I = \frac{\partial I}{\partial h}\delta h = \frac{3I}{h}\delta h$, and $z'_3 = \frac{\partial z_3}{\partial h}\delta h$. From the principle of virtual work, the solution of Eq. 4.151—that is, z'_3—belongs to the space S.

Using the same shape functions as in Eq. 4.143 to discretize the sensitivity equation (Eq. 4.151), we have on the left-hand side

$$\int_0^\ell EIz'_{3,11}\bar{z}_{3,11}\,dx = \bar{\mathbf{Z}}^T\mathbf{K}_g\mathbf{Z}' \tag{4.152}$$

where

$$\mathbf{Z}' = \frac{\partial \mathbf{Z}}{\partial h}\delta h = \left[\begin{array}{cccc} \dfrac{\partial z_{3_i}}{\partial h}\delta h & \dfrac{\partial \theta_i}{\partial h}\delta h & \dfrac{\partial z_{3_j}}{\partial h}\delta h & \dfrac{\partial \theta_j}{\partial h}\delta h \end{array}\right]^T \tag{4.153}$$

On the right-hand side of Eq. 4.151, we have

$$-\int_0^\ell E\delta Iz_{3,11}\bar{z}_{3,11}\,dx = -\frac{3EI}{h}\bar{\mathbf{Z}}^T\int_0^\ell \begin{bmatrix} \dfrac{-6}{\ell^2} + 12\dfrac{x}{\ell^3} \\[2mm] \dfrac{-4}{\ell} + 6\dfrac{x}{\ell^2} \\[2mm] \dfrac{6}{\ell^2} - 12\dfrac{x}{\ell^3} \\[2mm] \dfrac{-2}{\ell} + 6\dfrac{x}{\ell^2} \end{bmatrix}\left[\frac{q\ell}{12EI}(5\ell - 6x)\right]dx\delta h = \frac{q\ell}{4h}\begin{bmatrix} -18 \\ 5\ell \\ -6 \\ \ell \end{bmatrix}\delta h \tag{4.154}$$

By imposing the boundary conditions and removing the virtual displacement $\bar{\mathbf{Z}}$ and δh from both sides, we have $\mathbf{KZ}' = \mathbf{F}_{fic}$:

$$\frac{EI}{\ell^3}\begin{bmatrix} 12 & -6\ell \\ -6\ell & 4\ell^2 \end{bmatrix}\begin{bmatrix} \dfrac{\partial z_{3_j}}{\partial h} \\[3mm] \dfrac{\partial \theta_j}{\partial h} \end{bmatrix} = \frac{q\ell}{4h}\begin{bmatrix} -6 \\ \ell \end{bmatrix} \tag{4.155}$$

The right-hand side becomes a vector of a fictitious load:

$$\mathbf{F}_{\text{fic}} = \frac{q\ell}{4h} \begin{bmatrix} -6 \\ \ell \end{bmatrix} \tag{4.156}$$

By solving the sensitivity vector in the same way as solving for $\mathbf{Z}$ of the structure (see Eq. 4.148), we have

$$\begin{bmatrix} \dfrac{\partial z_{3_j}}{\partial h} \\ \dfrac{\partial \theta_j}{\partial h} \end{bmatrix} = \mathbf{K}^{-1}\mathbf{F}_{\text{fic}} = \frac{\ell}{12EI} \begin{bmatrix} 4\ell^2 & 6\ell \\ 6\ell & 12 \end{bmatrix} \frac{q\ell}{4h} \begin{bmatrix} -6 \\ \ell \end{bmatrix} = \begin{bmatrix} -\dfrac{3q\ell^4}{8EIh} \\ -\dfrac{q\ell^3}{2EIh} \end{bmatrix} \tag{4.157}$$

4.4.3.4 Sensitivity of the bending stress

Now, we calculate the sensitivity of the bending stress σ_{11} with respect to the height h of the beam cross-section. The bending stress σ_{11} can be calculated using the displacement results obtained from FEA as $\sigma_{11} = -Ex_3 z_{3,11}$, and the maximum tensile stress is at the top fiber ($x_3 = -h/2$) of the beam:

$$\sigma_{11} = E\frac{h}{2}z_{3,11} \tag{4.158}$$

Note that we follow the sign conventions shown in Figure 4.12 for the bending moment and stress. The sensitivity of the maximum bending stress in tensile σ_{11} with respect to height h can be computed by taking the derivative of Eq. 4.158 as

$$\frac{\partial \sigma_{11}}{\partial h} = \frac{E}{2}z_{3,11} + \frac{Eh}{2}\frac{\partial z_{3,11}}{\partial h} \tag{4.159}$$

Here, using finite element solution $z_{3,11}$ can be obtained by taking second derivative of z_3 obtained in Eq. 4.150 with respect to x, as

$$z_{3,11}(x) = \frac{q\ell}{12EI}(5\ell - 6x)$$

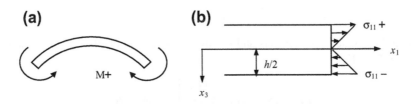

(a)   **(b)**

FIGURE 4.12

Sign conventions of beam stress. (a) Positive bending moment. (b) Positive stress in tensile and negative stress in compression.

The sensitivity of $z_{3,11}$ can be interpolated using element shape functions, similar to that of displacement z_3 in Eq. 4.150, as

$$\frac{\partial z_{3,11}(x)}{\partial h} = \mathbf{N}_{,11}^{\mathrm{T}}\frac{\partial \mathbf{Z}}{\partial h} = \begin{bmatrix} N_{3,11} & N_{4,11} \end{bmatrix} \begin{bmatrix} \dfrac{\partial z_{3_j}}{\partial h} \\[2mm] \dfrac{\partial \theta_j}{\partial h} \end{bmatrix} = \begin{bmatrix} \dfrac{6}{\ell^2} - \dfrac{12x}{\ell^3}, & -\dfrac{2}{\ell} + \dfrac{6x}{\ell^2} \end{bmatrix} \begin{bmatrix} -\dfrac{3q\ell^4}{8EIh} \\[2mm] -\dfrac{q\ell^3}{2EIh} \end{bmatrix} = \frac{q\ell}{4EIh}(6x - 5\ell)$$

Hence, Eq. 4.159 becomes

$$\frac{\partial \sigma_{11}}{\partial h} = \frac{q\ell}{24I}(5\ell - 6x) + \frac{q\ell}{8I}(6x - 5\ell) = -\frac{q\ell}{12I}(5\ell - 6x) \tag{4.160}$$

At the root when $x = 0$, we have

$$\frac{\partial \sigma_{11}}{\partial h} = -\frac{5q\ell^2}{12I} < 0 \tag{4.161}$$

This implies that increasing the height h will decrease the maximum bending stress in tensile at the top fiber of the beam.

Note that in theory, as mentioned earlier, the stress sensitivity at the tip $x = \ell$ is zero because the bending moment—and hence, the bending stress—is zero at the tip. However, as indicated in Eq. 4.160, this is not the case because the solution we used for evaluating the sensitivity information is obtained from FEA. As we are aware, the finite element solution is not exact.

If we use the exact solution to calculate the sensitivity using the continuum-analytical method, the stress sensitivity is exact. We illustrate this point in the next example.

EXAMPLE 4.13

The exact solution of the cantilever beam shown in Figure 4.11a was given in Eq. 4.149 and is restated below:

$$z_3^{\text{Exact}}(x) = \frac{q}{24EI}\left(6x^2\ell^2 - 4x^3\ell + x^4\right) \tag{4.162}$$

We use the exact solution to calculate the sensitivity of the maximum bending stress of the beam defined in Eq. 4.158 using the continuum formulation. Because no finite element discretization is involved, we are exercising the continuum-analytical approach in this example.

Solutions

We first carry out two integrations by parts on Eq. 4.151 to factor out the virtual displacement $\bar{z}$. We are getting back to the differential equation so that we can solve for z_3' when a fictitious load is calculated. The left-hand side of Eq. 4.151 becomes

$$\int_0^\ell EIz_{3,11}'\bar{z}_{3,11}\,dx = EIz_{3,11}'\bar{z}_{3,1}\Big|_0^\ell - EIz_{3,111}'\bar{z}_3\Big|_0^\ell + \int_0^\ell EIz_{3,1111}'\bar{z}_3\,dx \tag{4.163}$$

The first two terms vanish because $\bar{z}_3(0) = \bar{z}_{3,1}(0) = 0$ and z_3' belongs to the same space as z_3, in which $EIz_{3,11}(\ell) = 0$, $(EIz_{3,11})_{,1}(\ell) = 0$, as discussed earlier in this subsection. You may come back to check if $EIz_{3,11}'(\ell) = 0$ and $(EIz_{3,11}')_{,1}(\ell) = 0$ once we obtain z_3'.

Continued

EXAMPLE 4.13—cont'd

The right-hand side of Eq. 4.151 becomes

$$-\int_0^\ell E\delta I z_{3,11}\bar{z}_{3,11}\mathrm{d}x = -E\delta I z_{3,11}\bar{z}_{3,1}\Big|_0^\ell + E\delta I z_{3,111}\bar{z}_3\Big|_0^\ell - \int_0^\ell E\delta I z_{3,1111}\bar{z}_3\mathrm{d}x \tag{4.164}$$

Again, the first two terms of Eq. 4.164 vanish. We remove the integral and delete the virtual displacement $\bar{z}_3$ on both sides of the equation to obtain the sensitivity equation in differential form as

$$EIz'_{3,1111} = -E\delta I z_{3,1111} = -\frac{3EI}{h}z_{3,1111}\delta h = -\frac{3EI}{h}\frac{q}{EI}\delta h = -\frac{3q}{h}\delta h \tag{4.165}$$

Hence, the solution of Eq. 4.165 can be obtained as

$$z'_3(x) = -\frac{3q}{24EIh}\left(6x^2\ell^2 - 4x^3\ell + x^4\right)\delta h \tag{4.166}$$

Its second derivative with respect to x is

$$z'_{3,11}(x) = -\frac{3q}{2EIh}\left(\ell^2 - 2\ell x + x^2\right)\delta h \tag{4.167}$$

At the tip, $z'_{3,11}(\ell) = 0$ and $z'_{3,111}(\ell) = 0$ as expected. From Eq. 4.159, the sensitivity of bending stress σ_{11} is

$$\frac{\partial \sigma_{11}}{\partial h} = \frac{E}{2}z_{3,11} + \frac{Eh}{2}\frac{\partial z_{3,11}}{\partial h} = \frac{q}{4I}\left(\ell^2 - 2x\ell + x^2\right) - \frac{3q}{4I}\left(\ell^2 - 2\ell x + x^2\right) = -\frac{q}{2I}(\ell - x)^2 \tag{4.168}$$

Note that $\frac{\partial \sigma_{11}}{\partial h} < 0$ except at the tip, implying that increasing the height h reduces the maximum bending stress. At the tip, $\frac{\partial \sigma_{11}}{\partial h} = 0$ as expected. The first term of Eq. 4.168 represents the explicit dependence of the stress on the height variable, which is always positive (except at the tip) because increasing h—that is, raising the stress measurement along the height of the beam cross-section—increases bending stress.

Another point worth mentioning is that the stress sensitivity shown in Eq. 4.159 assumes that the stress measurement point is changing as the height h varies. If we assume the stress measurement point stays, the first term of Eq. 4.159 (and Eq. 4.168) vanishes. Therefore, the stress sensitivity becomes

$$\frac{\partial \sigma_{11}}{\partial h} = \frac{Eh}{2}\frac{\partial z_{3,11}}{\partial h} = -\frac{3q}{4I}\left(\ell^2 - 2\ell x + x^2\right) = -\frac{3q}{4I}(\ell - x)^2 \tag{4.169}$$

This implies that the stress decreases more as the height h increases if the measurement point does not move together with the height h.

4.4.3.5 Adjoint variable method of the continuum-discrete approach

We stay with the cantilever beam example to discuss the adjoint variable method of the continuum-discrete approach for a general performance measure, and then use the stress performance measure of the cantilever beam example for illustration.

In the continuum approach, the performance measure is expressed in an integral form as

$$\psi = \int_\Omega g(\mathbf{z}(\mathbf{b}), \nabla\mathbf{z}(\mathbf{b}); \mathbf{b})\mathrm{d}\Omega \tag{4.170}$$

where Ω is the structural domain. For a one-dimensional problem such as the cantilever beam, $\Omega = [0, \ell]$. The displacement $\mathbf{z}$ is function of design variables $\mathbf{b}$, and $\nabla\mathbf{z}$ is the gradient of $\mathbf{z}$ with respect to spatial variables $\mathbf{x}$. For simplicity, we assume a single design variable b. Note that the displacement $\mathbf{z}$ depends on both the spatial variable $\mathbf{x}$ and design variable b. For a one-dimensional problem, such as the cantilever beam in bending, vector $\mathbf{z}$ becomes a scalar quantity z_3, and $\mathbf{x}$ becomes x_1.

Taking the variation of the performance measure ψ with respect to design variable b, we have

$$
\psi' \equiv \frac{\mathrm{d}}{\mathrm{d}\tau}\left[\int_\Omega g(\mathbf{z}(\mathbf{x}, b + \tau\delta b), \nabla\mathbf{z}(\mathbf{x}, b + \tau\delta b),\ b + \tau\delta b)\mathrm{d}\Omega \right]\Bigg|_{\tau=0} \tag{4.171}
$$

$$
= \int_\Omega \left[g_{,\mathbf{z}}\mathbf{z}' + g_{,\nabla\mathbf{z}}\nabla\mathbf{z}' + g_{,b}\delta b \right]\mathrm{d}\Omega
$$

Here, ∇ is the gradient operator, and is defined as

$$
\nabla \equiv \begin{bmatrix} \dfrac{\partial}{\partial x_1} \\[2mm] \dfrac{\partial}{\partial x_2} \\[2mm] \dfrac{\partial}{\partial x_3} \end{bmatrix} \tag{4.172}
$$

for three-dimensional structures. If we simplify the formulation by considering only one-dimensional problems, we have $\nabla = \frac{\partial}{\partial x_1}$ and Eq. 4.171 becomes

$$
\psi' = \int_0^\ell \left[g_{,z}\mathbf{z}' + g_{,\mathbf{z}_{,1}}\mathbf{z}'_{,1} + g_{,b}\delta b \right]\mathrm{d}x \tag{4.173}
$$

In Eq. 4.171 (and Eq. 4.173), the third term in the integrand $g_{,b}\delta b$ represents the explicit dependence of g in design b, and the first two terms are implicit terms that need to be substituted by introducing the adjoint equation.

Note that some terms in the integrand of Eq. 4.173 are zero depending on the type of performance measures. For example, for a displacement measure, $g_{,b}\delta b$ is zero because no explicit dependence exists, and $g_{,\mathbf{z}_{,1}}$ is zero because no derivative is involved in a displacement measure. For a stress performance measure, as seen in previous examples (e.g., Example 4.13), $g_{,b}\delta b$ is nonzero, as well as $g_{,\mathbf{z}_{,1}}$. For beams in bending, we have $g_{,z_{3,11}}$ instead of $g_{,\mathbf{z}_{,1}}$.

The adjoint equation is written as follows, replacing the terms involved in variations in Eq. 4.173:

$$
a_b(\lambda, \overline{\lambda}) = \int_0^\ell \left[g_{,z}\overline{\lambda} + g_{,\mathbf{z}_{,1}}\overline{\lambda}_{,1} \right]\mathrm{d}x, \quad \forall\overline{\lambda}\in S \tag{4.174}
$$

Note that the left-hand side of Eq. 4.174 is the energy bilinear form in terms of adjoint response λ and virtual adjoint response $\overline{\lambda}$. The adjoint response can be obtained in the same way as the structural response $\mathbf{z}$ (e.g., using finite element equations), as long as the load vector, called adjoint load, on the right-hand side can be calculated.

We first evaluate Eq. 4.174 at $\overline{\lambda} = \mathbf{z}'$:

$$a_b(\lambda, \mathbf{z}') = \int_0^\ell \left[g_{,\mathbf{z}}\mathbf{z}' + g_{,\mathbf{z}_1}\mathbf{z}'_{,1} \right] dx , \quad \forall \mathbf{z}' \in S \tag{4.175}$$

Recall the sensitivity equation of the direct differentiation method in Eq. 4.131. We evaluate Eq. 4.131 at $\overline{\mathbf{z}} = \lambda$:

$$a_b(\mathbf{z}', \lambda) = -a'_{\delta b}(\mathbf{z}, \lambda) + \ell'_{\delta b}(\lambda) \tag{4.176}$$

Because the energy bilinear form is symmetric—that is, $a_b(\lambda, \mathbf{z}') = a_b(\mathbf{z}', \lambda)$, Eq. 4.175 equals Eq. 4.176. Therefore,

$$\int_0^\ell \left[g_{,\mathbf{z}}\mathbf{z}' + g_{,\mathbf{z}_1}\mathbf{z}'_{,1} \right] dx = -a'_{\delta b}(\mathbf{z}, \lambda) + \ell'_{\delta b}(\lambda) \tag{4.177}$$

Bringing Eq. 4.177 back to Eq. 4.173, we have

$$\psi' = \int_0^\ell g_{,b}\delta b dx - a'_{\delta b}(\mathbf{z}, \lambda) + \ell'_{\delta b}(\lambda) \tag{4.178}$$

This can be calculated once the original structural response $\mathbf{z}$ and adjoint response λ are available. Note that for beams, we have

$$a'_{\delta b}(\mathbf{z}, \lambda) = \int_0^\ell EIz_{3,11}\lambda_{3,11} dx \tag{4.179}$$

and

$$\ell'_{\delta b}(\lambda) = \int_0^\ell f\lambda_3 dx \tag{4.180}$$

EXAMPLE 4.14

The stress measure at the top fiber of the cantilever beam shown in Figure 4.11a at the root ($x = 0$) is written in an integral form as

$$\psi = \sigma_{11}(0) = E\frac{h}{2}z_{3,11}(0) = \int_0^\ell \left[E\frac{h}{2}z_{3,11}(x)\delta(x - 0) \right] dx \tag{4.181}$$

in which $\delta(x - 0)$ is a direct delta function defined as

$$\delta(x - 0) = \begin{cases} \infty, & x = 0 \\ 0, & x \neq 0 \end{cases} \tag{4.182}$$

EXAMPLE 4.14—cont'd

Calculate the sensitivity of the stress measure with respect to the height design variable h using the adjoint variable method of the continuum-discrete approach.

Solutions

Taking the variation of the performance measure ψ in Eq. 4.181, we have

$$\psi' = \int_0^\ell \left[E\frac{\delta h}{2}z_{3,11}(x)\delta(x-0) + E\frac{h}{2}z'_{3,11}(x)\delta(x-0) \right] dx$$

(4.183)

$$= E\frac{\delta h}{2}z_{3,11}(0) + \int_0^\ell \left[E\frac{h}{2}z'_{3,11}(x)\delta(x-0) \right] dx$$

in which the first term on the right-hand side shows the explicit dependence of the performance measure on design—that is, $\int_0^\ell g_{,b}\delta b dx$ in Eq. 4.173—which is straightforward to solve. The second term involves z'_3, which must be solved by introducing the adjoint structure, defined as

$$a_b(\lambda,\overline{\lambda}) = \int_0^\ell \left[E\frac{h}{2}\overline{\lambda}_{3,11}(x)\delta(x-0) \right] dx$$

(4.184)

Discretizing Eq. 4.184 in the same way as the original structure using the shape functions in Eq. 4.143, the left-hand side becomes

$$a_b(\lambda,\overline{\lambda}) = \int_0^\ell EI\lambda_{3,11}\overline{\lambda}_{3,11}dx = \overline{\lambda}_g^T\mathbf{K}_g\lambda_g$$

(4.185)

and the right-hand side is

$$\overline{\lambda}_g^T\int_0^\ell \left[E\frac{h}{2}\mathbf{N}_{,11}(x)\delta(x-0) \right] dx$$

$$= \overline{\lambda}_g^T\int_0^\ell E\frac{h}{2}\begin{bmatrix} \left[\dfrac{6}{\ell^2} + 12\dfrac{x}{\ell^3}\right] \\[2ex] \left[\dfrac{-4}{\ell} + 6\dfrac{x}{\ell^2}\right] \\[2ex] \dfrac{6}{\ell^2} - 12\dfrac{x}{\ell^3} \\[2ex] \dfrac{-2}{\ell} + 6\dfrac{x}{\ell^2} \end{bmatrix}\delta(x-0)\,dx = \overline{\lambda}_g^T\begin{bmatrix} \dfrac{3Eh}{\ell^2} \\[2ex] \dfrac{-2Eh}{\ell} \\[2ex] \dfrac{3Eh}{\ell^2} \\[2ex] \dfrac{-Eh}{\ell} \end{bmatrix} = \overline{\lambda}_g^T\mathbf{F}_{\mathrm{adj}_g}$$

(4.186)

By imposing the boundary conditions and removing the virtual adjoint displacement $\overline{\lambda}$ from both sides, we have $\mathbf{K}\lambda = \mathbf{F}_{\mathrm{adj}}$. That is, the adjoint load vector $\mathbf{F}_{\mathrm{adj}}$ is

$$\mathbf{F}_{\mathrm{adj}} = \begin{bmatrix} \dfrac{3Eh}{\ell^2} \\[2ex] \dfrac{-Eh}{\ell} \end{bmatrix}$$

(4.187)

Continued

EXAMPLE 4.14—cont'd

which is a point load $\frac{3Eh}{\ell^2}$ applied at the tip of the beam in the downward direction and a point moment $\frac{Eh}{\ell}$ applied at the tip of the beam in the negative direction (counterclockwise), as shown in the figure below.

$$\frac{3Eh}{\ell^2}$$

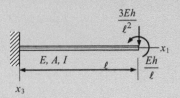

Solve the adjoint response by using the finite element equation, similar to Eq. 4.148:

$$\boldsymbol{\lambda} = \mathbf{K}^{-1}\mathbf{F}_{\text{adj}} = \frac{\ell}{12EI}\begin{bmatrix} 4\ell^2 & 6\ell \\ 6\ell & 12 \end{bmatrix}\begin{bmatrix} \dfrac{3Eh}{\ell^2} \\[2ex] -\dfrac{Eh}{\ell} \end{bmatrix} = \begin{bmatrix} \dfrac{h\ell}{2I} \\[2ex] \dfrac{h}{2I} \end{bmatrix} \tag{4.188}$$

Hence

$$\lambda_{3,11} = \begin{bmatrix} N_{3,11} & N_{4,11} \end{bmatrix}\begin{bmatrix} \dfrac{h\ell}{2I} \\[2ex] \dfrac{h}{2I} \end{bmatrix} = \begin{bmatrix} \dfrac{6}{\ell^2} - \dfrac{12x}{\ell^3} & -\dfrac{2}{\ell} + \dfrac{6x}{\ell^2} \end{bmatrix}\begin{bmatrix} \dfrac{h\ell}{2I} \\[2ex] \dfrac{h}{2I} \end{bmatrix} = \dfrac{h}{I\ell}\left(2 - \dfrac{3x}{\ell}\right) \tag{4.189}$$

By bringing the original responses $z_{3,11}$ and adjoint responses $\lambda_{3,11}$ to the sensitivity equation (Eq. 4.178), we have

$$\psi' = \int_0^\ell g_{,b}\delta b\, dx - a'_{\delta b}(\mathbf{z}, \boldsymbol{\lambda}) + \ell'_{\delta b}(\boldsymbol{\lambda})$$

$$= \int_0^\ell \left[E\frac{\delta h}{2}z_{3,11}(x)\delta(x-0)\right]dx - \int_0^\ell E\delta I z_{3,11}\lambda_{3,11}dx + 0 \tag{4.190}$$

$$= E\frac{\delta h}{2}z_{3,11}(0) - \int_0^\ell E\frac{3I}{h}\left[\frac{q\ell}{12EI}(5\ell - 6x)\right]\frac{h}{I\ell}\left(2 - \frac{3x}{\ell}\right)dx\delta h$$

$$= \frac{5q\ell^2}{24I}\delta h - \frac{5q\ell^2}{8I}\delta h = -\frac{5q\ell^2}{12I}\delta h$$

4.4.3.6 Adjoint variable method of the continuum-analytical approach

We define a compliance performance measure and derive the sensitivity expression for the measure using the adjoint variable method in the continuum form. We then use the analytical solution of the cantilever beam example again to illustrate a few details of the continuum-analytical approach.

A compliance performance measure of a structure is defined as

$$\psi = \int_\Omega f(\mathbf{x})z\mathrm{d}\Omega \tag{4.191}$$

where $f(\mathbf{x})$ is the external load per length (or per area and per volume for two- and three-dimensional problems, respectively) applied to the structure. Compliance is a global measure; a smaller compliance implies that the structure is more rigid. As shown in Eq. 4.191 for a less compliant structure, a smaller displacement is expected for a given external load $f(\mathbf{x})$.

If we assume a uniformly distributed load applied to the cantilever beam, as shown in Figure 4.11a, $f(x)$ is a uniformly distributed load per length; that is, $f(x) = q$. If we also include the self-weight of the beam, the force $f(x)$ per length can be written as

$$f(x) = q + \gamma A \tag{4.192}$$

where γ is the weight density. Note that for the cantilever beam shown in Figure 4.11a, the displacement is z_3. Hence, Eq. 4.191 becomes

$$\psi = \int_0^\ell (q + \gamma A)z_3\mathrm{d}x \tag{4.193}$$

Taking the variation of the performance measure defined in Eq. 4.193, we have

$$\psi' = \int_0^\ell (\gamma A)_{,b}\delta b z_3\mathrm{d}x + \int_0^\ell (q + \gamma A)z_3'\mathrm{d}x \tag{4.194}$$

If the height h is the design variable, then $(\gamma A)_{,h}\delta h = \gamma(wh)_{,h}\delta h = \gamma w\delta h$. We create an adjoint structure as

$$a_b(\lambda, \overline{\lambda}) = \int_0^\ell (q + \gamma A)\overline{\lambda}_3\mathrm{d}x \tag{4.195}$$

in which the right-hand side is identical to the load linear form $\ell_b(\overline{\mathbf{z}}) = \int_0^\ell (q + \gamma A)\overline{z}_3\mathrm{d}x$. Hence, solving Eq. 4.195 for λ_3 is identical to solving the original structure for z_3, implying that the adjoint response is identical to the original structural response; that is, $\lambda_3 = z_3$. This is called self-adjoint and is only true for a compliance performance measure. The sensitivity equation becomes

$$\psi' = \int_0^\ell \left[(\gamma A)_{,b}\delta b \right] z_3\mathrm{d}x - a_{\delta b}'(\mathbf{z}, \mathbf{z}) + \ell_{\delta b}'(\mathbf{z}) \tag{4.196}$$

EXAMPLE 4.15

The displacement of the cantilever beam due to the distributed load q and self-weight γA is

$$z_3^{\text{Exact}}(x) = \frac{q + \gamma A}{24EI}\left(6x^2\ell^2 - 4x^3\ell + x^4\right) \tag{4.197}$$

Calculate the sensitivity of the compliance performance measure for the area design variable A and modulus design variable E using the continuum-analytical approach.

Solutions

Twice taking the derivatives of Eq. 4.197 with respect to x, we have

$$z_{3,11}(x) = \frac{q + \gamma A}{2EI}\left(\ell^2 - 2x\ell + x^2\right) = \frac{q + \gamma A}{2EI}(\ell - x)^2 \tag{4.198}$$

From Eq. 4.196, we have

$$
\begin{aligned}
\psi &= \int_0^\ell (\gamma A)_{,b}\,\delta b z_3 \,dx - d'_{\delta b}(\mathbf{z}, \mathbf{z}) + \ell'_{\delta b}(\mathbf{z}) \\[2mm]
&= \int_0^\ell (\gamma w) z_3 \,dx\delta h - \int_0^\ell \left(E\frac{3I}{h}\delta h + I\delta E\right)z_{3,11}^2 \,dx\delta h + \int_0^\ell (\gamma w) z_3 \,dx\delta h \\[2mm]
&= \left\{2\int_0^\ell (\gamma w) z_3 \,dx - \int_0^\ell E\frac{3I}{h}z_{3,11}^2 \,dx\right\}\delta h + \left\{-\int_0^\ell I z_{3,11}^2 \,dx\right\}\delta E \\[2mm]
&= \frac{(q+\gamma A)\ell^5}{20EI}\left(2\gamma w - 3\frac{q+\gamma A}{h}\right)\delta h - \frac{(q+\gamma A)^2\ell^5}{20E^2 I}\delta E
\end{aligned}
\tag{4.199}
$$

The sensitivity coefficient of design variable E is negative, implying that increasing the modulus E decreases the compliance of the structure (the cantilever beam becomes more rigid). On the other hand, the sign of the sensitivity coefficient of design variable h depends on the changes in self-weight and change in the moment of inertia due to the change in height h. Increasing h increases the compliance due to the increment in self-weight. Increasing h makes the beam more rigid, hence decreasing the compliance of the structure.

4.4.4 NUMERICAL IMPLEMENTATION

As mentioned in Section 4.3.4, sensitivity equations derived from continuum-discrete approach, either direct differentiation method or adjoint variable method, can be implemented external to commercial FEA codes that are employed for structural analysis.

Numerical computation of Eq. 4.131 (direct differentiation method) or Eq. 4.178 (adjoint variable method) requires knowledge of the original structural response, $\mathbf{z}$, and/or the adjoint structural response, $\boldsymbol{\lambda}$. The solution $\mathbf{z}$ of Eq. 4.106 is obtained by structural finite element analysis. Using the adjoint variable method, the solution $\boldsymbol{\lambda}$ of the adjoint equation (Eq. 4.174) can be obtained by restarting the finite element analysis code that was used for the analysis model but with additional loading vectors that are defined by the right-hand side of Eq. 4.174. Notice that Eq. 4.174 has to be solved for each displacement or stress performance measure that has a corresponding adjoint load. For other performance measures, such as compliance, natural frequency, buckling load, volume, and mass performance measures, no adjoint structural analysis is necessary.

Using direct differentiation method, fictitious loads must be calculated by carrying out numerical integration of the terms on the right-hand side of Eq. 4.131 over the entire structural domain. Note that Eq. 4.131 has to be solved for each design variable per load case.

For implementation with an existing FEA code, there are four software programs that need to be developed: (1) interface program that retrieves results z and/or λ from database of the FEA code, (2) program that generates input data file of the FEA code for reanalysis with either adjoint load or fictitious load, (3) program that performs numerical integration to evaluate the terms on the right-hand side of Eq. 4.131 and/or Eq. 4.174, and (4) script that integrates programs to carry out batch mode computation. Note that computation for the fictitious load or sensitivity coefficients must be carried out numerically by using, for example, Gauss integration (Atkinson 1989) since the FEA results are given in numerical data instead of notations as seen in the examples of this section.

4.5 SHAPE SENSITIVITY ANALYSIS

Shape sensitivity analysis characterizes the influence of structural geometric change to its performance. For frame structures consisting of bars or beams, a change in dimension variables causes changes in the length and orientation angle of individual bar or beam members in the structure. Length change is referred to as domain shape change, and angle change is referred to as configuration change. For 2-D planar and 3-D solid structures, shape sensitivity analysis only involves domain shape change. In this section, we discuss domain shape sensitivity analysis. For those who are interested in learning more about configuration design for frame structures, Choi and Kim (2006a) offer excellent details. To avoid the complex underlying mathematical formulation involved in shape sensitivity analysis theory, our discussion is focused more on the practical aspects. However, because the basic concepts cannot be introduced without discussing the theory, we use a simple cantilever beam example to minimize complex mathematical formulations.

4.5.1 DOMAIN SHAPE SENSITIVITY ANALYSIS

In the shape design of 2-D planar or 3-D solid structures, the component under consideration is typically a continuum, and its shape is defined by geometric boundaries. Usually, a portion of the geometric boundary is designated as the design boundary for shape sensitivity analysis (and optimization). For a 2-D planar structure, the design boundary is a curve (or composite curve). For 3-D problems, the design boundary is a surface (or composite surface). In order to carry out shape sensitivity analysis, the design boundary of the structural component must be parameterized in a mathematical form that is compatible to computer-aided design (CAD) so that follow-up engineering assignments involved in product design, such as machining or manufacturing process planning, can be readily carried out. The design boundary can be represented by freeform curves or surfaces (e.g., a Bézier curve or a B-spline surface) or by regular geometric curves and surfaces (e.g., an elliptical arc, a cylindrical surface) commonly found in CAD. Shape parameterization of freeform surfaces as well as regular CAD geometric features, will be discussed in Section 4.5.3.

The second key issue in shape optimization is design velocity field computation (Choi and Chang 1994). The design velocity field characterizes the movement of material points of the structural domain due to changes at the boundary. In practice, the design velocity field is also used to update finite element mesh during shape optimization. Updating mesh using design velocity retains the topology of the finite element mesh throughout the design iterations; therefore, the consistency of performance measures evaluated using FEA throughout design iterations is ensured. This mesh update is especially critical for tracing local performance measures, such as displacements or stresses, during design iterations. The design velocity field must comply with the geometric shape of the boundary; that is, all finite element nodes at the design boundary must stay on the boundary when the design is varied. It is critical that one single geometric representation at the design boundary supports structural analysis, sensitivity analysis, and follow-on engineering tasks, such as machining. Design velocity field computation and related issues are briefly explained in Section 4.5.4.

After the design boundary is parameterized and design velocity field is calculated, shape sensitivity analysis can be carried out. The easiest approach for shape sensitivity analysis is probably the overall finite difference. The semianalytical method is more efficient than the overall finite difference. Like sizing sensitivity analysis, these methods require an adequate design perturbation size for accurate sensitivity coefficients. On the other hand, the continuum approach is efficient and does not require any design perturbations. We discuss the finite difference method and continuum approach in Sections 4.5.5 and 4.5.7, respectively. Before getting into these subjects, in the next subsection we use a simple cantilever beam to illustrate some of the basic but essential concepts involved in shape sensitivity analysis.

4.5.2 A SIMPLE CANTILEVER BEAM EXAMPLE

Shape design involves altering the geometric shape of the structure for improved or optimal performance. For a simple case, such as the cantilever beam discussed in Section 4.4.3 and shown in Figure 4.13a again, the displacement obtained from FEA is

$$z_3(x) = \frac{q\ell}{24EI}\left(5x^2\ell - 2x^3\right) \tag{4.200}$$

At the tip, the displacement is

$$z_3^\ell = z_3(\ell) = \frac{q\ell}{24EI}\left(5\ell^3 - 2\ell^3\right) = \frac{q\ell^4}{8EI} \tag{4.201}$$

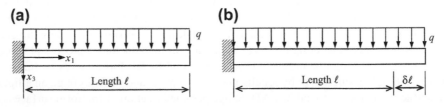

(a)   **(b)**

q q

x_1

x_3 Length ℓ Length ℓ $\delta\ell$

FIGURE 4.13

Cantilever beam example: (a) current design and (b) perturbed design with length $\ell + \delta\ell$, in which the overall force is increased from $q\ell$ to $q(\ell + \delta\ell)$.

For a cantilever beam, the only shape design variable is its length. The easiest way of calculating shape sensitivity coefficient for the displacement is by taking the derivative of z_3^ℓ with respect to length ℓ:

$$\frac{\partial z_3^\ell}{\partial \ell} = \frac{\partial}{\partial \ell} \frac{q\ell^4}{8EI} = \frac{q\ell^3}{2EI} \tag{4.202}$$

which is nothing but the analytical derivative method discussed in Section 4.3.1. Because the sign of the derivative in Eq. 4.202 is positive, it is implied that increasing the length of beam increases its displacement at the tip. Such an increment is due to three factors: the increment in beam length, the increment in the overall load from $q\ell$ to $q(\ell + \delta\ell)$, and the movement of the displacement measurement point that stays at the tip of the beam (i.e., moving with the length change).

There are different approaches to calculating the shape sensitivity of the beam. The analytical approach is shown in Eq. 4.201. Another simple way for calculating the sensitivity coefficient is the discrete-analytical approach. Similar to Eq. 4.40 formulated for sizing design variables, for the length design variable, we have

$$\mathbf{K}\frac{\partial \mathbf{Z}}{\partial \ell} = \frac{\partial \mathbf{F}}{\partial \ell} - \frac{\partial \mathbf{K}}{\partial \ell}\mathbf{Z} \tag{4.203}$$

in which

$$\frac{\partial \mathbf{K}}{\partial \ell} = \frac{\partial}{\partial \ell}\left\{ EI \begin{bmatrix} \dfrac{12}{\ell^3} & -\dfrac{6}{\ell^2} \\[2mm] -\dfrac{6}{\ell^2} & \dfrac{4}{\ell} \end{bmatrix} \right\} = EI \begin{bmatrix} -\dfrac{36}{\ell^4} & \dfrac{12}{\ell^3} \\[2mm] \dfrac{12}{\ell^3} & -\dfrac{4}{\ell^2} \end{bmatrix} \tag{4.204}$$

and

$$\frac{\partial \mathbf{F}}{\partial \ell} = \frac{\partial}{\partial \ell}\begin{bmatrix} \dfrac{1}{2}q\ell \\[2mm] -\dfrac{1}{12}q\ell^2 \end{bmatrix} = \begin{bmatrix} \dfrac{1}{2}q \\[2mm] -\dfrac{1}{6}q\ell \end{bmatrix} \tag{4.205}$$

Collecting the terms on the right-hand side of Eq. 4.203, we have the fictitious load vector:

$$\mathbf{F}_{\text{fic}} = \frac{\partial \mathbf{F}}{\partial \ell} - \frac{\partial \mathbf{K}}{\partial \ell}\mathbf{Z} = \begin{bmatrix} \dfrac{1}{2}q \\[2mm] -\dfrac{1}{6}q\ell \end{bmatrix} - EI \begin{bmatrix} -\dfrac{36}{\ell^4} & \dfrac{12}{\ell^3} \\[2mm] \dfrac{12}{\ell^3} & -\dfrac{4}{\ell^2} \end{bmatrix} \begin{bmatrix} \dfrac{q\ell^4}{8EI} \\[2mm] \dfrac{q\ell^3}{6EI} \end{bmatrix} = \begin{bmatrix} 3q \\[2mm] -q\ell \end{bmatrix} \tag{4.206}$$

which is a point load $3q$ acting at the tip in the downward direction, and a point moment $q\ell$ at the tip acting in a counterclockwise direction. Solving for Eq. 4.203, we have

$$\begin{bmatrix} \dfrac{\partial z_3^\ell}{\partial \ell} \\[3mm] \dfrac{\partial \theta}{\partial \ell} \end{bmatrix} = \mathbf{K}^{-1}\mathbf{F}_{\text{fic}} = \frac{\ell}{12EI}\begin{bmatrix} 4\ell^2 & 6\ell \\[1mm] 6\ell & 12 \end{bmatrix}\begin{bmatrix} 3q \\[1mm] -q\ell \end{bmatrix} = \begin{bmatrix} \dfrac{q\ell^3}{2EI} \\[3mm] \dfrac{q\ell^2}{2EI} \end{bmatrix} \tag{4.207}$$

in which $\frac{\partial z_3^\ell}{\partial \ell} = \frac{q\ell^3}{2EI}$, which is same as that of Eq. 4.202.

Now, let us try a slightly different way to calculate the displacement sensitivity. Instead of taking the derivative of the displacement at the tip z_3^ℓ with respect to length ℓ, we take derivative of the function $z_3(x)$ with respect to length ℓ, then evaluate the derivative at the tip $x = \ell$. Taking the derivative of $z_3(x)$ with respect to length ℓ, we have

$$\frac{\partial z_3(x)}{\partial \ell} = \frac{\partial}{\partial \ell}\left[\frac{q}{24EI}\left(5x^2\ell^2 - 2x^3\ell\right)\right] = \frac{qx^2}{12EI}\left(5\ell - x\right) \tag{4.208}$$

At the tip of the beam, the displacement sensitivity is

$$\frac{\partial z_3}{\partial \ell} = \left.\frac{\partial z_3(x)}{\partial \ell}\right|_{x=\ell} = \frac{q\ell^3}{3EI} \tag{4.209}$$

Compared Eqs. 4.209 to 4.202, the two sensitivity coefficients $\frac{\partial z_3}{\partial \ell}$ and $\frac{\partial z_3^\ell}{\partial \ell}$ are not identical. Why? What do these two sensitivity coefficients mean? To answer this question, let us first rewrite Eq. 4.202 into a finite difference form:

$$\frac{\partial z_3^\ell}{\partial \ell} \approx \frac{\Delta z_3^\ell}{\Delta \ell} = \frac{z_3^\ell(\ell + \delta\ell) - z_3^\ell(\ell)}{\delta\ell} = \frac{\dfrac{q(\ell + \delta\ell)^4}{8EI} - \dfrac{q\ell^4}{8EI}}{\delta\ell}$$

$$= \frac{q}{8EI}\frac{4\ell^3\delta\ell + 6\ell^2\delta\ell^2 + 4\ell\delta\ell^3 + \delta\ell^4}{\delta\ell} = \frac{q\ell^3}{2EI} + \frac{q}{8EI}\left(6\ell^2\delta\ell + 4\ell\delta\ell^2 + \delta\ell^3\right) \tag{4.210}$$

in which $z_3^\ell(\ell + \delta\ell) = \frac{q(\ell+\delta\ell)^4}{8EI}$ represents the displacement at the tip of the beam after a design change; that is, at $x = \ell + \delta\ell$, as illustrated in Figure 4.14. Certainly, only the first term survives in Eq. 4.210 when we let the design change $\delta\ell$ approach zero.

Now, we rewrite Eq. 4.208 in the finite difference form:

$$\frac{\partial z_3(x)}{\partial \ell} \approx \frac{\Delta z_3(x)}{\Delta \ell} = \frac{z_3(x; \ell + \delta\ell) - z_3(x; \ell)}{\delta\ell}$$

$$= \frac{\dfrac{q}{24EI}\left(5x^2(\ell + \delta\ell)^2 - 2x^3(\ell + \delta\ell)\right) - \dfrac{q}{24EI}\left(5x^2\ell^2 - 2x^3\ell\right)}{\delta\ell} \tag{4.211}$$

$$= \frac{q}{24EI}\frac{2x^2(5x - \ell)\delta\ell + 5x^2\delta\ell^2}{\delta\ell} = \frac{qx^2}{12EI}(5x - \ell) + \frac{q}{24EI}\left(5x^2\delta\ell\right)$$

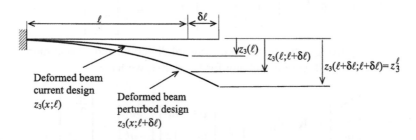

FIGURE 4.14

Deformed cantilever beam in the current and perturbed designs.

in which $z_3(x; \ell + \delta\ell)$ represents displacement function z_3 at the perturbed design:

$$z_3(x; \ell + \delta\ell) = \frac{q}{24EI}\left(5x^2(\ell + \delta\ell)^2 - 2x^3(\ell + \delta\ell)\right) \tag{4.212}$$

If we evaluate Eq. 4.212 at the tip of the perturbed design—that is, at $x = \ell + \delta\ell$—then

$$z_3(\ell + \delta\ell; \ell + \delta\ell) = \frac{q}{24EI}\left(5(\ell + \delta\ell)^4 - 2(\ell + \delta\ell)^4\right) = \frac{q}{8EI}(\ell + \delta\ell)^4 \tag{4.213}$$

which is $z_3^\ell(\ell + \delta\ell)$. However, in Eq. 4.208, we assume $x = \ell$; that is, the measurement point is stationary, not moving with the design—hence, the result of Eq. 4.209.

As illustrated above, different shape sensitivity coefficients are obtained depending on whether the measurement point is stationary.

Another important observation to point out is the measurement point inside the beam—that is, $0 < x < \ell$, for example, at the midpoint $x = \ell/2$. How does this measurement point move with design? It is moving half the length change $\delta\ell/2$, assuming the interior material moves proportionally with design, as shown in Figure 4.15a? Or, can the midpoint movement be determined in another way, such as by a quadratic function, as shown in Figure 4.15b?

In fact, in shape design, the material point movement is determined by the design velocity field, which plays an important role in shape sensitivity analysis (and optimization). The sensitivity we calculated predicts the performance measure of the material point whose location is determined by the velocity field employed in the calculation. The material point location at the next design iteration is also determined by the design velocity field. These key concepts illustrated using the beam example will be extended to more complex and general problems in the following discussion.

Although shape sensitivity analysis can be carried out easily for this cantilever beam example, the application of shape design for one-dimensional structures is limited. As mentioned at the beginning of the section, for a frame structure, when its geometric shape changes, individual components (beams or bars) experience not only length changes but also changes in orientation angles—the so-called configuration changes. In this section, we focus on domain shape design. Domain shape designs are often seen in two-dimensional planar or three-dimensional solid structural components.

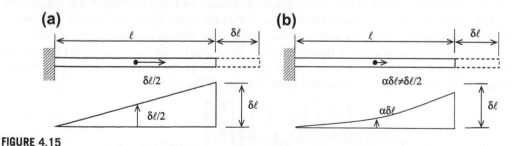

FIGURE 4.15

Midpoint movement with design change $\delta\ell$: (a) $\delta\ell/2$: linear velocity and (b) $\alpha\delta\ell \neq \delta\ell/2$, $\alpha \neq 0$: other than linear velocity.

4.5.3 SHAPE DESIGN PARAMETERIZATION

Shape design variables govern the geometric shape of the structural boundary, usually represented by parametric curves and surfaces for 2-D and 3-D applications, respectively. It is important to select a parameterization scheme that properly reflects the design intent. If you are not familiar with parametric curves and surfaces, you are encouraged to review Chapter 2 of *Product Design Modeling using CAD/CAE* (Chang 2014), one of the modules of the *Computer Aided Engineering Design* book series, before reading this subsection. The following explains parameterization of a few selected geometric entities for 2-D and 3-D structures. These geometric entities are supported in CAD in addition to widely accepted geometric modeling tools, such as MSC/PATRAN (www.mscsoftware.com/product/patran) and HyperMesh (www.altairhyperworks.com).

4.5.3.1 2-D Planar structures

For 2-D structures, the design boundaries are planar curves. To stay focused, we assume the design boundary is represented by a cubic curve. There are eight degrees of freedom for a planar parametric cubic curve:

$$\mathbf{P}(u) = \begin{bmatrix} u^3 & u^2 & u & 1 \end{bmatrix}_{1\times 4} \begin{bmatrix} a_{3x} & a_{3y} \\ a_{2x} & a_{2y} \\ a_{1x} & a_{1y} \\ a_{0x} & a_{0y} \end{bmatrix}_{4\times 2} = \mathbf{U}_{1\times 4}\mathbf{A}_{4\times 2}, \quad u \in [0, 1] \qquad (4.214)$$

where u is the parametric coordinate of the curve. The 4×2 matrix $\mathbf{A}$ contains algebraic coefficients $\mathbf{a}_i$ that determine the geometric shape of the curves. Equation (4.214) is called the algebraic format. In practice, boundary curves are not parameterized using these algebraic coefficients because the change of coefficients $\mathbf{a}_i$ to curve geometry is not clear.

There are several common formats by which the curves can be better parameterized, such as the spline curve, Hermit cubic curve, Bézier curve, and B-spline curve, where the influences of the parameter change to the geometric shape of the curve are obvious. Mathematically, these curves are represented as a set of basis functions $\mathbf{B}(u)$ and geometric control matrix $\mathbf{G}$ that does not depend on the parametric coordinate u:

$$\mathbf{P}(u) = \mathbf{U}(u)\mathbf{A} = \mathbf{U}(u)\mathbf{N}\mathbf{G} = \mathbf{B}(u)\mathbf{G} \qquad (4.215)$$

Here, the basic functions can be written as a multiplication of the vector $\mathbf{U}$ and a matrix $\mathbf{N}$ that is a constant 4×4 matrix determined by the respective curve format. For a cubic curve, $\mathbf{G}$ is a 4×2 matrix that controls the geometric shape of the curve. The entries of the matrix $\mathbf{G}$ are selected as shape design variables.

We assume a cubic Bézier curve to illustrate further. The cubic Bézier curve shown in Figure 4.16a is determined by the position of its four control points $\mathbf{G}$, which form a control polygon. Mathematically, a Bézier curve can be written in a form like that of Eq. 4.215:

$$\mathbf{P}(u) = \begin{bmatrix} u^3 & u^2 & u & 1 \end{bmatrix} \begin{bmatrix} -1 & 3 & -3 & 1 \\ 3 & -6 & 3 & 0 \\ -3 & 3 & 0 & 0 \\ 1 & 0 & 0 & 0 \end{bmatrix} \begin{bmatrix} \mathbf{P}_0 \\ \mathbf{P}_1 \\ \mathbf{P}_2 \\ \mathbf{P}_3 \end{bmatrix} = \mathbf{U}\mathbf{N}^{\mathrm{B}}\mathbf{G}^{\mathrm{B}} = \mathbf{B}^{\mathrm{B}}\mathbf{G}^{\mathrm{B}} \qquad (4.216)$$

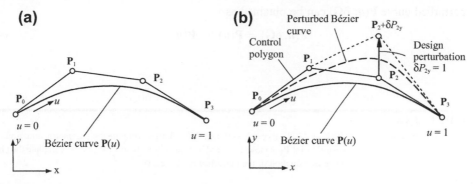

FIGURE 4.16

(a) A cubic Bézier curve parameterized by four control points that form a control polygon. (b) The perturbed curve by moving control point $\mathbf{P}_2$.

where

$$
\mathbf{G}^{\mathrm{B}} = \begin{bmatrix} \mathbf{P}_0 \\ \mathbf{P}_1 \\ \mathbf{P}_2 \\ \mathbf{P}_3 \end{bmatrix} = \begin{bmatrix} P_{0x} & P_{0y} \\ P_{1x} & P_{1y} \\ P_{2x} & P_{2y} \\ P_{3x} & P_{3y} \end{bmatrix}_{4\times 2}
\tag{4.217}
$$

and

$$
\mathbf{N}^{\mathrm{B}} = \begin{bmatrix} -1 & 3 & -3 & 1 \\ 3 & -6 & 3 & 0 \\ -3 & 3 & 0 & 0 \\ 1 & 0 & 0 & 0 \end{bmatrix}_{4\times 4}
\tag{4.218}
$$

which is symmetric. Hence, the basis functions, called Bernstein polynomials, can be obtained as

$$
\mathbf{U}\mathbf{N}^{\mathrm{B}} = \begin{bmatrix} -u^3 + 3u^2 - u + 1, & u^3 - 6u^2 + 3u, & -u^3 + 3u^2, & u^3 \end{bmatrix}_{1\times 4}
\tag{4.219}
$$

Movement of any of the four control points in the x- or y-direction can be chosen as a shape design variable. Any change in the position of the control points will result in a change of the geometric shape of the curve through the basis functions. For example, if the control point $\mathbf{P}_2$ moves vertically up to a new position by δP_{2y}, as shown in Figure 4.16b, the change in curve can be obtained as

$$
\delta \mathbf{P}(u) = \mathbf{U}\mathbf{N}^{\mathrm{B}}\delta \mathbf{G}^{\mathrm{B}} = \begin{bmatrix} u^3 & u^2 & u & 1 \end{bmatrix} \begin{bmatrix} -1 & 3 & -3 & 1 \\ 3 & -6 & 3 & 0 \\ -3 & 3 & 0 & 0 \\ 1 & 0 & 0 & 0 \end{bmatrix} \begin{bmatrix} 0 & 0 \\ 0 & 0 \\ 0 & \delta P_{2y} \\ 0 & 0 \end{bmatrix}
\tag{4.220}
$$

The perturbed curve $\mathbf{P}(u; \delta\mathbf{G})$ can be obtained as

$$\mathbf{P}(u; \delta\mathbf{G}) = \mathbf{P}(u) + \delta\mathbf{P}(u) \tag{4.221}$$

EXAMPLE 4.16

Given four control points—$\mathbf{P}_0 = [0, 0]$, $\mathbf{P}_1 = [1, 3]$, $\mathbf{P}_2 = [2, -2]$, and $\mathbf{P}_3 = [3, 0]$—compute the parametric equation of the Bézier curve formed by them. If the control point $\mathbf{P}_2$ is moved upward by two units—that is, $\delta P_{2y} = 2$—calculate the change in curve $\delta\mathbf{P}(u)$, and calculate the parametric equation of the perturbed Bézier curve.

Solutions
Using Eq. 4.216, we have

$$\mathbf{P}(u) = \begin{bmatrix} u^3 & u^2 & u & 1 \end{bmatrix} \begin{bmatrix} -1 & 3 & -3 & 1 \\ 3 & -6 & 3 & 0 \\ -3 & 3 & 0 & 0 \\ 1 & 0 & 0 & 0 \end{bmatrix} \begin{bmatrix} 0 & 0 \\ 1 & 3 \\ 2 & -2 \\ 3 & 0 \end{bmatrix} = \begin{bmatrix} 3u, & 15u^3 - 24u^2 + 9u \end{bmatrix} \tag{4.222}$$

The change in the curve due to $\delta P_{2y} = 2$ can be obtained using Eq. 4.220 as

$$\delta\mathbf{P}(u) = \mathbf{U}\mathbf{N}^B\delta\mathbf{P}^B = \begin{bmatrix} u^3 & u^2 & u & 1 \end{bmatrix} \begin{bmatrix} -1 & 3 & -3 & 1 \\ 3 & -6 & 3 & 0 \\ -3 & 3 & 0 & 0 \\ 1 & 0 & 0 & 0 \end{bmatrix} \begin{bmatrix} 0 & 0 \\ 0 & 0 \\ 0 & 2 \\ 0 & 0 \end{bmatrix} = \begin{bmatrix} u^3 & u^2 & u & 1 \end{bmatrix} \begin{bmatrix} 0 & -6 \\ 0 & 6 \\ 0 & 0 \\ 0 & 0 \end{bmatrix}$$

$$= \begin{bmatrix} 0, & -6u^3 + 6u^2 \end{bmatrix} \tag{4.223}$$

The perturbed curve $\mathbf{P}(u; \delta\mathbf{G})$ can be obtained using Eq. 4.221 as

$$\mathbf{P}(u; \delta\mathbf{G}) = \mathbf{P}(u) + \delta\mathbf{P}(u) = \begin{bmatrix} 3u, & 9u^3 - 18u^2 + 9u \end{bmatrix} \tag{4.224}$$

The Bézier curves before and after perturbation are shown below:

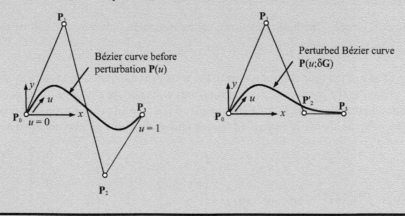

4.5.3.2 3-D Solid structures—freeform surfaces

For a 3-D solid structure, its design boundary can be modeled as a spatial parametric surface or CAD-generated surface if a CAD solid model is considered. A spatial parametric surface, referred to as freeform parametric surfaces in this subsection, such as a Coons patch, ruled surface, Bézier surface, and B-spline (or NURB) surface (Mortenson 2006), are the popular choices for shape parameterization. A parametric surface can be represented by a parametric vector equation $\mathbf{S}$ and a parametric area $A \in R^2$, usually $A = [0,1] \times [0,1]$, such that the surface consists of the set of points $\{\mathbf{S}(u,w)|(u,w) \in A\}$, as illustrated in Figure 4.17. Note that $\mathbf{S}$ is sometimes called the evaluation function and A is called the domain of evaluation.

We assume that the design boundary is represented by a bicubic parametric surface, cubic in both its parametric coordinates u and w. In general, there are 48 degrees of freedom for a parametric bicubic surface. The mathematical expressions in algebraic format for a bicubic parametric surface are as follows:

$$\mathbf{S}_x(u,w) = \sum_{i,j=0}^{3} a_{ijx}u^i w^j = \begin{bmatrix} u^3 & u^2 & u & 1 \end{bmatrix} \begin{bmatrix} a_{33x} & a_{32x} & a_{31x} & a_{30x} \\ a_{23x} & a_{22x} & a_{21x} & a_{20x} \\ a_{13x} & a_{12x} & a_{11x} & a_{10x} \\ a_{03x} & a_{02x} & a_{01x} & a_{00x} \end{bmatrix}_{4\times4} \begin{bmatrix} w^3 \\ w^2 \\ w \\ 1 \end{bmatrix}$$

$$= \mathbf{U}\mathbf{A}_x\mathbf{W}^T, \quad (u,w) \in [0,1] \times [0,1] \tag{4.225a}$$

$$\mathbf{S}_y(u,w) = \sum_{i,j=0}^{3} a_{ijy}u^i w^j = \begin{bmatrix} u^3 & u^2 & u & 1 \end{bmatrix} \begin{bmatrix} a_{33y} & a_{32y} & a_{31y} & a_{30y} \\ a_{23y} & a_{22y} & a_{21y} & a_{20y} \\ a_{13y} & a_{12y} & a_{11y} & a_{10y} \\ a_{03y} & a_{02y} & a_{01y} & a_{00y} \end{bmatrix}_{4\times4} \begin{bmatrix} w^3 \\ w^2 \\ w \\ 1 \end{bmatrix}$$

$$= \mathbf{U}\mathbf{A}_y\mathbf{W}^T, \quad (u,w) \in [0,1] \times [0,1] \tag{4.225b}$$

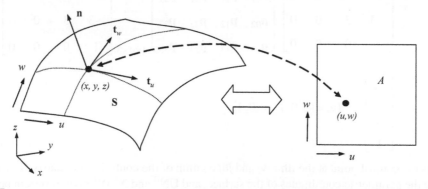

FIGURE 4.17

Parametric surface $\mathbf{S}$ and its parametric area A.

and

$$\mathbf{S}_z(u, w) = \sum_{i,j=0}^{3} a_{ijz}u^i w^j = \begin{bmatrix} u^3 & u^2 & u & 1 \end{bmatrix} \begin{bmatrix} a_{33z} & a_{32z} & a_{31z} & a_{30z} \\ a_{23z} & a_{22z} & a_{21z} & a_{20z} \\ a_{13z} & a_{12z} & a_{11z} & a_{10z} \\ a_{03z} & a_{02z} & a_{01z} & a_{00z} \end{bmatrix}_{4 \times 4} \begin{bmatrix} w^3 \\ w^2 \\ w \\ 1 \end{bmatrix}$$

$$= \mathbf{U}\mathbf{A}_z\mathbf{W}^T, \quad (u, w) \in [0, 1] \times [0, 1] \tag{4.225c}$$

For convenience, we have combined the above equations as

$$\mathbf{S}(u, w) = \sum_{i,j=0}^{3} \mathbf{a}_{ij}u^i w^j = \begin{bmatrix} u^3 & u^2 & u & 1 \end{bmatrix} \begin{bmatrix} \mathbf{a}_{33} & \mathbf{a}_{32} & \mathbf{a}_{31} & \mathbf{a}_{30} \\ \mathbf{a}_{23} & \mathbf{a}_{22} & \mathbf{a}_{21} & \mathbf{a}_{20} \\ \mathbf{a}_{13} & \mathbf{a}_{12} & \mathbf{a}_{11} & \mathbf{a}_{10} \\ \mathbf{a}_{03} & \mathbf{a}_{02} & \mathbf{a}_{01} & \mathbf{a}_{00} \end{bmatrix}_{4 \times 4 \times 3} \begin{bmatrix} w^3 \\ w^2 \\ w \\ 1 \end{bmatrix}$$

$$= \mathbf{U}\mathbf{A}\mathbf{W}^T, \quad (u, w) \in [0, 1] \times [0, 1] \tag{4.226}$$

where $\mathbf{S}(u,w) = [S_x, S_y, S_z]$, $\mathbf{a}_{ij} = [\mathbf{a}_{ijx}, \mathbf{a}_{ijy}, \mathbf{a}_{ijz}]$, 4×4 matrix each, are the algebraic coefficients of the surface. Similar to the parametric curves, in practice, the algebraic format of the parametric surface is not suitable for shape parameterization. Two surfaces that are commonly employed to parameterize design surfaces are Bézier and B-spline (or NURB) surfaces. We use a bicubic Bézier surface to illustrate further.

A bicubic Bézier surface, shown in Figure 4.18, is determined by the position of its 16 control points **G**, which form a control polyhedron. Mathematically, a Bézier surface can be written in a form like that of Eq. 4.226 as

$$\mathbf{S}(u, w) = \mathbf{U}\mathbf{N}^B\mathbf{G}\mathbf{N}^{B^T}\mathbf{W}^T$$
$$= \begin{bmatrix} u^3 & u^2 & u & 1 \end{bmatrix}$$

$$\times \begin{bmatrix} -1 & 3 & -3 & 1 \\ 3 & -6 & 3 & 0 \\ -3 & 3 & 0 & 0 \\ 1 & 0 & 0 & 0 \end{bmatrix} \begin{bmatrix} \mathbf{P}_{00} & \mathbf{P}_{10} & \mathbf{P}_{20} & \mathbf{P}_{30} \\ \mathbf{P}_{01} & \mathbf{P}_{11} & \mathbf{P}_{21} & \mathbf{P}_{31} \\ \mathbf{P}_{02} & \mathbf{P}_{12} & \mathbf{P}_{22} & \mathbf{P}_{32} \\ \mathbf{P}_{03} & \mathbf{P}_{13} & \mathbf{P}_{23} & \mathbf{P}_{33} \end{bmatrix}_{4 \times 4 \times 3} \begin{bmatrix} -1 & 3 & -3 & 1 \\ 3 & -6 & 3 & 0 \\ -3 & 3 & 0 & 0 \\ 1 & 0 & 0 & 0 \end{bmatrix}$$

$$\times \begin{bmatrix} w^3 \\ w^2 \\ w \\ 1 \end{bmatrix}, \quad (u, w) \in [0, 1] \times [0, 1] \tag{4.227}$$

where $\mathbf{P}_{ij}$ is the control point at the ith row and jth column of the control point matrix **G** of $4 \times 4 \times 3$, u and w are the parametric coordinates of the surface, and $\mathbf{U}\mathbf{N}^B$ and $\mathbf{N}^B\mathbf{W}^T$ give Bernstein polynomials in u and w, respectively.

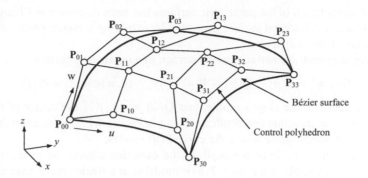

FIGURE 4.18

A bicubic Bézier surface determined by a control polyhedron of 4 × 4 control points.

Similar to the Bézier curve, movement of any of the 16 control points in the x-, y-, or z-direction can be chosen as the shape design variable. Any change in the position of the control points will result in the change of the geometric shape of the surface. For example, if control point $\mathbf{P}_{22}$ moves vertically up to a new position in the y-direction by δP_{22y}, the change in the surface can be obtained as

$$\delta S_y(u, w) = \mathbf{U}\mathbf{N}^B \delta\mathbf{G}_y \mathbf{N}^B \mathbf{W}^T$$

$$= \begin{bmatrix} u^3 & u^2 & u & 1 \end{bmatrix} \begin{bmatrix} -1 & 3 & -3 & 1 \\ 3 & -6 & 3 & 0 \\ -3 & 3 & 0 & 0 \\ 1 & 0 & 0 & 0 \end{bmatrix} \begin{bmatrix} 0 & 0 & 0 & 0 \\ 0 & 0 & 0 & 0 \\ 0 & 0 & \delta P_{22y} & 0 \\ 0 & 0 & 0 & 0 \end{bmatrix} \begin{bmatrix} -1 & 3 & -3 & 1 \\ 3 & -6 & 3 & 0 \\ -3 & 3 & 0 & 0 \\ 1 & 0 & 0 & 0 \end{bmatrix}$$

$$\times \begin{bmatrix} w^3 \\ w^2 \\ w \\ 1 \end{bmatrix}, \quad (u, w) \in [0, 1] \times [0, 1]$$

(4.228)

Here, $\delta\mathbf{G}_x = \delta\mathbf{G}_z = \mathbf{0}$ because the change only takes place in the y-direction. Then, the perturbed surface $\mathbf{S}(u, w; \delta\mathbf{G})$ can be obtained as

$$\mathbf{S}(u, w; \delta\mathbf{G}) = \mathbf{S}(u, w) + \delta\mathbf{S}(u, w)$$

(4.229)

4.5.3.3 3-D Solid sructures—CAD-generated surfaces

In CAD, we sketch an open profile and protrude it for a surface or protrude a closed profile for a solid feature (or a surface feature). The basic protrusion capabilities commonly available in CAD include extrusion, blend (or loft), revolve, and sweep. The boundary surface of a solid feature generated in CAD involves moving a curve segment $\mathbf{P}(u)$ (representing the profile of a section sketch) along a certain path $\mathbf{Q}(w)$, where u and w are the parametric coordinates of the resulting parametric surface.

The mathematical formulation of the parametric surfaces has been discussed in Chang (2014). In this subsection, we assume a cylindrical surface generated by extruding a sketch profile along a direction that is perpendicular to the sketch plane.

Mathematically, a cylindrical surface can be written in a parametric form as

$$\mathbf{S}(u, w) = \mathbf{P}(u) + \mathbf{Q}(w) = \mathbf{P}(u) + w\mathbf{r}, \quad (u, w) \in [0, 1] \times [0, 1] \tag{4.230}$$

in which $\mathbf{P}(u)$ is a curve in the sketch profile, and $\mathbf{Q}(w) = w\mathbf{r}$, $\mathbf{r}$ is the vector of the straight line representing the extrusion direction and depth, as shown in Figure 4.19. In this case, the straight line $\mathbf{r}$ is perpendicular to the sketch plane where the curve $\mathbf{P}(u)$ resides.

A change in the sketch profile or the depth of the extrusion affects the geometric shape of the cylindrical surface. For example, if the curve $\mathbf{P}(u)$ is modeled as a Bézier curve, then the movement of a control point on the sketch plane (e.g., the x–y plane shown in Figure 4.19) alters the geometry of the cylindrical surface by

$$\delta\mathbf{S}(u, w) = \delta\mathbf{P}(u) = \delta\mathbf{P}(u), \quad (u, w) \in [0, 1] \times [0, 1] \tag{4.231}$$

Similarly, if the extrusion depth is changed, the change in the cylindrical surface is

$$\delta\mathbf{S}(u, w) = \delta\mathbf{Q}(w) = w\delta\mathbf{r}, \quad (u, w) \in [0, 1] \times [0, 1] \tag{4.232}$$

Then, the perturbed surface $\mathbf{S}(u, w; \delta\mathbf{G})$ can be obtained as in Eq. 4.229.

EXAMPLE 4.17

Find the parametric equation of the cylindrical surface generated by extruding a cubic Bézier curve on the x–y plane along the positive z-direction for 5 units, as shown below (left). Note that the four points that control the Bézier curve are identical to those of Example 4.16.

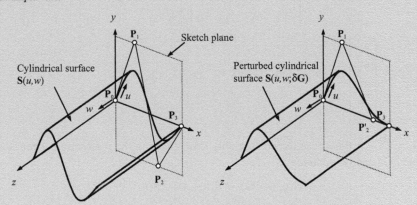

If the control point $\mathbf{P}_2$ is moved upward by two units—that is, $\delta P_{2y} = 2$—calculate the change in surface $\delta\mathbf{S}(u, w)$, and calculate the parametric equation of the perturbed cylindrical surface $\mathbf{S}(u, w; \delta\mathbf{G})$.

Solutions

From Example 4.16, we have the parametric equation of the Bézier curve as

$$\mathbf{P}(u) = \begin{bmatrix} 3u, & 15u^3 - 24u^2 + 9u \end{bmatrix} \tag{4.233}$$

EXAMPLE 4.17—cont'd

Using Eq. 4.230, the cylindrical surface can be written as

$$S(u,w) = P(u) + w\mathbf{r} = [3u, \ 15u^3 + 24u^2 + 9u, \ 0] + w[0,0,5]$$
$$= [3u, \ 15u^3 - 24u^2 + 9u, \ 5w], \quad (u,w) \in [0,1] \times [0,1] \tag{4.234}$$

The change in curve due to $\delta P_{2y} = 2$ was obtained in Example 4.16 as

$$\delta P(u) = [0, -6u^3 + 6u^2]$$

Hence, the change in surface is

$$\delta S(u,w) = \delta P(u) = [0, \ -6u^3 + 6u^2, \ 0], \quad (u,w) \in [0,1] \times [0,1] \tag{4.235}$$

Then, the perturbed cylindrical surface $S(u,w; \delta G)$ can be obtained using Eq. 4.229 as

$$S(u,w; \delta G) = S(u,w) + \delta S(u,w) = [3u, \ 15u^3 - 24u^2 + 9u, \ 5w] + [0, \ -6u^3 + 6u^2, \ 0]$$
$$= [3u, \ 9u^3 - 18u^2 + 9u, \ 5w], \quad (u,w) \in [0,1] \times [0,1] \tag{4.236}$$

The cylindrical surface after perturbation is shown above (previous page) to the right.

Note that Eq. 4.230 is not the only way to represent a cylindrical surface mathematically. Another widely employed formulation is based on a polar coordinate system, such as the circular cylindrical surface shown in Figure 4.20. The surface is created by extruding a circular arc along a straight path $\mathbf{d} = w\mathbf{e}_3$, yielding a circular cylindrical surface:

$$S(u,w) = r \cdot [\cos(u) \cdot \mathbf{e}_1 + \sin(u) \cdot \mathbf{e}_2] + w \cdot \mathbf{e}_3 + \mathbf{o}, \quad (u,w) \in A = [0,\varphi] \times [0,h], \quad \mathbf{o} \in R^3 \tag{4.237}$$

where $\mathbf{e}_1$, $\mathbf{e}_2$, and $\mathbf{e}_3$ are orthogonal unit vectors in R^3 and $\mathbf{o}$ is the origin, as shown in Figure 4.20. For this cylindrical surface, design variables can be radius r and height h.

4.5.4 DESIGN VELOCITY FIELD COMPUTATION

When design variables vary, the geometric shape of the structural boundary—and therefore, the location of material points inside the structural domain—must change accordingly. The design velocity field governs the movement of material points both on the boundary and inside the structural domain based on the changes in shape design variables. The design velocity field provides a systematic scheme that maps the location of material points from original design to the updated design. Naturally, design velocity field supports finite element mesh updates while maintaining mesh topology. Maintaining mesh topology is critical in terms of tracking local measures, such as stress and displacement, during design iterations.

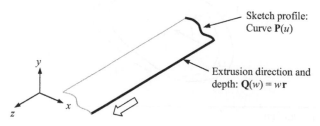

FIGURE 4.19

Extrusion of a sketch profile for a cylindrical surface.

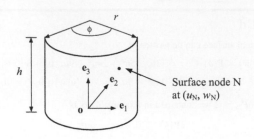

FIGURE 4.20

A parametric cylindrical surface.

We first define the design velocity mathematically and briefly discuss its theoretical and practical requirements. We present the velocity computation for design boundary, and then domain velocity.

4.5.4.1 Design velocity field

Consider a structural domain Ω with its boundary Γ as a continuous medium at the current design $\tau = 0$ shown in Figure 4.21 (solid lines). Suppose only one parameter τ defines the transformation **T** that changes the structural domain from Ω to Ω_τ (dotted lines). The transformation mapping **T** that represents this process can be defined as

$$\mathbf{T}: \mathbf{x} \to \mathbf{x}_\tau(\mathbf{x}), \quad \mathbf{x} \in \Omega \tag{4.238}$$

where **x** is a material point (Haug et al. 1986).

We define the design velocity field **V** as

$$\mathbf{V}(\mathbf{x}_\tau, \tau) \equiv \frac{d\mathbf{x}_\tau}{d\tau} = \frac{d\mathbf{T}(\mathbf{x}, \tau)}{d\tau} \tag{4.239}$$

where τ plays the role of design time (or the design iteration in practice). In the neighborhood of the current design $\tau = 0$, assume the mapping function **T** is smooth. Ignoring higher-order terms, **T** can be approximated by

$$\mathbf{T}(\mathbf{x}, \tau) = \mathbf{T}(\mathbf{x}, 0) + \tau \frac{d\mathbf{T}(\mathbf{x}, 0)}{d\tau} + \mathrm{O}(\tau^2) \approx \mathbf{x} + \tau \mathbf{V}(\mathbf{x}) \tag{4.240}$$

where $\mathbf{x} \equiv \mathbf{T}(\mathbf{x},0)$ and $\mathbf{V}(\mathbf{x}) \equiv \mathbf{V}(\mathbf{x},0)$. Note that only the linear term is retained in Eq. 4.240.

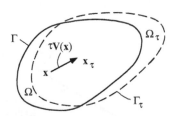

FIGURE 4.21

Changing of the structural domain.

In shape optimization, the design velocity field is calculated first at the design boundary, called the boundary velocity field. The mapping $\mathbf{T}$ is characterized by the parametric equations employed for representing the design boundary. Therefore, the boundary velocity field can be calculated by varying the parametric equations of the design boundary through changes in design variables:

$$\mathbf{V}_i(u) = \delta\mathbf{P}(u; \mathbf{b}) = \frac{\partial\mathbf{P}}{\partial b_i}\,\delta b_i \tag{4.241a}$$

and

$$\mathbf{V}_i(u, w) = \delta\mathbf{S}(u, w; \mathbf{b}) = \frac{\partial\mathbf{S}}{\partial b_i}\,\delta b_i \tag{4.241b}$$

where $\mathbf{P}(u)$ and $\mathbf{S}(u,w)$ are the parametric curves and surfaces representing the design boundary of 2-D and 3-D structures, respectively; and $\mathbf{b}$ is the vector of shape design variables. Note that the design perturbation for the ith design variable, δb_i, is usually set to 1 for convenience in practice.

The change of structural boundary also causes the movement of material points in the domain of the structural component, which is characterized by the so-called domain velocity field. Both boundary and domain velocity fields must be calculated. Before discussing the computation methods, there are a few requirements for the design velocity that are worth mentioning (Choi and Chang 1994).

First, the velocity field must depend linearly on the variation of shape design variables, as required by its definition shown in Eq. 4.240. Linear dependence requires that if control point $\mathbf{P}_1$ moves a distance $\delta b_1 = 1$ in the x-direction, producing domain design velocity $\mathbf{V}^1(\mathbf{x}^1)$ at node 1, as shown in Figure 4.22a, then node 1 must move $k\mathbf{V}^1(\mathbf{x}^1)$ along the same direction when $\mathbf{P}_1$ moves $k\,\delta b_1 = k$, $k \neq 0$, in the x-direction, as depicted in Figure 4.22b. This linear dependence must be true for all boundary and interior nodes of the structure.

For a boundary velocity field, as long as the derivatives $\partial\mathbf{P}/\partial b_i$ and $\partial\mathbf{S}/\partial b_i$ shown in Eqs 4.241a and 4.241b, respectively, are constant at a given u and (u,w), the linearity requirement is satisfied.

For a finite element model, design sensitivity coefficients predict structural performance measures of the perturbed design with finite element mesh updated by moving nodal points along the direction of the design velocity field using

$$\mathbf{x}^k(\mathbf{b} + \delta\mathbf{b}) = \mathbf{x}^k(\mathbf{b}) + \delta\mathbf{x}^k(\mathbf{b}) = \mathbf{x}^k(\mathbf{b}) + \sum_{i=1}^{n}\mathbf{V}_i^k\delta b_i \tag{4.242}$$

where $\mathbf{x}^k(\mathbf{b} + \delta\mathbf{b})$ and $\mathbf{x}^k(\mathbf{b})$ are the locations of the kth node of the perturbed and the current designs, respectively; $\delta\mathbf{x}^k(\mathbf{b})$ is the nodal point movement due to design changes; $\mathbf{V}_i^k$ and δb_i are the design

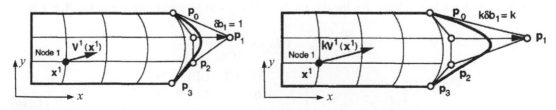

FIGURE 4.22

Illustration of the linearity requirement of the design velocity field. (a) Control point $\mathbf{P}_1$ moves $\delta b_1 = 1$ in the x-direction. (b) Control point $\mathbf{P}_1$ moves $k\delta b_1 = k$, $k \neq 0$ in the x-direction.

velocity of the ith design variable at the kth node and the change of the ith design variable b_i, respectively; and n is the total number of design variables. It is important to note that shape sensitivity coefficients do not predict performance measures at the new design if the finite element mesh is not updated using the design velocity field employed for sensitivity analysis.

For practical applications using FEM, a velocity field computation method must retain the topology of the original finite element mesh (i.e., no elements or nodes are added or removed) so that local performance measures, such as displacement or stress, at a specific node are retained consistently throughout the design process. It is also desirable that the finite element mesh that is updated using the design velocity computed does not get distorted in the design process.

4.5.4.2 Boundary velocity computation

We discuss boundary velocity computation for 2-D planar and 3-D solid structures.

4.5.4.2.1 2-D Planar structures

As discussed earlier, the design boundary of a 2-D planar structure is represented using parametric curves. If FEM is employed for structural analysis in shape design, the finite element nodes at the design boundary will have to move to the new geometric boundary at the new design. The movement (i.e., the boundary velocity field) can be calculated by plugging the parametric coordinate u of a boundary node (e.g., u_j of the jth node) on the boundary curve into Eq. 4.241a:

$$\mathbf{V}_i^j = \mathbf{V}_i(u_j) = \delta \mathbf{P}(u_j; \mathbf{b}) = \frac{\partial \mathbf{P}}{\partial b_i} \delta b_i \tag{4.243}$$

For example, if a cubic Bézier curve is parameterized for the design boundary, the boundary velocity field can be calculated by bringing Eq. 4.216 into Eq. 4.243 as

$$\mathbf{V}_i^j = \mathbf{V}_i(u_j) = \frac{\partial \mathbf{P}(u_j)}{\partial b_i} \delta b_i = \mathbf{U}(u_j) \mathbf{N}^{\mathrm{B}} \frac{\partial \mathbf{G}^{\mathrm{B}}}{\partial b_i} \delta b_i \tag{4.244}$$

Here, $\frac{\partial \mathbf{G}^{\mathrm{B}}}{\partial b_i}$ is $\mathbf{0}$, except for the entry corresponding to a design variable, in which the value is 1. For the design change shown in Figure 4.16b, in which P_{2y} is varied, we have

$$\frac{\partial \mathbf{G}^{\mathrm{B}}}{\partial b_i} = \frac{\partial \mathbf{G}^{\mathrm{B}}}{\partial P_{2y}} = \frac{\partial}{\partial P_{2y}} \begin{bmatrix} P_{0x} & P_{0y} \\ P_{1x} & P_{1y} \\ P_{2x} & P_{2y} \\ P_{3x} & P_{3y} \end{bmatrix}_{4\times 2} = \begin{bmatrix} 0 & 0 \\ 0 & 0 \\ 0 & 1 \\ 0 & 0 \end{bmatrix}_{4\times 2} \tag{4.245}$$

Hence, the boundary velocity for the design variable P_{2y} can be calculated using Eq. 4.244 as

$$\mathbf{V}^j = \mathbf{V}(u_j) = \frac{\partial \mathbf{P}}{\partial P_{2y}} \delta P_{2y} = \mathbf{U}(u_j) \mathbf{N}^{\mathrm{B}} \frac{\partial \mathbf{G}^{\mathrm{B}}}{\partial P_{2y}} \delta P_{2y}$$

$$= \begin{bmatrix} u_j^3 & u_j^2 & u_j & 1 \end{bmatrix} \begin{bmatrix} -1 & 3 & -3 & 1 \\ 3 & -6 & 3 & 0 \\ -3 & 3 & 0 & 0 \\ 1 & 0 & 0 & 0 \end{bmatrix} \begin{bmatrix} 0 & 0 \\ 0 & 0 \\ 0 & 1 \\ 0 & 0 \end{bmatrix} \delta P_{2y} \tag{4.246}$$

EXAMPLE 4.18

Given four control points—$\mathbf{P}_0 = [0, 0]$, $\mathbf{P}_1 = [1, 1]$, $\mathbf{P}_2 = [2, 0.5]$, and $\mathbf{P}_3 = [3, 0]$—create a Bézier curve that represents a design boundary of a planar structure meshed with 5×3 finite elements, as shown below. There are six nodes meshed on the design boundary with parametric coordinates $u = 0, 0.2, 0.4, 0.6, 0.8$, and 1.0, respectively. If the vertical movement of the control point $\mathbf{P}_2$ is defined as a design variable—that is, $\delta b = \delta P_{2y} = 1$—then calculate the boundary velocity for nodes 4 and 5 on the design boundary.

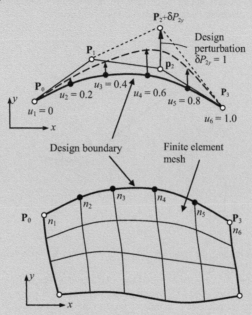

Solutions

Using Eq. 4.246, we have

$$\mathbf{V}^j = \mathbf{V}(u_j) = \frac{\partial \mathbf{P}}{\partial P_{2y}} \delta P_{2y} = \mathbf{U}(u_j) \mathbf{N}^\mathbf{B} \frac{\partial \mathbf{G}^\mathbf{B}}{\partial P_{2y}} \delta P_{2y} = \begin{bmatrix} u_j^3 & u_j^2 & u_j & 1 \end{bmatrix} \begin{bmatrix} -1 & 3 & -3 & 1 \\ 3 & -6 & 3 & 0 \\ -3 & 3 & 0 & 0 \\ 1 & 0 & 0 & 0 \end{bmatrix} \begin{bmatrix} 0 & 0 \\ 0 & 0 \\ 0 & 1 \\ 0 & 0 \end{bmatrix} \delta P_{2y} \qquad (4.247)$$

The parametric coordinate of node 4 is $u_4 = 0.6$. Therefore, the boundary velocity at the node is

$$\mathbf{V}^4 = \mathbf{V}(u_4) = \begin{bmatrix} 0.6^3 & 0.6^2 & 0.6 & 1 \end{bmatrix} \begin{bmatrix} -1 & 3 & -3 & 1 \\ 3 & -6 & 3 & 0 \\ -3 & 3 & 0 & 0 \\ 1 & 0 & 0 & 0 \end{bmatrix} \begin{bmatrix} 0 & 0 \\ 0 & 0 \\ 0 & 1 \\ 0 & 0 \end{bmatrix} = [0.432] \qquad (4.248)$$

Similarly, $\mathbf{V}^5$ for node 5, where $u_5 = 0.8$, and $\mathbf{V}^5 = [0, 0.384]$.

In practice, implementing Eq. 4.243 for the boundary velocity computation requires knowledge of the parametric coordinates of individual boundary nodes. Usually, only Cartesian coordinates (x, y) of a node are available, such as those found in the finite element input data file. How can we convert x- and y-coordinates into the parametric coordinate u? One straightforward technique is using MATLAB's root finding functions, such as `solve` or `fzero`. We use the following example to illustrate further.

EXAMPLE 4.19

Given four control points—$P_0 = [10, 2]$, $P_1 = [6, 1]$, $P_2 = [4, 5]$, and $P_3 = [0, 3]$—create a Bézier curve that represents a design boundary of a planar structure. The parametric equation of the Bézier curve is given as

$$P(u) = [-4u^3 + 6u^2 - 12u + 10, \quad 0, \quad -11u^3 + 15u^2 - 3u + 2], \quad u \in [0, 1]$$

The Cartesian coordinates of a nodal point on the curve are found in the finite element input data file as $x = 7.6564$ and $y = 1.9393$. Find the parametric coordinate u of the nodal point on the curve.

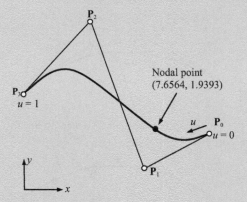

Solutions

We use two MATLAB functions to show the solutions, `solve` and `fzero`. We are solving the x-component of the curve equation, as follows:

Find u by solving: $-4u^3 + 6u^2 - 12u + 10 = 7.6564$ (or $-4u^3 + 6u^2 - 12u + 10 - 7.6564 = 0$).

The function `solve`, syntax `solve (eq,x)`, returns the set of all complex solutions of an equation `eq` with respect to x. We pick the real solution as the parametric coordinate for the point. Enter the following in MATLAB:

```
[u]=solve('-4*u^3+6*u^2-12*u+10 = 7.6564')
```

MATLAB returns the following:

```
u =
               0.21511995841548803773588873485
0.64244002079225598113205563257 5 + 1.5201537680675251178484289817741*i
0.64244002079225598113205563257 5 - 1.5201537680675251178484289817741*i
```

The only real solution is `u = 0.21512`.

The function `fzero`, syntax `x=fzero(fun,x0)`, tries to find a point x where `fun(x) = 0`, `x0` is the initial point the user enters. Enter the following in MATLAB:

```
x=fzero('-4*x^3+6*x^2-12*x+10-7.6564',0)
```

MATLAB returns the following:

```
x =
0.2151
```

which is the parametric coordinate we are solving.

4.5.4.2.2 3-D Solid structures—freeform surfaces

Similar to the 2-D applications, boundary velocity field can be calculated by plugging the parametric coordinates at the nodes, such as (u_j, w_j) of the jth node, on the boundary surface into Eq. 4.241b:

$$\mathbf{V}_i^j = \mathbf{V}_i(u_j, w_j) = \delta \mathbf{S}(u_j, w_j; \mathbf{b}) = \frac{\partial \mathbf{S}}{\partial b_i} \delta b_i \tag{4.249}$$

For example, if a bicubic Bézier surface is parameterized for the design boundary, the boundary velocity field can be calculated by bringing Eq. 4.227 into Eq. 4.249:

$$\mathbf{V}_i^j = \mathbf{V}_i(u_j, w_j) = \frac{\partial \mathbf{S}}{\partial b_i} \delta b_i = \mathbf{U}(u_j) \mathbf{N}^B \frac{\partial \mathbf{G}^B}{\partial b_i} \mathbf{N}^B \mathbf{W}(w_j)^T \delta b_i \tag{4.250}$$

in which $\frac{\partial \mathbf{G}^B}{\partial b_i}$ is **0**, except for the entry corresponding to a design variable, in which the value is 1. For example, if the y-coordinate of the control point $\mathbf{P}_{22}$—that is, P_{22y}—is chosen as a design variable, we have

$$\frac{\partial \mathbf{G}_y^B}{\partial b_i} = \frac{\partial \mathbf{G}_y^B}{\partial P_{22y}} = \frac{\partial}{\partial P_{22y}} \begin{bmatrix} p_{00y} & p_{10y} & p_{20y} & p_{30y} \\ p_{01y} & p_{11y} & p_{21y} & p_{31y} \\ p_{02y} & p_{12y} & p_{22y} & p_{32y} \\ p_{03y} & p_{13y} & p_{23y} & p_{33y} \end{bmatrix}_{4\times 4} = \begin{bmatrix} 0 & 0 & 0 & 0 \\ 0 & 0 & 0 & 0 \\ 0 & 0 & 1 & 0 \\ 0 & 0 & 0 & 0 \end{bmatrix}_{4\times 4} \tag{4.251}$$

where $\frac{\partial \mathbf{G}_x^B}{\partial b_i} = \frac{\partial \mathbf{G}_z^B}{\partial b_i} = 0$ because the change only takes place in the y-direction. Hence, the boundary velocity for the design variable P_{22y} can be calculated using Eq. 4.250:

$$\mathbf{V}_i^j = \mathbf{V}_i(u_j, w_j) = \frac{\partial \mathbf{S}}{\partial P_{22y}} \delta P_{22y} = \mathbf{U}(u_j) \mathbf{N}^B \frac{\partial \mathbf{G}_y^B}{\partial p_{2y}} \mathbf{N}^B \mathbf{W}(w_j)^T \delta P_{2y}$$

$$= \begin{bmatrix} u_j^3 & u_j^2 & u_j & 1 \end{bmatrix} \begin{bmatrix} -1 & 3 & -3 & 1 \\ 3 & -6 & 3 & 0 \\ -3 & 3 & 0 & 0 \\ 1 & 0 & 0 & 0 \end{bmatrix} \begin{bmatrix} 0 & 0 & 0 & 0 \\ 0 & 0 & 0 & 0 \\ 0 & 0 & 1 & 0 \\ 0 & 0 & 0 & 0 \end{bmatrix} \begin{bmatrix} -1 & 3 & -3 & 1 \\ 3 & -6 & 3 & 0 \\ -3 & 3 & 0 & 0 \\ 1 & 0 & 0 & 0 \end{bmatrix} \begin{bmatrix} w_j^3 \\ w_j^2 \\ w_j \\ 1 \end{bmatrix} \delta P_{2y} \tag{4.252}$$

An important role that freeform surfaces play in shape optimization is that they are suitable for integrating topology optimization (Bendsøe and Sigmund 2003) with shape optimization because the structural boundary obtained from topology optimization is unsmooth and highly irregular. Freeform surfaces approximate the irregular structural boundary within a given error bound and smooth the boundary for shape optimization. In addition, for follow-up engineering assignments, such as machining simulation using computer-aided manufacturing (CAM) software, the surface parameterization must be compatible with CAD so that the geometry of the component can be imported into CAD for machining simulation and toolpath generation. However, not all freeform surfaces are compatible with CAD systems. In fact, only very few are supported. Among them are Bézier and the B-spline surfaces. Between the two, B-spline (or NURB) surface is probably the safer choice because it is supported by several major CAD software programs, including SolidWorks. This point is further illustrated in a case study presented in Section 4.7.1.

4.5.4.2.3 3-D Solid structures—CAD-generated surfaces

Similar to the freeform surfaces, the boundary velocity field for a CAD-generated surface can be calculated using Eq. 4.249.

For example, the cylindrical surface shown in Figure 4.19 is parameterized as a design boundary. The boundary velocity field can be calculated by bringing Eq. 4.230 into Eq. 4.249:

$$
\mathbf{V}_i(u_j, w_j) = \frac{\partial \mathbf{S}}{\partial b_i}\delta b_i = \frac{\partial}{\partial b_i}\left(\mathbf{P}(u_j) + \mathbf{Q}(w_j)\right)\delta b_i = \frac{\partial \mathbf{P}(u_j)}{\partial b_i}\delta b_i + \frac{\partial \mathbf{Q}(w_j)}{\partial b_i}\delta b_i
$$

$$
= \mathbf{U}(u_j)\mathbf{N}^{\mathrm{B}}\frac{\partial \mathbf{G}^{\mathrm{B}}}{\partial b_i}\delta b_i + w_j\frac{\partial \mathbf{r}}{\partial b_i}\delta b_i, \quad (u_j, w_j)\in[0,1]\times[0,1]
$$

(4.253)

Here, we assume that $\mathbf{P}(u)$ is, for example, a Bézier curve. As with the boundary curve, $\frac{\partial \mathbf{G}^{\mathrm{B}}}{\partial b_i}$ is $\mathbf{0}$, except for the entry corresponding to a design variable, in which the value is 1. If the extrusion depth is the design variable, then $\frac{\partial \mathbf{r}}{\partial b_i} = [0, 0, 1]$.

EXAMPLE 4.20

Revisiting Example 4.17, the parametric equation of the cylindrical surface shown below is given as follows:

$$
\mathbf{S}(u, w) = \mathbf{P}(u) + w\mathbf{r} = \left[3u,\ 15u^3 - 24u^2 + 9u,\ 5w\right], \quad (u, w)\in[0,1]\times[0,1]
$$

(4.254)

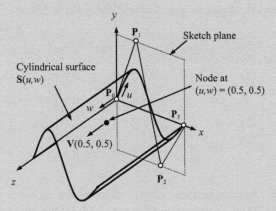

If the extrusion depth is defined as the design variable, calculate the change in surface $\delta \mathbf{S}(u, w)$, and calculate the design velocity at a node whose parametric coordinates are (0.5, 0.5).

Solutions

From Eq. 4.232, the change in surface $\delta \mathbf{S}(u, w)$ can be calculated as

$$
\delta \mathbf{S}(u, w) = \delta \mathbf{Q}(w) = w\delta \mathbf{r} = w[0, 0, 1]\delta b_i, \quad (u, w)\in[0,1]\times[0,1]
$$

(4.525)

Then, the velocity at surface node $(u, w) = (0.5, 0.5)$ can be obtained using Eq. 4.253:

$$
\mathbf{V}(0.5, 0.5) = 0.5\frac{\partial \mathbf{r}}{\partial b_i}\delta b_i = 0.5[0, 0, 1]\delta b_i = [0, 0, 0.5]
$$

(4.256)

Here, δb_i is set to 1, as discussed before.

Similar to the curve boundary, for a surface design boundary, only the Cartesian coordinates (x,y,z) of a node are available, such as those found in the finite element input data file. One may use MATLAB script (or other numerical tools or techniques) to convert (x,y,z) to (u,w). The following illustrates such an example.

EXAMPLE 4.21

The Cartesian coordinates of the surface node $(u,w) = (0.5,0.5)$ on the surface in Example 4.20 are $(1.5, 0.375, 2.5)$. Convert the Cartesian coordinates of the node back to the parametric coordinates (u,w).

Solutions

One possible way to solve for (u,w) from (x,y,z) is using the MATLAB function `fsolve`. We solve (u,w) from the following equations:

$$3u = x = 1.5$$
$$15u^3 - 24u^2 + 9u = y = 0.375$$
$$5w = 2.5$$

This problem is certainly trivial to solve; they are simple, and u and w are decoupled. We pick two equations, $3u$ $-1.5 = 0$ and $5w-2.5 = 0$, to illustrate the solution steps.

We first write a file that computes function F, the values of the equations at x, which is a vector containing u and w:

```
function F = myfun(x)
F = [3*x(1)-1.5;
     5*x(2)-2.5];
```

Save this function file as `myfun.m` somewhere on your MATLAB path. Next, set up the initial point (e.g., $x0 = [0; 0]$) and options and call `fsolve`:

```
x0 = [0; 0]; % Make a starting guess at the solution
options = optimoptions('fsolve','Display','iter'); % Option to display output
[x,fval] = fsolve(@myfun,x0,options) % Call solver
```

MATLAB returns the following:

Iteration	Func-count	f(x)	Norm of step	First-order optimality	Trust-region radius
0	3	8.5		12.5	1
1	6	0	0.707107	0	1

```
Equation solved.

...

x =
      0.5000
      0.5000
 fval =
      0
      0
```

These are the correct parametric coordinates $(u,w) = (0.5, 0.5)$.

4.5.4.2.4 Implementation with CAD software

The boundary velocity field computation methods discussed above are straightforward to implement because the parametric equations of the boundary curves or surfaces are explicitly expressed in design variables, and numerical data required for evaluating the curves or surfaces are available.

In some situations, however, the design variable is not explicit in the geometric representation for a design surface and the parametric equation and parametric area A of the perturbed design surface cannot be readily anticipated. Note that in CAD surfaces, (u,w) may not be bound by the parametric area $[0,1] \times [0,1]$. Therefore, an analytical solution for the velocity at a node is not available. In this case, it is easier to let the CAD tool regenerate the part based on the perturbed design variable than try to trace the influence of the design variable based on surface's geometric representation. In this situation, a viable solution is to use finite difference method to calculate the boundary design velocity. The use of the CAD application protocol interface (API) allows this regeneration to be done automatically. The API also allows one to retrieve the data needed for carrying out velocity field computation.

In the following, we assume a circular cylindrical surface created in Pro/ENGINEER similar to that of Figure 4.20 for discussion. Geometric representations of curves and surfaces in Pro/ENGINEER can be retrieved via the API. Pro/ENGINEER offers API functions that are needed for velocity field computation; for example, pro_get_surface returns a data structure containing the geometric representation of the specified surface, and pro_get_face_params provides the parameter values u and w corresponding to a given point on the surface. These functions are essential for supporting the velocity computation with CAD software, as discussed next.

Suppose the radius r is chosen as a design variable. The parametric equation for the surface is written as $\mathbf{S} = \mathbf{S}(u,w;r)$ to indicate that r is varying. When r is perturbed to $r + \delta r$, the perturbed surface can be represented by

$$\mathbf{S}(u, w; r + \delta r) = (r + \delta r) \cdot [\cos(u) \cdot \mathbf{e}_1 + \sin(u) \cdot \mathbf{e}_2] + (w) \cdot \mathbf{e}_3 + \mathbf{o}, \quad (u, w) \in A = [0, \varphi] \times [0, h] \tag{4.257}$$

On the other hand, suppose h (the height) is chosen as a design variable. Let h be perturbed to $h + \delta h$. Then, the perturbed surface can be represented by

$$\mathbf{S}(u, w; h + \delta h) = r \cdot [\cos(u) \cdot \mathbf{e}_1 + \sin(u) \cdot \mathbf{e}_2] + w \cdot \mathbf{e}_3 + \mathbf{o}, \quad (u, w) \in A' = [0, \varphi] \times [0, h + \delta h] \tag{4.258}$$

where A' denotes the perturbed parametric area. Equivalently, the parametric area may be left unchanged, and the perturbed face represented by

$$\mathbf{S}\left(u, \left(1 + \frac{\delta h}{h}\right) \cdot w; h\right) = r[\cos(u) \cdot \mathbf{e}_1 + \sin(u) \cdot \mathbf{e}_2] + \left(1 + \frac{\delta h}{h}\right) \cdot w \cdot \mathbf{e}_3 + \mathbf{o}, \quad (u, w) \in A$$

$$= [0, \varphi] \times [0, h]. \tag{4.259}$$

Let N be a node on the cylindrical surface with parametric coordinates (u_N, w_N), as shown in Figure 4.20. If r is the design variable, we can calculate the design velocity at N using finite differences:

$$\mathbf{V}^N = \frac{\partial \mathbf{S}}{\partial r} \cong \frac{[\mathbf{S}(u_N, w_N; r + \delta r) - \mathbf{S}(u_N, w_N; r)]}{\delta r} \tag{4.260}$$

Note that, in this case, $\mathbf{V}^N = \cos(u_N)\,\mathbf{e}_1 + \sin(u_N)\,\mathbf{e}_2$. If h is the design variable, the design velocity at N will be

$$\mathbf{V}^N = \frac{\partial \mathbf{S}}{\partial h} \cong \frac{\left[\mathbf{S}\left(u_N, \left(1 + \frac{\delta h}{h}\right) \cdot w_N; r\right) - \mathbf{S}(u_N, w_N; r)\right]}{\delta h} \qquad (4.261)$$

In this case, $\mathbf{V}^N = (w_N/h)\,\mathbf{e}_3$. This is the basic approach to computing the design velocity for all nodes on a design surface.

Note that the design perturbation for velocity field computation in CAD must be determined based on the size of the solid features and components. No topological change is allowed in such a perturbation. In this cylindrical surface example, the boundary design velocity field for the two types of perturbations is linearly dependent upon the variation of the design variable. This is so because the parametric equation $\mathbf{S}$ was linear with respect to r and to h. In many situations, the linearity may not be possible (Hardee et al. 1999).

4.5.4.3 Domain velocity computation

Perturbation of shape design variables results in the movement of the design boundary, as discussed in the previous section. Interior material points, or finite element nodes, in the structural domain must also move following certain rules. Among numerous approaches for domain velocity field computations, we briefly discuss two representative methods: the boundary displacement method and the isoparametric mapping method (Choi and Chang 1994).

4.5.4.3.1 Boundary displacement method

In the boundary displacement method, movements of nodal points at the design boundary are treated as the prescribed displacements. The domain velocity field that corresponds to the design perturbation can be obtained by solving an auxiliary elasticity problem with prescribed displacements specified at design boundary nodes as well as prescribed rollers at the nondesign boundary, as depicted in Figure 4.23. Note that rollers are added to constrain the nodal point movements at the nondesign boundary, as shown in Figure 4.23b, in accordance with the design intent.

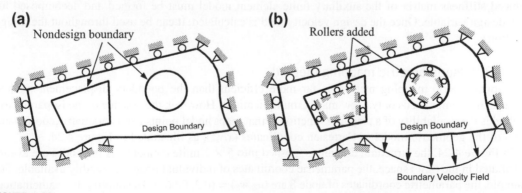

FIGURE 4.23

Illustration of the boundary displacement method for domain velocity field computation. (a) The original structure with boundary conditions. (b) The boundary velocity field added to the design boundary and additional rollers added to the nondesign boundary.

To form the auxiliary problem for the domain velocity computation, both the velocity field of the design boundary and the displacement constraints (e.g., rollers) that define nodal point movements on the nondesign boundary must be imposed to the finite element model. The discretized equilibrium equation of the auxiliary finite element model can be written as

$$\mathbf{KV} = \mathbf{f} \tag{4.262}$$

where $\mathbf{K}$ is the reduced stiffness matrix of the auxiliary structure (e.g., the structure shown in Figure 4.23b, which is different from that of the original structure (e.g., the one shown in Figure 4.23a. $\mathbf{V}$ is the design velocity vector and $\mathbf{f}$ is the unknown vector of boundary forces, which produce the prescribed boundary velocity at both the design and nondesign boundary. In a partitioned form, Eq. 4.262 can be written as

$$\begin{bmatrix} \mathbf{K}_{bb} & \mathbf{K}_{db} \\ \mathbf{K}_{db} & \mathbf{K}_{dd} \end{bmatrix} \begin{bmatrix} \mathbf{V}_b \\ \mathbf{V}_d \end{bmatrix} = \begin{bmatrix} \mathbf{f}_b \\ 0 \end{bmatrix} \tag{4.263}$$

where $\mathbf{V}_b$ is the prescribed velocity of nodes on the boundary, $\mathbf{V}_d$ is the nodal velocity vector in the interior domain (domain velocity), and $\mathbf{f}_b$ is the unknown boundary force that acts on the structural boundary. Equation 4.263 can be rearranged as

$$\mathbf{K}_{dd}\mathbf{V}_d = -\mathbf{K}_{db}\mathbf{V}_b \tag{4.264}$$

This defines a linear relationship between the boundary and domain velocity fields.

The boundary displacement method is independent of the way in which the boundary velocity field is computed. The same finite element code used for analysis can be used to compute the displacement field of the auxiliary model; yielding a domain velocity field that naturally satisfies the linearity requirement (because the finite element matrix equations are a system of linear equations). An important characteristic of the elasticity problem is that the solution trajectory tends to maintain orthogonality. As a result, the updated mesh obtained using the velocity field, as stated in Eq. 4.242, tends to be more regular (see Figure 4.26 as an example).

One drawback of the boundary displacement method is that finite element matrix equations must be formulated and solved to generate the velocity field for each shape design variable. For this, the reduced stiffness matrix of the auxiliary finite element model must be formed and decomposed for each design variable. Once the design velocity field is calculated, it can be used throughout the design iterations.

4.5.4.3.2 Isoparametric mapping method

The isoparametric mapping method is far more efficient than the boundary displacement method because the former needs only a few matrix multiplications. However, the essence of the isoparametric method is the availability of a mapping function $\mathbf{N}$ that maps nodal points from parametric coordinates (u, w) of the geometric model to Cartesian coordinates (x, y, z), as illustrated in Figure 4.24.

In Figure 4.24, the structural domain is meshed into 5×3 finite elements. If the nodes are evenly distributed in the (u, w) space, the parametric coordinates of individual nodes are readily available. For example, the parametric coordinates of node 8 are $(u_8, w_8) = (0.2, 0.667)$. To simplify the mathematical expressions in our discussion, we assume a 2-D planar structure for discussing the parametric mapping method.

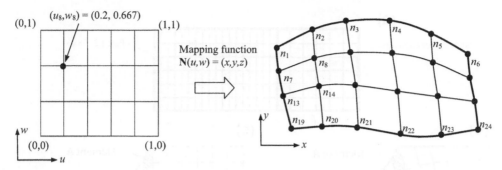

FIGURE 4.24

Concept of isoparametric mapping.

Finding an appropriate mapping function **N** for accurate velocity field computation is difficult when the geometric shape of the structure is complicated. However, for a simple geometric model or a smaller subdomain of the structure, for which an accurate mapping function **N** can be found, the isoparametric mapping method is attractive. For such a model or subdomain that can be modeled by a single geometric entity, velocity fields can be computed using the isoparametric mapping method.

Geometric modeling software, such as MSC/PATRAN and HyperMesh, employs a standard patch to represent geometric surfaces. For example, in PATRAN, a surface patch is modeled in Coons patch (Mortenson 2006). A Coons patch, as shown in Figure 4.25a, is a bicubic parametric surface in terms of parametric coordinates u and w, where u and $w \in [0,1]$. Obviously, one single patch is not able to represent a complicate geometric entity. Therefore, a 2-D structural domain is often decomposed into several smaller patches in the modeling process.

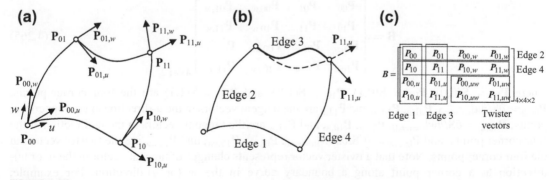

FIGURE 4.25

A bicubic parametric patch. (a) Corner points and tangent vectors. (b) Adjusting tangent vector $\mathbf{P}_{11,u}$. (c) Data formulated in a $4 \times 4 \times 2$ B-matrix that defines the patch.

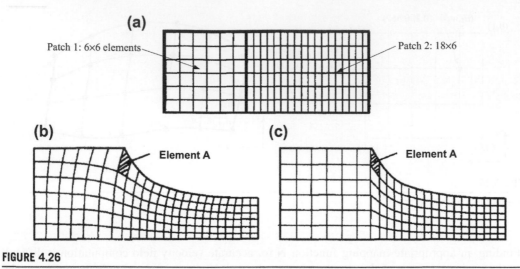

FIGURE 4.26

The two-dimensional fillet example modeled in two patches. (a) Original mesh. (b) Remesh using the boundary displacement method. (c) Remesh the using mapping method (Choi and Chang 1994).

The edges of a Coons patch are cubic curves in u and w, respectively. The shape of the cubic curve can be controlled by adjusting tangent vectors in both direction and magnitude. For example, changing the direction of the tangent vector $\mathbf{P}_{11,u}$ will alter the geometry of edge 3 and, therefore, the geometry of the patch, as illustrated in Figure 4.25b. A mesh of quad-elements can be generated by specifying the number of elements along the u- and w-parametric directions, such as the mesh shown in Figure 4.24, in which the number of elements along the u- and w-parametric directions are 5 and 3, respectively.

The data representation of a Coons patch is in a so-called B-matrix, as shown in Figure 4.25c, defined as

$$
\mathbf{B} = \begin{bmatrix}
\mathbf{P}_{00} & \mathbf{P}_{01} & \mathbf{P}_{00,w} & \mathbf{P}_{01,w} \\
\mathbf{P}_{10} & \mathbf{P}_{11} & \mathbf{P}_{10,w} & \mathbf{P}_{11,w} \\
\mathbf{P}_{00,u} & \mathbf{P}_{01,u} & \mathbf{P}_{00,uw} & \mathbf{P}_{01,uw} \\
\mathbf{P}_{10,u} & \mathbf{P}_{11,u} & \mathbf{P}_{10,uw} & \mathbf{P}_{11,uw}
\end{bmatrix}_{4 \times 4 \times 2}
\tag{4.265}
$$

where $\mathbf{P}_{00} = \mathbf{S}(0,0)$, $\mathbf{P}_{01} = \mathbf{S}(0,1)$, $\mathbf{P}_{10} = \mathbf{S}(1,0)$, and $\mathbf{P}_{11} = \mathbf{S}(1,1)$ are the four corner points; $\mathbf{P}_{00,u} = \partial\mathbf{S}/\partial u|_{u=w=0}$, $\mathbf{P}_{01,u}$, $\mathbf{P}_{10,u}$, and $\mathbf{P}_{11,u}$ are the tangent vectors in the u-direction at the four corner points; $\mathbf{P}_{00,w} = \partial\mathbf{S}/\partial w|_{u=w=0}$, $\mathbf{P}_{01,w}$, $\mathbf{P}_{10,w}$, and $\mathbf{P}_{11,w}$ are the tangent vectors in the w-direction at the four corner points; and $\mathbf{P}_{00,uw} = \partial^2\mathbf{S}/\partial u \partial w|_{u=w=0}$, $\mathbf{P}_{01,uw}$, $\mathbf{P}_{10,uw}$, and $\mathbf{P}_{11,uw}$ are the twister vectors at the four corner points. Note that a twister vector represents changes of tangent vector in the u (or w)-direction at a corner point along a boundary curve in the w (or u)-direction. For example, $\mathbf{P}_{10,uw} = \partial/\partial w(\partial\mathbf{S}/\partial u)|_{u=1,\,w=0} = \partial\mathbf{P}_{10,u}/\partial w$ is the derivative of the tangent vector along the u-direction at $\mathbf{P}_{10}$ with respect to w. The same twister vector can also be interpreted as $\mathbf{P}_{10,uw} = \partial/\partial u(\partial\mathbf{S}/\partial w)|_{u=1,\,w=0} = \partial\mathbf{P}_{10,w}/\partial u$, which is the derivative of the tangent vector along the w-direction at

P_{10} with respect to u. Note that for a planar patch, all twister vectors vanish. Also, the first two columns of the **B** matrix are boundary edges 1 and 3, respectively; and rows 1 and 2 are boundary edges 2 and 4, respectively.

Recall that the mathematical expression (algebraic format) for a bicubic parametric surface was given in Eq. 4.226. Similar to a Bézier surface, a Coons patch can also be written in a form like that of Eq. 4.227:

$$\mathbf{S}(u, w) = \mathbf{U}\mathbf{N}^C\mathbf{B}\mathbf{N}^{C^T}\mathbf{W}^T$$

$$= \begin{bmatrix} u^3 & u^2 & u & 1 \end{bmatrix} \begin{bmatrix} 2 & -2 & 1 & 1 \\ -3 & 3 & -2 & -1 \\ 0 & 0 & 1 & 0 \\ 1 & 0 & 0 & 0 \end{bmatrix} \begin{bmatrix} \mathbf{P}_{00} & \mathbf{P}_{01} & \mathbf{P}_{00,w} & \mathbf{P}_{01,w} \\ \mathbf{P}_{10} & \mathbf{P}_{11} & \mathbf{P}_{10,w} & \mathbf{P}_{11,w} \\ \mathbf{P}_{00,u} & \mathbf{P}_{01,u} & \mathbf{P}_{00,uw} & \mathbf{P}_{01,uw} \\ \mathbf{P}_{10,u} & \mathbf{P}_{11,u} & \mathbf{P}_{10,uw} & \mathbf{P}_{11,uw} \end{bmatrix}_{4 \times 4 \times 3}$$

$$\times \begin{bmatrix} 2 & -3 & 0 & 1 \\ -2 & 3 & 0 & 0 \\ 1 & -2 & 1 & 0 \\ 1 & -1 & 0 & 0 \end{bmatrix} \begin{bmatrix} w^3 \\ w^2 \\ w \\ 1 \end{bmatrix}$$

(4.266)

Here, $(u,w) \in [0,1] \times [0,1]$, and

$$\mathbf{N}^C = \begin{bmatrix} 2 & -2 & 1 & 1 \\ -3 & 3 & -2 & -1 \\ 0 & 0 & 1 & 0 \\ 1 & 0 & 0 & 0 \end{bmatrix}$$

(4.267)

The boundary edges of a Coons patch are Hermit cubic curves, defined as

$$\mathbf{P}(u) = \mathbf{U}\mathbf{N}^C\mathbf{H} = \begin{bmatrix} u^3 & u^2 & u & 1 \end{bmatrix} \begin{bmatrix} 2 & -2 & 1 & 1 \\ -3 & 3 & -2 & -1 \\ 0 & 0 & 1 & 0 \\ 1 & 0 & 0 & 0 \end{bmatrix} \begin{bmatrix} \mathbf{P}_0 \\ \mathbf{P}_1 \\ \mathbf{P}_{0,u} \\ \mathbf{P}_{1,u} \end{bmatrix}$$

(4.268)

Here, $\mathbf{H} = [\mathbf{P}_0, \mathbf{P}_1, \mathbf{P}_{0,u}, \mathbf{P}_{1,u}]^T$, in which $\mathbf{P}_0$ and $\mathbf{P}_1$ are the end points and $\mathbf{P}_{0,u}$ and $\mathbf{P}_{1,u}$ are tangent vectors at the end points. Entries in the matrix **H** can be used for shape design variables, depending on which boundary edge the curve represents. Any change in **H** alters the shape of the Coons patch:

$$\delta\mathbf{S}(u, w) = \mathbf{U}\mathbf{N}^C\delta\mathbf{B}\mathbf{N}^{C^T}\mathbf{W}^T$$

(4.269)

where

$$\delta\mathbf{B} = \begin{bmatrix} \delta\mathbf{P}_{00} & \delta\mathbf{P}_{01} & \delta\mathbf{P}_{00,w} & \delta\mathbf{P}_{01,w} \\ \delta\mathbf{P}_{10} & \delta\mathbf{P}_{11} & \delta\mathbf{P}_{10,w} & \delta\mathbf{P}_{11,w} \\ \delta\mathbf{P}_{00,u} & \delta\mathbf{P}_{01,u} & 0 & 0 \\ \delta\mathbf{P}_{10,u} & \delta\mathbf{P}_{11,u} & 0 & 0 \end{bmatrix}_{4 \times 4 \times 2}$$

(4.270)

For example, if the y-component of the tangent vector along the u-direction at $(u,w) = (1,1)$—that is, $P_{11y,u}$—is varied by $\delta P_{11y,u} = 1$, then $\delta\mathbf{B}$ is

$$\delta\mathbf{B}_y = \begin{bmatrix} 0 & 0 & 0 & 0 \\ 0 & 0 & 0 & 0 \\ 0 & 0 & 0 & 0 \\ 0 & 1 & 0 & 0 \end{bmatrix}_{4\times4} \tag{4.271}$$

In this case, $\delta\mathbf{B}_x(u,w) = 0$ because no change occurs in the x-direction. The domain velocity can be calculated by plugging the (u,w) of the node into Eq. 4.269. Note that in Eq. 4.270, we set the variation of the twister vectors to zero for planar structures.

EXAMPLE 4.22

Assume that $\delta P_{11y,u} = 1$. Calculate the velocity for node 8 (n_8) shown in Figure 4.24, where $(u_8, w_8) = (0.2, 0.667)$.

Solutions

Using Eq. 4.269 with $\delta\mathbf{B}_y$ shown in Eq. 4.271, we have

$$V_y^8 = \delta S_y(u_8, w_8) = \delta S_y(0.2, 0.667)$$

$$= \begin{bmatrix} 0.2^3 & 0.2^2 & 0.2 & 1 \end{bmatrix} \begin{bmatrix} 2 & -2 & 1 & 1 \\ -3 & 3 & -2 & -1 \\ 0 & 0 & 1 & 0 \\ 1 & 0 & 0 & 0 \end{bmatrix} \begin{bmatrix} 0 & 0 & 0 & 0 \\ 0 & 0 & 0 & 0 \\ 0 & 0 & 0 & 0 \\ 0 & 1 & 0 & 0 \end{bmatrix}_{4\times4} \begin{bmatrix} 2 & -3 & 0 & 1 \\ -2 & 3 & 0 & 0 \\ 1 & -2 & 1 & 0 \\ 1 & -1 & 0 & 0 \end{bmatrix} \begin{bmatrix} 0.667^3 \\ 0.667^2 \\ 0.667 \\ 1 \end{bmatrix} = -0.0237$$

Hence, the velocity is $\mathbf{V}^8 = [0, -0.0237]$.

If the boundary curve is not parameterized using a Hermit cubic curve, the change in the curve must be transformed to a Hermit cubic form and then the perturbed B matrix $\delta\mathbf{B}$ is created accordingly. For example, a cubic curve defined in Eq. 4.216 can be written as Bézier curve or a Hermit cubic curve, as follows:

$$\mathbf{P}(u) = \mathbf{U}(u)\mathbf{A} = \mathbf{U}\mathbf{N}^B\mathbf{G}^B = \mathbf{U}\mathbf{N}^C\mathbf{H} \tag{4.272}$$

Here, $\mathbf{G}^B$ contains the four control points of the cubic Bézier curve, and $\mathbf{H}$ was defined in Eq. 4.268, consisting of the end points and tangent vectors of a Hermit cubic curve. Therefore, control point vector of a Bézier curve can be transformed into an $\mathbf{H}$ vector of a Hermit cubic curve as

$$\mathbf{H} = (\mathbf{N}^C)^{-1}\mathbf{N}^B\mathbf{G}^B \tag{4.273}$$

Then, the changes in a Bézier curve can be transformed into a change in the Hermit cubic curve:

$$\delta\mathbf{H} = (\mathbf{N}^C)^{-1}\mathbf{N}^B\delta\mathbf{G}^B \tag{4.274}$$

EXAMPLE 4.23

Assume that the design boundary of the patch that contains nodes n_1 to n_6 shown in Figure 4.24 is parameterized by a Bézier curve identical to that of Example 4.18. As stated in Example 4.18, the design change is $\delta b = \delta P_{2y} = 1$; that is, the design variable is the y-direction movement of control point $\mathbf{P}_2$. Note that the design boundary is edge 3 of a Coons patch shown in Figure 4.25b. Calculate the velocity for node 8 inside the patch.

Solutions

From Example 4.18, the $\partial \mathbf{G}_y^B$ of the Bézier curve is

$$\delta \mathbf{G}_y^B = \begin{bmatrix} 0 \\ 0 \\ 1 \\ 0 \end{bmatrix} \delta P_{2y} = \begin{bmatrix} 0 \\ 0 \\ 1 \\ 0 \end{bmatrix} \tag{4.275}$$

From Eq. 4.274, we have

$$\delta \mathbf{H}_y = (\mathbf{N}^C)^{-1} \mathbf{N}^B \delta \mathbf{G}_y^B = \begin{bmatrix} 2 & -2 & 1 & 1 \\ -3 & 3 & -2 & -1 \\ 0 & 0 & 1 & 0 \\ 1 & 0 & 0 & 0 \end{bmatrix}^{-1} \begin{bmatrix} -1 & 3 & -3 & 1 \\ 3 & -6 & 3 & 0 \\ -3 & 3 & 0 & 0 \\ 1 & 0 & 0 & 0 \end{bmatrix} \begin{bmatrix} 0 \\ 0 \\ 1 \\ 0 \end{bmatrix} = \begin{bmatrix} 0 \\ 0 \\ -3 \\ 0 \end{bmatrix} \tag{4.276}$$

Using Eq. 4.269 with the second column of $\delta \mathbf{B}_y$ in Eq. 4.271 (with edge 3 as the design boundary in this case) replaced by the vector $\delta \mathbf{H}_y$ in Eq. 4.276, we have

$$\mathbf{V}_y^8 = \delta \mathbf{S}_y(u_8, w_8) = \delta \mathbf{S}_y(0.2, 0.667)$$

$$= \begin{bmatrix} 0.2^3 & 0.2^2 & 0.2 & 1 \end{bmatrix} \begin{bmatrix} 2 & -2 & 1 & 1 \\ -3 & 3 & -2 & -1 \\ 0 & 0 & 1 & 0 \\ 1 & 0 & 0 & 0 \end{bmatrix} \begin{bmatrix} 0 & 0 & 0 & 0 \\ 0 & 0 & 0 & 0 \\ 0 & -3 & 0 & 0 \\ 0 & 0 & 0 & 0 \end{bmatrix}_{4\times4} \begin{bmatrix} 2 & -3 & 0 & 1 \\ -2 & 3 & 0 & 0 \\ 1 & -2 & 1 & 0 \\ 1 & -1 & 0 & 0 \end{bmatrix} \begin{bmatrix} 0.667^3 \\ 0.667^2 \\ 0.667 \\ 1 \end{bmatrix} = -0.2846$$

Hence, the domain velocity at node 8 is obtained as $\mathbf{V}^8 = [0, -0.2846]$.

As mentioned before, the isoparametric mapping method is much more efficient than the boundary displacement method; however, the velocity field obtained using the boundary displacement method tends to keep the finite element mesh more regular because tracing of the solution to an elastic problem, governed by partial differential equations, is more orthogonal. Figure 4.26a shows a 2-D fillet example, in which the finite element mesh updated using the velocity obtained from the boundary displacement method (Figure 4.26b) is of better quality than those updated using isoparametric mapping method. Figure 4.26c shows that, using isoparametric mapping, the finite element mesh could be severely distorted after a large design change (e.g., element A).

4.5.5 SHAPE SENSITIVITY ANALYSIS USING FINITE DIFFERENCE OR SEMIANALYTICAL METHOD

As with the sensitivity analysis for sizing or material design variables, shape DSA can be carried out using the overall finite difference or semianalytical method, in which a design variable is perturbed and a finite element model is regenerated with the design change. These methods are general and widely employed in support of gradient-based optimization, despite the drawbacks pointed out previously.

For shape sensitivity analysis using the finite difference method or semianalytical method, it is desirable to update the finite element model (using Eq. 4.242) at the perturbed design using a velocity field obtained from the methods discussed above. In the overall finite difference method, an FEA is carried out for the FE model at the perturbed design, then the sensitivity coefficient is approximated using Eq. 4.35. In the semianalytical method, the stiffness matrix and load vector of the perturbed design must be generated, and Eq. 4.41 is followed to calculate the sensitivity coefficients.

In many situations, the design velocity field may not be readily available. Overall, the finite difference method is widely implemented for shape sensitivity analysis using CAD software equipped with a mesh generator and FEA solver, such as SolidWorks Simulation. A new dimension value with a design perturbation can be entered in CAD and the solid model can be rebuilt and remeshed. Thereafter, an FEA model for the perturbed design can be created (usually meshed by an automatic mesh generator), and an FEA can be carried out. The results obtained from FEA of the current and perturbed designs can then be used for sensitivity calculations using Eq. 4.35.

Although the approach is easy to implement, it has two potential pitfalls in addition to the problems of step size determination and computation efficiency, as discussed in Section 4.3. For shape design, the added pitfalls are altering topology of finite element mesh or geometric features at the perturbed design, as illustrated in Figure 4.27.

The pitfalls affect local performance measures, such as the stress measure defined at node A shown in Figure 4.27a. We assume a design change pulls the holes further apart. As a result, mesh is regenerated due to a design perturbation, as shown in Figure 4.27b. Due to the change in mesh topology, there is no clear trace in locating node A in the perturbed design: Is it at node A1, A2, or A3 that the stress performance is to be measured for sensitivity calculation using finite difference method? Another such example is the disappearing of nodes where performance measures are defined. As shown in Figure 4.27c, a performance measure, such as displacement, is defined at node B. Due to a design perturbation, the hole is moved to the right, such that it intersects the right edge of the rectangular structural boundary. As a result, part of the right edge disappears, as well as node B. Such a topology change in geometric features cause problems in calculating shape sensitivity coefficients using finite difference method with CAD software.

4.5.6 MATERIAL DERIVATIVES

Besides finite difference and semianalytical methods, the continuum approach offers options for shape sensitivity analysis with a rigorous mathematical basis and better computational efficiency. Moreover, no step size is required for design perturbation, such as in the finite difference and semianalytical methods.

The key theory in continuum approach for shape sensitivity analysis is the material derivative. In general, the material derivative describes the time rate of change of a physical quantity, such as heat or momentum, for a material element subjected to a space- and time-dependent velocity field. For structural shape design, time is replaced by design change, and the time rate of change of a physical quantity becomes the rate of a performance measure change with respect to design (i.e., gradient or sensitivity coefficient).

Recall the energy bilinear form and load linear form discussed in Section 4.4.1, which governs the behavior of the structure under static load:

$$a_\Omega(\mathbf{z}, \overline{\mathbf{z}}) = \ell_\Omega(\overline{\mathbf{z}}), \quad \forall \overline{\mathbf{z}} \in S \tag{4.277}$$

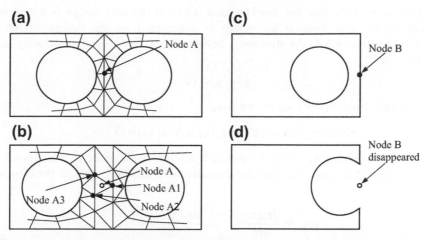

FIGURE 4.27

Pitfalls in shape sensitivity analysis using finite difference or semianalytical methods with the mesh generator in CAD software: (a) mesh of current design, (b) mesh topology altered after design change, (c) Node B at the right edge of the current design, and (d) topology of geometric features altered eliminating Node B.

Note that the subscript Ω is added to the energy and load forms in Eq. 4.277 to emphasize their dependence on the structural domain Ω, which varies. In Eq. 4.277, $\mathbf{z}$ is the solution, or displacement, of the governing equation of the structure at the current design, in which the structural domain is Ω, as illustrated in Figure 4.28.

Let $\mathbf{z}_\tau(\mathbf{x}_\tau)$ be the solution of the same governing equation but at a perturbed domain Ω_τ due to a design change:

$$a_{\Omega_\tau}(\mathbf{z}_\tau, \overline{\mathbf{z}}_\tau) = \ell_{\Omega_\tau}(\overline{\mathbf{z}}), \quad \forall \overline{\mathbf{z}}_\tau \in S_\tau \tag{4.278}$$

Here, the energy bilinear form and load linear form are formulated at the new structural domain Ω_τ, and S_τ is the space of kinematically admissible virtual displacement at the new design. Note that, in general, $S = S_\tau$ because the essential boundary conditions do not change with design and the smoothness of the solution function is not affected by the shape change. Again, τ is a scalar parameter that defines a shape change. In a material derivative for a physical problem, τ plays the role of time.

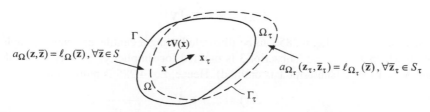

FIGURE 4.28

Illustration of material derivative concept.

It is important to note that the displacement $z_\tau(x_\tau)$ at the new design is a new function z_τ and measured at the new location x_τ, which is determined by the design velocity field V, as discussed in Section 4.5.4. As discussed before, for a linear design velocity field, x_τ is determined by

$$x_\tau = x + \tau V \tag{4.279}$$

Expanding $z_\tau(x_\tau)$ using Taylor's series, we have

$$z_\tau(x_\tau) = z_\tau(x + \tau V) \approx z_\tau(x) + \nabla[z_\tau^T(x)](\tau V) + \dots \tag{4.280}$$

Here, the gradient operator is employed—that is, $\nabla = [\partial/\partial x^1, \partial/\partial x^2, \partial/\partial x^3]^T$, which is a 3×1 vector for three-dimensional structures. The material derivative of a function, such as the displacement, is defined as

$$\dot{z}(x) \equiv \left.\frac{dz_\tau(x_\tau)}{d\tau}\right|_{\tau=0} = \left.\frac{dz_\tau(x + \tau V)}{d\tau}\right|_{\tau=0} \tag{4.281}$$

in which the dot on top of the function z denotes the material derivative on z. Equation 4.281 can be rewritten in a limiting form as

$$\dot{z}(x) = \left.\frac{dz_\tau(x + \tau V)}{d\tau}\right|_{\tau=0} = \lim_{\tau \to 0} \frac{z_\tau(x + \tau V) - z(x)}{\tau} = \lim_{\tau \to 0} \frac{z_\tau(x + \tau V) - z_\tau(x)}{\tau} + \lim_{\tau \to 0} \frac{z_\tau(x) - z(x)}{\tau} \tag{4.282}$$

The first term on the right-hand side of Eq. 4.282 defines the change of the same function z_τ evaluated at different locations (x_τ and x respectively). This can be further derived using the expansion shown in Eq. 4.280 as

$$\lim_{\tau \to 0} \frac{z_\tau(x + \tau V) - z_\tau(x)}{\tau} = \lim_{\tau \to 0} \frac{\nabla[z_\tau^T(x)](\tau V) + \dots}{\tau} = \nabla(z^T)V = \nabla z^T V \tag{4.283}$$

The second term on the right-hand side of Eq. 4.282 defines the change due to different functions (z_τ and z) evaluated at the same location x, which is nothing but the variation of z from its definition discussed in Section 4.4.2:

$$\lim_{\tau \to 0} \frac{z_\tau(x) - z(x)}{\tau} = z' \tag{4.284}$$

Hence, Eq. 4.282 becomes

$$\dot{z}(x) = z' + \nabla z^T V \tag{4.285}$$

The physical meaning of Eq. 4.285 can be illustrated using a cantilever beam example shown in Figure 4.29, in which $z = z_3 = z$ (subscript 3 is omitted for simplicity), $x = x_1 = x$ (subscript 1 is omitted), and $V = V_1 = V$ (subscript 1 is omitted). Hence, Eq. 4.285 is reduced to

$$\dot{z}(x) = z' + z_{,1}V \tag{4.286}$$

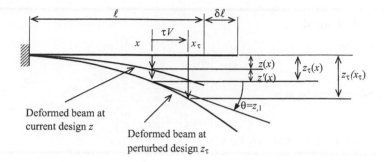

FIGURE 4.29

Deformed cantilever beam at current and perturbed designs due to a length change.

in which V is the design velocity that determines the material point movement due the beam length change. As shown in Figure 4.29, the difference between $z(x)$ and $z_\tau(x_\tau)$ consists of two parts, $z'(x) \approx z_\tau(x) - z(x)$ and $\theta \cdot (x_\tau - x) = z_{,1} \tau V$.

Recall that the FEA solution for the cantilever beam with an evenly distributed load shown in Figure 4.13 is obtained in Eq. 4.200 as

$$z_3(x) = \frac{q\ell}{24EI}\left(5x^2\ell - 2x^3\right) \tag{4.287}$$

In the next example, we use this solution to further illustrate the concept of material derivatives for the cantilever beam example shown in Figure 4.13.

EXAMPLE 4.24

Calculate $\dot{z}$, z', and $z_{,1}V$ at $x = \ell$ for the cantilever beam shown in Figure 4.13, using the finite element solution of Eq. 4.200.

Solutions

We first evaluate z' from the definition stated in Eq. 4.284. The displacement at the perturbed design due to the change $\delta\ell$ in the length of beam can be found as

$$z_\tau(x) = \frac{q(\ell + \tau\delta\ell)}{24EI}\left(5x^2(\ell + \tau\delta\ell) - 2x^3\right) = \frac{q}{24EI}\left(5x^2(\ell + \tau\delta\ell)^2 - 2x^3(\ell + \tau\delta\ell)\right) \tag{4.288}$$

Hence from Eq. 4.284, z' can be calculated as

$$z' = \lim_{\tau \to 0} \frac{z_\tau(x) - z(x)}{\tau} = \frac{\frac{q}{24EI}\left(5x^2(\ell + \tau\delta\ell)^2 - 2x^3(\ell + \tau\delta\ell)\right) - \frac{q}{24EI}\left(5x^2\ell^2 - 2x^3\ell\right)}{\tau} \tag{4.289}$$

$$= \lim_{\tau \to 0} \frac{q}{24EI} \frac{2x^2(5\ell - x)\tau\delta\ell + 5x^2(\tau\delta\ell)^2}{\tau} = \frac{qx^2}{12EI}(5\ell - x)\delta\ell$$

For simplicity in shape sensitivity analysis, we set $\delta\ell = 1$. Note that Eq. 4.289 is nothing but $\frac{\partial z_3(x)}{\partial \ell}$ shown in Eq. 4.208. Evaluating Eq. 4.289 at $x = \ell$, we have

$$z' = \frac{q\ell^3}{3EI} \tag{4.290}$$

which is identical to Eq. 4.209. Now, we take the derivative of Eq. 4.200 with respect to x:

$$z_{,1}(x) = \frac{q\ell}{12EI}\left(5x\ell - 3x^2\right) \tag{4.291}$$

Continued

EXAMPLE 4.24—cont'd

At $x = \ell$, $V = \delta\ell$. Hence,

$$z_{,1}V = \frac{q\ell^3}{6EI}\delta\ell = \frac{q\ell^3}{6EI} \qquad (4.292)$$

Adding Eqs 4.290 and 4.292, we have

$$\dot{z}(x) = z' + z_{,1}V_1 = \frac{q\ell^3}{3EI} + \frac{q\ell^3}{6EI} = \frac{q\ell^3}{2EI} \qquad (4.293)$$

which is $\frac{\partial z_1^{\ell}}{\partial \ell} = \frac{q\ell^3}{2EI}$, as obtained in Eq. 4.202 (and Eq. 4.207).

As demonstrated in Example 4.24, for the cantilever beam example, the sensitivity coefficient obtained using the discrete approach discussed at the beginning of this section is the same as the result of material derivative, in which the measurement point is assumed to be moving with the design. On the other hand, the result obtained in Eq. 4.209 is nothing but z', in which the measurement point is assumed to be stationary. As discussed, the movement of the material point or the measurement point is determined by the design velocity field. At the boundary, the velocity field is $\delta\ell$. How do the interior points move due to the design change $\delta\ell$? How does the domain velocity field affect the shape sensitivity coefficient?

Now, we discuss further on the design velocity field. At the tip, the boundary velocity is defined as $\delta\ell$. The interior or domain velocity can be defined in different ways, depending on the design intent. For example, let us consider two design velocity fields V^A and V^B, shown in Figures 4.30a and 4.30b, respectively. V^A is a linear function, and V^B is a quadratic function. Both assume a length change at the right end by $\delta\ell$. Mathematically, they are defined as

$$V^A = \frac{x}{\ell}\delta\ell \qquad (4.294)$$

$$V^B = \left(\frac{x}{\ell}\right)^2\delta\ell \qquad (4.295)$$

No matter how the velocity field is defined, the z' term is unchanged because the functions z_τ and z are evaluated at the same point. The same is true for $z_{,1}$. The only term that is affected is the velocity field V in the second term of Eq. 4.287. We assume the displacement is measured at midpoint $x = \ell/2$, the design velocities at the midpoint are $\delta\ell/2$ and $\delta\ell/4$, respectively, using V^A and V^B. Hence, the material derivatives of the displacement at the midpoint become

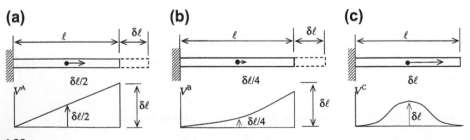

(a) **(b)** **(c)**

FIGURE 4.30

Midpoint movement with design change $\delta\ell$. (a) $\delta\ell/2$: linear velocity. (b) $\alpha\delta\ell \neq \delta\ell/2$: other than linear velocity. (c) $\delta\ell$: beam length change is zero.

$$\dot{z}_{\ell/2}^{A} = z' + z_{,1}V_1 = \frac{3q\ell^3}{32EI}\delta\ell + \frac{7q\ell^3}{48EI}\left(\frac{\delta\ell}{2}\right) = \frac{q\ell^3}{6EI}\delta\ell \tag{4.296}$$

and

$$\dot{z}_{\ell/2}^{B} = z' + z_{,1}V_1 = \frac{3q\ell^3}{32EI}\delta\ell + \frac{7q\ell^3}{48EI}\left(\frac{\delta\ell}{4}\right) = \frac{25q\ell^3}{192EI}\delta\ell \tag{4.297}$$

in which $\dot{z}_{\ell/2}^{A}$ predicts the displacement at the midpoint of the beam if the movement of the displacement measurement point is determined by velocity field V^A when a beam length is changed by $\delta\ell$. Similarly, $\dot{z}_{\ell/2}^{B}$ predicts the displacement at the midpoint of the beam if the movement of the displacement measurement point is determined by velocity field V^B. Therefore, in general, domain velocity is not unique, as long as it satisfies the requirements discussed in Section 4.5.4 and adequately captures the design intent.

We now take a look at a design velocity field V^C shown in Figure 4.30c, which is defined as a quadratic function with zero at both ends and maximum at the midpoint. Mathematically, V^C is defined as

$$V^C = \frac{4x}{\ell^2}(\ell - x)\delta\ell \tag{4.298}$$

In this case, the shape sensitivity coefficient at the tip is zero because $V = 0$ at $x = 1$. However, at the midpoint, the sensitivity is not zero; instead

$$\dot{z}_{\ell/2}^{C} = z' + z_{,1}V = 0 + \frac{7q\ell^3}{48EI}(\delta\ell) = \frac{7q\ell^3}{48EI}\delta\ell \tag{4.299}$$

Here, $z' = 0$ because the displacement function is unchanged (length stays constant). The second term of Eq. 4.299 is nonzero, which characterizes the change of displacement due to the change of the measurement point.

With a basic understanding of the concept of shape design, we move on to introduce the shape sensitivity analysis method based on the continuum approach. To simplify the mathematical derivations, we focus only on the beam example.

4.5.7 SHAPE SENSITIVITY ANALYSIS USING THE CONTINUUM APPROACH

We start by stating one of the major material derivative equations that are essential for the continuum approach. Readers are referred to Choi and Kim (2006a) for details and mathematical proof of the equation.

For an integral function defined as

$$\Phi = \int_{\Omega_\tau} f_\tau(\mathbf{x}_\tau)\Omega_\tau \tag{4.300}$$

in which f is a scalar function, the material derivative of Φ is given as

$$\dot{\Phi} = \left(\int_{\Omega_\tau} f(\mathbf{x}_\tau)\Omega_\tau\right)^{\cdot} = \int_{\Omega} (f'(\mathbf{x}) + \mathrm{div}(f\mathbf{V}))\Omega \tag{4.301}$$

in which the divergence is defined as $\mathrm{div}(f\mathbf{V}) = (f\mathbf{V})_{,1} + (f\mathbf{V})_{,2} + (f\mathbf{V})_{,3}$ for 3-D structures.

This equation is applied to the energy bilinear form and load linear form of Eq. 4.277. We rewrite the energy bilinear form as

$$a_{\Omega_\tau}(\mathbf{z}_\tau, \bar{\mathbf{z}}_\tau) = \int_{\Omega_\tau} c(\mathbf{z}_\tau, \bar{\mathbf{z}}_\tau)\Omega_\tau \tag{4.302}$$

For a beam structure, $c(\mathbf{z}, \bar{\mathbf{z}}) = EIz_{3,11}\bar{z}_{3,11}$. The material derivative of $a_{\Omega_\tau}(\mathbf{z}_\tau, \bar{\mathbf{z}}_\tau)$ can be written using Eq. 4.301:

$$
\begin{aligned}
[a_{\Omega_\tau}(\mathbf{z}_\tau, \bar{\mathbf{z}}_\tau)]^{\boldsymbol{\cdot}} &= \int_\Omega [c(\mathbf{z}, \bar{\mathbf{z}})]'\Omega + \int_\Omega \mathrm{div}[c(\mathbf{z}, \bar{\mathbf{z}})\mathbf{V}]\Omega \\
&= \int_\Omega [c(\mathbf{z}', \bar{\mathbf{z}}) + c(\mathbf{z}, \bar{\mathbf{z}}')]\Omega + \int_\Omega \mathrm{div}[c(\mathbf{z}, \bar{\mathbf{z}})\mathbf{V}]\Omega \\
&= \int_\Omega [c(\dot{\mathbf{z}} - \nabla\mathbf{z}^{\mathrm{T}}\mathbf{V}, \bar{\mathbf{z}}) + c(\mathbf{z}, \dot{\bar{\mathbf{z}}} - \nabla\bar{\mathbf{z}}^{\mathrm{T}}\mathbf{V})]\Omega + \int_\Omega \mathrm{div}[c(\mathbf{z}, \bar{\mathbf{z}})\mathbf{V}]\Omega \\
&= \int_\Omega c(\dot{\mathbf{z}}, \bar{\mathbf{z}})\Omega - \int_\Omega [c(\nabla\mathbf{z}^{\mathrm{T}}\mathbf{V}, \bar{\mathbf{z}}) + c(\mathbf{z}, \nabla\bar{\mathbf{z}}^{\mathrm{T}}\mathbf{V})]\Omega + \int_\Omega \mathrm{div}[c(\mathbf{z}, \bar{\mathbf{z}})\mathbf{V}]\Omega \\
&= a_\Omega(\dot{\mathbf{z}}, \bar{\mathbf{z}}) + a'_V(\mathbf{z}, \bar{\mathbf{z}}) \tag{4.303}
\end{aligned}
$$

in which we employed the relation $\mathbf{z}' = \dot{\mathbf{z}}(x) - \nabla\mathbf{z}^{\mathrm{T}}\mathbf{V}$ and set $\dot{\bar{\mathbf{z}}}(x) = 0$ because $\bar{\mathbf{z}}$ is virtual displacement. We define

$$a_\Omega(\dot{\mathbf{z}}, \bar{\mathbf{z}}) \equiv \int_\Omega c(\dot{\mathbf{z}}, \bar{\mathbf{z}})\Omega \tag{4.304}$$

and

$$a'_V(\mathbf{z}, \bar{\mathbf{z}}) \equiv -\int_\Omega [c(\nabla\mathbf{z}^{\mathrm{T}}\mathbf{V}, \bar{\mathbf{z}}) + c(\mathbf{z}, \nabla\bar{\mathbf{z}}^{\mathrm{T}}\mathbf{V})]\Omega + \int_\Omega \mathrm{div}[c(\mathbf{z}, \bar{\mathbf{z}})\mathbf{V}]\Omega \tag{4.305}$$

Repeating the same for the load linear form, we have

$$[\ell_{\Omega_\tau}(\bar{\mathbf{z}})]^{\boldsymbol{\cdot}} = \int_\Omega (f\bar{z})^{\boldsymbol{\cdot}}\Omega = \int_\Omega (f\bar{z})'\Omega + \int_\Omega \mathrm{div}(f\bar{z}V)\Omega = -\int_\Omega f\nabla\bar{z}^{\mathrm{T}}V\Omega + \int_\Omega \mathrm{div}(f\bar{z}V)\Omega \equiv \ell'_V(\bar{\mathbf{z}}) \tag{4.306}$$

in which we assume $f' = 0$. Hence, equating $[a_{\Omega_\tau}(\mathbf{z}_\tau, \bar{\mathbf{z}}_\tau)]^{\boldsymbol{\cdot}}$ and $[\ell_{\Omega_\tau}(\bar{\mathbf{z}}_\tau)]^{\boldsymbol{\cdot}}$, we have

$$a_\Omega(\dot{\mathbf{z}}, \bar{\mathbf{z}}) = -a'_V(\mathbf{z}, \bar{\mathbf{z}}) + \ell'_V(\bar{\mathbf{z}}), \quad \forall \bar{\mathbf{z}} \in S \tag{4.307}$$

Similar to sizing sensitivity analysis, in implementation, the right-hand side of Eq. 4.307 is calculated as a fictitious load, which is employed to solve for $\dot{\mathbf{z}}$. Equation 4.307 represents the direct differentiation method of the continuum approach. For the adjoint variable method of the continuum approach, readers are referred to Choi and Kim (2006a).

4.5.7.1 Shape sensitivity analysis for a cantilever beam

For one-dimensional beam bending problems, we have $c(\mathbf{z}, \bar{\mathbf{z}}) = EIz_{3,11}\bar{z}_{3,11}$, $\mathbf{z} = z_3$, $\mathbf{V}^T\mathbf{n} = V$, $\nabla\mathbf{z} = z_{3,1}$, and div $f = f_{,1}$. Hence, from Eqs 4.304, 4.305, and 4.306, we have

$$a_\Omega(\dot{\mathbf{z}}, \bar{\mathbf{z}}) = \int_0^\ell EI\dot{z}_{3,11}\bar{z}_{3,11}dx \tag{4.308}$$

$$a_V'(\mathbf{z}, \bar{\mathbf{z}}) = -\int_0^\ell \left[EI(z_{3,1}V)_{,11}\bar{z}_{3,11} + EIz_{3,11}(\bar{z}_{3,1}V)_{,11}\right]dx + \int_0^\ell (EIz_{3,11}\bar{z}_{3,11}V)_{,1}dx$$

$$= -3\int_0^\ell (EIz_{3,11}\bar{z}_{3,11}V_{,1})dx - \int_0^\ell \left[EI(z_{3,11}\bar{z}_{3,1} + z_{3,1}\bar{z}_{3,11})V_{,11}\right]dx \tag{4.309}$$

and

$$\ell_V'(\bar{\mathbf{z}}) = -\int_\Omega f\nabla\bar{\mathbf{z}}^T\mathbf{V}\Omega + \int_\Omega \mathrm{div}(f\bar{\mathbf{z}}V)\Omega = -\int_0^\ell f\bar{z}_{3,1}Vdx + \int_0^\ell (f\bar{z}_3V)_{,1}dx = \int_0^\ell (f_{,1}V + fV_{,1})\bar{z}_3dx \tag{4.310}$$

Plugging these terms into Eq. 4.307, we have

$$\int_0^\ell EI\dot{z}_{3,11}\bar{z}_{3,11}dx = 3\int_0^\ell (EIz_{3,11}\bar{z}_{3,11}V_{,1})dx + \int_0^\ell \left[EI(z_{3,11}\bar{z}_{3,1} + z_{3,1}\bar{z}_{3,11})V_{,11}\right]dx$$

$$+ \int_0^\ell (f_{,1}V + fV_{,1})\bar{z}_3dx \tag{4.311}$$

Using finite element shape functions, such as those shown in Eq. 4.143, to discretize the sensitivity equation (Eq. 4.311), we have on the left-hand side

$$\int_0^\ell EI\dot{z}_{3,11}\bar{z}_{3,11}dx = \bar{\mathbf{Z}}_g^T\mathbf{K}_g\dot{\mathbf{Z}}_g \tag{4.312}$$

where

$$\dot{\mathbf{Z}}_g = \begin{bmatrix} \dot{z}_{3_i} & \dot{\theta}_i & \dot{z}_{3j} & \dot{\theta}_j \end{bmatrix}^T \tag{4.313}$$

The right-hand side of Eq. 4.311 can be computed as $\overline{\mathbf{Z}}^T\mathbf{F}_{\text{fic}_g}$, in which $\mathbf{F}_{\text{fic}_g}$ is the global fictitious load vector. By imposing the boundary conditions and removing the virtual displacement $\overline{\mathbf{Z}}$ from both sides, we have

$$\mathbf{K}\dot{\mathbf{Z}} = \mathbf{F}_{\text{fic}} \tag{4.314}$$

which can be solved using the same decomposed stiffness matrix $\mathbf{K}$ as the original structure considering the fictitious load $\mathbf{F}_{\text{fic}}$ as additional load cases. Equation 4.311 represents the direct differentiation method of the continuum-discrete approach for shape sensitivity analysis. We illustrate a few more details in the following example.

EXAMPLE 4.25

Calculate shape sensitivity for the cantilever beam shown in Figure 4.13, assuming a linear design velocity of Eq. 4.294—that is, $V = V^A = \frac{x}{\ell}$ (set $\delta\ell = 1$)—using the continuum-discrete method.

Solutions

We first rewrite Eq. 4.311 for the cantilever beam with $f = q$ (hence $f_{,1} = q_{,1} = 0$) and a linear velocity field (hence $V_{,11} = 0$) as

$$\int_0^\ell EI\dot{z}_{3,11}\overline{z}_{3,11}\,dx = 3\int_0^\ell \left(EIz_{3,11}\overline{z}_{3,11}V_{,1}\right)dx + \int_0^\ell qV_{,1}\overline{z}_3\,dx \tag{4.315}$$

Using the shape functions of Eq. 4.144, we have

$$z_3 = \mathbf{N}^T\mathbf{Z} = \begin{bmatrix} N_3 & N_4 \end{bmatrix}\begin{bmatrix} z_{3j} \\ \theta_j \end{bmatrix} = \begin{bmatrix} \dfrac{3x^2}{\ell^2} - \dfrac{2x^3}{\ell^3} & -\dfrac{x^2}{\ell} + \dfrac{x^3}{\ell^2} \end{bmatrix}\begin{bmatrix} \dfrac{q\ell^4}{8EI} \\ \dfrac{q\ell^3}{6EI} \end{bmatrix} = \dfrac{q\ell}{24EI}\left(5x^2\ell - 2x^3\right) \tag{4.316}$$

$$z_{3,1} = \left(\mathbf{N}^T\mathbf{Z}\right)_{,1} = \begin{bmatrix} N_{3,1} & N_{4,1} \end{bmatrix}\begin{bmatrix} z_{3j} \\ \theta_j \end{bmatrix} = \begin{bmatrix} \dfrac{6x}{\ell^2} - \dfrac{6x^2}{\ell^3} & -\dfrac{2x}{\ell} + \dfrac{3x^2}{\ell^2} \end{bmatrix}\begin{bmatrix} \dfrac{q\ell^4}{8EI} \\ \dfrac{q\ell^3}{6EI} \end{bmatrix} = \dfrac{q\ell}{12EI}\left(5x\ell - 3x^2\right) \tag{4.317}$$

$$z_{3,11} = \left(\mathbf{N}^T\mathbf{Z}\right)_{,11} = \begin{bmatrix} N_{3,11} & N_{4,11} \end{bmatrix}\begin{bmatrix} z_{3j} \\ \theta_j \end{bmatrix} = \begin{bmatrix} \dfrac{6}{\ell^2} - \dfrac{12x}{\ell^3} & -\dfrac{2}{\ell} + \dfrac{6x}{\ell^2} \end{bmatrix}\begin{bmatrix} \dfrac{q\ell^4}{8EI} \\ \dfrac{q\ell^3}{6EI} \end{bmatrix} = \dfrac{q\ell}{12EI}\left(5\ell - 6x\right) \tag{4.318}$$

Bringing Eqs 4.316 to 4.318 to the right-hand side of Eq. 4.315, we have

$$\int_0^\ell EIz_{3,11}\overline{z}_{3,11}V_{,1}\,dx = \overline{\mathbf{Z}}^T\int_0^\ell \begin{bmatrix} \dfrac{6}{\ell^2} - \dfrac{12x}{\ell^3} \\ -\dfrac{2}{\ell} + \dfrac{6x}{\ell^2} \end{bmatrix}\left[\dfrac{q\ell}{12}(5\ell - 6x)\right]\dfrac{1}{\ell}\,dx = \overline{\mathbf{Z}}^T\dfrac{q}{12}\int_0^\ell \begin{bmatrix} \dfrac{6}{\ell^2} - \dfrac{12x}{\ell^3} \\ -\dfrac{2}{\ell} + \dfrac{6x}{\ell^2} \end{bmatrix}(5\ell - 6x)\,dx = \overline{\mathbf{Z}}^T\begin{bmatrix} \dfrac{q}{2} \\ -\dfrac{q\ell}{12} \end{bmatrix} \tag{4.319}$$

EXAMPLE 4.25—cont'd

and

$$\int_0^\ell q\bar{z}_3 V_{,1} \mathrm{d}x = \overline{\mathbf{Z}}^\mathrm{T} \int_0^\ell q \begin{bmatrix} \dfrac{3x^2}{\ell^2} - \dfrac{2x^3}{\ell^3} \\[2mm] -\dfrac{x^2}{\ell} + \dfrac{x^3}{\ell^2} \end{bmatrix} \dfrac{1}{\ell} \mathrm{d}x = \overline{\mathbf{Z}}^\mathrm{T} \dfrac{q}{\ell} \int_0^\ell \begin{bmatrix} \dfrac{3x^2}{\ell^2} - \dfrac{2x^3}{\ell^3} \\[2mm] -\dfrac{x^2}{\ell} + \dfrac{x^3}{\ell^2} \end{bmatrix} \mathrm{d}x = \overline{\mathbf{Z}}^\mathrm{T} \begin{bmatrix} \dfrac{q}{2} \\[2mm] \dfrac{q\ell}{12} \end{bmatrix} \tag{4.320}$$

Bringing Eqs 4.319 and 4.320 to 4.315, we have

$$\overline{\mathbf{Z}}^\mathrm{T} \left\{ 3 \begin{bmatrix} \dfrac{q}{2} \\[2mm] -\dfrac{q\ell}{12} \end{bmatrix} + \begin{bmatrix} \dfrac{q}{2} \\[2mm] -\dfrac{q\ell}{12} \end{bmatrix} \right\} = \overline{\mathbf{Z}}^\mathrm{T} \begin{bmatrix} 2q \\[2mm] \dfrac{q\ell}{3} \end{bmatrix} = \overline{\mathbf{Z}}^\mathrm{T} \mathbf{F}_{\mathrm{fic}} \tag{4.321}$$

By solving the sensitivity vector in the same way as solving **Z** of the structure (see Eq. 4.149), we have

$$\begin{bmatrix} \dot{z}_{3j} \\[2mm] \dot{\theta}_j \end{bmatrix} = \mathbf{K}^{-1} \mathbf{F}_{\mathrm{fic}} = \dfrac{\ell}{12EI} \begin{bmatrix} 4\ell^2 & 6\ell \\[1mm] 6\ell & 12 \end{bmatrix} \begin{bmatrix} 2q \\[2mm] -\dfrac{q\ell}{3} \end{bmatrix} = \begin{bmatrix} \dfrac{q\ell^3}{2EI} \\[4mm] \dfrac{2q\ell^2}{3EI} \end{bmatrix} \tag{4.322}$$

4.6 TOPOLOGY OPTIMIZATION

Sizing optimization varies the sizes of the structural elements, such as the diameter of a bar or beam, as shown in Figure 4.31a, or the thickness of a sheet metal. In sizing optimization, the shape of the structure is known and unchanged. For shape optimization, the topology of geometric features in the structure, such as the number of holes, is known. The optimal shape is confined to the topology of the initial structural geometry, as illustrated in Figure 4.31b. No additional holes can be created during the shape optimization process. For both sizing and shape optimization, the topology of the structural geometry is unchanged in the design process.

Another important type of optimization problem is topology optimization, which solves the basic engineering problem of distributing a prescribed amount of material in a design space. Topology optimization is sometimes referred to as layout optimization. In general, the goal of topology optimization is to find the best use of material for a structural body that is subject to either a single load or multiple load distributions. The best use of material in the case of topology optimization represents a maximum stiffness (or minimum compliance), maximum buckling load, or maximum first vibration frequency design. With topology optimization, the resulting shape or topology is not known; the number of holes, structural members (for frame structures), etc. are not decided upon, as illustrated in Figure 4.31c.

In topology optimization, we start from a given design domain and proceed to find the optimum distribution of material and voids. In general, a design domain is discretized by using the FEM into discrete mesh. Individual finite elements in the discretized design domain are then evaluated to determine if they are staying or removed from the structure using optimization methods. This results in a so-called 0–1 problem—the elements either exist or do not.

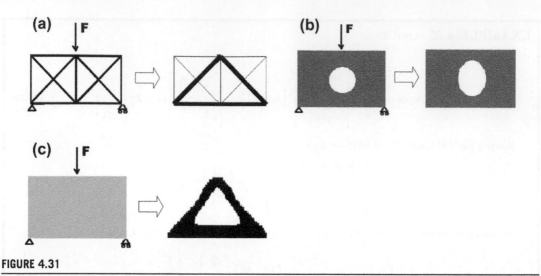

FIGURE 4.31

Different types of structural optimization. (a) Sizing optimization. (b) Domain shape optimization. (c) Topology optimization.

The two main solution strategies for solving the topology optimization problem are the density method and the homogenization method. In this section, we briefly discuss the density method. Those interested in learning the homogenization method are referred to Bendsøe and Sigmund (2003).

4.6.1 BASIC CONCEPT AND PROBLEM FORMULATION

In general, the objective of a topology optimization problem is either to minimize the compliance, which is equivalent to maximizing the stiffness, maximize the lowest frequency, or maximize the lowest buckling load. A volume constraint is imposed to limit material usage. The optimization problem can be formulated as

$$\text{Minimize:} \quad \phi(c(\mathbf{x})) \tag{4.323a}$$

$$\text{Subject to:} \quad \int_{\Omega} c(\mathbf{x})\rho_0 d\Omega \leq m_0 \tag{4.323b}$$

$$\psi_i - \psi_i^u \leq 0, \quad i = 1, m \tag{4.323c}$$

$$0 \leq c(\mathbf{x}) \leq 1 \tag{4.323d}$$

where ϕ is a structural performance measure to be minimized (or maximized), such as compliance; ψ_i is ith structural performance constraint, such as displacement; m is the total number of constraints; Ω is the structural domain; m_0 is the mass limit; $c(\mathbf{x})$ is the design variable; and ρ_0 is material mass density. In implementation, the design variable $c(\mathbf{x})$ is assigned to individual finite elements; therefore,

$c(\mathbf{x}) = c_i$, $i = 1$, NE (number of elements). In each element, the density becomes $\rho_i = c_i\,\rho_0$, and Eq. 4.323b becomes

$$\int_{\Omega} c(\mathbf{x})\rho_0 d\Omega = \sum_{i=1}^{NE} c_i\rho_0 V_i \leq m_0 \tag{4.324}$$

where V_i is the volume (or area for planar problems) of the ith finite element. Also, Eq. 4.323d becomes

$$0 \leq c_i \leq 1, \quad i = 1, \quad NE \tag{4.325}$$

It is desired that the solution of topology optimization only consists of solid or empty elements—that is, $c_i = 0$ or 1. One popular method to suppress or penalize intermediate densities is by letting the stiffness of the material be expressed as

$$E = c_i^p E_0, \quad i = 1, \quad NE \tag{4.326}$$

where E_0 is the modulus of elasticity of the material and p is a penalization factor that is greater than zero, typically 2 to 5 in implementation. For $p > 1$, local stiffness for values of $c_i < 1$ is lowered, as illustrated in Figure 4.32, thus making it "uneconomical" to have intermediate densities in the optimal design. In the literature, the density method together with this penalization is often called the SIMP (solid isotropic microstructures with penalization) method (Rozvany 2000).

Although the density method is straightforward to implement for topology optimization, one must be aware of numerical instabilities. These can manifest as checkerboard, mesh dependence, and local minimum. More detailed discussion and remedies to treat these instabilities can be found in Sigmund and Petersson (1998).

Topology optimization often converges to a design that contains a checkerboard pattern—that is, the alternately solid and void elements, as shown in Figure 4.33b. This is a typical result in topology optimization using finite element methods.

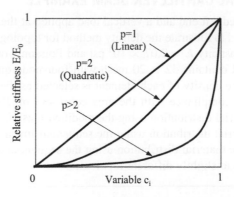

FIGURE 4.32

Relative stiffness as a function of variable c_i with different penalization factors p (Eschenauer and Olhoff 2001).

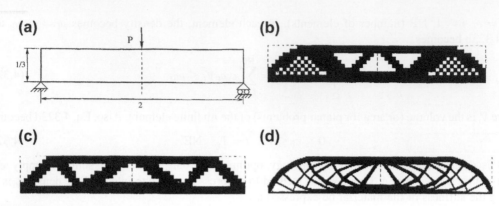

FIGURE 4.33

Numerical instabilities in topology optimization. (a) Design problem. (b) Example of checkerboards. (c) Solution for 600-element discretization. (d) Solution for 5400-element discretization (Sigmund and Petersson 1998).

The quality of topology optimization result is dependent on the discretization of the finite element model, the so-called mesh-dependence problem. Figure 4.33c shows the topology optimization result for the discretization using 600 elements, and Figure 4.33d shows the result using 5400 elements for the same physical problem. The result in Figure 4.33d is much more detailed than that in Figure 4.33c; however, the two topologies are different in nature.

Different solutions to the same discretized problem are observed when choosing different parameters, such as finite element type (triangular vs quadratic elements), optimization algorithms, convergence criteria, and so forth. These numerical instabilities are all related to whether the continuous constrained optimization problems are well posed or not—that is, whether they have a solution or not.

4.6.2 TWO-DIMENSIONAL CANTILEVER BEAM EXAMPLE

A cantilever beam with a fixed left end and a vertical load applied at the midpoint of the free end, as shown in Figure 4.34a, is used to illustrate the density method for topology optimization. The material properties are modulus of elasticity $E = 2.07 \times 10^5$ psi and Poisson's ratio $\nu = 0.3$.

The finite element model contains 32×20 mesh with four-node quad elements. There are 640 elements in the model and the density of each element is selected as the design variable. The design problem is to minimize the compliance with the area constraint of 25% imposed on the domain. Figure 4.34b shows the material distribution using the modified feasible direction after 10 iterations. Figure 4.34c shows the material distribution using the sequential linear programming after 36 iterations. Figure 4.34d shows the material distribution using the sequential quadratic programming after 28 iterations. They converge to slightly different topologies.

4.7 CASE STUDIES

Two case studies are presented in this section. A design for tracked vehicle roadarm demonstrates a practical application of topology optimization and shape optimization as well as a geometric

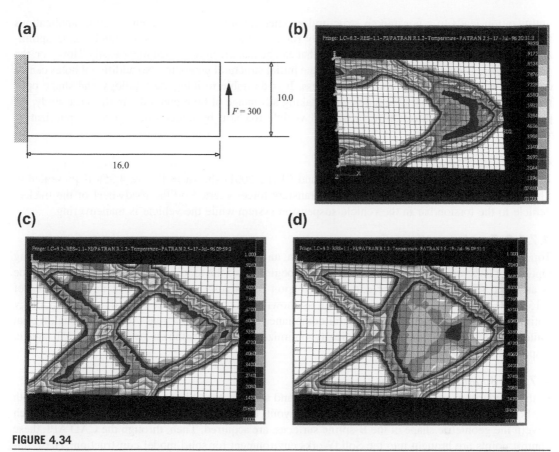

FIGURE 4.34

The two-dimensional topology optimization example. (a) Analysis model (dimensions are in inches and force is in pounds). (b) Optimal topology design using modified feasible direction. (c) Optimal topology design using sequential linear programming. (d) Optimal topology design using sequential quadratic programming.

smoothing technique that converts the result of topology optimization into smooth parametric geometric surfaces for shape optimization. The second case demonstrates shape sensitivity analysis for a three-dimensional multiscale dynamic crack propagation problem, in which performance measures are crack propagation speed defined at atomistic level obtained by solving coupled atomistic/continuum structures using the bridging scale method (BSM) (Wagner and Liu 2003). The goal is to quantify the impact of geometric shape changes on crack speed at the atomistic level.

4.7.1 TOPOLOGY AND SHAPE OPTIMIZATION

Topology optimization has drawn significant attention in the recent development of structural optimization. This method has been proven very effective in determining the initial geometric shape or layout for structural design. The main drawback of the method, however, is that the topology optimization leads to a nonsmooth structural geometry, while most of the engineering applications require a smooth geometric shape, especially for manufacturing. On the other hand, shape optimization starts with a smooth geometric model that can be manufactured much more easily. However, the optimal shape is confined to the topology of the initial structural geometry. No additional holes can be created during the shape optimization process. It is desirable to integrate topology and shape optimizations in support of structural design by taking advantage of both methods. In this case study, we present structural optimization for a tracked vehicle roadarm by integrating topology optimization with shape optimization.

4.7.1.1 The tracked vehicle roadarm

A tracked vehicle roadarm example (Tang and Chang 2001) shown in Figure 4.35a is presented to illustrate the design process. The roadarm transfers forces exerted on the roadwheel of the tracked vehicle to the torsion bar in the vehicle suspension system while the vehicle is maneuvering.

4.7.1.2 Topology optimization

Topology optimization is carried out to obtain an optimal structural layout, in this case using OptiStruct (www.altairhyperworks.com). At the beginning of the topology optimization, a bulk shape is assumed (Figure 4.35b). The end of the torsion bar is fully constrained and an axial force of 2.15×10^4 lb. is applied at the wheel shaft, as shown in Figure 4.35b. The objective of the topology optimization is to minimize the structural compliance subject to a 45% mass reduction. The original finite element model and its topologically optimized layout are shown in Figures 4.35b and c, respectively.

4.7.1.3 Boundary smoothing

In this roadarm example, cubic B-spline curves and surfaces are utilized to approximate and smooth the boundary points of the irregular structural layout. During this process, the control points, which govern the geometric shape of the B-spline surfaces, are acquired. Then, through the CAD API, these control points are brought into the SolidWorks environment for solid model construction. These imported control points also parameterize the boundary surfaces of the reconstructed solid model and later serve as design variables for shape optimization.

As an example, the geometric points of five representative sections of the roadarm (Figure 4.35c) are selected and fitted with B-spline curves (Figure 4.36). Following the surface skinning technique

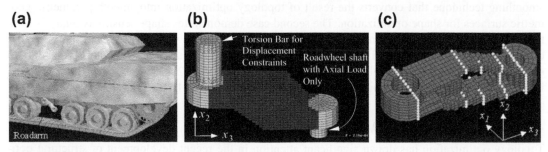

FIGURE 4.35

The tracked vehicle roadarm. (a) Physical model. (b) Initial finite element model. (c) Topologically optimized model.

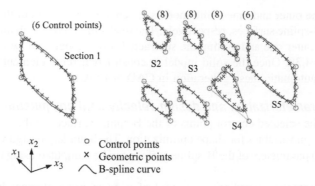

FIGURE 4.36

Section profiles and B-spline curves.

(Tang and Chang 2001), an outer polygon surface formed by 6×5 control points and the enclosed B-spline surface are created (Figure 4.37a). Similarly, an inner B-spline surface (4×3 control points) that represents the hole in the roadarm is created (Figure 4.37b). Note that the B-spline surface constructed is C^2-continuous in both u- and w-parametric directions because cubic basis functions are employed. The control points and basis functions of the B-spline surface can be imported into CAD tools, in this case SolidWorks, to support solid modeling and shape optimization.

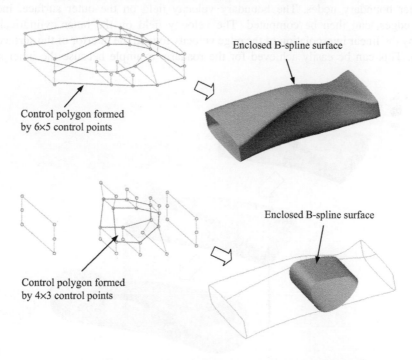

FIGURE 4.37

B-spline surfaces: (a) outer surface and (b) inner surface.

In SolidWorks, the outer and inner solid models are created by filling out the cavities enclosed by the outer and inner B-spline surfaces, respectively. The final solid model is obtained by subtracting the inner solid from the outer one and uniting the subtracted solid model with two end half cylinders, as illustrated in Figure 4.38. Once the solid model is constructed, finite element mesh can be created using, for example, automatic mesh generation in CAD or FEA.

4.7.1.4 Shape parameterization and design velocity field computation

The movements of the selected control points of the B-spline surfaces in the x_1, x_2, and x_3 directions can be defined as design variables for shape optimization. The boundary design velocity field of a node N defined by (u_N, w_N) parameters of the B-spline surface can be computed using the methods discussed in Section 4.5.4.

Note that the parametric coordinates (u_N, w_N) of a given finite element boundary node N are determined by CAD via its API function. For example, in SolidWorks, the API function GetClosestPointsOn(x_1, x_2, x_3), where (x_1, x_2, x_3) are the Cartesian coordinates of the finite element node, can be used to retrieve the parametric coordinates (u_N, w_N) of the node.

According to the design variables shown in Figure 4.39, the outer and inner B-spline surfaces will vary due to design changes. When the control points shared by the outer and inner B-spline surfaces move, the material points on the intersection edges can move following either the outer or inner surface. Such movements are not unique. To alleviate the problem, the boundary velocity field of both the outer and inner surfaces is determined by the outer surface only. The parametric locations of the finite element nodes of the intersection edges on the outer B-spline surface are first identified along with the other boundary nodes. The boundary velocity field on the outer surface, including the intersection edges, can then be computed. The velocity field on the inner cylindrical surface is determined by the linear interpolation of the edge velocity toward the interior of the surface (along the x_1-direction). This can be easily achieved for the roadarm example because the inner surface is a

FIGURE 4.38

Reconstruction of roadarm solid model.

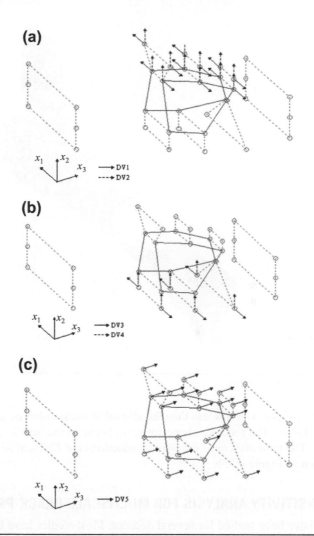

FIGURE 4.39

Shape design parameterization. (a) Design variables DV1 and DV2. (b) Design variables DV3 and DV4. (c) Design variable DV5.

cylindrical surface. Once the boundary velocity field is computed, the domain velocity can be calculated using the boundary displacement method discussed in Section 4.5.4.

4.7.1.5 Shape design optimization

In shape optimization, the objective function is structural mass, and constraint functions are the structural compliance measure obtained from topology optimization and stress measures. Note that the stress upper bound is defined as 37.5 ksi, and the material is SAE 1045 carbon steel with a yield strength of 45 ksi. Hence, a safety factor of 1.2 is employed for the design.

The optimal shape was obtained in five design iterations. The von Mises stress distribution of the optimal design is shown in Figure 4.40. The high stress areas are distributed around the upper and lower branch of the roadarm. The highest stress is found located at the top surface of the lower branch

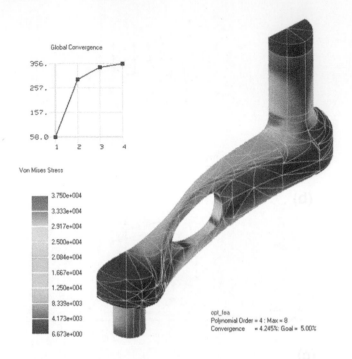

FIGURE 4.40

von Mises stress of the optimal roadarm.

with a value of 3.69×10^4 psi, which is less than the allowable stress—that is, material yield strength divided by the safety factor. The roadarm shape variation between the initial and optimal designs is shown in Figure 4.41. The total mass and compliance reductions are 47% and 81%, respectively, from the initial shape shown in Figure 4.35b.

4.7.2 SHAPE SENSITIVITY ANALYSIS FOR MULTISCALE CRACK PROPAGATION

Fatigue and fracture have been studied for several decades. Most studies have been conducted at the macroscopic level. Theories and computational methods developed at the macroscopic level involve assumptions, approximations, and uncertainties, which are not able to accurately predict and explain the fatigue and fracture. The approaches presented in this case study are to simulate the fatigue and fracture behavior at the atomistic level using BSM (Wagner and Liu 2003) and analyze the influence of shape changes at the macroscopic level to the crack growth at the atomistic level. Going from macroscopic to atomistic levels involves order-of-magnitude refinements in length and time scales, as well as an understanding of the mechanics and material constitutive law at the atomistic level.

In this case study, we present a shape sensitivity analysis method and apply it to a three-dimensional multiscale dynamic crack propagation problem, in which performance measures are crack propagation speed (or crack speed) defined at atomistic level obtained by solving coupled atomistic/continuum structures. Our goal is to quantify the impact of geometric shape changes on crack speed at the atomistic level.

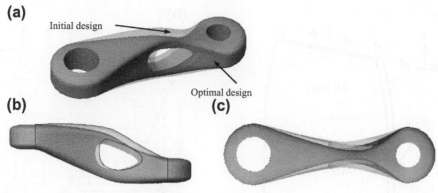

(a)

Initial design

Optimal design

(b)   **(c)**

FIGURE 4.41

Comparison of the 3-D roadarm at initial and optimal designs. (a) Isometric view (light color: initial design, dark color: optimal design). (b) Side view. (c) Top view.

4.7.2.1 Interatomic potential

In this case study, the interatomic potential function chosen to be implemented in the MD simulation is the Lennard-Jones (LJ) 6–12 potential (Jones 1924), which takes the following form:

$$\Phi(r) = 4\varepsilon \left(\left(\frac{\sigma}{r} \right)^{12} - \left(\frac{\sigma}{r} \right)^{6} \right) \tag{4.327}$$

Moreover, the parameters σ and ε are set to unity, while the atomic mass chosen is $m_A = 1$ for all atoms. Although the choice of normalized units cannot lead to quantitative representations of the behavior of a specific material, it allows the drawing of generic conclusions about the fundamental and general material behaviors in fracture. The atomic structure to be utilized in the 3-D numerical example represents a perfect FCC (face-centered cubic) crystal oriented along the (001) direction, as illustrated in Figure 4.42.

4.7.2.2 Simulation model

The structure under consideration is a nanoscale solid beam depicted schematically in Figure 4.43. As can be seen, the beam has a uniform cross-section along the thickness (y-direction) and is symmetric in

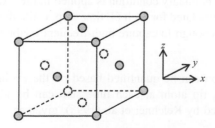

FIGURE 4.42

A unit cell of an face-centered cubic atomic lattice.

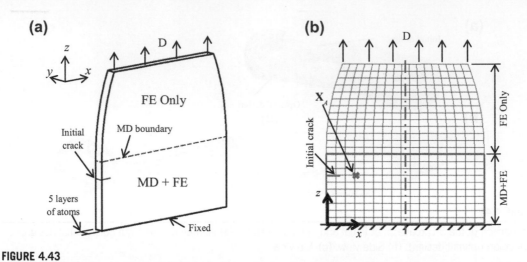

FIGURE 4.43

A 3-D nanoscale beam model. (a) Schematic illustration. (b) FE mesh in the x–z plane.

the x–z plane with respect to the midplane (dashed vertical axis in Figure 4.43b). For a bridging scale simulation, the beam is modeled with finite elements everywhere, while the MD region is confined to a rectangular area at the bottom portion. Specifically, the entire domain contains 782 hexahedral eight-node isoparametric finite elements (two layers along the thickness). The 340 elements within the MD region are of the same rectangular shape, with width $8\sqrt{2}h_a$, height $6\sqrt{2}h_a$, and thickness $\sqrt{2}h_a$ (normalized units), where h_a is the equilibrium distance between atoms. In the FE-only region, the elements are trapezoidal in the x–z plane with the same height of $6\sqrt{2}h_a$, whereas their dimensions in x-direction are subject to the curved shape of the boundary. The MD domain is comprised of 82,583 atoms in total, with five layers of atoms in the y-direction. A predefined horizontal edge crack of length $12\sqrt{2}h_a$ is created in the MD region by blocking the interaction between two adjacent layers of atoms.

During simulation, the structure is fixed at the bottom and pulled at the top face by a displacement boundary condition shown in Figure 4.44. This boundary displacement corresponds to a strain in the z-direction that grows from zero to approximately 2.5% at $t = 6000\Delta t_m$. Thereafter, it keeps stretching the beam at a lower strain rate to prevent the crack faces from contacting each other. To mimic a plane-strain configuration, a periodic boundary condition is applied in the y-direction. Following Park et al. (2005), 10 MD time steps are simulated for each FE time step, while the MD time step was taken to be $\Delta t_m = 0.0075$, which is small enough to capture high-frequency vibrations of atoms.

4.7.2.3 Crack propagation

The speed of crack propagation can be quantified based on the evolution of the crack tip location through time, while the crack tip atom at each time step can be identified based on the centrosymmetry parameter P proposed by Kelchner et al. (1998), given as

$$P = \sum_i |\mathbf{R}_i + \mathbf{R}_{i+n}|^2 \tag{4.328}$$

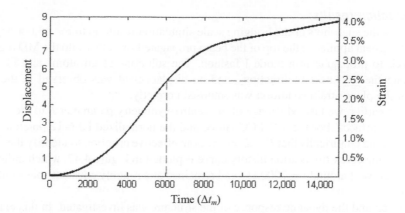

FIGURE 4.44

Displacement boundary condition applied at the top face of the beam.

where $\mathbf{R}_i$ and $\mathbf{R}_{i+n}$ are vectors corresponding to the n pairs of opposite atomic bonds surrounding a given atom. As an example, Figure 4.45a illustrates two of the six pairs of bonds ($n = 6$) around the atom α in a 3-D FCC lattice. Note that $P = 0$ when the lattice is undisturbed or deformed in a symmetric manner, and P becomes large near defects or free surfaces. Figure 4.45b depicts a scenario in which the four nearest neighbors above atom α have been pulled far away so that atom α becomes a surface atom.

Due to the fact that the P parameters of surface atoms are in most cases much larger than those of atoms near defects, such as dislocations or stacking faults (Kelchner et al. 1998), a critical value of P can be defined to distinguish all atoms located on free surfaces of the crack provided that the lattice structure and the interatomic potential model are available. Therefore, if the crack grows along the $+x$ direction, for example, then the crack tip position can be determined by finding the x location of the crack interior surface atom with the maximum x coordinate.

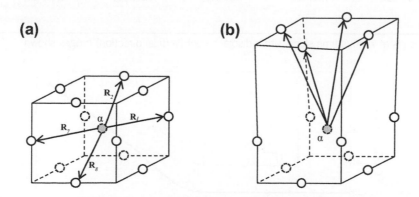

FIGURE 4.45

Criterion for determining atom separation. (a) Atom α within a 3-D FCC lattice. (b) Deformed lattice in which atom α becomes a surface atom.

4.7.2.4 Simulation results

Figure 4.46 gives the snapshots of the bridging scale simulation result up to $t = 15,000\Delta t_{\mathrm{m}}$. As can be seen, the displacement applied at the top of the beam propagated smoothly into the MD region, causing the initial crack to propagate in a mode I fashion. The solutions of all atoms and FE nodes were consistent along the thickness (y-direction), and no displacement was observed in the y-direction, indicating that the plane strain condition was imposed correctly.

As discussed earlier, we take advantage of the centro–symmetry parameter P to identify the atoms located on crack surfaces. For the 3-D FCC lattice and the normalized LJ 6-12 potential used in our simulation, we found empirically that $P > 2$ serves as an effective criterion to identify the crack surface atoms. The resulting crack tip location history curve is plotted in Figure 4.47, which indicates that the crack started to grow at around $t = 7000\Delta t_{\mathrm{m}}$ and remained at a roughly constant speed until the end of the simulation.

The physics behind the dynamic response of the structure was investigated. In this example, at the macroscopic level, the speed of the z-direction elastic stress wave was measured to be 11.7 (Wang 2014), which agrees well with the theoretical longitudinal wave speed $c_{\mathrm{l}} = 12$ (Buehler 2008) for the FCC lattice and LJ potential used in our simulation. At the atomistic level, the atomic vibration period in z-direction was found to be approximately $48\Delta t_{\mathrm{m}}$ (Wang 2014), corresponding to a frequency of 2.78, which is very close to 2.86 (Wang 2014)—the theoretical vibration frequency along the x, y, and z directions of the lattice structure shown in Figure 4.42. As for crack propagation, it can be observed from Figure 4.48 that, during most of the simulation period, the crack propagated in a straight line and

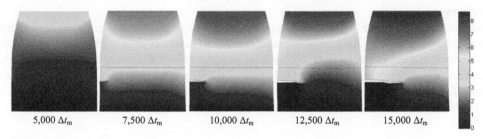

| 5,000 Δt_{m} | 7,500 Δt_{m} | 10,000 Δt_{m} | 12,500 Δt_{m} | 15,000 Δt_{m} |

FIGURE 4.46

Simulation results at various time steps with z-displacement (vertical direction) fringes shown.

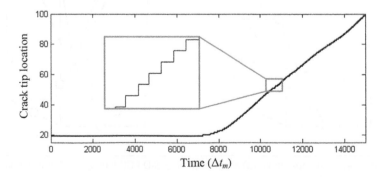

FIGURE 4.47

Crack tip location history curve.

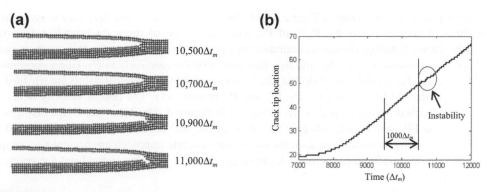

FIGURE 4.48

Instability of brittle crack propagation. (a) Crack surface roughened at instability, where atoms near the crack are plotted in blue, except for crack surface atoms being highlighted in red. (b) Calculation of crack speed before instability. (For interpretation of the references to color in this figure legend, the reader is referred to the online version of this book.)

left "mirror" cleaved surfaces, except for that the crack was roughened at around $t = 10,500\Delta t_m$ by the instability behavior in brittle crack propagation (Figure 4.48a). The crack speed before the onset of instability was measured by averaging the crack tip locations from $9500\Delta t_m$ to $10,500\Delta t_m$ using least square fitting (Figure 4.48b), and the calculated local crack speed turned out to be 1.76 (Wang 2014)— approximately 32% of the Rayleigh wave speed $c_R = 5.6$ (Buehler 2008) for the FCC crystal modeled in our simulation. This is in good agreement with the experimental and simulation results reported in the literature (Buehler 2008; Abraham et al. 1994); that is, dynamic instability in brittle crack propagation occurs when the crack speed approaches one-third of the Rayleigh wave speed c_R. Based on the observations above, it is clear that the bridging scale simulation accurately captured the essential physics of brittle crack propagation. More details regarding the physical interpretations of the simulation results can be found in Wang (2014).

4.7.2.5 Analytical shape sensitivity analysis

A shape sensitivity analysis method of continuum-discrete approach was developed for coupled atomistic and continuum problems using the bridging scale method. An energy formulation similar to what was discussed in Section 4.5 for the bridging scale decomposition is developed based on the Hamilton's principle. The energy formulation provides a uniform and generalized system of equations from which the differential equations can be obtained naturally. The sensitivity expressions are formulated in a continuum setting using both the adjoint variable method and direct differentiation method. The sensitivity formulation based on direct differentiation method is implemented numerically for the displacement, velocity, and acceleration for individual atoms and finite element nodes because the number of design variables is small. We skipped the sensitivity formulations in this case study. Readers who are interested in the formulation are referred to Wang (2014).

The shape sensitivity analysis method was applied to the nanobeam example. The geometry of the curved boundary face above the MD region on each side of the beam is modeled as a Bézier curve with three control points. As illustrated in Figure 4.49a, the shape of the left boundary face is controlled by the locations of control points P_1, P_2, and P_3, while the shape of the right boundary face varies accordingly during a design change to maintain the symmetry of the beam. In this study, we define three shape design variables—b_1, b_2, and b_3—that correspond respectively to P_{2y} (the y coordinate of point P_2), P_{3y}, and P_{3x}. At the current design, the locations of the three control points are $(0, 66\sqrt{2}h_a)$, $(0, 102\sqrt{2}h_a)$, and $(13.6h_a, 138h_a)$ for P_1, P_2, and P_3, respectively. The design velocity fields associated with individual design variables are illustrated in Figure 4.49b. Note that shape changes only take place in the FE-only region, whereas the shape of the MD region will remain untouched. The shape sensitivity analysis results based on the continuum-discrete approach are accurate as compared with the overall finite difference method. A convergence study was also carried out for verification (Wang 2014).

4.7.2.6 Crack speed performance measure

For the nanobeam example, we define the performance measure of crack speed by taking the average of all crack tip locations between $t = 9000\Delta t_m$ and $t = 15{,}000\Delta t_m$ (where the crack grows at a roughly constant rate), as shown in Figure 4.50. The goal is to obtain one single performance measure that quantifies the speed of crack propagation. We use least square fitting to identify a straight line whose slope can be taken as the crack speed. At the current design, the crack speed is calculated to be 1.466 (in normalized units).

Note that the performance measure of the crack propagation speed is verified to be continuous and differentiable in the design space (Wang 2014). As a result, the sensitivity of these measures exists in theory. However, in numerical implementation, due to the finite size of integration time steps, these measures become nondifferentiable. To overcome the difficulty of nondifferentiability, we use a regression hybrid method that calculates the shape sensitivity coefficients of crack speed through polynomial regression analysis based on the sensitivity of atomistic responses. In the proposed method, the sensitivity of atomistic responses is calculated using continuum-discrete formulation, which is fully generalized for three-dimensional problems.

4.7.2.7 Regression hybrid method

The sensitivity coefficients of crack speed computed for individual design variables are obtained using a regression hybrid method (Wang 2014), as listed in Table 4.1. The signs of the sensitivity values

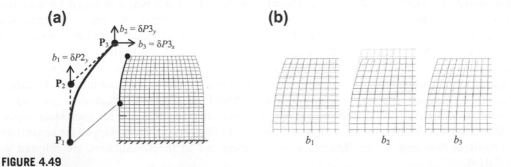

FIGURE 4.49

Design parameterization for the nanobeam. (a) Parametric boundary curve. (b) Design velocity fields for individual design variables.

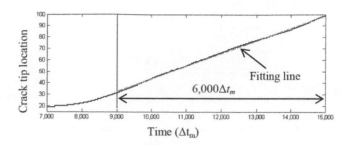

FIGURE 4.50

Performance measure defined for the nanobeam example. A straight line (red dashed line) is fitted to the crack tip locations within $9000\Delta t_m \sim 15{,}000\Delta t_m$. (For interpretation of the references to color in this figure legend, the reader is referred to the e-version of this book.)

Table 4.1 Sensitivity coefficients of crack propagation speed

Design Variable	b_1	b_2	b_3
Crack speed sensitivity	9.7097E-04	−4.3939E-02	−2.6893E-02

provide a clear indication that within the simulation period of $9{,}000\Delta t_m \sim 15{,}000\Delta t_m$, crack propagation will slow down due to an increment of either b_2 or b_3, or a decrement of b_1.

The sensitivity results can be interpreted using basic laws of physics. As with most MD simulations reported in literature, in our nanobeam example, the applied boundary displacement corresponds to an extremely high strain rate compared to one that can be implemented in laboratory experiments. Also, only small design perturbations are considered during sensitivity accuracy verification. Under these circumstances, the impact of shape change on crack propagation is mainly through the variation of elastic stress wave. It is found that when either b_2 or b_3 increases due to the design change, the elastic stress wave delays at the very beginning of the simulation when the boundary displacement propagates into the MD region, and this delay escalates as the simulation moves forward, resulting in a slower crack propagation. In the meantime, compared to b_3, increasing b_1 causes the opposite shape change of the beam, and therefore accelerates crack propagation. Moreover, as can be seen from Figure 4.49b, with the same amount of variation in three design variables, the amount of beam shape change is far less for b_1 than for the other two design variables. This explains the much smaller sensitivity value for b_1 compared to b_2 and b_3, as listed in Table 4.1.

4.7.2.8 Accuracy verification of crack speed sensitivity

To verify the accuracy of the crack speed sensitivity coefficients calculated using the regression hybrid method, we compare the first-order predictions of crack speeds $\tilde{V}^*$ with reanalyzed results V. In Figure 4.51, the blue data points are crack speeds V obtained through re-analyses, while the red dashed lines represent the first-order predictions of crack speeds.

As seen in Figure 4.51, using the regression hybrid method, the sensitivity coefficients of crack speed with respect to all three design variables are accurately computed. In addition, due to the close-to-linear behavior of the crack speed in design space, the first-order predictions of crack speed are

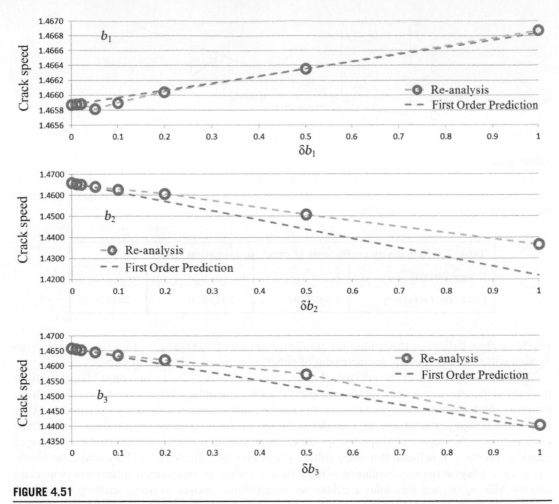

FIGURE 4.51

Accuracy verification of crack speed sensitivity for (a) shape design variable b_1, (b) shape design variable b_2, and (c) shape design variable b_3. (For interpretation of the references to color in this figure legend, the reader is referred to the e-version of this book.)

accurate for much larger design perturbations than those of atomic responses (Wang 2014), which is desirable.

4.8 SUMMARY

Efficient and accurate sensitivity analysis is essential to the success of gradient-based optimization, especially for large-scale applications that require substantial computational effort for function evaluations. We discussed three major approaches for sensitivity analysis—the brute-force finite difference, discrete, and continuum methods. We pointed out the pros and cons of individual approaches. In practical implementation, the semianalytical and continuum-discrete methods are general and efficient. In addition, considering accuracy, the continuum-discrete method is, in our opinion, the best approach for support of DSA in general and large-scale structural problems. However, it requires a more in-depth study for the reader to become proficient in implementing the method and integrating it with commercial FEA software. We hope the discussions provided in this chapter offer a gateway for those who intend to get into a more in-depth study in the sensitivity analysis area. We also briefly discussed topology optimization, which has gained lots of attention in recent years in support of structural layout design. For those who are interested in pursuing graduate study, topology optimization and relevant subjects can be an interesting thesis topic. In addition, research in studying mechanics at the atomistic level, using techniques such as MD or the bridging scale method presented in the second case study, also provides viable topics for further research.

We hope that this chapter has helped readers gain a general understanding of the concept and computation methods for DSA, in order to be able to implement some of the methods for their own applications. We hope, at this point, that software offering sensitivity analysis capabilities for gradient-based optimization, such as MSC/NASTRAN, is no longer seen as a black box.

QUESTIONS AND EXERCISES

1. The governing equation of the bar shown below is given as

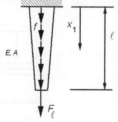

$$-(EAz_{1,1})_{,1} = f, \ x_1 \in [0,\ell]$$
with $z_1(0) = 0, f(\ell) = F_\ell$

where

f: body force, $f = \rho g A$
E: modulus of elasticity
A: cross-sectional area of the bar defined as

$$A(x_1) = A_0\left(1 - \frac{x_1}{\ell}\right) + \frac{x_1}{\ell}A_\ell$$

    A_0 and A_ℓ are the areas at $x = 0$ and $x = \ell$, respectively
ρ: mass density
g: gravitational acceleration
ℓ: length of the bar
F_ℓ: point force applied at the bottom end of the bar

a. Define the space $\tilde{S}$ for the solution z_1 of the above partial differential equation, similar to that of Eq. 4.7.

b. Derive the energy bilinear form and load linear form for the bar.

c. Show that the energy bilinear form is bilinear and load linear form is linear.

d. Define the space of kinematically admissible virtual displacement S (as in Eq. 4.12) for the bar structure.

e. State the principle of virtual work for the bar structure. Identify trial and test functions, as well as trial and test function spaces.

f. Solve for z_1 of the bar as a function of x_1.

2. Continue with Problem 1.
   a. Discretize the bar by using two equal-length finite elements of the same shape function shown in Eq. 4.17. Formulate the finite element matrix equations.
   b. Solve the matrix equations for the nodal displacements and use the same shape function to write the solution z_1 as functions of x_1 for individual finite elements.
   c. Compare the finite element solution with the exact solution obtained in Problem 1(f), and comment on the differences between these two solutions.
3. Continue with Problem 2. Assume the modulus of elasticity E and cross-sectional area A_2 of the second element of the bar as design variables.
   a. Calculate sensitivity of the displacement at the bottom end of the bar for the two design variables using the direct differentiation method of the discrete-analytical approach.
   b. Repeat (a) by using the adjoint variable method of the discrete-analytical approach. Show the adjoint loads on the bar structure.
4. Continue with Problem 3.
   a. Calculate the sensitivity of the stress at the top end of the bar for the two design variables using the direct differentiation method of the continuum-discrete approach.
   b. Repeat (a) by using the adjoint variable method of the continuum-discrete approach. Show the adjoint loads on the bar structure.
5. For the following clamped–clamped beam, calculate the displacement sensitivity for both modulus E and area A using the direct differential method of the continuum-analytical approach. Note that no shape function can be used.

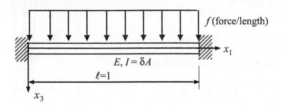

6. Answer the following questions for the beam structure shown below.

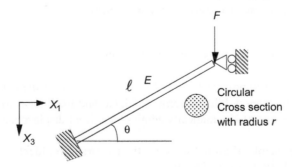

a. Compute and identify the maximum stress along the beam (bending plus axial) using the FEM (one finite element, linear functions for truss effect, and cubic shape functions for bending effect).

b. Define the maximum stress obtained in (a) as the performance measure and radius r as the design variable. Compute the adjoint load using the continuum-discrete approach, and identify the adjoint load on the beam.

c. Compute the adjoint load for the same problem using the discrete approach.

d. Compute stress sensitivity with respect to the radius defined in (b) using the adjoint variable method of the continuum-discrete approach.

e. Explain the meaning of the term $\int_0^\ell g_{,b}\delta b dx$, as seen in Eq. 4.173, in this problem. Why is it nonzero? What does it mean if we force it to be zero?

f. Compute stress sensitivity with respect to the radius design variable defined in (b) using the direct differentiation method of the discrete-analytical approach, and identify the fictitious loads on the beam.

7. Four control points on the x–y plane are given as follows:
$\mathbf{P}_0 = [0,0]$, $\mathbf{P}_1 = [1,4]$, $\mathbf{P}_2 = [2, -5]$, $\mathbf{P}_3 = [3,8]$.

a. Construct a Bézier curve enclosed by the control polygon formed by the four given points.

b. If the control point $\mathbf{P}_3$ is moved upward by two units (i.e., $\delta P_{3y} = 2$), calculate the change in curve $\delta\mathbf{P}(u)$, and calculate the parametric equation of the perturbed Bézier curve.

c. For a node with parametric coordinate $u = 0.5$, calculate its velocity field.

8. Continue with Problem 7. Assume the Bézier curve is edge 4 of a Coons patch. Calculate the velocity for a node inside the patch with $(u,w) = (0.5, 0.5)$.

9. Consider the simple support beam loaded with a point load F in the middle, as shown below.

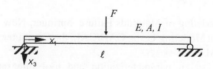

a. Compute the sensitivity of displacement z_3 at the middle of the beam with respect to its length ℓ, assuming $\partial F/\partial\ell = 0$ and that the measuring point is changing with $\delta\ell$.

b. What is the meaning of $\dot{z}_3$ in (a), what is $\dot{z}_3$ predicting?

c. What is are the boundary and loading conditions of the perturbed design in (a)?

d. What is $\partial z_3/\partial\ell$, if F does not change with $\delta\ell$?

e. What is $\partial z_3/\partial\ell$, if z_3 is measured at the fixed point?

10. This problem is a continuation of Problem 9.

A simple support beam loaded with a point load F in the middle is shown below. Answer the following questions for designs A, B, and C shown below:

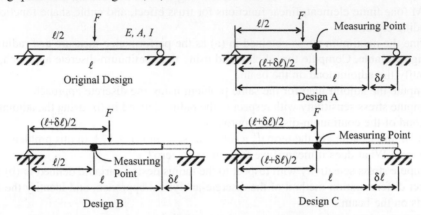

a. Is the sensitivity $\partial z_3/\partial \ell$ identical for all three designs? Why or why not?

b. In (a), $\partial z_3/\partial \ell = \dot{z} = z' + \nabla z^T V$, calculate z' for all three designs.

c. Calculate sensitivity $\dot{z}$ at $x = \ell/4$ for Design B, and at $x = (\ell + \delta\ell)/4$ for designs A and C.

REFERENCES

Abraham, F.F., Brodbeck, D., Rafey, R.A., Rudge, W.E., 1994. Instability dynamics of fracture: a computer simulation investigation. Physical Review Letters 73 (2), 272–275.

Atkinson, K., 1989. An Introduction to Numerical Analysis, second ed. Wiley, New York.

Bendsøe, M.P., Sigmund, O., 2003. Topology Optimization, Theory, Methods, and Applications, second ed. Springer, New York.

Buehler, M.J., 2008. Atomistic Modeling of Materials Failure. Springer, New York.

Chang, K.H., 2014. Product Design Modeling Using CAD/CAE, the Computer Aided Engineering Design Series. Academic Press, Burlington, MA.

Chang, K.H., Joo, S.H., 2006. Design parameterization and tool integration for CAD-based mechanism optimization. Advances in Engineering Software 37, 779–796.

Choi, K.K., Chang, K.H., 1994. A study of design velocity field computation for shape optimal design. Finite Elements in Analysis and Design 15, 317–341.

Choi, K.K., Kim, N.H., 2006a. Structural Sensitivity Analysis and Optimization 1: Linear Systems (Mechanical Engineering Series). Springer, New York.

Choi, K.K., Kim, N.H., 2006b. Structural Sensitivity Analysis and Optimization 2: Nonlinear Systems and Applications (Mechanical Engineering Series). Springer, New York.

Edke, M., Chang, K.H., 2011. Shape optimization for 2-D mixed mode fractures using extended FEM (XFEM) and level set method (LSM). Structural and Multidisciplinary Optimization 44 (2), 165–181.

Eschenauer, H.A., Olhoff, N., 2001. Topology optimization of continuum structures: a review. Applied Mechanics Reviews 54 (4), 331–390.

Hardee, E., Chang, K.H., Tu, J., Choi, K.K., Grindeanu, I., Yu, X., 1999. CAD-based shape design sensitivity analysis and optimization. Advances in Engineering Software 30 (3), 153–175.

Haug, E.J., Choi, K.K., Komkov, V., 1986. Design Sensitivity Analysis of Structural Systems. Academic Press, New York.

Jones, J.E., 1924. On the determination of molecular fields. II. From the equation of state of a gas. Proceedings of the Royal Society of London A 106 (738), 463–477.

Kelchner, C.L., Plimton, S.J., Hamilton, J.C., 1998. Dislocation nucleation and defect structure during surface indentation. Physical Review B 58 (17), 11085–11088.

Mortenson, M.E., 2006. Geometric Modeling, third ed. Industrial Press.

Park, H.S., Karpov, E.G., Klein, P.A., Liu, W.K., 2005. Three-dimensional bridging scale analysis of dynamic fracture. Journal of Computational Physics 207, 588–609.

Rozvany, G.I.N., 2000. Aims, scope, methods, history and unified terminology of computer aided topology optimization in structural mechanics. Structural Multidisciplinary Optimization 21, 90–108.

Sigmund, O., Petersson, J., 1998. Numerical instabilities in topology optimization: a survey on procedures dealing with checkerboards, mesh-dependencies and local minima. Structural Optimization 16, 68–75.

Tang, P.S., Chang, K.H., 2001. Integration of topology and shape optimizations for design of structural components. Journal of Structural Optimization 22 (1), 65–82.

Twu, S., Choi, K.K., 1993. Configuration design sensitivity analysis of built-up structures part II: numerical method. International Journal for Numerical Methods in Engineering 36, 4201–4222.

Wagner, G.J., Liu, W.K., 2003. Coupling of atomistic and continuum simulations using a bridging scale decomposition. Journal of Computational Physics 190, 249–274.

Wang, Y., 2014. Shape Sensitivity Analysis for Three-Dimensional Multi-scale Crack Propagation Problems Using the Bridging Scale Method (Ph.D. thesis). University of Oklahoma, Norman, OK.

MULTIOBJECTIVE OPTIMIZATION AND ADVANCED TOPICS

CHAPTER OUTLINE

Multiobjective optimization (also known as multiobjective programming, vector optimization, multicriteria optimization, multiattribute optimization, or Pareto optimization) is an area of multiple-criteria decision making, concerning mathematical optimization problems involving more than one objective functions to be optimized simultaneously. Multiobjective optimization has been applied to many fields of science and engineering, where optimal decisions need to be taken in the presence of trade-offs between two or more objectives that may be in conflict. Indeed, in many practical engineering applications, designers are making decisions between conflict objectives, such as maximizing performance while minimizing fuel consumption and emission of pollutants of a vehicle. In these cases, a multiobjective optimization study should be performed, which provides multiple solutions representing the trade-offs among the objective functions.

Even for a trivial multiobjective optimization problem, it is unlikely that a single solution that simultaneously optimizes each objective exists. In many cases, the objective functions are said to be in conflict, and there exist (a possibly infinite number of) Pareto optimal solutions. A solution is called nondominated if none of the objective functions can be improved in value without degrading some of the other objective function values. Such solutions are called Pareto optimal. Without additional preference information, all Pareto optimal solutions are considered equally good.

Francis Y. Edgeworth (1845–1926) and Vilfredo Pareto (1848–1923) are credited with first introducing the concept of noninferiority in the context of economics (de Weck 2004). Since then, multiobjective optimization has permeated engineering and design. The translation of Pareto's work into English in 1971 spurred the development of multiobjective methods in applied mathematics and engineering. The growth of this field manifested itself particularly strongly in the United States, with pioneering contributions made by Stadler (1979) and Steuer (1985), as well as many others. Another major development, particularly in the theoretical aspects of multiobjective optimization, can be found in Japan (Sawaragi et al. 1985). Over the last three decades, the applications of multiobjective optimization have grown steadily in many areas of engineering and design.

Researchers have studied multiobjective optimization problems from different viewpoints; thus, there exist different solution philosophies and goals when setting and solving them. The goal may be to find a representative set of Pareto optimal solutions, quantify the trade-offs in satisfying the different objectives, and/or find a single solution that satisfies the subjective preferences of a designer. There are numerous ways to categorize the solution methods, such as scalarization versus Pareto methods, a priori versus a posteriori articulation of preferences, and so on. Although there have been many solution techniques developed in the past decades, none of these methods are perfect, and selecting among them depends on the requirements of a particular design situation.

In this chapter, we introduce the basic concepts, the nature of multiobjective problems, and solution techniques for readers who have not been exposed to these topics. We do not intend to provide a comprehensive review nor offer presentations for research-style discussions. We include only sufficient mathematical details to explain concepts and solution techniques, and then use simple examples for illustration. We offer MATLAB scripts for most examples so that readers are able to learn to solve their own multiobjective problems by creating similar scripts. We use similar examples throughout the chapter while discussing different solution techniques. By doing so, we hope readers are able to gain a better understanding of these methods and see the pros and cons between them. We also revisit the decision theories discussed in Chapter 2 and discuss the application of them to engineering design in the context of multiobjective optimization. Also provided is a short review of the available software tools for use. For a more thorough and comprehensive discussion of multiobjective optimization, readers are referred to excellent books and articles (e.g., Deb 2001; Miettinen 1999; Marler and Arora 2004; Branke et al. 2008).

In addition, we include advanced topics that are relevant to design optimization. We present reliability-based design optimization that incorporates uncertainty of physical parameters and design variables into design optimization, leading to designs with less probability of failure. Also, we present a case of design optimization that takes product cost including manufacturing as objective function. This case represents a more realistic design scenario that eventually leads to less expensive designs. These topics are open and under active research.

Although we do not get into a discussion of multidisciplinary optimization, we include a single-piston engine example as a tutorial project to illustrate a design scenario that involves multiple engineering disciplines. We use the suite of computer-aided design (CAD), engineering (CAE), and manufacturing (CAM) software in Pro/ENGINEER and SolidWorks and we employ the interactive process discussed in Chapter 3 to proceed with the design.

The objectives of this chapter include (1) introducing readers to the basic concepts and solution techniques of multiobjective optimization, (2) pointing out the pros and cons of different solution techniques and their applicability to practical engineering problems, and (3) offering sample MATLAB scripts as references so that readers are be able to write their own to solve similar problems. We also provide a short discussion on software tools, both academic and commercial, in the hope of offering readers basic ideas of selecting proper software tools that are suitable to specific needs. Certainly, the advanced topics discussed at the end of the chapter aim to provide readers with the flavor of those under active research.

5.1 INTRODUCTION

So far, we have considered problems with only one objective function, the so-called single-objective optimization. In practice, many engineering design problems involve more than one objective function. In many situations, these objective functions may be in conflict with one another. For example, it is desirable for a design team to minimize weight while maximizing the strength of a particular structural component, or maximize performance of a vehicle while minimizing fuel consumption and emission of pollutants. Both are multiobjective optimization (MOO) problems involving two or more objectives. Mathematically, a MOO problem can be formulated as follows:

$$\text{Minimize}: \quad \mathbf{f}(\mathbf{x}) \tag{5.1a}$$

$$\text{Subject to}: \quad g_i(\mathbf{x}) \leq 0, \quad i = 1, m \tag{5.1b}$$

$$h_j(\mathbf{x}) = 0, \quad j = 1, p \tag{5.1c}$$

$$x_k^\ell \leq x_k \leq x_k^u, \quad k = 1, n \tag{5.1d}$$

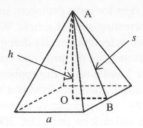

FIGURE 5.1

The pyramid example.

This is identical to that of Eq. 3.3 discussed in Chapter 3, except that in Eq. 5.1a, $\mathbf{f}(\mathbf{x})$ is the vector of objective functions, $\mathbf{f}(\mathbf{x}) = [f_1(\mathbf{x}), f_2(\mathbf{x}), \ldots, f_q(\mathbf{x})]^T$, and q is the number of the objective functions.

Before entering formal discussion, we present a simple pyramid example to illustrate the basic concept of formulating a multiobjective optimization problem. As shown in Figure 5.1, the base width and height of the pyramid are a and h, respectively. The chord length of the triangle OAB is s. The objective functions of the pyramid design problem are minimizing the lateral surface area A and minimizing the total surface area T by varying two design variables, base width a and height h. The volume of the pyramid must be greater than 1500 in.^{3}, and the upper bounds of the base width a and height h are 30 in. Mathematically, the optimization problem is defined as follows:

$$\text{Minimize} : A(a, h) = 2as = 2a\sqrt{\left(\frac{a}{2}\right)^2 + h^2}$$

(5.2a)

$$T(a, h) = S + a^2 = 2a\sqrt{\left(\frac{a}{2}\right)^2 + h^2} + a^2$$

$$\text{Subject to} : V(a, h) = \frac{a^2 h}{3} \geq 1500$$

(5.2b)

$$0 < a \leq 30$$

(5.2c)

$$0 < h \leq 30$$

(5.2d)

How do we solve this problem? Is there one single solution that minimizes both $A(a, h)$ and $T(a, h)$ and satisfies all the constraints? How do we solve a multiobjective optimization problem, such as the pyramid example defined in Eq. 5.2?

To illustrate the nature of the problem involved in MOO, we present the following simpler example (Example 5.1), in which the problem and solutions can be visualized graphically. We return to the pyramid example in Section 5.2.3 after introducing a few basic solution concepts of MOO.

Note that for a simple problem such as Example 5.1, the Pareto optimum can be found easily in the design space. Unfortunately, this is not the case even for a simple problem like the pyramid example. In general, it is insufficient to discuss the Pareto optimum in the design space. To understand the concept of MOO problems and solution techniques, we will have to convert the MOO problem to its criterion (or objective) space. We discuss the criterion space together with concepts and basic terminologies in

EXAMPLE 5.1

Solve the following MOO problem, defined as

$$\text{Minimize} : f_1(x_1, x_2) = x_1^2 + x_2^2 \tag{5.3a}$$

$$f_2(x_1, x_2) = (x_1 - 2)^2 + x_2^2 \tag{5.3b}$$

$$\text{Subject to} : x_1 \geq 0, x_2 \geq 0 \tag{5.3c}$$

Solutions:

The two objective functions f_1 and f_2 can be graphed in the design space, in this case the x_1-x_2 plane, as shown below.

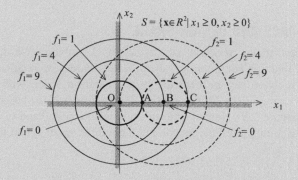

The iso-lines of the objective functions (or objective function contours) f_1 and f_2 are circles with center points $(0, 0)$ and $(2, 0)$, respectively. They are graphed above in solid and dotted lines, respectively. The feasible set (or feasible region) of this problem is defined as:

$$S = \left\{ \mathbf{x} \in R^2 \middle| x_1 \geq 0, x_2 \geq 0 \right\}$$

which is the first quadrant of the x_1-x_2 plane shown in the figure above. It is apparent that the minimum of f_1 is at point $O = (0, 0)$, and the minimum of f_2 is at point $B = (2, 0)$. However, at point $O, f_2 = 4$, and at point $B, f_1 = 4$, which are not their respective minimum. It is obvious that there is no single point that offers both f_1 and f_2 at their minimum. A best possible compromised solution could be at point $A = (1, 0)$, at which $f_1 = f_2 = 1$, which is a compromised solution. Moving away from this point may reduce one objective but result in increasing the other objective. As a matter of fact, any solutions between points O and B may be considered acceptable because reducing one objective function cannot be accomplished without increasing the other objective function. A point that is outside the line segment between points O and B (that is, $x_1 \notin (0, 2)$ and $x_2 = 0$) can be changed to reduce both objective functions simultaneously. For example, at point $C = (3, 0)$, we have $f_1 = 9$ and $f_2 = 1$. By moving from point C to B, both objectives f_1 and f_2 can be reduced. Note that points between O and B, defined as $S_p = \{ \mathbf{x} \in R^2 | 0 \leq x_1 \leq 2, x_2 = 0 \}$, are called the Pareto optimal set of the MOO problem. A Pareto optimal set usually consists of infinite number of solutions that are considered legitimate. More about the Pareto optimum is discussed in Section 5.2.

Section 5.2. We then discuss solution techniques to MOO problems in Section 5.3. Recall that we discussed design examples of multiple objectives in Chapter 2 using both utility and game theories. We will revisit these examples in Section 5.4. In Section 5.5, we present a brief overview on the software tools for solving MOO problems, both academic and commercial. We then present two advanced topics relevant to design optimization in Section 5.6. They are reliability-based design optimization and design optimization for structural performance and manufacturing cost.

5.2 BASIC CONCEPT

The scalar concept of "optimality" of single-objective optimization problems does not apply directly in the multiobjective setting. To understand the concept and solution techniques, the notion of Pareto optimality has to be introduced. In this section, we start by introducing the criterion space by revisiting the simple MOO example of Example 5.1. We then introduce basic terminologies to facilitate our later discussion. With the basic concept discussed, we revisit the pyramid example to reinforce the concept.

5.2.1 CRITERION SPACE AND DESIGN SPACE

In Example 5.1, we depicted the design space of the MOO problem. We defined its feasible set S, plotted the objective function contours, and identified its Pareto optimal set S_p in design space. Alternatively, a MOO problem can be depicted in the criterion space with axes represented by objective functions. For example, the MOO problem of Example 5.1 can be depicted in its criterion space f_1-f_2, as shown in Figure 5.2b. Note that the side constraints $x_1 \geq 0$ and $x_2 \geq 0$ are translated into the criterion space in this simple example by solving them in terms of f_1 and f_2:

$$
\begin{aligned}
x_1(f_1, f_2) &= 0.25 \times (f_1 - f_2) + 1 \geq 0 \\
x_2(f_1, f_2) &= \sqrt{f_1 - x_1^2} = \sqrt{f_1 - (0.25 \times (f_1 - f_2) + 1)^2} \geq 0
\end{aligned}
\tag{5.4}
$$

The feasible criterion space Z is defined simply as the set of objective function values corresponding to the feasible points in the design space:

$$
Z = \{f(\mathbf{x}) | \mathbf{x} \in S\}
\tag{5.5}
$$

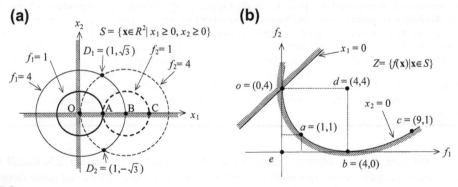

FIGURE 5.2

The example MOO problem depicted in (a) design space and (b) criterion space.

As shown in Figure 5.2a, the points O, A, and B in the design space are mapped into the criterion space at o, a, and b, respectively. The Pareto optimum in the design space $S_p = \{\mathbf{x} \in R^2 | \ 0 \leq x_1 \leq 2, x_2 = 0\}$ is converted into the curve segment oab in the criterion space, defined as $Z_p = \{\mathbf{f} = (f_1, f_2) \in R^2 | \ 0 \leq f_1 \leq 4, 0 \leq f_2 \leq 4, \sqrt{f_1 - (0.25 \times (f_1 - f_2) + 1)^2} = 0\}$. The curve oab in the criterion space is called the Pareto solution, Pareto optimum, Pareto front, or Pareto set, representing the solutions of the MOO problem. It is clearly shown in the criterion space that the minimum of the objective function f_1 is 0 and is located at point $o = (0, 4)$. On the other hand, the minimum of the objective function f_2 is 0 and is located at point $b = (4, 0)$. Moreover, an objective function (f_1 or f_2) cannot be further reduced without increasing the other objective function for any point on the Pareto front. For a point not on the Pareto front, such as point $d = (4, 4)$, it is possible to reduce both objective function values simultaneously by moving the point toward the Pareto front.

A concept related to the feasibility of design points in the design space is that of attainability in criterion space. As discussed in Chapter 3, the feasibility of a design implies that no constraint is violated in the design space. Attainability implies that a point in the criterion space can be related to a point in the design space. It is important to note that each point in the design space can be mapped to the point in the criterion space. However, the reverse may not be true; that is, every point in the criterion space does not necessarily correspond to points or a single point in the design space. For example, point $e = (0, 0)$ in the criterion space does not map back to any point in the design space. Also, point $d = (4, 4)$ in the criterion space maps two points $D_1 = (1, \sqrt{3})$ and $D_2 = (1, -\sqrt{3})$ in the design space, and $D_1 \in S$ and $D_2 \notin S$. We are only interested in the points in the criterion space that are attainable.

5.2.2 PARETO OPTIMALITY

The concept in defining solutions for MOO problems is that of Pareto optimality. A point $\mathbf{x}^*$ in the feasible design space S is called Pareto optimal if there is no other point in the set S that reduces at least one objective function without increasing another one. In Example 5.1, the Pareto optimal is $\mathbf{x}^* \in S_p = \{\mathbf{x} \in R^2 | \ 0 \leq x_1 \leq 2, x_2 = 0\}$. Pareto optimal is defined more precisely as follows.

Definition 1: Pareto Optimal. A point $\mathbf{x}^* \in S$ is Pareto optimal iff (i.e., if and only if) $\nexists$ (there does not exist) another point $\mathbf{x} \in S$ such that $f_i(\mathbf{x}) \leq f_i(\mathbf{x}^*)$ $\forall$ (for all) i and $f_i(\mathbf{x}) < f_i(\mathbf{x}^*)$ for least one i.

To illustrate the above statement, we assume that $\mathbf{x}^* = (3, 0)$ is a Pareto optimal (in fact, it is point C in Figure 5.2a and we know point C is not Pareto optimal). At this point, $f_1(\mathbf{x}^*) = 9$ and $f_2(\mathbf{x}^*) = 1$. Let us see if $\mathbf{f}(\mathbf{x}) \leq \mathbf{f}(\mathbf{x}^*)$ and $f_i(\mathbf{x}) < f_i(\mathbf{x}^*)$ for at least one i are true. We can easily find many points to test the conditions. For example, we pick point A, where $\mathbf{x} = (1, 0)$, and $f_1(\mathbf{x}) = f_2(\mathbf{x}) = 1$. Hence, $\mathbf{f}(\mathbf{x}) \leq \mathbf{f}(\mathbf{x}^*)$ with $f_1(\mathbf{x}) = 1 < f_1(\mathbf{x}^*) = 9$. Therefore, from Definition 1, $\mathbf{x}^* = (3, 0)$ is not Pareto optimal. On the other hand, if we pick $\mathbf{x}^* = (1, 0)$ (point A), $f_1(\mathbf{x}^*) = 1$ and $f_2(\mathbf{x}^*) = 1$. There does not exist any other point where f_1 and f_2 are less than or equal to those of $(1, 0)$. Hence, $\mathbf{x}^* = (1, 0)$ is a Pareto optimal. In fact, all points in S_p satisfy Definition 1; hence, all are Pareto optimal.

It is important to note that the Pareto optimal set Z_p is always on the boundary of the feasible criterion space Z. When there are just two objective functions, as shown in Example 5.1, the minimum points of individual objective functions define the endpoints of the Pareto front (i.e., points o and b in Figure 5.2b), assuming the minima to be unique.

Although the Pareto optimal set is always on the boundary of Z, it is not necessarily defined by the constraints. If the MOO problem of Example 5.1 is redefined as a nonconstrained problem by removing the side constraints $x_1 \geq 0, x_2 \geq 0$, how do we find the Pareto optimal? In this case, the Pareto optimal is defined by the relationship between the gradients of the objective functions. For cases of two objective functions, the gradients of the objective functions point in opposite directions at all Pareto optimal points (Arora 2012). For the problem in Example 5.1 without side constraints, the Pareto optimal is the line connecting the two centers of the objective function contours, which is unchanged from the constrained problem. From Figure 5.2b, the Pareto optimal can be easily identified on the curve segment oab.

A concept closely related to Pareto optimality is that of weak Pareto optimality. At the weak Pareto optimal points, it is possible to improve some objective functions without penalizing others. A weak Pareto optimal point is defined as follows.

Definition 2: Weak Pareto Optimal. A point $\mathbf{x}^* \in S$ is weakly Pareto optimal iff $\nexists$ another point $\mathbf{x} \in S$ such that $f_i(\mathbf{x}) < f_i(\mathbf{x}^*) \ \forall \ i$.

In another word, a point is weakly Pareto optimal if there is no other point that improves all objective functions simultaneously. However, there may be points that improve some of the objectives while keeping others unchanged.

The concept of weak Pareto optimality is illustrated in Figure 5.3. We minimize two objectives f_1 and f_2. Note that lines AB and BC are the boundary of the feasible criterion space Z and are respectively horizontal and vertical. In this case, all points on line $A–B–C$ are weakly Pareto optimal. However, only points A and C are Pareto optimal.

If we take any point on AB (not including A), such as point E, we cannot find another point $\mathbf{x}$ in the feasible criterion space such that $f_1(\mathbf{x}) < f_1(\mathbf{x}_E)$ and $f_2(\mathbf{x}) < f_2(\mathbf{x}_E)$; therefore, point E is a weakly Pareto optimal. On the other hand, we can find at least one point $\mathbf{x}$ (such as all points between AE) such that $f_1(\mathbf{x}) < f_1(\mathbf{x}_E)$ and $f_2(\mathbf{x}) = f_2(\mathbf{x}_E)$; therefore, point E is not Pareto optimal. Similarly, all points on BC (not including C) are weakly Pareto optimal but not Pareto optimal. Pareto optimal is weakly Pareto optimal, but a weakly Pareto optimal is not necessarily Pareto optimal.

Another common concept is that of nondominated and dominated points, which is defined as follows.

Definition 3: Nondominated and Dominated. A vector of objective functions $\mathbf{f}^* = \mathbf{f}(\mathbf{x}^*) \in Z$ is nondominated iff $\nexists$ another vector $\mathbf{f} \in Z$ such that $f_i \leq f_i^* \ \forall \ i$ and $f_i < f_i^*$ for least one i. Otherwise, $\mathbf{f}^*$ is dominated.

Pareto optimality generally refers to both the design and the criterion spaces. In numerical algorithms, the idea of nondomination in the criterion space is often used for a subset of points; one point

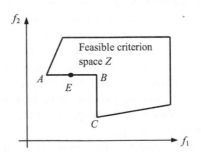

FIGURE 5.3

Illustration of weak Pareto optimality.

may be nondominated compared with other points in the subset. A Pareto optimal point has no other point that improves at least one objective without detriment to another; that is, it is nondominated.

A unique point, called utopia point or ideal point, in the criterion space is defined as follows.

Definition 4: Utopia Point. A point $\mathbf{f}^o$ in the criterion space is called the utopia point if $f_i^o = \min \{f_i(\mathbf{x}) \,|\, \mathbf{x} \in S\}$, $i = 1$ to q, and q is the number of the objective functions.

The utopia point is obtained by minimizing each objective function without regard for other objective functions. Each minimization yields a design point in the design space and the corresponding value for the objective function. As shown in Figure 5.2b, point e is the utopia point of Example 5.1.

In general, it is rare that each minimization will end up at the same point in the design space. That is, one design point cannot simultaneously minimize all of the objective functions. Thus, the utopia point exists only in the criterion space and, in general, it is not attainable, such as the utopia point e in Figure 5.2b.

The next best thing to a utopia point is a Pareto solution that is as close as possible to the utopia point. Such a solution is called a compromise solution, for example, point a in Figure 5.2b. The term closeness can be defined in several different ways. Usually, it implies that one minimizes the Euclidean distance $D(\mathbf{x})$ from the utopia point in the criterion space, which is defined as follows:

$$D(\mathbf{x}) = \|\mathbf{f}(\mathbf{x}) - \mathbf{f}^o\| = \sqrt{\sum_{i=1}^{q} \left[f_i(\mathbf{x}) - f_i^o\right]^2} \tag{5.6}$$

where f_i^o represents a component of the utopia point in the criterion space. Compromise solutions are Pareto optimal.

5.2.3 GENERATION OF PARETO OPTIMAL SET

For a simple problem such as Example 5.1, we can easily solve it by translating the problem from design space to criterion space and identifying its Pareto optimal set in the criterion space as $Z_p = \{\mathbf{f} = (f_1, f_2) \in R^2 \,|\, 0 \le f_1 \le 4, 0 \le f_2 \le 4, \sqrt{f_1} - (0.25 \times (f_1 - f_2) + 1)^2 = 0\}$. The Pareto optimal set can also be translated back to its design space $S_p = \{\mathbf{x} \in R^2 \,|\, 0 \le x_1 \le 2, x_2 = 0\}$. For problems just a bit more complicated, such as the pyramid design, none of the above is possible. So, how do we solve MOO problems in general?

We first use a brute-force approach—the generative method similar to that discussed in Chapter 3—to solve for the pyramid problem. We will then make a few comments before moving into the next section for the discussion of general solution techniques.

Recall the problem definition of the pyramid problem in Eq. 5.2. The feasible set of the problem is defined in its design space (in this case, the a-h plane) as

$$S = \left\{ (a, h) \in R^2 \,|\, V(a, h) = \frac{a^2 h}{3} \ge 1,500; \text{ and } 0 < a \le 30, 0 < h \le 30 \right\} \tag{5.7}$$

The feasible set S is graphed in Figure 5.4, with optimal points $\mathbf{x}_t^* = (14.7, 20.8)$ and $\mathbf{x}_a^* = (18.5, 13.1)$ for objective functions T and A, respectively. Note that the optimal points $\mathbf{x}_t^*$ and $\mathbf{x}_a^*$ are obtained by solving the two respective single-objective problems using MATLAB (Script 1 in Appendix A).

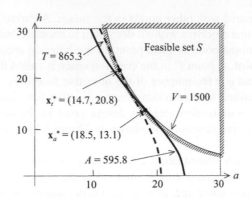

FIGURE 5.4

The design space of the pyramid example with feasible set and optimal points $\mathbf{x}_t^*$ and $\mathbf{x}_a^*$.

Translating this simple problem from design space to its criterion space is not straightforward. In fact, it is very difficult, if not entirely impossible, to write the design variables a and h as functions of objective functions A and T analytically. Therefore, we use the generative method to graph feasible and infeasible solutions in criterion space.

We first generate 1,000,000 random points for a and h, respectively. We then calculate the numerical values of the two objective functions A and T for each point (a, h). In the meantime, we calculate volume V. If V is greater or equal to 1500, the design represented in the point (a, h) is feasible; otherwise, it is infeasible. We store feasible and infeasible points separately and graph them on the criterion space in different colors. The points at the junction of the areas of different colors are the boundary of the feasible criterion space Z. Once the feasible criterion space is identified, we locate the minimum points of functions A and T and identify the Pareto front. The MATLAB script that graphs the criterion space and Pareto front shown in Figure 5.5 can be found in Appendix A (Script 2). The MATLAB script also found that the minima of A and T are, respectively, 594.8 and 865.5. The minimum A occurs at $a = 18.63$ and $h = 12.97$, which is very close to the $\mathbf{x}_a^*$ found earlier (see Figure 5.4). The minimum T occurs at $a = 14.56$ and $h = 21.23$, which is again very close to $\mathbf{x}_t^*$.

In fact, we ran six cases with sample points ranging from 100 to 10,000,000, as shown in Table 5.1. The results show that for sample points more than 10,000, the results do not vary significantly. However, if we graph the criterion space with a sample size of 10,000, the boundary of the feasible criterion set, hence the Pareto front, cannot be clearly identified (see Figure 5.6).

Although the generative method is simple, it works well only for simple problems, such as the pyramid example. In general, engineering design problems involve significant computation efforts for function evaluations, and it is impractical to find Pareto front using the generative method in general. On the other hand, do we really need to solve for an entire set of the Pareto front in order to make adequate design decisions? Instead of the generative method, are there plausible methods that are practical for support of engineering design involving multiple objectives?

We discuss solution techniques for multiobjective optimization problems in the next section. Note that a key characteristic of MOO solution techniques is the nature of the solutions that they provide. Some methods always yield Pareto optimal solutions but may skip certain points in the Pareto optimal set; that is, they may not be able to yield all or most of the Pareto optimal points. Other methods are

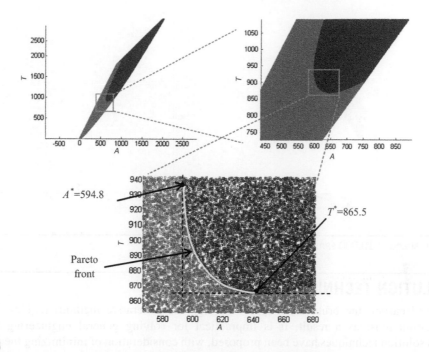

FIGURE 5.5

The criterion space of the pyramid example shown for 1,000,000 sample points.

Table 5.1 Cases of the Pyramid Example with Different Sample Sizes Ranging from 100 to 10,000,000

Sample Points	Surface Area A			Total Area T		
	Minimum A	a	h	Minimum T	a	h
10,000,000	594.82	18.486	13.168	865.35	14.675	20.895
1,000,000	594.85	18.627	12.970	865.54	14.559	21.232
100,000	595.68	18.924	12.577	866.26	14.999	20.021
10,000	595.57	18.368	13.360	866.23	14.724	20.787
1000	612.16	19.199	12.728	876.89	15.021	20.336
100	643.25	20.499	11.878	901.44	15.416	20.100

able to capture most points in the Pareto optimal set, but they may also provide non-Pareto optimal points. In any case, the primary goal of solving a multiobjective optimization problem is to support the designer to make adequate design decision, which may involve the designer's preferences in ordering or setting the relative importance of individual objective functions. A designer may articulate these preferences before solving the MOO problem or wait until sufficient solutions become available and then set the preferences. In some cases, a designer may not have any preferences at all. Different solution techniques are suitable for the respective situations.

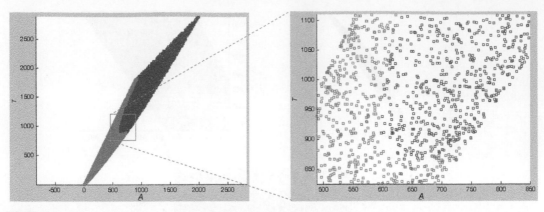

FIGURE 5.6

The criterion space of 10,000 sample points.

5.3 SOLUTION TECHNIQUES

As discussed earlier, the brute-force approach (e.g., the generative method) requires too many function evaluations; as a result, it is impractical for solving general engineering problems. Numerous solution techniques have been proposed, with consideration of minimizing the number of function evaluations, among other factors. Because a primary goal of multiobjective optimization is to model a designer's preferences (ordering or relative importance of objectives and goals), methods are categorized depending on how the designer articulates these preferences: methods with a priori articulation of preferences, methods with a posteriori articulation of preferences, and methods with no articulation of preferences. Section 5.3.2 contains methods that involve a priori articulation of preferences, which implies that the designer indicates the relative importance of the objective functions or desired goals before running the optimization algorithm. With the preferences specified, the MOO is converted to a single optimization problem, leading to a single solution. Section 5.3.3 describes methods with a posteriori articulation of preferences, which offer a set of solutions that allow the designer to choose. In Section 5.3.4, methods that require no articulation of preferences are addressed. Although methods based on genetic algorithms (GA) generate multiple solutions for designers to choose, the concept and solution techniques are somehow different than those of conventional methods with a posteriori articulation of preferences. Therefore, we discuss GA-based methods separately in Section 5.3.5.

5.3.1 NORMALIZATION OF OBJECTIVE FUNCTIONS

Many multiobjective optimization methods involve comparing and making decisions about different objective functions. However, values of different functions may have different units and/or significantly different orders of magnitude, making comparisons difficult. Thus, it is usually necessary to transform the objective functions such that they all have similar orders of magnitude. Although there are different approaches proposed for such a purpose, one of the simplest approaches is to normalize individual objective functions by their respective absolute function values at current (or initial) designs:

$$f_i^{\text{norm}}(\mathbf{x}) = \frac{f_i(\mathbf{x})}{|f_i(\mathbf{x}^0)|} \tag{5.8}$$

where f_i^{norm} is the ith normalized objective function, and $\mathbf{x}^0$ is the vector of design variables at current or initial design. This method ensures that all objective functions are normalized to 1 or -1 to start with. Certainly, we assume that $f_i(\mathbf{x}^0)$ is not zero or close to zero at the initial design or throughout the optimization process. Note, however, that if all of the objective functions have similar values, such as the pyramid example, normalization may not be needed.

In the following sections, we assume that the objective functions have been normalized in a certain way if necessary.

5.3.2 METHODS WITH A PRIORI ARTICULATION OF PREFERENCES

The methods to be discussed in this section allow the designer to specify preferences, which may be articulated in terms of goals or the relative importance of different objectives. Most of these methods incorporate parameters, which are coefficients, exponents, constraint limits, and so on, that can either be set to reflect designer preferences or be continuously altered in an attempt to find multiple solutions that roughly represent the Pareto optimal set.

5.3.2.1 Weighted-sum method

A multiobjective optimization problem is often solved by combining its multiple objectives into one single-objective scalar function. The simplest and most common approach is the weighted-sum or scalarization method, defined as

$$\underset{\mathbf{x} \in S}{\text{Minimize}} \quad u(\mathbf{x}) = \sum_{i=1}^{q} w_i f_i(\mathbf{x}) \tag{5.9}$$

which represents a new optimization problem with a unique objective function $u(\mathbf{x})$. Note that the weights w_i's are typically set by the decision maker, such that $\sum_{i=1}^{q} w_i = 1$ and $w_i \geq 0 \;\forall\; i$. Graphically, the new objective function $u(\mathbf{x})$ is a hyperplane in the criterion space of q dimensions. For a two-objective problem ($q = 2$), the new objective function $u(\mathbf{x})$ is a straight line in the f_1-f_2 plane. Note that the slope of the straight line is determined by weights w_1 and w_2. The optimal solution is the tangent point of the straight line intersecting with the Pareto front of the feasible criterion space, as shown in Figure 5.7.

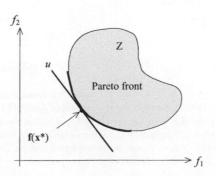

FIGURE 5.7

Geometrical representation of the weighted-sum approach in the case of a convex Pareto front.

We introduce a simple example in Example 5.2 to illustrate the method. We use multiobjective linear programming (MOLP) problems to simplify the mathematical calculations, and we use two design variables to facilitate graphical presentation.

EXAMPLE 5.2

Solve the following MOLP problem using the weighted-sum method,

$$\text{Minimize}: f_1 = 2x_1 - 3x_2 \tag{5.10a}$$

$$f_2 = 2x_1 + x_2 \tag{5.10b}$$

$$\text{Subject to}: g_1 = 7 - x_1 - 5x_2 \leq 0 \tag{5.10c}$$

$$g_2 = 10 - 4x_1 - x_2 \leq 0 \tag{5.10d}$$

$$g_3 = -7x_1 + 6x_2 - 9 \leq 0 \tag{5.10e}$$

$$g_4 = -x_1 + 6x_2 - 24 \leq 0 \tag{5.10f}$$

Solutions:

We define the vectors of steepest descent direction (negative gradients) of the objective functions f_1 and f_2 in the design space as $\mathbf{C}_1 = (-2, 3)$ and $\mathbf{C}_2 = (-2, -1)$, respectively. Note that $\mathbf{C}_1$ is a vector that defines the direction on the x_1-x_2 plane that minimizes f_1, as is $\mathbf{C}_2$ for function f_2. The feasible set S, as well as the vectors $\mathbf{C}_1$ and $\mathbf{C}_2$, are shown in the left figure below.

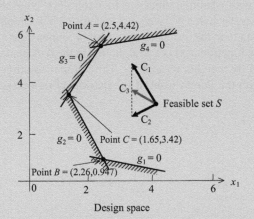

Design space

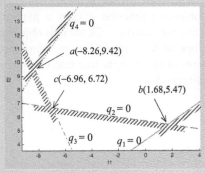

Criterion space

It is obvious that the minimum of $f_1 = -8.26$ (as if $w_1 = 1$, $w_2 = 0$) is found at point $A = (2.5, 4.42)$ (the intersecting point of $g_3 = 0$ and $g_4 = 0$; and $f_2 = 9.42$ at this point), and minimum of $f_2 = 5.47$ (as if $w_1 = 0$, $w_2 = 1$) is at point $B = (2.26, 0.947)$ (the intersecting point of $g_1 = 0$ and $g_2 = 0$; and $f_1 = 1.68$ at this point).

For any $0 < w_i < 1$ and $w_1 + w_2 = 1$, the gradient $\mathbf{C}_3$ of the weighted-sum function $u(\mathbf{x})$ points in a direction that is between $\mathbf{C}_1$ and $\mathbf{C}_2$ (e.g., $w_1 = w_2 = 0.5$, as shown in the left figure above). Therefore, the minimum of the weighted-sum function is at point $C = (1.65, 3.42)$ (the intersecting point of $g_2 = 0$ and $g_3 = 0$), at which $f_1 = -6.96$, $f_2 = 6.72$.

EXAMPLE 5.2—cont'd

The MOLP problem can be easily translated into its criterion space as shown in the right figure above. Note that the constraint functions in the criterion space are, respectively,

$$q_1 = 9f_1 - 13f_2 + 56 \leq 0 \tag{5.11a}$$

$$q_2 = -f_1 - 7f_2 + 40 \leq 0 \tag{5.11b}$$

$$q_3 = -19f_1 - 9f_2 - 72 \leq 0 \tag{5.11c}$$

$$q_4 = -13f_1 + 9f_2 - 192 \leq 0 \tag{5.11d}$$

The feasible criterion space is graphed. The three vertices of the feasible criterion space $a = (-8.26, 9.42)$, $b = (1.68, 5.47)$, and $c = (-6.96, 6.72)$ are the intersection points of $q_3 = 0$ and $q_4 = 0$, $q_1 = 0$ and $q_2 = 0$, as well as $q_2 = 0$ and $q_3 = 0$, respectively. The Pareto front consists of all points on the line segments ac and cb. All the three solutions obtained (for $w_1 = 1$, $w_2 = 0$; $w_1 = 0$, $w_2 = 1$; and $w_1 = w_2 = 0.5$) are Pareto optimum.

If we set $w_1 = 19/28$ and $w_2 = 9/28$, then $\mathbf{C}_3 = w_1\mathbf{C}_1 + w_2\mathbf{C}_2 = (-2, 12/7)$, which is the gradient of $g_3 = 0$. In this case, the solutions contain points on the line segment between A and C on $g_3 = 0$ in the design space. The optimum is not unique. In the criterion space, the solutions are line segment ac, which is part of the Pareto front.

The relative value of the weights generally reflects the relative importance of the objectives. The weights can be used in two ways. The designer may either set w_i to reflect preferences before the problem is solved or systematically alter weights to yield different Pareto optimal points. In fact, most methods that involve weights can be used in both of these capacities—to generate a single solution or multiple solutions. We demonstrate this statement in Example 5.3.

EXAMPLE 5.3

Solve the following MOO problem.

$$\text{Minimize}: f_1 = x_1 \tag{5.12a}$$

$$f_2 = x_2 \tag{5.12b}$$

$$\text{Subject to}: x_1^2 + x_2^2 - 1 = 0 \tag{5.12c}$$

Solutions:

This problem has identical design space and criterion space, as shown below.

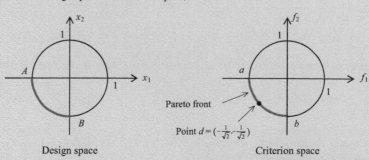

Design space

Criterion space

Continued

EXAMPLE 5.3—cont'd

It is obvious that the Pareto front is the circular arc *ab* (including points *a* and *b*) in the criterion space. The negative gradients of the objective functions f_1 and f_2 are $\mathbf{C}_1 = (-1, 0)$ and $\mathbf{C}_2 = (0, -1)$, respectively. If we choose $w_1 = 1/2$ and $w_2 = 1/2$, then $\mathbf{C}_3 = (-1/2, -1/2)$. If we solve for this single-objective optimization problem, we should obtain a solution at point $d = \left(-\frac{1}{\sqrt{2}}, -\frac{1}{\sqrt{2}}\right)$. In general, the Pareto optimal set can be obtained by choosing different combinations of weights w_1 and w_2 as many times as desired. For this simple example, when the weights are properly selected—for example, $(w_1, w_2) = (0.1, 0.9), (0.2, 0.8), ..., (0.9, 0.1)$—the Pareto optimal points are evenly distributed, which is desirable. This is because that the Pareto front in this case is a 90° circular arc, whose curvature is constant. Note that, in general, evenly distributed Pareto points may not be possible using the weighted-sum method when the curvature of the front is varying.

Examples 5.2 and 5.3 are solved graphically with minimum calculations. We revisit the pyramid problem discussed in Sections 5.1 and 5.2 to illustrate a more realistic problem-solving scenario in Example 5.4.

EXAMPLE 5.4

Solve the pyramid problem using the weighted-sum method.

Solutions:

Although the Pareto front of the pyramid problem has been solved using the generative method and is shown in Figure 5.5, we assume this front is unknown.

We re-sketch below (left) Figure 5.4 of the design space of the pyramid example with feasible set S and optimal points $\mathbf{x}_t^*$ and $\mathbf{x}_a^*$. The two optimal points of the respective single-objective optimization problems are $\mathbf{x}_t^* = (14.7, 20.8)$ and $\mathbf{x}_a^* = (18.5, 13.1)$, respectively. At $\mathbf{x}_t^*$, the two function values are $A(\mathbf{x}_t^*) = A(14.7, 20.8) = 648.5$ and $T(\mathbf{x}_t^*) = T(14.7, 20.8) = 865.3$. Similarly, at $\mathbf{x}_a^*$, the two function values are $A(\mathbf{x}_a^*) = A(18.5, 13.1) = 595.8$ and $T(\mathbf{x}_a^*) = T(18.5, 13.1) = 935.6$. The two optimal points $a = (A(\mathbf{x}_t^*), T(\mathbf{x}_t^*))$ and $b = (A(\mathbf{x}_a^*), T(\mathbf{x}_a^*))$ are pointed out in the criterion space shown below (right) for illustration.

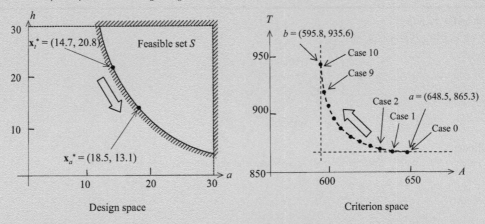

Design space   Criterion space

Using the weighted-sum method, the MOO problem is converted into a single-objective optimization as

$$\underset{(a,h)\in S}{\text{Minimize}} \ \ u(a,h) = w_a A(a,h) + w_t T(a,h) = w_a \left[2a\sqrt{\left(\frac{a}{2}\right)^2 + h^2}\right] + w_t \left[2a\sqrt{\left(\frac{a}{2}\right)^2 + h^2} + a^2\right] \qquad (5.13a)$$

where the feasible space S is defined in Eq. 5.7. The single-objective problem can be solved, for example, using MATLAB function fmincon (Script 3 in Appendix A). We choose $w_a = 0.0, 0.1, 0.2, ..., 1.0$ (and $w_t = 1 - w_s$) and solve for the 11 cases. The results are listed in the following table.

EXAMPLE 5.4—cont'd

Case No.	Weights		Design Point		Surface Area	Total Area	Weighted Objective
	w_a	w_t	a	h	A	T	u
0	0.0	1.0	14.71	20.80	649.0	865.3	865.3
1	0.1	0.9	14.99	20.02	641.0	865.8	843.3
2	0.2	0.8	15.29	19.24	633.3	867.1	820.4
3	0.3	0.7	15.61	18.47	626.0	869.6	796.7
4	0.4	0.6	15.95	17.69	619.0	873.4	771.6
5	0.5	0.5	16.31	16.92	612.6	878.6	745.6
6	0.6	0.4	16.69	16.15	606.9	885.6	718.4
7	0.7	0.3	17.10	15.38	602.0	894.6	689.8
8	0.8	0.2	17.54	14.62	598.2	906.1	659.8
9	0.9	0.1	18.01	13.86	595.7	920.4	628.2
10	1.0	0.0	18.53	13.10	594.8	938.2	594.8

From the solutions of the 11 cases, it is clear that in the design space, as w_a increases from 0 to 1, the design point moves from $\mathbf{x}_t^*$ to $\mathbf{x}_a^*$. In the criterion space, the function values of A and T are plotted. A dotted curve that connects these points approximates Pareto front accurately. Also, as w_a increases from 0 to 1, the objective point moves from points a to b, indicating the influence of the weight w_a to the solutions of the problem using the weighted-sum method. Essentially, increasing w_a results in smaller values of the objective function A, pushing the solutions in the criterion space from points a to b. Although, as shown in the figure above (right), the Pareto points seem to be fairly evenly distributed, in general, it may not be the case if the Pareto front is of significantly varying curvature.

As shown in the above examples, the weighted-sum method is easy to use. If all weights are positive, the minimum of Eq. 5.9 is always a Pareto optimal. However, there are a few recognized difficulties with the weighted-sum method (Arora 2012). First, even with some of the methods discussed in the literature for determining weights, a satisfactory a priori weight selection does not necessarily guarantee that the final solution will be acceptable; one may have to resolve the problem with different weights. In fact, this is true of most weighted methods. The second problem is that it is impossible to obtain points on nonconvex portions of the Pareto optimal set in the criterion space. This is illustrated in Figure 5.8, which shows feasible criterion space and Pareto front of a two-objective problem. In this case, the Pareto front is a concave curve. Using the weighted-sum method, as discussed earlier, the converted objective function $u(\mathbf{x})$ is a straight line in the f_1-f_2 plane. We select two sets of weights to create two new objective functions $u_A(\mathbf{x})$ and $u_B(\mathbf{x})$; both are

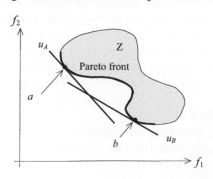

FIGURE 5.8

Geometric illustration of the weighted-sum approach for a nonconvex Pareto front.

straight lines with slopes determined by the respective weights, as shown in Figure 5.8. The two straight lines intersect the Pareto front at points a and b, respectively, resulting in two Pareto points. However, it is apparent that a straight line would never be able to reach the concave portion of the Pareto front. Therefore, the weighted-sum method is not able to find any solution located in the concave Pareto front. Another difficulty with the weighted-sum method is that varying the weights consistently and continuously may not necessarily result in an even distribution of Pareto optimal points and an accurate, complete representation of the Pareto optimal set.

An improvement to the weighted-sum is called the weighted-exponential sum, in which exponential p is added to objective functions as

$$\min_{\mathbf{x} \in S} \ u(\mathbf{x}) = \sum_{i=1}^{q} w_i (f_i(\mathbf{x}))^p \tag{5.14}$$

where $\sum_{i=1}^{q} w_i = 1$ and $w_i \geq 0 \ \forall \ i$, and $p > 0$. In this case, p can be thought of as a compensation parameter; that is, a larger p implies that one prefers solutions with both very high and very low objective values rather than those with averaged values. In general, p may need to be very large to capture Pareto points in the nonconvex regions.

5.3.2.2 Weighted min–max method

The idea of the weighted min–max method (also called the weighted Tchebycheff method) is to minimize $u(\mathbf{x})$, which represents the "distance" to an ideal utopia point in criterion space and is given as follows:

$$\text{Minimize} \ \ u(\mathbf{x}) = \max_{i} \{ w_i [f_i(\mathbf{x}) - f_i^o] \} = \underset{\mathbf{x} \in S}{\text{Min}} \left(\max_{i} \{ w_i [f_i(\mathbf{x}) - f_i^o] \} \right) \tag{5.15}$$

A common approach to the treatment of Eq. 5.15 is to introduce an additional unknown parameter λ as follows:

$$\underset{\mathbf{x} \in S}{\text{Minimize:}} \ \ \lambda \tag{5.16a}$$

$$\text{Subject to:} \ \ w_i [f_i(\mathbf{x}) - f_i^o] - \lambda \leq 0, \ \ \forall i \tag{5.16b}$$

The concept of solving the MOO problem defined Eq. 5.16 is illustrated in Figure 5.9 with two objective functions f_1 and f_2. First, the first term of Eq. 5.16b represents the length of a line originating from the utopia point and pointing in a direction, determined by the weights w_1 and w_2, toward the feasible criterion set. By minimizing λ, and hence the length of the line segment, the solution leads to a point on the Pareto front while keeping the design to be feasible.

The key advantage of the weighted min–max method is that it is able to provide almost all the Pareto optimal points, even for a nonconvex Pareto front. It is relatively well suited for generating the representative Pareto optimal front (with variation in the weights). However, this method requires the minimization of individual single-objective optimization problems to determine the utopia point, which can be computationally expensive. We demonstrate the weighted min–max method in Example 5.5.

5.3.2.3 Lexicographic method

With the lexicographic method, preferences are imposed by ordering the objective functions according to their importance or significance, rather than by assigning weights. After we arrange the objective functions by importance, the most important objective is solved first as a single-objective problem. The

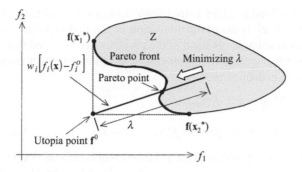

FIGURE 5.9

Geometric illustration of the weighted min–max method.

EXAMPLE 5.5

Solve the same MOO problem of Example 5.3 using the weighted min–max method.

Solutions:

Following the weighted min–max method, we need to find the utopia point first. In this case, the point is found at $\mathbf{f}^0 = (-1, -1)$, as shown in the figure below, in which we recall that points $a = (-1, 0)$ and $b = (0, -1)$ represent the ends of the Pareto front—in this case, a 90° circular arc.

If we choose $w_1 = 1/2$ and $w_2 = 1/2$, then the constraint function of Eq. 5.16b becomes

$$w_i\left[f_i(\mathbf{x}) - f_i^o\right] - \lambda = w_1\left(f_1(\mathbf{x}) - f_1^o\right) + w_2\left(f_2(\mathbf{x}) - f_2^o\right) - \lambda = \frac{1}{2}(x_1 + 1) + \frac{1}{2}(x_2 + 1) - \lambda \leq 0$$

Hence, using the weighted min–max method, we are essentially solving the following single-objective optimization problem:

$$\underset{\mathbf{x} \in S}{\text{Minimize:}} \quad \lambda \tag{5.17a}$$

$$\text{Subject to:} \quad \frac{1}{2}(x_1 + 1) + \frac{1}{2}(x_2 + 1) - \lambda \leq 0 \tag{5.17b}$$

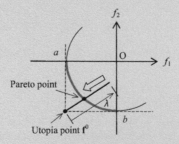

Criterion space

The single-objective problem can be solved, for example, using MATLAB (Script 4 in Appendix A). The solutions are obtained as $x_1 = x_2 = -0.7071$, and $\lambda = 0.2929$, which is on the Pareto front. We may adjust the weights to obtain other Pareto points. For example, if we choose $w_1 = 0.1$ and $w_2 = 0.9$, Eq. 5.17b becomes $0.1x_1 + 0.9x_2 + 1 - \lambda \leq 0$, and solutions are $x_1 = -0.1104$, $x_2 = -0.9939$, and $\lambda = 0.0945$, which is another Pareto point.

second objective is then solved as again a single-objective problem with an added constraint, defined as $f_1(\mathbf{x}) \leq f_1(\mathbf{x}_1^*)$, in which $\mathbf{x}_1^*$ is the optimal solution of the first objective function. The process is repeated, in which optimal solution obtained in the previous step is added as a new constraint, and the sequence of single-objective optimization problems is solved, one problem at a time. Mathematically, the lexicographic method is defined as

$$\text{Minimize: } f_i(\mathbf{x}) \qquad \qquad (5.18a)$$
$$\mathbf{x} \in S$$

$$\text{Subject to: } f_j(\mathbf{x}) \leq f_j(\mathbf{x}_j^*); \quad j = 1, i-1; \quad i = 1, q \qquad (5.18b)$$

where i represents a function's position in the preferred sequence, and $f_j(\mathbf{x}_j^*)$ represents the minimum value for the jth objective function, found in the jth optimization problem. Note that after the first iteration, $(j > 1)$, $f_j(\mathbf{x}_j^*)$ is not necessarily the same as the independent minimum of $f_j(\mathbf{x})$ because new constraints are introduced for each problem. The algorithm terminates once a unique optimum is determined. Generally, this is indicated when two consecutive optimization problems yield the same solution point. However, determining if a solution is unique (within the feasible design space S) can be difficult, especially with gradient-based solution techniques.

For this reason, often with continuous problems, this approach terminates after simply finding the optimum of the first objective $f_1(\mathbf{x})$. Thus, it is best to use a non-gradient-based solution technique with this approach. In any case, the solution is, theoretically, always Pareto optimal. We use an example problem (Example 5.6) similar to that of Example 5.2 to illustrate the concept of the method.

EXAMPLE 5.6

Solve the following MOLP problem using the lexicographic method, defined as

$$\text{Minimize}: f_1 = 4x_1 + x_2 \qquad \qquad (5.19a)$$

$$f_2 = 2x_1 + x_2 \qquad \qquad (5.19b)$$

$$f_3 = x_1 + x_2 \qquad \qquad (5.19c)$$

$$\text{Subject to}: g_1 = 7 - x_1 - 5x_2 \leq 0 \qquad \qquad (5.19d)$$

$$g_2 = 10 - 4x_1 - x_2 \leq 0 \qquad \qquad (5.19e)$$

$$g_3 = -7x_1 + 6x_2 - 9 \leq 0 \qquad \qquad (5.19f)$$

$$g_4 = -x_1 + 6x_2 - 24 \leq 0 \qquad \qquad (5.19g)$$

Solutions:

Note that the feasible set of the problem defined by Eqs. 5.19d–5.19g is identical to that of Example 5.2; that is, $S = \{\mathbf{x} \in R^2 | g_i(\mathbf{x}) \leq 0, i = 1, 4\}$. In this problem, we have three objective functions f_1, f_2, and f_3. We arrange the objective functions by importance as f_1, f_2, and f_3. Therefore, we first optimize a single-objective problem for objective f_1 as

$$\text{Minimize } f_1(\mathbf{x}) = 4x_1 + x_2 \qquad \qquad (5.19h)$$
$$\mathbf{x} \in S$$

EXAMPLE 5.6—cont'd

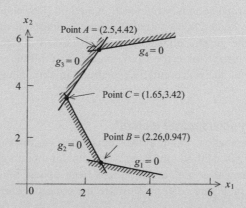

For this simple problem, we found the solution to this problem is all points in line segment between points B and C, as shown in the figure above. The function value is $f_1(\mathbf{x}) = 4x_1 + x_2 = 10.02$. Because the optimum solution is not unique, we continue by minimizing the second objective function $f_2 = 2x_1 + x_2$ while adding the result of the first optimization problem to the constraint set:

$$\underset{\mathbf{x} \in S}{\text{Minimize:}} \quad f_2(\mathbf{x}) = 2x_1 + x_2 \tag{5.19i}$$

$$\text{Subject to:} \quad f_1(\mathbf{x}) \le f_1(\mathbf{x}_1^*) = 10.02 \tag{5.19j}$$

Note that Eq. 5.19j is the additional constraint from the result of the first optimization. For the second optimization problem, we found the solution at point B, where $f_2 = 5.47$, while f_1 is unchanged satisfying the constraint function defined in Eq. 5.19j. Because the solution is unique, the process stops. The solution is found at point B, where $f_1 = 10.02$, $f_2 = 5.47$, and $f_3 = 3.21$. Note that f_3 was not even considered in the solution process because it is the least important objective among the three. If we change the importance order of the objective functions, we will most likely reach a different solution.

The advantages of the method include that it offers a unique approach to specifying preferences, it does not require that the objective functions be normalized, and it always provides a Pareto optimal solution. A few disadvantages on this method include that it can require the solution of many single-objective problems to obtain just one solution point, and it requires that additional constraints to be imposed. When there are more objective functions, the constraint set becomes large toward the end of the solution process.

5.3.3 METHODS WITH A POSTERIORI ARTICULATION OF PREFERENCES

In some cases, it is difficult for a designer to express preferences a priori. Therefore, it can be effective to allow the designer to choose from a palette of solutions. To this end, a number of methods aim at producing almost all the Pareto optimal solutions or a representative subset of the Pareto optimal. Such methods incorporate a posteriori articulation of preferences; they are called cafeteria or generate-first-choose-later approaches (Messac and Mattson 2002). Several methods belong to this category, such as the normal boundary intersection method (Das and Dennis 1998) and adaptive weighted-sum (Kim and de Weck 2006), to name a few. In addition, methods based on evolution algorithms (EA), such as the genetic algorithm to be discussed in Section 5.4, are considered a posteriori methods, although they solve the MOO problems in a much different way. In this sub-section, we introduce one of the representative methods in this category, the normal boundary intersection (NBI) method.

5.3.3.1 *Normal boundary intersection method*

The normal boundary intersection method was developed in response to deficiencies in the weighted-sum approach. This method provides a means for obtaining an even distribution of Pareto optimal points, even with a nonconvex Pareto front of varying curvature. We use a two-objective problem shown in Figure 5.10 to illustrate the concept. The approach is formulated as follows:

$$\text{Maximize:} \quad \beta \qquad \qquad (5.20a)$$
$$\text{x} \in S$$

$$\text{Subject to:} \quad \boldsymbol{\alpha} + \beta \mathbf{n} = f(\mathbf{x}) \qquad \qquad (5.20b)$$

where $\boldsymbol{\alpha}$ is a point on the line segment AB, called the convex hull of individual minima (CHIM), which is also called the utopia line. Points A and B are the optimal points of the objective functions f_1 and f_2, respectively. $\mathbf{n}$ is a vector perpendicular to CHIM and pointing toward the Pareto front. The parameter β is a scalar to be maximized. Essentially, the concept of NBI is identifying a point α on the CHIM and searching a Pareto point along the $\mathbf{n}$ direction by maximizing parameter β. Because the constraint equation (Eq. 5.20b) ensures an attainable design point $\mathbf{x}$ in the feasible set S, maximizing β pushes the vector $\mathbf{n}$ to eventually intersect the Pareto front and yields a Pareto solution. If α points are chosen uniformly along the CHIM, a set of fairly evenly distributed Pareto points can be reasonably expected.

We use the same MOO problem of Example 5.3 to further illustrate the NBI method.

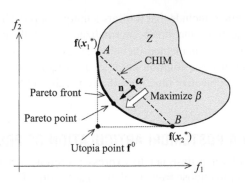

FIGURE 5.10

Geometrical illustration of the NBI method.

EXAMPLE 5.7

Solve the same MOO problem of Example 5.3 using the NBI method.

Solutions:

In this case, we recall that the utopia point is found at $\mathbf{f}^0 = (-1, -1)$, and points $A = (-1, 0)$ and $B = (0, -1)$. The CHIM line connecting points A and B is shown below. For this simple example, $\mathbf{n}$ can be found as $\mathbf{n} = [-1, -1]^T$. Therefore, using the NBI method, we are essentially solving the following single-objective optimization subproblem:

$$\text{Maximize:} \quad \beta \tag{5.21a}$$
$$\underset{x \in S}{}$$

$$\text{Subject to:} \quad [\alpha_1, \alpha_2]^T + \beta[-1, -1]^T = [x_1, x_2]^T \tag{5.21b}$$

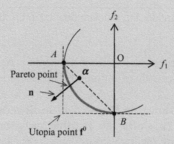

Criterion space

The single-objective problem can be solved, for example, using MATLAB (Script 5 in Appendix A). If we choose $\alpha = [-0.7, -0.3]^T$, the solution is obtained as $x_1 = -0.8782$, $x_2 = -0.4782$, and $\beta = 0.1782$, which is on the Pareto front. Note that β is the distance between α and the Pareto point found. We may pick another α point to obtain another Pareto point. For example, if we choose $\alpha_1 = -0.5$ and $\alpha_2 = -0.5$, the first term in Eq. 5.21b becomes $[-0.5, -0.5]^T$, and solutions are $x_1 = -0.7071$, $x_2 = -0.7071$, and $\beta = 0.2071$.

In fact, for this simple problem, β can be solved from Eqs. 5.21a and 5.21b as

$$\beta = \frac{1}{2}\left[(\alpha_1 + \alpha_2) + \sqrt{2 - (\alpha_1^2 - \alpha_2^2)}\right] \tag{5.21c}$$

which does not require solving the single-objective optimization problem of Eqs. 5.21a and 5.21b. Derivation of Eq. 5.21c is left as an exercise.

For problems with more than two objectives, the NBI method can be formulated in a more general form as

$$\text{Maximize:} \quad \beta \tag{5.22a}$$
$$\underset{x \in S}{}$$

$$\text{Subject to:} \quad \mathbf{\Phi}\mathbf{w} + \beta\mathbf{n} = \mathbf{f}(\mathbf{x}) - \mathbf{f}^0 \tag{5.22b}$$

Here, $\mathbf{\Phi}$ is a $q \times q$ payoff matrix with ith column composed of the vector $\mathbf{f}(\mathbf{x}_i^*) - \mathbf{f}^0$, where $\mathbf{f}(\mathbf{x}_i^*)$ is the vector of objective functions evaluated at the minimum of the ith objective function. As a result, the diagonal elements of $\mathbf{\Phi}$ are zeros. $\mathbf{w}$ is a vector of scalars such that $\sum_{i=1}^{q} w_i = 1$ and $w_i \geq 0 \; \forall \; i$. $\mathbf{n} = -\mathbf{\Phi}\mathbf{e}$, where $\mathbf{e} \in R^q$ is a column vector of ones in the criterion space. $\mathbf{n}$ is called a quasi-normal vector. Because each component of $\mathbf{\Phi}$ is positive, the negative sign ensures that $\mathbf{n}$ points toward the origin of the criterion space. $\mathbf{n}$ gives the NBI method the property that for any $\mathbf{w}$, a solution

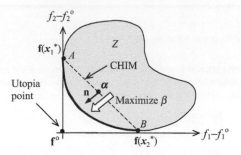

FIGURE 5.11

Geometric illustration of the NBI method for the problem defined in Eq. 5.22.

point is independent of how the objective functions are scaled. As $\mathbf{w}$ is systematically modified, the solution to Eq. 5.22 yields an even distribution of Pareto optimal points well representing the Pareto set.

We use a two-objective case shown in Figure 5.11 to illustrate the formulation geometrically. The right-hand side of Eq. 5.22b shifts the feasible criterion space Z such that the utopia point coincides with the origin of the two coordinate axes. The matrix $\mathbf{\Phi}$ for this example is constructed as

$$\mathbf{\Phi} = \left[\mathbf{f}(\mathbf{x}_1^*) - \mathbf{f}^o \ \ \mathbf{f}(\mathbf{x}_2^*) - \mathbf{f}^o \right] = \begin{bmatrix} f_1(\mathbf{x}_1^*) - f_1^o & f_1(\mathbf{x}_2^*) - f_1^o \\ f_2(\mathbf{x}_1^*) - f_2^o & f_2(\mathbf{x}_2^*) - f_2^o \end{bmatrix} = \begin{bmatrix} 0 & f_1(\mathbf{x}_2^*) - f_1^o \\ f_2(\mathbf{x}_1^*) - f_2^o & 0 \end{bmatrix}$$

$$(5.23)$$

The vector $\mathbf{n}$ can be written as

$$\mathbf{n} = -\mathbf{\Phi e} = - \begin{bmatrix} 0 & f_1(\mathbf{x}_2^*) - f_1^o \\ f_2(\mathbf{x}_1^*) - f_2^o & 0 \end{bmatrix} \begin{bmatrix} 1 \\ 1 \end{bmatrix} = - \begin{bmatrix} f_1(\mathbf{x}_2^*) - f_1^o \\ f_2(\mathbf{x}_1^*) - f_2^o \end{bmatrix} \qquad (5.24)$$

Hence, the constraint Eq. 5.22b is

$$\mathbf{\Phi w} + \beta \mathbf{n} = \begin{bmatrix} w_2 \left[f_1(\mathbf{x}_2^*) - f_1^o \right] \\ w_1 \left[f_2(\mathbf{x}_1^*) - f_2^o \right] \end{bmatrix} - \beta \begin{bmatrix} f_1(\mathbf{x}_2^*) - f_1^o \\ f_2(\mathbf{x}_1^*) - f_2^o \end{bmatrix} = \mathbf{f}(\mathbf{x}) - \mathbf{f}^o = \begin{bmatrix} f_1(\mathbf{x}) - f_1^o \\ f_2(\mathbf{x}) - f_2^o \end{bmatrix} \qquad (5.25)$$

Note that the first term on the left-hand side of Eq. 5.25 ($\mathbf{\Phi w}$) is nothing but a point on the CHIM line in the criterion space with axes $f_1 - f_1^o$ and $f_2 - f_2^o$. The vector in the second term is the normal vector $\mathbf{n}$ in Eq. 5.20b. Therefore, the constraint equations in Eqs. 5.22b and 5.20b are very similar. The major difference is that in Eq. 5.22b the α points are generated by adjusting the vector $\mathbf{w}$. For MOO problems with more than two objectives ($q > 2$), in which the CHIM becomes an utopia-hyperplane, Eq. 5.22b is more general in determining a set of uniformly distributed α points, as well as the normal vector $\mathbf{n}$. As a result, uniformly distributed Pareto solutions can be expected.

In the Example 5.8, we revisit the pyramid example to illustrate more details on the NBI method formulated in Eq. 5.22.

EXAMPLE 5.8

Solve the pyramid problem using the NBI method.

Solutions:

From Example 5.4, we have solved the two respective single-objective optimization problems for the optimal points $\mathbf{x}_t^*$ and $\mathbf{x}_a^*$ in the design space. The objective functions at these two design points are located in the criterion space $\mathbf{f}(\mathbf{x}_t^*) = (A(\mathbf{x}_t^*), T(\mathbf{x}_t^*))$ and $\mathbf{f}(\mathbf{x}_a^*) = (A(\mathbf{x}_a^*), T(\mathbf{x}_a^*))$ in the figure shown below. Hence, the utopia point is found at $\mathbf{f}^o = (A(\mathbf{x}_a^*), T(\mathbf{x}_t^*)) = (595.8, 865.3)$.

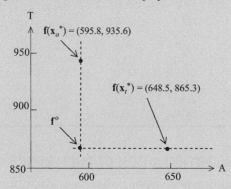

Using the NBI method, the constraint equation of Eq. 5.22b can be found, using Eq. 5.25, as

$$\mathbf{\Phi}\mathbf{w} + \beta\mathbf{n} = \begin{bmatrix} w_2\left[A(\mathbf{x}_t^*) - A(\mathbf{x}_a^*)\right] \\ w_1\left[T(\mathbf{x}_a^*) - T(\mathbf{x}_t^*)\right] \end{bmatrix} - \beta\begin{bmatrix} \left[A(\mathbf{x}_t^*) - A(\mathbf{x}_a^*)\right] \\ \left[T(\mathbf{x}_a^*) - T(\mathbf{x}_t^*)\right] \end{bmatrix} = \mathbf{f}(\mathbf{x}) - \mathbf{f}^o = \begin{bmatrix} A(\mathbf{x}) - A(\mathbf{x}_a^*) \\ T(\mathbf{x}) - T(\mathbf{x}_t^*) \end{bmatrix} \quad (5.26a)$$

Plugging in the numbers, we have

$$\begin{bmatrix} w_2(648.5 - 595.8) \\ w_1(935.6 - 865.3) \end{bmatrix} - \beta\begin{bmatrix} (648.5 - 595.8) \\ (935.6 - 865.3) \end{bmatrix} = \begin{bmatrix} 2a\sqrt{\left(\dfrac{a}{2}\right)^2 + h^2} - 595.8 \\ \left[2a\sqrt{\left(\dfrac{a}{2}\right)^2 + h^2} + a^2\right] - 865.3 \end{bmatrix} \quad (5.26b)$$

Hence, using the NBI method, the problem is converted into a single-objective optimization as

$$\underset{\mathbf{x}\in S}{\text{Maximize:}} \quad \beta \quad (5.26c)$$

Subject to: $\quad w_2(648.5 - 595.8) - \beta(648.5 - 595.8) - 2a\sqrt{\left(\dfrac{a}{2}\right)^2 + h^2} + 595.8 = 0$

$$w_1(935.6 - 865.3) - \beta(935.6 - 865.3) - \left[2a\sqrt{\left(\dfrac{a}{2}\right)^2 + h^2} + a^2\right] + 865.3 = 0 \quad (5.26d)$$

where S is the feasible space defined in Eq. 5.7. The single-objective problem can be solved, for example, using MATLAB (Script 6 in Appendix A).

We choose $w_1 = 0.0, 0.1, \ldots, 1.0$ (and $w_2 = 1 - w_1$) and solved for these 11 cases. The results are listed in the table next page. Because the α points are selected uniformly along the CHIM by specifying a uniform increment 0.1 in w_1 from 0 to 1, the Pareto points found from NBI are more evenly distributed. In this example, however, because the Pareto front is mild in geometry without significant changes in curvature, the advantage of the NBI is not apparent, and the results shown in the table below are similar to those in Example 5.4, in which weighted-sum method was employed.

Continued

EXAMPLE 5.8—cont'd

Case No.	Weights		Design Point		Surface Area	Total Area	β
	w_1	w_2	a	h	A	T	
0	0.0	1.0	14.72	20.76	648.5	865.4	0.0007
1	0.1	0.9	15.08	19.78	638.6	866.1	0.0887
2	0.2	0.8	15.45	18.86	629.6	868.2	0.1586
3	0.3	0.7	15.81	18.00	621.7	871.7	0.2090
4	0.4	0.6	16.18	17.19	614.8	876.6	0.2399
5	0.5	0.5	16.55	16.43	608.9	882.8	0.2514
6	0.6	0.4	16.92	15.72	604.1	890.4	0.2434
7	0.7	0.3	17.29	15.05	600.2	899.3	0.2160
8	0.8	0.2	17.67	14.41	597.4	909.6	0.1692
9	0.9	0.1	18.05	13.82	595.6	921.3	0.1032
10	1.0	0.0	18.42	13.26	594.9	934.3	0.0180

5.3.4 METHODS WITH NO ARTICULATION OF PREFERENCE

Sometimes the designer cannot concretely define what he or she prefers. Consequently, this section describes methods that do not require any articulation of preferences. One of the simplest methods is the min–max method, formulated as

$$\underset{\mathbf{x} \in S}{\text{Minimize}} \ \max_{i} \left[f_i(\mathbf{x}) \right] \qquad (5.27)$$

The concept and formulation are straightforward. However, implementing Eq. 5.27 is not obvious. One possible approach is to introduce a new parameter β, which is to be minimized while requiring that all objective functions are no greater than β. Mathematically, the min–max problem can be formulated as

$$\underset{\mathbf{x} \in S}{\text{Minimize:}} \ \beta \qquad (5.28a)$$

$$\text{Subject to:} \ f_i(\mathbf{x}) \le \beta, \quad i = 1, q \qquad (5.28b)$$

The following example (Example 5.9) illustrates the min–max method.

EXAMPLE 5.9

Solve the pyramid problem using the min–max method.

Solutions:

Using the min–max method, the pyramid problem is formulated as

$$\underset{\mathbf{x} \in S}{\text{Minimize:}} \ \beta \qquad (5.29a)$$

$$\text{Subject to:} \ 2a\sqrt{\left(\frac{a}{2}\right)^2 + h^2} - \beta \le 0$$

$$\qquad (5.29b)$$

$$2a\sqrt{\left(\frac{a}{2}\right)^2 + h^2} + a^2 - \beta \le 0$$

EXAMPLE 5.9—cont'd

where S is the feasible region defined in Eq. 5.7. The single-objective problem can be solved, for example, using MATLAB (Script 7 in Appendix A). The solution found is design point $\mathbf{x}_t^*$, at which $a = 14.71$, $h = 20.80$, surface area $A = 649.0$, total area $T = 865.3$, and $\beta = 865.3$. The parameter $\beta = 865.3$ is indeed the minimum because it is the minimum of the total area T. Although the solution $\mathbf{x}_t^*$ does not minimize surface area A, reducing A cannot be done without increasing the total area T, in which β is no longer the minimum.

5.3.5 MULTIOBJECTIVE GENETIC ALGORITHMS

The solution techniques we discussed so far are mainly searching a single best solution representing the best compromise given the information from the designer. These techniques are called aggregating approaches because they combine (or "aggregate") all the objectives into a single one. Such techniques approximate the Pareto front by repeating the solution process by adjusting parameters that redefine the search, such as the weights in the weighted-sum method.

In contrast to these techniques, methods based on evolution algorithms, such as multiobjective genetic algorithms, support direct generation of the Pareto front by simultaneously optimizing the individual objectives. As in the single-objective optimization problem discussed in Chapter 3, genetic algorithms emulate the biological evolution process. A population of individuals representing different solutions is evolving to find the optimal solutions. The fittest individuals are chosen, with mutation and crossover operations applied, thus yielding a new generation (offspring).

Evolutionary algorithms seem particularly suitable to solving multiobjective optimization problems because they deal simultaneously with a set of possible solutions (or a population). This allows designers to find several members of the Pareto optimal set in a single run of the algorithm, instead of having to perform a series of separate runs. Additionally, genetic algorithms are less susceptible to the shape or continuity of the Pareto front (e.g., they can easily deal with discontinuous or concave Pareto fronts), which are the two issues for mathematical programming techniques.

In general, for multiobjective problems, genetic algorithms have the advantage of evaluating multiple potential solutions in a single iteration. They offer greater flexibility for the designer, mainly in cases where no a priori information is available, as is the case for most real-life multiobjective problems. However, the challenge is how to guide the search toward the Pareto optimal set, and how to maintain a diverse population in order to prevent premature convergence. In addition, as discussed in Chapter 3, genetic algorithms require a large amount of function evaluations. For complex problems that require significant computation time for function evaluations, such as problems involving nonlinear finite element analysis, methods other than genetic algorithms are more practical.

One of the first treatments of multiobjective genetic algorithms, called the vector-evaluated genetic algorithm (VEGA), was presented by Schaffer (1985). Although this method was surpassed by others soon after it was proposed, it has provided a foundation for later development. Many prominent algorithms have been developed in the past three decades, in which ranking was explicitly used in order to determine the probability of replication of an individual. Such methods are so-called Pareto-based approaches.

In this section, we first provide a general discussion on the Pareto-based approaches. Thereafter, we zoom in to one of the most popular Pareto-based approaches, called the nondominated sorting genetic algorithm (NSGA). NSGA was proposed by Srinivas and Deb (1994) and revised to NSGA II (Deb et al. 2002), with significant improvements on the efficiency of the sorting algorithm as well as on its niche technique. We also present an example of NSGA II implementation in MATLAB, followed by the pyramid example for illustration. Note that some aspects of the method presented are shared by many other multiobjective genetic algorithms.

5.3.5.1 Pareto-based approaches

Pareto-based approaches incorporate three major elements—ranking, niching mechanism, and elitist strategy—in addition to the common mutation and crossover techniques seen in genetic algorithms. Ranking determines the probability of replication of an individual. The basic idea is to find the set of nondominated individuals in the population. These are assigned the highest rank and eliminated from further contention. The process is then repeated with the remaining individuals until the entire population is ranked and assigned a fitness value. In conjunction with Pareto-based fitness assignment, a niching mechanism is used to prevent the algorithm from converging to a single region of the Pareto front. The elitist strategy provides a means for ensuring that Pareto optimal solutions are not lost. Mutation and crossover operations are then performed to create the next generation of individuals.

5.3.5.1.1 Ranking

A simple and efficient method assumes that the fitness value of an individual is proportional to the number of other individuals it dominates, as illustrated in Figure 5.12. In Figure 5.12a, point 2 dominates six other individuals (points 3, 5, 6, 7, 9, and 10) because the objective function values at this point are less than those of the six points it dominates. Therefore, point 2 is assigned a higher fitness, and hence the probability of selecting this point is higher than, for example point 8, which only dominates two individuals (points 9 and 10).

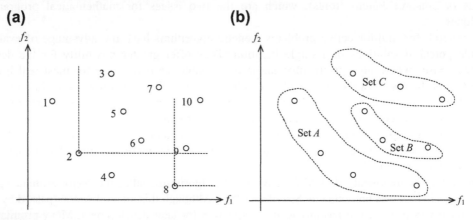

FIGURE 5.12

Illustration of fitness assignments: (a) ranking and (b) fitness computation for NSGA.

Another version is the nondominated sorting genetic algorithm, which uses a layered classification technique, as illustrated in Figure 5.12b. In Figure 5.12b, a layered classification technique is used whereby the population is incrementally sorted using Pareto dominance. Individuals in set A have the same fitness value, which is higher than the fitness of individuals in set B, which in turn are superior to individuals in set C. All nondominated individuals are assigned the same fitness value. The process is repeated for the remainder of the population, with a progressively lower fitness value assigned to the nondominated individuals.

5.3.5.1.2 Niche techniques

A niche in genetic algorithms is a group of points that are close to each other, typically in the criterion space. Niche techniques (also called niche schemes, niche mechanism, or the niche-formation method) are methods for ensuring that a set of designs does not converge to a niche. Thus, these techniques foster an even spread of points in the criterion space. Genetic multiobjective algorithms tend to create a limited number of niches; they converge to or cluster around a limited set of Pareto optimal points. This phenomenon is known as genetic or population drift; niche techniques force the development of multiple niches while limiting the growth of any single niche.

Fitness sharing is a common niche technique. The basic idea of which is to penalize the fitness of points in crowded areas, thus reducing the probability of their survival for the next iteration. The fitness of a given point is divided by a constant that is proportional to the number of other points within a specified distance in the criterion space, thus reducing its fitness value and lowering its chance to survive for the next iteration. In this way, the fitness of all points in a niche is shared in some sense—thus the term "fitness sharing."

5.3.5.1.3 Elitist strategy

Elitist strategy provides a means for ensuring that Pareto optimal solutions are not lost. It functions independently of the ranking scheme. Two sets of solutions are stored: a current population and a tentative set of nondominated solutions, which is an approximate Pareto optimal set. In each iteration, all points in the current population that are not dominated by any points in the tentative set are added to the tentative set. Then, the dominated points in the tentative set are discarded. After crossover and mutation operations are applied, a user-specified number of points from the tentative set are reintroduced into the current population. These are called elite points. In addition, q points with the best values for each objective function can be regarded as elite points and preserved for the next generation. Recall that q is the number of objective functions.

5.3.5.2 Nondominated sorting genetic algorithm II

With the understanding of the basic elements in the GA-based approaches for MOO problems, we now discuss NSGA II.

In NSGA II, before selection is performed, the population is ranked on the basis of non-domination: all nondominated individuals are classified into one category to provide an equal reproductive potential for these individuals. Because the individuals in the first front have the maximum fitness value, they always get more copies than the rest of the population when the selection is carried out. Additionally, the NSGA II estimates the density of solutions surrounding a particular solution in the population by computing the average distance of two points on either side of this point along each of the objectives of the problem. This value is called the crowding distance. During selection, NSGA II uses a crowded-comparison scheme that takes into consideration both the nondomination rank of an individual in the population and its crowding distance. The nondominated solutions are preferred over dominated solutions; however, between two solutions with the same nondomination rank, the one that resides in the less crowded region is preferred. The NSGA II uses the elitist mechanism that consists of combining the best parents with the best offspring obtained. More details are described in the next section.

5.3.5.2.1 Nondominated sorting

One of the major computation efforts in performing the nondominated sorting involves comparisons. Each solution can be compared with every other solution in the population to find out if it is dominated. This requires $q \times (N - 1)$ comparisons for each solution, where q is the number of objectives and N is the population size. The first round of sorting involves $\sum_{i=1}^{N-1} q(N - i) = \frac{1}{2} qN(N - 1)$ comparisons, which is in the order of $O(qN^2)$.

The results of the sorting are usually stored in an $N \times N$ matrix, in which each term indicates the dominance relation between two individuals in the population. In general, four scenarios need to be considered in terms of the feasibilities of the two individuals x and y being compared. First, if both x and y are feasible, then objectives at x and y are compared. If x dominates y, then the domination matrix donMat$(x, y) = 1$; else, if y dominates x, then donMat$(x, y) = -1$. Second, if x is feasible and y is infeasible, then donMat$(x, y) = 1$. Third, if x is infeasible and y is feasible, then donMat$(x, y) = -1$. Fourth, if both x and y are infeasible, then the constraint violations of x and y are compared. If violation at x is less than that of y, then donMat$(x, y) = 1$. Otherwise, donMat$(x, y) = -1$.

To further illustrate the idea, we use the example of population size $N = 10$ shown in Figure 5.12a. To simplify the discussion, we assume a nonconstrained problem. The matrix is set to zero initially. We pick point 2 to illustrate the process. As shown in Figure 5.12a, point 2 dominates points 3, 5, 6, 7, 9, and 10. Therefore, in the following matrix, entries $(2, 3)$, $(2, 5)$, $(2, 6)$, $(2, 7)$, $(2, 9)$, and $(2, 10)$ are set to 1. The entries $(3, 2)$, $(5, 2)$, $(6, 2)$, $(7, 2)$, $(9, 2)$, and $(10, 2)$ are set to -1. By going over the comparisons, the domination matrix can be constructed, as shown in Figure 5.13.

As proposed by Deb et al. (2002), for each individual we calculate two entities: (1) domination count n_p, the number of solutions that dominate the solution p, $p = 1, N$; and (2) D_p, a set of individuals that the solution dominates. The count n_p can be found by adding the number of -1 of the pth column in the matrix. The set D_p can be created by collecting individuals with -1 along the pth row in the matrix. For example, for point 2, $n_2 = 0$ and $D_2 = \{3, 5, 6, 7, 9, \text{and } 10\}$. By checking the columns and rows of the domination matrix, Table 5.2 can be constructed.

All solutions in the first nondominated front will have their domination count as zero ($n_p = 0$). From Table 5.2, the first nondominated front is identified as $ND_1 = \{1, 2, 4, 8\}$. Now, for each solution in ND_1, we visit each member of its set D_p and reduce its domination count by 1. For example, for the second member of ND_1 (point 2), the counts of its members in D_2 become respectively, $n_3 = 3 - 1 = 2$, $n_5 = 1$, $n_6 = 1$, $n_7 = 4$, $n_9 = 2$, and $n_{10} = 6$. In doing so, if for any member the domination count becomes zero (if

	1	2	3	4	5	6	7	8	9	10
1	0	0	-1	0	0	0	-1	0	0	-1
2	0	0	-1	0	-1	-1	-1	0	-1	-1
3	1	1	0	1	0	0	0	0	0	0
4	0	0	-1	0	-1	-1	-1	0	-1	-1
5	0	1	0	1	0	0	-1	0	0	-1
6	0	1	0	1	0	0	-1	0	0	-1
7	1	1	0	1	1	1	0	0	0	0
8	0	0	0	0	0	0	0	0	-1	-1
9	0	1	0	1	0	0	0	1	0	-1
10	1	1	0	1	1	1	0	1	1	0

FIGURE 5.13

Domination matrix for the example shown in Figure 5.12a.

Table 5.2 Domination Count n_p and the Set D_p Derived from the Domination Matrix

Solution Point p	Solution p Dominates (D_p)	Solutions Dominate p	n_p
1	3, 7, 10	None	0
2	3, 5, 6, 7, 9, 10	None	0
3	None	1, 2, 4	3
4	5, 6, 7, 9, 10	None	0
5	7, 10	2, 4	2
6	7, 10	2, 4	2
7	None	1, 2, 4, 5, 6	5
8	9, 10	None	0
9	10	2, 4, 8	3
10	None	1, 2, 4, 5, 6, 8, 9	7

none is 0, we pick the ones with the lowest integer), these members belong to the second nondominated front ND_2. In this example, solutions 5 and 6 of D_2 have the lowest count 1. After repeating the same process for the remaining solutions 1, 4, and 8 in ND_1, the second nondominated front is $ND_2 = \{5, 6\}$. Now, the above procedure is continued with each member of ND_2, and the third front is identified as $ND_3 = \{7\}$. This process continues until all fronts are identified. Note that some individuals may not get counted, for example, 3, 9, and 10 in this case, which are dominated points. It is important that the first few nondominated fronts are identified accurately. In this example, the first two fronts, ND_1 and ND_2, are accurately identified as can be seen in Figure 5.12a.

5.3.5.2.2 Niche technique for diversity preservation

NSGA II employs a different niche technique to preserve the diversity of the Pareto solutions. This technique involves two factors: density-estimation metric and crowded-comparison operator.

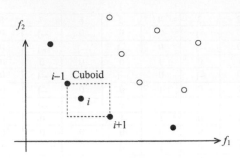

FIGURE 5.14

Crowding-distance calculation. Points marked in filled circles are solutions of the same nondominated front.

Density estimation calculates an estimate of the density of points surrounding a particular solution in the population. The estimate calculates the average distance of two points on either side of this point along each of the objectives. This quantity serves as an estimate of the perimeter of the cuboid formed by using the nearest neighbors as the vertices (this is called the crowding distance). For example, in Figure 5.14, the crowding distance of the ith solution in its front (marked with solid circles) is the average side length of the cuboid (shown with a dashed box). The crowding-distance computation requires sorting the population according to each objective function value in ascending order of magnitude. Thereafter, for each objective function, the boundary solutions (solutions with smallest and largest function values) are assigned an infinite distance value. All other intermediate solutions are assigned a distance value equal to the absolute normalized difference in the function values of two adjacent solutions. This calculation is continued with other objective functions. The overall crowding-distance value is calculated as the sum of individual distance values corresponding to each objective. Note that each objective function is normalized before calculating the crowding distance.

The crowded-comparison operator ($\prec$) guides the selection process at the various stages of the algorithm toward a uniformly spread-out Pareto optimal front. Assume that every individual in the population has two attributes:

1. Nondomination rank (i_{rank})
2. Crowding distance ($i_{distance}$).

We now define a partial order as

$i \prec j$ if ($i_{rank} < j_{rank}$) or if ($i_{rank} = j_{rank}$, $i_{distance} > j_{distance}$).

That is, between two solutions with different nondomination ranks, we prefer the solution with the lower (better) rank. Otherwise, if both solutions belong to the same front, then we prefer the solution that is located in a lesser crowded region.

At the end of nondominated sorting, the individuals of the population of size N are sorted based on their rank. For the individuals of same rank, they are sorted by crowding distances.

5.3.5.2.3 Overall process

The overall NSGA II process consists of the creation of initial population and design iterations. Initially, a random parent population P_0 of size N is created. Function evaluations of objectives and constraints are carried out for each of the N individuals. The population is sorted based on the non-domination and crowding distance, as discussed above.

A. Assume that the current generation P is [①, ②, ③, ④, ⑤, ⑥], so the size of the population is $N = 6$.

B. Assume that ② > ③ > ① > ⑥ > ④ > ⑤, where ② > ③ means the individual ② is better than ③. The comparison between each two individuals is based on their ranks (if ranks are the same, then compare crowding distance).

C. The next step is to create a new population Q_0 of size N. The procedure is shown below

Create $2N$ (in this case 12) random integers between 1 and 6, for example:

[⑤, ④, ①, ③, ②, ②, ⑥, ⑤, ①, ④, ③ ⑤]

Carry out binary tournament selection, crossover, and mutation:

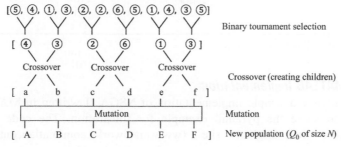

FIGURE 5.15

An example of the selection of new population of size $N = 6$.

At first, the usual binary tournament selection, recombination, and mutation operators are used to create an offspring population Q_0 of size N. The procedure is as follows. Function evaluations and nondominated (and crowding distance) sorting are performed. Then, a binary tournament selection is carried out to select the best N individuals. These N individuals go through crossover and mutation to generate a new population Q_0 of size N. An example of these steps is shown in Figure 5.15 for illustration, in which $N = 6$ is assumed.

Because elitism is introduced by comparing the current population with the previously found best nondominated solutions, the procedure is different after the initial generation. We describe the ith generation of the algorithm. First, a combined population $R_i = P_i \cup Q_i$ is formed. The population is of size $2N$. Then, the population is sorted according to nondomination (and crowding distance). Because all previous and current population members are included in R_i, elitism is ensured. Now, solutions belonging to the best nondominated set F_1 are of best solutions in the combined population (see Figure 5.16) and must be emphasized more than any other solution in the combined population. If the size of F_1 is smaller than N, we choose all members of the set F_1 for the new population P_{i+1}. The remaining members of the population P_{i+1} are chosen from subsequent nondominated fronts in the order of their ranking. Thus, solutions from the set F_2 are chosen next, followed by solutions from the set F_3, and so on. This procedure is continued until no more sets can be accommodated. Say that the set F_ℓ is the last nondominated set beyond which no other set can be accommodated. In general, the count of solutions in all sets from F_1 to F_ℓ would be larger than the population size. To choose exactly N population members, we sort the solutions of the last front using the crowded-comparison operator in descending order and choose the best solutions needed to fill all population slots. This procedure is also shown in Figure 5.16. The new population P_{i+1} of size N is now used for selection, crossover, and mutation to create a new population Q_{i+1} of size N.

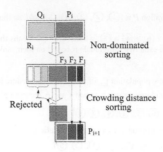

FIGURE 5.16

Procedure of selecting a new population P_{i+1}.

5.3.5.3 Sample MATLAB implementation

In this section, we review a sample implementation of NSGA II written in MATLAB, and then we use this code to solve the pyramid example for illustration. The code is available for download at either its original site (www.mathworks.com/matlabcentral/fileexchange/ 31166-ngpm-a-nsga-ii-program-in-matlab-v1-4) or from the book's companion site. The zipped file consists of a set of MATLAB files and three folders. The doc folder includes a license file and a user manual. The other two folders, TP_NSGA2 and TP_R-NSGA2, include test problems that come with the code. Readers may download the code to go over some of the exercises at the end of this chapter.

There are two MATLAB files that users will have to create: one contains the input parameters, such as the number of objectives, whereas the other defines the objective and constraint functions. Once the two MATLAB files are created, we run the file with input parameters in MATLAB. The code searches for solutions at the Pareto front following the NSGA II algorithm discussed above. During the solution process, the code shows a list of generations, plots the solutions, and provides a summary on the status of the optimization, as shown in Figure 5.17.

To solve the pyramid example using the code, we first create the two needed files and name them, respectively, PYRAMID.m and PYRAMID_objfun.m. The contents of these two files are shown in Figure 5.18. Note that the name of the function needs to match that defined in PYRAMID.m. In this case, the name of the function must be options.objfun = @PYRAMID_objfun, as circled in red in Figure 5.18a.

As shown in Figure 5.18a, the inputs of the pyramid example include the following:

- Population size: 200
- Number of generation: 500
- Number of objective functions: 2
- Number of design variables: 2
- Number of constraints: 1
- Lower bounds of design variables: 0 for the two design variables
- Upper bounds of design variables: 30 for the two design variables

The MATALB script PYRAMID.m is called to start the optimization process. Results of a number of iterations are graphed in Figure 5.19, in which the red dotted line represents the Pareto front. As seen

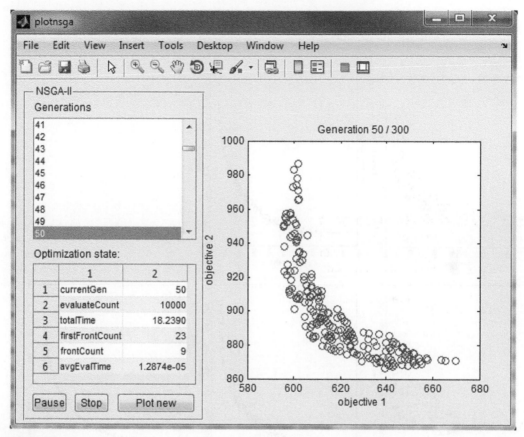

FIGURE 5.17

Screen capture of an intermediate iteration in MATLAB.

(a)

```
%***********************************************************
% Test Problem : 'PYRAMID'
%***********************************************************

options = nsgaopt();                % create default options structure
options.popsize = 200;              % populaion size
options.maxGen  = 500;              % max generation

options.numObj = 2;                 % number of objectives
options.numVar = 2;                 % number of design variables
options.numCons = 1;                % number of constraints
options.lb = [0 0];                 % lower bound of x
options.ub = [30 30];               % upper bound of x
options.objfun = @PYRAMID_objfun;   % objective function handle
options.plotInterval = 5;           % interval between two calls of "plotnsga".
result = nsga2(options);            % begin the optimization!
```

(b)

```
function [y, cons] = PYRAMID_objfun(x)
% Objective function : Test problem 'PYRAMID'.
%y(1): S    y(2): T
%x(1): a    x(2): h
%***********************************************************

y = [0,0];
cons = 0;

y(1) = 2*x(1)*sqrt(x(1)^2/4+x(2)^2);
y(2) = 2*x(1)*sqrt(x(1)^2/4+x(2)^2)+x(1)^2;

% calculate the constraint violations
c = x(1)^2*x(2)/3 - 1500;
if(c<0)
    cons(1) = abs(c);
end
```

FIGURE 5.18

Contents of the two user-created files: (a) file: PYRAMID and (2) file: PYRAMID_objfun.

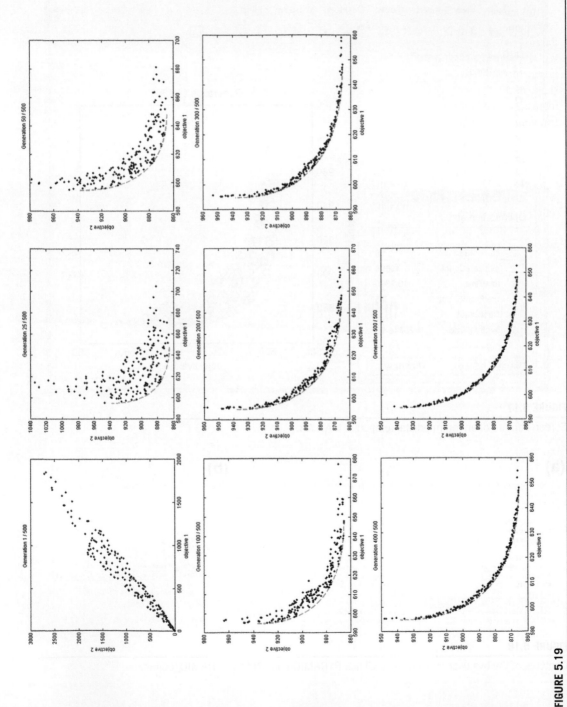

FIGURE 5.19

Result graphs of pyramid example at selected iterations.

in Figure 5.19, the solutions converge rapidly to the Pareto front at the 200th iteration. At the last iteration (500), the Pareto front is well captured.

In general, the genetic algorithms successfully address the limitations of the classical approaches when generating the Pareto front. Because they allow concurrent exploration of different points of the Pareto front, they can generate multiple solutions in a single run. The optimization can be performed without a priori information about objectives relative importance. These techniques can handle ill-posed problems with incommensurable or mixed-type objectives. They are not susceptible to the shape of the Pareto front. Their main drawback is performance degradation as the number of objectives increases. Furthermore, they require additional parameters such as crossover fraction and mutation fraction, which need to be tweaked to improve the performance of the algorithms.

5.4 DECISION-BASED DESIGN

In general, engineering design is viewed as a problem-solving process. Such as in this chapter, a design problem is formulated mathematically as optimization problem and solved by meeting functional requirements subject to constraints at minimum objectives.

As mentioned in Chapter 2, engineering design is increasingly recognized as a decision-making process. From a product development perspective, design involves a series of decisions—some of which may be made sequentially and others that must be made concurrently. The term decision-based design (DBD) was introduced in 1990s. A formal definition introduced in (Hazelrigg 1998) states that decision-based design is a normative approach that prescribes a methodology to make unambiguous design alternative selections under uncertainty and risk wherein the design is optimized in terms of the expected utility.

Uncertainty and risk involved in decision making, as well as two prominent decision theories, utility theory and game theory, were introduced in Chapter 2. In addition, we presented examples to illustrate the application of these theories to engineering design. These examples involve multiple objectives. In this section, we revisit these examples and theories. Although we do not advocate bringing decision-based design into the framework of multiobjective optimization, we compare the methods of MOO and DBD for dealing with multiobjective problems. We provide closure in treating design as a problem-solving and decision-making process by bringing the decision theories discussed in Chapter 2 into the context of multiobjective optimization.

5.4.1 UTILITY THEORY AS A DESIGN TOOL: CANTILEVER BEAM EXAMPLE

In Chapter 2, the cantilever beam example shown in Figure 5.20 was employed to illustrate the application of utility theory as a design tool. Both constrained and nonconstrained problems were solved. In this section, we bring back the results of the constrained problem for discussion.

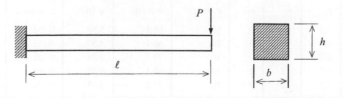

FIGURE 5.20

Cantilever beam example.

The constrained problem involves minimizing weight w and vertical displacement z and subject to bending stress σ. Mathematically, the MOO problem is formulated as

$$\text{Minimize}: \; f_w = w(b, h) = \rho g b h \ell, \; \text{and} \; f_z = z(b, h) = z = \frac{4P\ell^3}{Ebh^3} \tag{5.30a}$$

$$\text{Subject to}: \; g_\sigma = \sigma(b, h) - S_y \le 0 \tag{5.30b}$$

$$10 \le b \le 60 \, \text{mm}, \; \text{and} \; 20 \le h \le 80 \, \text{mm} \tag{5.30c}$$

where $\sigma = \frac{6P\ell}{bh^2}$ and the yield strength is $S_y = 27.6$ MPa.

As discussed in Chapter 2, the MOO problem was converted into a single-objective nonconstrained optimization problem, defined as

$$\text{Maximize}: \; u(w, z, \sigma) = [k_c + (1 - k_c)u(w, z)]u(\sigma) \tag{5.31}$$

in which $u(w, z)$ is the multiattribute utility (MAU) function defined in Eq. 2.51. We assume $0 \le k_w \le 1, 0 \le k_z \le 1$, and $k_w + k_z = 1$ (recall that k_w and k_z are the scaling constants representing the designer's preference between attributes w and z). Hence, the multiplicative MAU is reduced to an additive MAU, as in Eq. 5.32:

$$u(w, z) = k_w u_w + k_z u_z \tag{5.32}$$

which is similar to the weighted-sum method for solving MOO problems.

Also, in Eq. 5.31, the utility function of the stress constraint is defined in Eq. 5.33 as

$$u(\sigma) = \frac{1}{1 + e^{\frac{\sigma_{0.5} - \sigma}{s}}} \tag{5.33}$$

in which s is the slope of the utility function $u(\sigma)$ at $u = 0.5$, and we chose ($\sigma_{0.5} = S_y = 27.6$ MPa and $s = -0.01$ in the example.

We restate the results of the five cases below in Table 5.3.

We are now solving the MOO problem defined in Eq. 5.30. We plot the Pareto front using the generative method similar to that of the pyramid problem (MATLAB Script 8 in Appendix A). The boundary points of the front are identified as $\mathbf{x}_w^* = (10.01, 65.95)$ and $\mathbf{x}_z^* = (59.97, 79.94)$, as shown in Figure 5.21. At the points, the objective functions are $\mathbf{f}(\mathbf{x}_w^*) = (3.497 \, \text{N}, 0.1615 \, \text{mm})$ and $\mathbf{f}(\mathbf{x}_z^*) = (25.39 \, \text{N}, 0.01514 \, \text{mm})$, respectively, as shown in the zoomed-in figures A and B in Figure 5.21.

Table 5.3 Results Comparison for the Constrained Design Problems

Case No.	Problem Setup				Results					
	k_w	k_z	r_w	r_z	b (mm)	h (mm)	w (N)	z (mm)	σ (MPa)	u
Case 4	0.5	0.5	0	0	10	66.1	3.50	0.161	27.46	0.950
Case 5a	0.7	0.3	0	0	10	66.1	3.50	0.161	27.46	0.938
Case 5b	0.3	0.7	0	0	10	71.3	3.78	0.128	23.60	0.962
Case 6a	0.5	0.5	-2	0	10	66.1	3.50	0.161	27.46	0.906
Case 6b	0.5	0.5	2	0	10	72.8	3.85	0.120	22.64	0.977

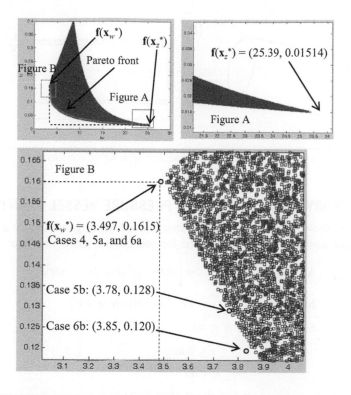

FIGURE 5.21

Pareto solutions of the beam example in criterion space.

The results obtained in Chapter 2 (see Table 5.3) show that the base case (Case 4, $k_w = k_z = 0.5$) is identical to $f(\mathbf{x}_w^*)$, which is a boundary point of the Pareto front shown in Figure 5.21. In Case 5a, k_w increases to 0.7, implying that the preference is given to weight; the design is identical to that of the base case. Adjusting the preference to weight did not alter the design. This is because that there is no more room to further minimize weight without violating the stress constraint. In Case 6a, r_w is reduced to -2, indicating that the designer is risk prone to the weight attribute; the design is identical to that of the base case. Adjusting the risk attitude toward the weight did not alter the design due to the same reason. In Case 5b, k_w decreases to 0.3, implying that the preference is given to displacement; the design moves away from $\mathbf{x}_w^*$, as shown in the zoomed-in figure B of Figure 5.21b. A similar result is found in Case 6b. However, all cases led to designs that are cluster to $\mathbf{x}_w^*$. The results indicate that there is a large portion of the Pareto front not being explored.

As seen in the beam example, the approach of using the utility theory as a design tool converts the MOO into a single-objective nonconstrained problem by the incorporating designer's preference and risk attitude. From a MOO perspective, this approach is similar to methods with a priori articulation of preferences discussed in Section 5.3.2 in the context of conventional MOO solution techniques.

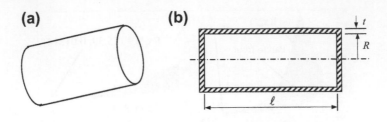

FIGURE 5.22

Cylindrical pressure vessel: (a) isometric view and (b) section view with dimensions.

5.4.2 GAME THEORY AS A DESIGN TOOL: PRESSURE VESSEL EXAMPLE

The pressure vessel shown in Figure 5.22 was investigated using game theory as a design tool in Chapter 2. The same design problem can also be formulated as a MOO problem:

$$\text{Minimize} : W(R,t) = \gamma \left[\pi(R+t)^2(\ell+2t) - \pi R^2 \ell \right] \tag{5.34a}$$

$$\text{Maximize} : V(R,t) = \pi R^2 \ell \, (\text{or Minimize} -V) \tag{5.34b}$$

$$\text{Subject to} : t + R \leq 40 \tag{5.34c}$$

$$0.5 \leq t \leq 6 \, \text{in.} \tag{5.34d}$$

$$5t \leq R \leq 8t \tag{5.34e}$$

$$2.5 < R \leq 5.55 \, \text{in.} \tag{5.34f}$$

We plot the Pareto front using the generative method (MATLAB Script 9 in Appendix A). Note that Eq. 5.34b was converted to minimize $-V$ in the MATLAB implementation. The boundary points of the Pareto front are points A, B, and C shown in Figure 5.23, in which $\mathbf{f}(A) = (f_W(2.5,0.5), f_V(2.5,0.5)) = (252.5, -1963)$, $\mathbf{f}(B) = (f_W(4,0.5), f_V(4,0.5)) = (395.9, -5027)$, and $\mathbf{f}(C) = (f_W(35.55,4.44), f_V(35.55,4.44)) = (42,440, -39,700)$.

Recall the solutions of Section 2.6.2, where we first found that the Nash solution in criterion space is curve BC, which is a subset of the Pareto front. In addition, point B is the solution to the sequential game, with Player W as the leader; point C is the solution to the sequential game with Player V as the leader. Both points are Pareto solutions.

As seen in the pressure vessel example, the approach of using game theory as a design tool converts a MOO problem into a series of single-objective problems by considering the objectives of the individual designers. The single-objective problems are then solved sequentially. The basic concept of handling MOO using game theory is different from those discussed in Section 5.3. In the former, multiple designers are making decisions. In the latter, a MOO problem is solved by a single designer. Practically, on many occasions, design decisions are made by individual groups or designers. In that regard, game theory as design tools may be more general and suitable for solving complex design problems of distributed design teams.

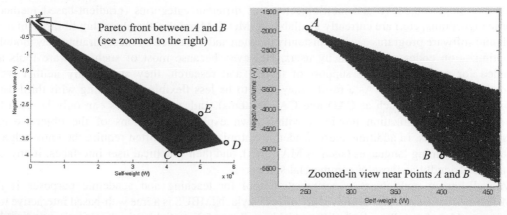

FIGURE 5.23

Pareto solutions of the beam example in criterion space.

5.5 SOFTWARE TOOLS

This section presents a general overview of academically and commercially available multiobjective optimization software. Instead of introducing detailed software capabilities, we offer only brief descriptions for software tools from both categories. For readers who are selecting software for conducting optimization tasks, you are encouraged to carry out further investigations and hands-on evaluations to find a candidate that best suits your needs. In line with the theme of this chapter, multicriteria decision-making software designed for choosing among discrete alternatives although many are readily available, will not be discussed in this section.

5.5.1 ACADEMIC CODES

A number of multiobjective optimization tools developed by universities and research laboratories are available. These software tools are usually compact and free for download from their websites. Most of them are essentially collections of multiobjective optimization algorithms implemented based on different solution techniques. For example, jMetal (jmetal.sourceforge.net) is a Java-based multiobjective optimizer; PISA (www.tik.ee.ethz.ch/pisa) is a C-based framework; interalg (openopt.org/interalg) and PaGMO/PyGMO (sourceforge.net/projects/pagmo) are programmed in Python; BENSLOVE (ito.mathematik.uni-halle.de/~loehne/index_en_dl.php) is implemented in MATLAB; and NIMBUS (wwwnimbus.it.jyu.fi) is a web-based optimization system. The capabilities of these academic tools vary from one to another due to the fact that different optimization algorithms are used. While most of them are general-purpose software, some are developed to deal with a specific group of problems; an example is the linear vector optimizer BENSLOVE, which can be used to solve only MOLP problems.

The advantage of academic codes is that they offer in general more choices of optimization techniques than commercial software packages; for example, more than 30 optimization algorithms (including single-objective and multiobjective) of different categories (gradient-based methods, genetic algorithms, etc.) are currently available in jMetal. Moreover, the algorithm libraries in most academic software programs can be constantly updated and expanded by incorporating new modules or optimization codes contributed by users. However, because most of such software tools are intended for academic use in support of testing and research, they are usually neither well-maintained nor advertised. As a result, they tend to be less flexible in coupling with third-party engineering software (such as CAD and CAE systems), and in most cases can only be used for solving simple optimization problems with known explicit expressions of the objective and constraint functions. In addition, these academic optimization tools often require the knowledge of certain programming languages (such as MATLAB), and their graphical user interfaces, if any, are not as user-friendly as those of commercial software.

A well-known multiobjective optimization tool for teaching and academic purposes is the NIMBUS system, developed by University of Jyväskylä. NIMBUS is a free web-based interactive tool that is capable of handling both differentiable and nondifferentiable single-objective and multi-objective optimization problems subject to nonlinear and linear constraints. The central idea of NIMBUS is to first find one Pareto optimum, and then generate additional solutions on the Pareto front based on the inputs from the user, including which of the objectives should be optimized most and what the limits of the objectives are. In other words, the optimization process is directed by the user. In NIMBUS, this interactive process is called a classification, and the optimal solutions obtained in this way effectively reflect the user's preferences.

The NIMBUS system is intuitive and easy for a novice to use. Because it operates via the Internet, no software needs to be downloaded. A simple tutorial example is available at wwwnimbus.it.jyu.fi/N4/tutorial/index.html. Now, we use the pyramid example to demonstrate the basic capabilities of NIMBUS. As shown in Figure 5.24a, we first enter the expressions of the objective and constraint functions, then identify the bounds of design variables ($x1$, $x2$, $f1$, and $f2$ correspond respectively to the two design variables a and h and the two objectives A and T in the pyramid example). Note that a starting point (initial design) is required (in this case, we choose $a = 15$ and $h = 15$) to initiate the optimization. The first solution obtained is shown in Figure 5.24b ($A = 608.3$, $T = 883.5$), which

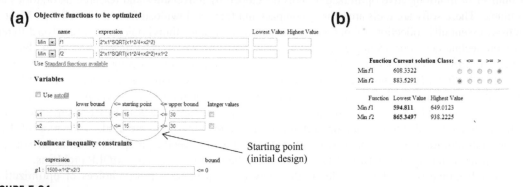

FIGURE 5.24

Screen shots of NIMBUS for the pyramid example. (a) Defining the optimization problem. (b) Interface of classification with the first Pareto solution and the minimum of individual objectives listed.

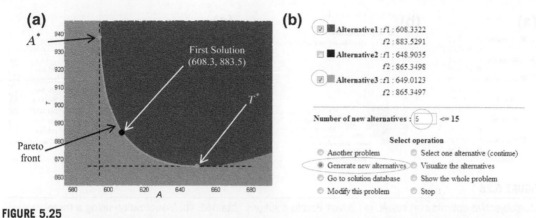

FIGURE 5.25

Sample windows of NIMBUS software. (a) Pareto front of the pyramid example with the first solution obtained. (b) New solutions generated based on the first classification iteration.

represents the black dot on the Pareto front (Figure 5.25a). The points A^* (594.8, 938.2) and T^* (649.0, 865.3) corresponding to the minimum of individual objectives are also calculated automatically, and are listed under the first solution, as shown in Figure 5.24b.

The next step is classification. As can be seen in Figure 5.24b, a group of buttons to the right of the first solution allow us to choose the "class" for each objective function. In this case, we choose ">" for $f1$ (objective function A) and "<" for $f2$ (objective function T), which implies our preference—further minimizing $f2$ based on the current solution while allowing $f1$ to change freely. Alternatively, if, for example, "$\geq$" is chosen for $f1$, then a value needs to be assigned as the maximum allowed value for $f1$ when minimizing $f2$. The two new solutions (alternatives 2 and 3) generated according to the classification information are shown in Figure 5.25b. Note that $f2$ has been minimized to its lower bound in alternative 3. We then choose alternatives 1 and 3 to create five more Pareto solutions between the two (Figure 5.25b), and the results are listed in Figure 5.26a as alternatives 2 to 6.

A number of solution visualization methods are available in NIMBUS, two of which are demonstrated in Figures 5.26b and c. In Figure 5.26c, each line represents a solution on the Pareto front (the seven solutions so far) between the first solution (alternative 1) and point T^* (alternative 7). We can certainly continue the optimization process to obtain additional Pareto solutions.

Note that for nonacademic users, an implementation of NIMBUS that operates under several operating systems (such as Linux and Windows) is also available, known as IND-NIMBUS (ind-nimbus.it.jyu.fi). IND-NIMBUS has been tested with several industrial problems and can be connected with different simulators or modeling tools, such as MATLAB.

5.5.2 COMMERCIAL TOOLS

There are several general-purpose multiobjective optimization software packages currently available in the commercial sector. In general, commercial optimization software tools provide intuitive user interfaces with shorter learning curves than academic codes. These software programs share similar

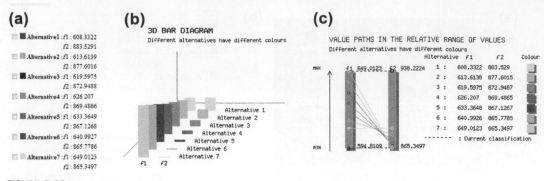

FIGURE 5.26

Multiobjective optimization result. (a) Seven Pareto solutions obtained. (b) Visualization using a three-dimensional bar diagram. (c) Visualization using value paths.

major characteristics, while some of them also offer unique capabilities. Examples of such software include the following:

- modeFRONTIER®, developed by ESTECO (www.esteco.com/modefrontier)
- OPTIMUS®, developed by Noesis Solutions (www.noesissolutions.com/Noesis/about-optimus)
- optiSLang®, developed by Dynardo (www.dynardo.de/en/software/optislang.html)
- BOSS Quattro®, developed by LMS (www.lmsintl.com/samtech-boss-quattro)
- Isight®, developed by Dassault Systèmes (www.modelon.com/products/isight)

One of the key advantages of commercial codes is their ability to seamlessly couple with multiple third-party (commercial and in-house) engineering tools. Unlike the academic codes, these commercial software packages are designed to work with a wide variety of modeling and simulation software programs. Each of them can be thought of as an integration platform that enables the automation of a complex design simulation, and in the meantime offers optimization capabilities to support decision making. For example, the workflow environment in modeFRONTIER® is able to formalize, drive, and manage individual steps, such as CAD modeling, computation fluid dynamics (CFD), finite element analysis (FEA), and so on, composing an engineering process through the integration of the most popular engineering solvers, while the data can be transferred automatically from one simulation to the next. Such communication between software is usually guaranteed by application protocol interfaces or automatic file exchange. Figure 5.27 shows a screen shot of the Workflow Editor in modeFRONTIER®, and some of the third-party software supported in modeFRONTIER®, including CAD, CAE, and generic applications, are listed in Table 5.4. Additional wizard style tools are also available in modeFRONTIER® to achieve the coupling with other commercial tools or in-house codes. Some commercial optimization software, such as OPTIMUS® and BOSS Quattro®, also provide native interfaces with major CAE and FEA systems, broadening the basic capabilities of those engineering tools.

In addition to a strong connection with third-party applications, each of the commercial software packages mentioned above also bundles a collection of advanced optimization techniques for both

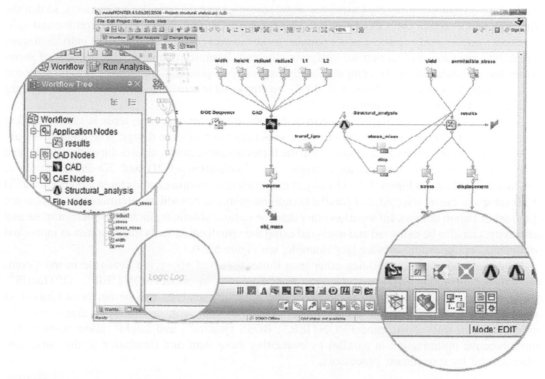

FIGURE 5.27

Workflow Editor in modeFRONTIER® (www.esteco.com/modefrontier/modefrontier-platform-overview).

Table 5.4 Examples of Third-Party Applications Supported in modeFRONTIER®	
CAD	SolidWorks, CATIA, Creo, NX, Spaceclaim
CAE	ANSYS Workbench, ABAQUS, Adams, Virtual.Lab, Image.Lab, ANSA
Generic applications	Excel, OpenOffice, SCIBAL, MATLAB, LABVIEW, MATHCAD

single and multiobjective problems, ranging from gradient-based methods to genetic algorithms. For example, the multiobjective algorithms available in modeFRONTIER® include multiobjective genetic algorithm (MOGA), adaptive range MOGA, multiobjective simulated annealing (MOSA), nondominated sorting genetic algorithm (NSGA II), multiobjective game theory, evolutionary strategies methodologies, and normal boundary intersection. Moreover, in modeFRONTIER®, different algorithms can even be combined by the user in order to obtain hybrid approaches. For example, it is possible to combine the robustness of a genetic algorithm together with the accuracy of a gradient-based method, using the former for initial screening and the latter for refinements, so that the efficiency of the design cycle can be further enhanced. During optimization, these commercial software codes continuously update product parameters and extract relevant outputs from individual steps within the workflow until an optimal design is obtained. Because running a large number of simulations for optimization can be computationally expensive in practical engineering design, response surface methods (Jones 2001; Poles et al. 2008) are employed in most commercial software packages to accelerate the optimization process.

Another advantage of commercial optimization software over academic tools is that they often provide interactive result viewers to help users select the most suitable design. The alternatives obtained during optimization can be displayed in the forms of tables, charts, or two-dimensional (2D) and three-dimensional (3D) plots. As an example, the visualization of 2D and 3D Pareto fronts in OPTIMUS® is shown in Figure 5.28. Usually, if the problem definition exceeds three dimensions, a 2D or 3D subspace can be selected, and parallel coordinate plots can be used to determine best designs out of the set of Pareto designs. Information other than the optimal solutions, such as the design space and sensitivity, can also be extracted and analyzed using the visualization tools incorporated in individual commercial optimization software (for example, see Figure 5.29).

Note that additional capabilities other than those discussed above are available in most commercial optimization software packages. For example, in modeFRONTIER®, OPTIMUS®, optiSLang®, and Isight®, robustness analysis and reliability analysis can be performed based on statistical methods, such as the Monte Carlo method, to deal with uncertainties that impact the optimization process. Also, modeFRONTIER®, BOSS Quattro®, and Isight® allow users to run multiobjective optimization in parallel by evaluating more than one simulation at the same time using several local or remote processors.

(a)

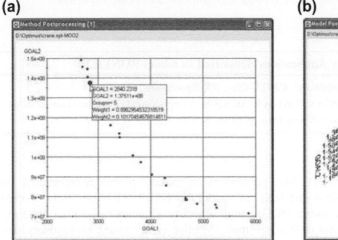

(b)

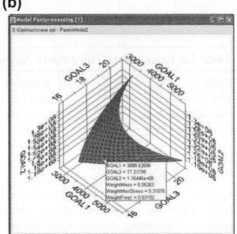

FIGURE 5.28

Optimization result visualization in OPTIMUS®. (a) A scatter plot showing points on the Pareto optimal set for two objectives. (b) A 3D plot of the Pareto optimal set (Poles et al. 2008).

(a)

(b)

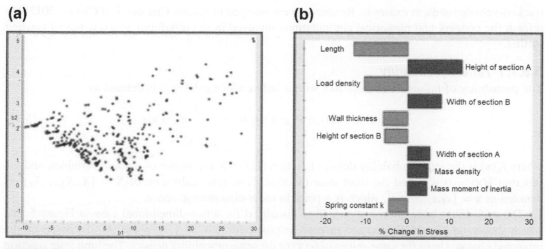

FIGURE 5.29

Optimization result visualization in Isight™. (a) A scatter plot of the design space. (b) Parameter effects on the performance measure (www.modelon.com/products/isight/).

5.6 **ADVANCED TOPICS**

Before closing out this chapter, we include two advanced topics that are relevant to design optimization. The first topic is reliability-based design optimization (RBDO), which incorporates variations in physical parameters and design variables into design optimization, leading to design with less probability of failure. Second, we introduce a case of design optimization that takes product cost, including manufacturing, as an objective function. This case represents a more realistic design optimization problem that leads to less expensive designs.

5.6.1 **RELIABILITY-BASED DESIGN OPTIMIZATION**

In Chapter 5 of Chang (2013a), we discussed reliability analysis. We introduced the concept of failure modes associated with certain critical product performance. We incorporated the variability (or uncertainty) of physical parameters or a manufacturing process that affects the performance (hence, failure modes) of the product to estimate a failure probability. We used a beam example for illustration, in which the failure probability of a stress failure mode predicts the percentage of the incidents when the maximum stress of the beam exceeds its material yield strength. The result offered by the reliability analysis is far more precise and effective than that of the safety factor approach.

In this subsection, we move one step further to discuss design optimization by incorporating failure modes (and failure probabilities) into optimization problem formulation. We first briefly review the basics of the reliability analysis, in particular, the most probable point (MPP) search for failure probability estimates. We then formulate the mathematic equations for a standard RBDO and its solution technique, and we present a sample case to illustrate the topic of RBDO using a tracked-vehicle roadarm example. Readers are encouraged to review Chapter 5 of Chang (2013a) to refresh the concept and numerical computations involved in the reliability analysis before reading further.

5.6.1.1 Failure probability

The probability of failure P_f of a product with a failure mode $g(\mathbf{X}) \leq 0$ is defined as

$$P_f = P(M \leq 0) = P(g(\mathbf{X}) \leq 0) = \int_{g(\mathbf{x}) \leq 0} \mathbf{f_X}(\mathbf{x})d\mathbf{x} \tag{5.35}$$

where $f_\mathbf{X}(\mathbf{x})$ is the joint probability density function (PDF), $\mathbf{X}$ is a vector of random variables, and the function $g(\mathbf{x}) = 0$ is called the limit state function. Note that realization of $\mathbf{X} = [X_1, X_2, \ldots, X_n]^T$ is denoted as $\mathbf{x} = [x_1, x_2, \ldots, x_n]^T$, which is a point in the n-dimensional space.

The probability integration in Eq. 5.35 is visualized for a two-dimensional case in Figure 5.30a, which shows the joint PDF $f_\mathbf{X}(\mathbf{x})$ and its contour projected onto the x_1-x_2 plane. All the points on the projected contours have the same values of $f_\mathbf{X}(\mathbf{x})$ or the same probability density. The limit state function $g(\mathbf{x}) = 0$ is also shown. The failure probability P_f is the volume underneath the surface of the joint PDF $f_\mathbf{X}(\mathbf{x})$ in the failure region $g(\mathbf{x}) \leq 0$. To show the integration more clearly, the contours of the joint PDF $f_\mathbf{X}(\mathbf{x})$ and the limit state function $g(\mathbf{x}) = 0$ are plotted on the x_1-x_2 plane, as shown in Figure 5.30b.

The direct evaluation of the probability integration of Eq. 5.35 is very difficult if not impossible. A number of methods have been developed. Monte Carlo simulation is simple and easy to

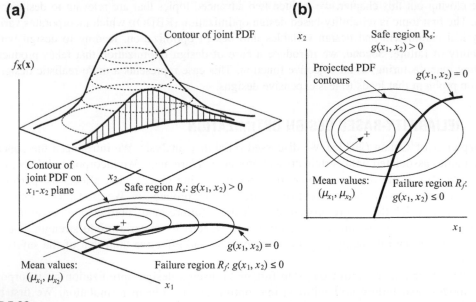

FIGURE 5.30

Probability integration illustrated using a two-dimensional example. (a) Isometric view. (b) Projected view on the x_1-x_2 plane.

implement. However, this brute-force approach requires tens of thousands analyses or more, which is impractical for engineering problems that often require significant computational effort. One of the widely employed methods to alleviate the computational issue to some extent, while still offering a sufficiently accurate estimate on the failure probability, is the first-order reliability method (FORM). This method aims at providing an acceptable estimate for the integral form of the failure probability P_f defined in Eq. 5.35, which can be achieved by two important steps. The first step involves the simplification of the joint probability density function $f_{\mathbf{X}}(\mathbf{x})$. The simplification is accomplished by transforming a given joint probability density function, which may be multidimensional, to a standard normal distribution function of independent random variables of the same dimensions. The second step approximates the limit state function $g(\mathbf{x}) = 0$ by a Taylor's series expansion and keeps up to the first-order terms for approximation. Note that if both linear and quadratic terms are included in the calculation, the method is called the second-order reliability method (SORM).

The space that contains the given set of random variables $X = [X_1, X_2, \ldots, X_n]^{\mathrm{T}}$ is called the X-space. These random variables are transformed to a standard normal space (U-space), where the transformed random variables $U = [U_1, U_2, \ldots, U_n]^{\mathrm{T}}$ follow the standard normal distribution (i.e., with mean value $\mu = 0$ and standard deviation $\sigma = 1$). Such a transformation is carried out based on the condition that the cumulative density functions (CDFs) of the random variables remain the same before and after transformation; that is,

$$F_{X_i}(x_i) = \Phi(u_i) \tag{5.36}$$

Here, $\Phi(\cdot)$ is the CDF of the standard normal distribution. The transformation can be written as

$$u_i = \Phi^{-1}(F_{X_i}(x_i)) \tag{5.37a}$$

Hence, the transformed random variable U_i can be written as

$$U_i = \Phi^{-1}(F_{X_i}(X_i)) \tag{5.37b}$$

If the random variables X_i are independent and normally distributed—that is, $F_{X_i}(x_i) = \Phi\left(\frac{x_i - \mu_i}{\sigma_i}\right)$—the transformation, as illustrated in Figure 5.31, can be obtained as

$$u_i = \Phi^{-1}\left(\Phi\left(\frac{x_i - \mu_i}{\sigma_i}\right)\right) = \frac{x_i - \mu_i}{\sigma_i} \tag{5.38}$$

Hence,

$$x_i = \mu_i + \sigma_i u_i \tag{5.39}$$

Note that, as shown in Figure 5.31c, the projected contours of the transformed PDF on the u_1-u_2 plane are circles centered at the origin.

The limit state function is also transformed into the U-space, as

$$g(\mathbf{x}) = g_u(\mathbf{u}) = 0 \tag{5.40}$$

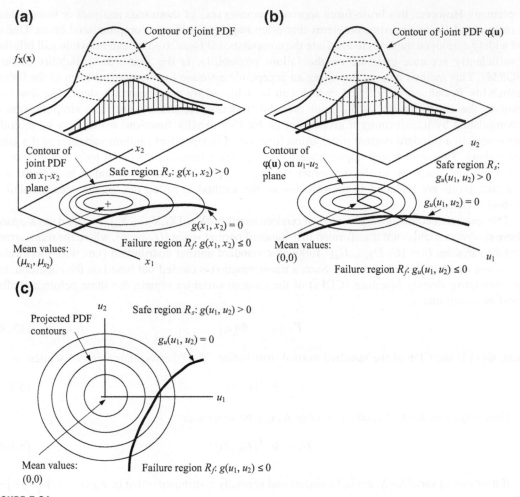

(a)
Contour of joint PDF
$f_X(x)$

Contour of
joint PDF
on x_1-x_2
plane

Safe region R_s: $g(x_1, x_2) > 0$

$g(x_1, x_2) = 0$

Failure region R_f: $g(x_1, x_2) \le 0$

Mean values:
(μ_{x_1}, μ_{x_2})

x_2

x_1

(b)
Contour of joint PDF $\varphi(\mathbf{u})$

Contour of
$\varphi(\mathbf{u})$ on u_1-u_2
plane

Safe region R_s:
$g_u(u_1, u_2) > 0$

$g_u(u_1, u_2) = 0$

u_2

Mean values:
$(0,0)$

u_1

Failure region R_f: $g_u(u_1, u_2) \le 0$

(c)
u_2

Safe region R_s: $g(u_1, u_2) > 0$

Projected PDF
contours

$g_u(u_1, u_2) = 0$

u_1

Mean values:
$(0,0)$

Failure region R_f: $g(u_1, u_2) \le 0$

FIGURE 5.31

Transformation of random variables from **X** to **U**, illustrated using a two-dimensional example. (a) PDF and projected contours in X-space. (b) Standard normal distribution in U-space. (c) Projected view on the u_1-u_2 plane.

The transformed limit state function separates the U-space into safe and failure regions, as illustrated in Figure 5.31b and c. After the transformation, the probability integration of Eq. 5.35 becomes

$$P_f = P(g_u(\mathbf{u}) \le 0) = \int_{g_u(\mathbf{u}) \le 0} \phi(\mathbf{u})d\mathbf{u} \tag{5.41}$$

Because all of the random variables **U** are independent, the joint PDF is the product of the individual PDFs of standard normal distribution:

$$\phi(\mathbf{u}) = \prod_{i=1}^{n} \frac{1}{\sqrt{2\pi}} e^{-\frac{1}{2}u_i^2} \tag{5.42}$$

Therefore, the probability integration becomes

$$P_f = \iint \cdots \int_{g_u(\mathbf{u}) \leq 0} \prod_{i=1}^{n} \frac{1}{\sqrt{2\pi}} e^{-\frac{1}{2}u_i^2} du_1 du_2 \ldots du_n \tag{5.43}$$

Although it is obvious that Eq. 5.43 is relatively easier to calculate than Eq. 5.35, calculating Eq. 5.43 is still difficult because the limit state function $g_u(\mathbf{u})$ is in general a nonlinear function of variables **u**.

If the nonlinearity of the limit state function is not too severe, the integration of Eq. 5.43 can be approximated by

$$P_f \approx \int_{-\infty}^{-\beta} \frac{1}{\sqrt{2\pi}} e^{-\frac{1}{2}u^2} du = \Phi(-\beta) \tag{5.44}$$

where Φ is the standard normal distribution function of a single dimension, and β is the shortest distance between the origin and the point on the limit state function $g_u(\mathbf{u}) = 0$. This point is depicted as **u*** in Figure 5.32 and is called the β-point, design point, or the most probable point. The β value is called reliability index.

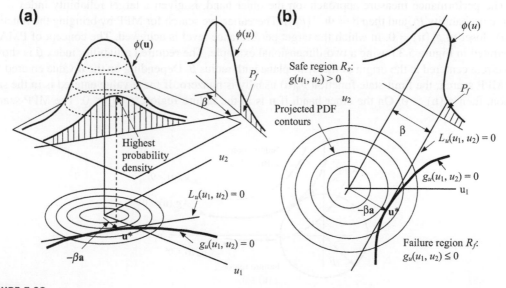

FIGURE 5.32

Approximation of failure probability integration, illustrated using a two-dimensional example. (a) PDF and projected contours in the U-space. (b) Projected view on the u_1-u_2 plane.

As illustrated in Figure 5.32a, the line that connects the origin and the MPP must be perpendicular to the tangent line $L_u(u_1,u_2) = 0$ at the MPP. Because the joint PDF $\phi(\mathbf{u})$ of the multidimension is axisymmetric, its projection on the plane that is normal to the tangent line is nothing but a probability density function of the standard normal distribution $\phi(u)$ of a single dimension. The probability integration of Eq. 5.35 can then be approximated by $\Phi(-\beta)$, as stated in Eq. 5.44.

As discussed above, the key idea in calculating the failure probability using FORM or SORM is to locate the MPP in the U-space. Many numerical approaches have been developed for the MPP search. These methods can be categorized in two major categories: the reliability index approach (RIA) and the performance measure approach (PMA). The reliability index approach employs a forward reliability analysis algorithm that computes failure probability for a prescribed performance level in the limit state function. The performance measure approach employs an inverse reliability analysis algorithm that computes response level for a prescribed failure probability. We briefly state both approaches. For more details, readers are referred to Chang (2013a).

The problem for MPP search using RIA can be formulated as follows:

$$\begin{aligned} \text{Minimize :} \quad & \|\mathbf{u}\| \\ \text{Subject to :} \quad & g_u(\mathbf{u}) = 0 \end{aligned} \tag{5.45}$$

in which the MPP is identified by searching a point on the limit state function $g_u(\mathbf{u}) = 0$, where the distance between the point to the origin of the U-space is minimum. Again, the distance β is called the reliability index, hence the name of this approach. Note that in Eq. 5.45, the performance level of the limit state function is prescribed. In this case, MPP can only be searched for the performance level prescribed. Once the MPP is found, the distance β (reliability index) can be used to approximate the failure probability as $P_f = \Phi(-\beta)$.

The performance measure approach, on the other hand, is given a target reliability index β (or failure probability P_f, and then $\beta = \Phi^{-1}(P_f)$). Thereafter, we search for MPP by bringing the function $g_u(\mathbf{u})$ closer to $g_u(\mathbf{u}) = 0$, in which the target performance level is achieved. The concept of PMA is illustrated in Figure 5.33 using a two-dimensional example. The required reliability index β is shown as a circle centered at the origin of the u_1-u_2 plane with radius β. Depending on the $\mathbf{u}$ value entered for the MPP search, the limit state function $g_u(\mathbf{u})$ usually is nonzero. If the $\mathbf{u}$ value entered is in the safe region, then $g_u(\mathbf{u}) > 0$. On the other hand, if $\mathbf{u}$ is in the failure region, $g_u(\mathbf{u}) < 0$. The MPP search

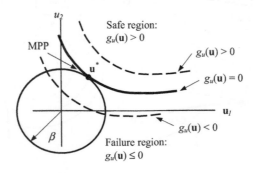

FIGURE 5.33

MPP search using the performance measure approach.

becomes finding $\mathbf{u}$ that brings the limit state function $g_u(\mathbf{u})$ to $g_u(\mathbf{u}) = 0$. Therefore, the problem for MPP search using PMA can be formulated as follows:

$$\text{Minimize}: \quad |g_u(\mathbf{u})|$$
$$\text{Subject to}: \quad \beta = \|\mathbf{u}\| \tag{5.46}$$

in which the reliability index $\beta = \Phi^{-1}(P_f)$, and P_f is the required failure probability. Also, in Eq. 5.46, the limit state function $g_u(\mathbf{u})$ can be rewritten as

$$g_u(\mathbf{u}) = g^t(\mathbf{u}) - g^t \tag{5.47}$$

where $g^t(\mathbf{u})$ is the performance measure corresponding to the failure mode, and g^t is the target performance level.

Next, we formulate the standard reliability-based design optimization problems, in which we assume the RIA for the MPP search.

5.6.1.2 RBDO problem formulation

The classical design optimization problem based on deterministic analysis is typically formulated as a nonlinear constrained optimization problem, as discussed in Chapter 3. Similarly, the RBDO problem can also be formulated as a nonlinear constrained optimization problem where reliability measures are included as constraint functions. Probabilistic constraints in RBDO ensure a more evenly distributed failure probability in the component of a product. In general, the RBDO model contains two types of design variables: distributional design variable $\boldsymbol{\theta}$ and conventional deterministic design variable $\mathbf{b}$.

Let $\boldsymbol{\theta} = [\theta_1, \theta_2, ..., \theta_{n1}]^T$ and $\mathbf{b} = [b_1, b_2, ..., b_{n2}]^T$ be the distributional and deterministic design variable vectors of dimensions n_1 and n_2, respectively. The RBDO problem can be formulated as follows:

$$\text{Minimize}: \quad f(\boldsymbol{\theta}, \mathbf{b}) \tag{5.48a}$$
$$\text{Subject to}: P_{f_i} = P\big(g_i(\boldsymbol{\theta}, \mathbf{b}) \le 0\big) - P_i^u \le 0, \quad i = 1, m \tag{5.48b}$$
$$\theta_j^\ell \le \theta_j \le \theta_j^u, \quad j = 1, n_1 \tag{5.48c}$$
$$b_k^\ell \le b_k \le b_k^u, \quad k = 1, n_2 \tag{5.48d}$$

where $f(\boldsymbol{\theta}, \mathbf{b})$ is the objective function, $P(\bullet)$ denotes the probability of the event $(\bullet)$, and P_i^u is the required upper bound of the probability of failure for the ith constraint function P_{f_i}. In Eqs (5.48c) and (5.48d), θ_j^ℓ and θ_j^u, and b_k^ℓ and b_k^u, are the lower and upper bounds of the jth distributional and kth deterministic design variables, respectively.

The reliability constraints defined in Eq. 5.48b are assumed to be independent and thus no correlation exists. As mentioned, it is almost impossible to calculate the probability of failure P_{f_i} in Eq. 5.48b by a multiple integration for general design applications. Consequently, the FORM or other more efficient reliability analysis methods are employed. The computational flow of RBDO using the FORM is illustrated in Figure 5.34. Note that at each design iteration, the FORM needs to be carried out several times for individual failure functions. As formulated in Eq. 5.45, each FORM is equivalent to a deterministic optimization, which is very computationally demanding. This is the reason why RBDO is mostly limited to academic problems. Furthermore, the first-order derivative of the failure

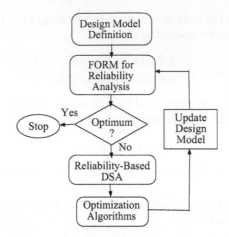

FIGURE 5.34

Computation flow for gradient-based RBDO.

probability with respect to both distributional and deterministic design variables must be computed to support gradient-based RBDO.

5.6.1.2.1 Reliability-based design sensitivity analysis

The sensitivity of the failure probability includes two parts: the sensitivity of the failure probability with respect to the distributional parameters $\boldsymbol{\theta}$ of random variables (e.g., mean value, standard deviation), and the sensitivity of the failure probability with respect to the deterministic design variables $\mathbf{b}$.

The derivative of the estimated failure probability P_f obtained using the FORM with respect to a design variable η, which can be either θ_j or b_k, is

$$\frac{\partial P_f}{\partial \eta} = \frac{\partial \Phi(-\beta)}{\partial \eta} = \frac{\partial \Phi(-\beta)}{\partial \beta}\frac{\partial \beta}{\partial \eta} = -\Phi(-\beta)\frac{\partial \beta}{\partial \eta} \tag{5.49}$$

where Φ is the standard normal density function. Therefore, to compute the sensitivity of the failure probability P_f, $\partial\beta/\partial\eta$ must be computed as

$$\frac{\partial \beta}{\partial \eta} = \frac{\partial \left(\mathbf{U}^{*\mathrm{T}}\mathbf{U}^*\right)^{1/2}}{\partial \eta} = \frac{1}{\beta}\mathbf{U}^{*\mathrm{T}}\frac{\partial \mathbf{U}^*}{\partial \eta} \tag{5.50}$$

where $\mathbf{U}^*$ is the MPP found in the U-space.

The sensitivity of the reliability index with respect to a distributional design variable θ_j can be obtained by substituting $\eta = \theta_j$ and $\mathbf{U}^* = \mathbf{T}(\mathbf{X}^*, \boldsymbol{\theta})$ in Eq. 5.50 as

$$\frac{\partial \beta}{\partial \theta_j} = \frac{1}{\beta}\mathbf{U}^{*\mathrm{T}}\left(\frac{\partial \mathbf{T}(\mathbf{X}^*, \boldsymbol{\theta})}{\partial \theta_j} + \frac{\partial \mathbf{T}(\mathbf{X}^*, \boldsymbol{\theta})}{\partial \mathbf{X}^*}\frac{\partial \mathbf{X}^*}{\partial \theta_j}\right) = \frac{1}{\beta}\mathbf{U}^{*\mathrm{T}}\frac{\partial \mathbf{T}(\mathbf{X}^*, \boldsymbol{\theta})}{\partial \theta_j} \tag{5.51}$$

For the normally distributed random variables, where $\mathbf{T}$ can be explicitly written as a transformation function of $\boldsymbol{\theta}$, Eq. 5.51 can be calculated analytically. For the non-normally distributed random variables, the transformation **T** cannot be obtained explicitly. In such a case, the finite difference method can be used to approximate the derivative of **T** with respect to θ_j.

As discussed, the reliability index β is the distance between the origin and the MPP in the U-space. The MPP vector **U*** on the failure surface can be written as

$$\mathbf{U}^* = -\beta \frac{\nabla g(\mathbf{U}^*, \mathbf{b})}{|\nabla g(\mathbf{U}^*, \mathbf{b})|} \tag{5.52}$$

in which, $\nabla g(\mathbf{U}^*, \mathbf{b})$ is the gradient of the failure function at the MPP:

$$\nabla g(\mathbf{U}^*, \mathbf{b}) = \frac{\partial g(\mathbf{U}^*, \mathbf{b})}{\partial \mathbf{U}} \tag{5.53}$$

From Eq. 5.52, the MPP vector **U*** is also a function of **b** because the failure function g depends on the deterministic design variables **b**. Substituting Eq. 5.52 into Eq. 5.50 yields

$$\frac{\partial \beta}{\partial \mathbf{b}} = -\frac{1}{|\nabla g(\mathbf{U}^*, \mathbf{b})|} \frac{\partial g^T(\mathbf{U}^*, \mathbf{b})}{\partial \mathbf{U}} \frac{\partial \mathbf{U}^*}{\partial \mathbf{b}} \tag{5.54}$$

By taking the derivative of $g(\mathbf{U}^*, \mathbf{b}) = 0$ with respect to **b**,

$$\frac{\partial g^T(\mathbf{U}^*, \mathbf{b})}{\partial \mathbf{U}} \frac{\partial \mathbf{U}^*}{\partial \mathbf{b}} + \frac{\partial g(\mathbf{U}^*, \mathbf{b})}{\partial \mathbf{b}} = 0 \tag{5.55}$$

Substituting Eq. 5.55 into Eq. 5.54 yields

$$\frac{\partial \beta}{\partial \mathbf{b}} = \frac{1}{|\nabla g(\mathbf{U}^*, \mathbf{b})|} \frac{\partial g^T(\mathbf{U}^*, \mathbf{b})}{\partial \mathbf{b}} \tag{5.56}$$

Note that evaluation of Eq. 5.56 needs only the first-order derivative of the failure function with respect to deterministic design variables.

To avoid the prohibitively expensive computational efforts required for a large number of reliability analyses during a batch-mode RBDO for design optimization, a mixed-design approach (Yu et al. 1997) that includes deterministic design optimization in batch mode and reliability-based design in an integrated mode, as discussed in Chapter 3, is employed. The mixed design approach starts with a deterministic design optimization, in which performance measures employed as failure modes are defined as constraint functions. After an optimal design is obtained, a reliability analysis is performed to ascertain if the deterministic optimal design is reliable. If the probability of the failure of the deterministic optimal design is found to be unacceptable, a reliability-based design approach that employs a set of interactive design steps, such as trade-off analysis and what-if study, is used to obtain a near-optimal design that is reliable with an affordable computational cost. A tracked-vehicle roadarm is employed to illustrate the approach next.

5.6.1.3 RBDO for a tracked-vehicle roadarm

A roadarm of the military tracked-vehicle shown in Figure 3.26a is employed to illustrate the mixed design approach for the RBDO. A deterministic design optimization is presented first. Then, reliability analysis using the FORM is discussed. The reliability-based design obtained using the interactive design process follows.

5.6.1.3.1 Deterministic design optimization

A 17-body dynamic simulation model (Chang 2013a) is created to drive the tracked-vehicle on a proving ground, at a constant speed of 20 miles per hour. A 20-s dynamic simulation is performed at a maximum integration time step of 0.05 s using dynamic simulation and design system or DADS (Haug and Smith 1990). The joint reaction forces applied at the wheel end of the roadarm, accelerations, angular velocities, and angular accelerations of the roadarm are obtained from the dynamic simulation. Four beam elements, STIF4, and 310 20-node isoparametric finite elements, STIF95, of ANSYS are used for the roadarm finite element model shown in Figure 5.35. The roadarm is made of S4340 steel and the length between the centers of the two holes is 20 in.

The fatigue life fringe plot is shown in Figure 5.36. At the initial design, the structural volume is 486.7 in.^{3}. The crack initiation lives at 24 critical points (with node IDs shown in Figure 5.36) are defined as the constraints with a lower bound of 9.63×106 blocks (20 s per block). Note that the lower bound defined is equivalent to 20 years service life, assuming the tracked-vehicle is operated 8 h per

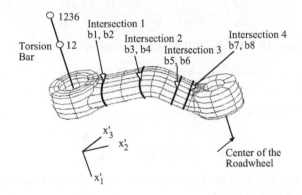

FIGURE 5.35

Roadarm finite element model.

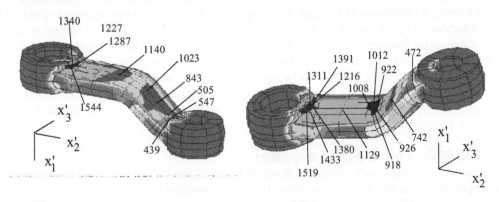

FIGURE 5.36

Fringe plots of crack initiation life.

Table 5.5 Objective and Critical Constraint Functions

Function	Description	Lower Bound	Current Design	Status
Objective	Volume		487.678 in.^{3}	
Constraint 1	Node 1216	9.63×10^6 (20 years)	9.631×10^6 bks	Active
Constraint 2	Node 926	9.63×10^6 (20 years)	8.309×10^7 bks	Inactive
Constraint 3	Node 1544	9.63×10^6 (20 years)	8.926×10^7 bks	Inactive
Constraint 4	Node 1519	9.63×10^6 (20 years)	1.447×10^8 bks	Inactive
Constraint 5	Node 1433	9.63×10^6 (20 years)	2.762×10^8 bks	Inactive

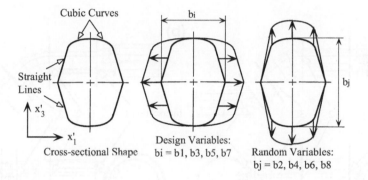

FIGURE 5.37

Design variable definition.

day, 5 days per week. Definitions of the objective function and five critical constraint functions are listed in Table 5.5.

For shape design parameterization, eight design variables are defined to characterize the geometric shapes of the four intersections, as shown in Figure 5.37. The profile of the crosssection shape is composed of four straight lines and four cubic curves. Side expansions (x_1' – direction) of cross-sectional shapes are defined using design variables b1, b3, b5, and b7 for intersections 1 to 4, respectively. Vertical expansions (x_3' – direction) of the cross-sectional shapes are defined using the remaining four design variables.

A deterministic optimal design is obtained in six design iterations using the modified feasible direction method in the design optimization tool (DOT). As shown in Table 5.6, at the deterministic optimal design, all fatigue lives are greater than the lower bound and objective function is reduced by 10.5%. The geometric shapes of the roadarm at initial and deterministic optimal designs are shown in Figure 5.38.

5.6.1.3.2 Probabilistic fatigue life predictions

The random variables and their statistical values for the crack initiation life predictions are listed in Table 5.7, including material and tolerance random variables. The eight tolerance

Table 5.6 Objective and Critical Constraint Function Values at Initial and Deterministic Optimal Designs

Function	Description	Initial Design	Deterministic Optimal Design	Changes
Objective	Volume	487.678 in.^{3}	436.722 in.^{3}	−10.5%
Constraint 1	Node 1216	9.631×10^6 bks	7.704×10^7 bks	699.9%
Constraint 2	Node 926	8.309×10^7 bks	9.631×10^6 bks	−88.4%
Constraint 3	Node 1544	8.926×10^7 bks	9.678×10^6 bks	−89.2%
Constraint 4	Node 1519	1.447×10^8 bks	4.698×10^7 bks	−67.5%
Constraint 5	Node 1433	2.762×10^8 bks	4.815×10^8 bks	74.3%

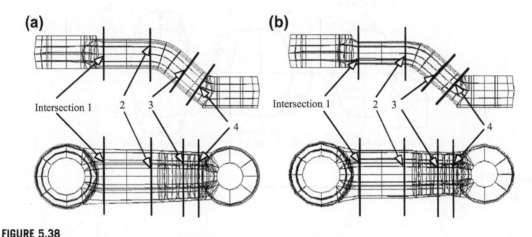

FIGURE 5.38

Geometric shape of the roadarm in front and top views. (a) Initial design. (b) Deterministic optimal design.

random variables b1 to b8 are defined corresponding to the eight shape design variables defined in Figure 5.37.

FORM is used to calculate the reliability of the crack initiation life at five critical points. The results shown in Table 5.8 indicate that the failure probability at nodes 926 and 1544 is greater than 3%. Because the failure probability of the roadarm at the deterministic optimal design is too high, a reliability-based design must be conducted to reduce the failure probability (to obtain a feasible design).

5.6.1.3.3 Reliability-based design

For the reliability-based design, the mean values of the eight shape parameters shown in Figure 5.37 are chosen as the design variables. The objective function is still the structural volume. The constraint functions are the failure probability of the fatigue life at the five critical points, with an upper bound of 1% (i.e., the required reliability of fatigue life larger than 20 years is 99%). Table 5.8 shows that the

Table 5.7 Definition of Random Variables

Random Variables	Mean Value	Standard Deviation	Distribution
Young's modulus E (psi)	30.0×10^6	0.75×10^6	LogNormal
Fatigue strength coefficient σ'_f	1.77×10^5	0.885×10^4	LogNormal
Fatigue ductility coefficient ε'_f	0.41	0.0205	LogNormal
Fatigue strength exponent b	−0.07300	0.00365	Normal
Fatigue ductility exponent c	−0.6	0.003	Normal
Tolerance b1 (in.)	2.889	0.032450	Normal
Tolerance b2 (in.)	1.583	0.019675	Normal
Tolerance b3 (in.)	2.911	0.031703	Normal
Tolerance b4 (in.)	1.637	0.019675	Normal
Tolerance b5 (in.)	2.870	0.031703	Normal
Tolerance b6 (in.)	2.420	0.026352	Normal
Tolerance b7 (in.)	2.801	0.032496	Normal
Tolerance b8 (in.)	4.700	0.050568	Normal

Table 5.8 Objective and Failure Function Values at Deterministic Optimal and Improved Designs

Function	Description	Deterministic Optimal Design	Improved Design (2 RBSOs)	Changes
Objective	Volume	436.722 in.^{3}	447.691 in.^{3}	2.5%
Constraint 1	Node 1216	0.476%	0.532%	
Constraint 2	Node 926	3.24%	0.992%	
Constraint 3	Node 1544	3.21%	0.998%	
Constraint 4	Node 1519	0.83%	0.721%	
Constraint 5	Node 1433	0.023%	0.018%	

initial design is infeasible because the second and third constraints are violated. The reliability-based design sensitivity analysis (DSA) method discussed above is used to calculate the sensitivity coefficients of the fatigue failure probability with respect to the design variables.

Because the current design is infeasible, a constraint correction algorithm is selected for the trade-off analysis. Using the sensitivity coefficients, a QP (quadratic programming) subproblem is employed to search a direction in which the reliability will quickly increase. Then, a what-if study is performed along the search direction suggested by the trade-off study, plus a step size.

Through two iterations, a feasible design is achieved, as shown in Table 5.9. The two design iterations took 10 FORMs and two reliability-based DSAs. At the improved design, failure

Table 5.9 Design Variable Values at Initial, Deterministic Optimal, and Improved Designs

Design Variables	Initial Design	Deterministic Optimal Design	Improved Design
b1 (in.)	3.250	2.889	2.902
b2 (in.)	1.968	1.583	1.593
b3 (in.)	3.170	2.911	2.925
b4 (in.)	1.968	1.637	1.687
b5 (in.)	3.170	2.870	2.904
b6 (in.)	2.635	2.420	2.442
b7 (in.)	3.170	2.801	2.881
b8 (in.)	5.057	4.700	4.700

probabilities at five critical points are less than 1%, with 2.5% increments in volume. However, the total volume savings starting from the initial design is 8%—that is, from 487 in.^{3} to 447 in.^{3}. The design variable values of the initial, deterministic optimal, and improved (after two interactive RBDOs) designs are listed in Table 5.9.

5.6.2 DESIGN OPTIMIZATION FOR STRUCTURAL PERFORMANCE AND MANUFACTURING COST

In mechanical and aerospace industries, engineers often confront the challenge of designing components that can sustain structural loads and meet the functional requirements (e.g., automotive suspension and engine components). It is imperative that these components contain the minimum material to reduce cost and increase efficiency of the mechanical system, such as fuel consumption. The geometry of these components is usually complicated due to strength and efficiency requirements, which often results in increased manufacturing time and cost. Some of these structural components are shown in Figure 5.39.

Although structural optimization has been widely used for many decades to solve such problems, the primary focus has been on the functionality aspects of the design. During the course of an

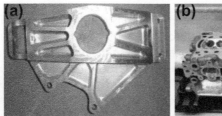

FIGURE 5.39

Mechanical components involving time-consuming and precision machining operations. (a) Upright in an automotive suspension. (b) Engine block.

optimization, the geometric complexity of the component may increase, making manufacturing difficult or uneconomical. Due to the increasing geometric complexity, conventional structural optimization problems that define mass as the objective function may not yield components with minimum cost from the manufacturing perspective. This especially holds true for machined components because machining cost is often the dominant constituent of product cost for such components.

In this subsection, we present a case study of structural shape optimization that incorporates machining and material costs as the objective function subject to structural performance constraints. Structural shape optimization reduces material, but it may be accompanied by increases in the geometric complexity due to changes in the design boundary, which ultimately increases manufacturing cost. In this case study, we present a design process that incorporates manufacturing costs into structural shape optimization and produces components that are cost effective and satisfy specified structural performance requirements.

5.6.2.1 Design problem definition and optimization process

As discussed in Chapter 3, a typical single-objective optimization problem is defined as follows:

$$\text{Minimize}: \ f(\mathbf{b}) \tag{5.57a}$$

$$\text{Subject to}: \ g_i(\mathbf{b}) \leq 0, \quad i = 1, m \tag{5.57b}$$

$$h_j(\mathbf{b}) = 0, \quad j = 1, p \tag{5.57c}$$

$$b_k^\ell \leq b_k \leq b_k^u, \quad k = 1, n \tag{5.57d}$$

where $f(\mathbf{b})$ is the objective function; $\mathbf{b}$ is the vector of design variables, $g_i(\mathbf{b})$ is the ith inequality constraint, and $h_j(\mathbf{b})$ is the jth equality constraint. The objective function $f(\mathbf{b})$ is the product cost for the component to be discussed shortly.

It is assumed that the designer has all of the required data, such as the initial shape of the component, boundary and loading conditions, material properties, and machining sequences. A solid model of the component is created using solid features in CAD software. Dimensions of the solid features also serve as design variables for the optimization problem. A virtual machining (VM) model is created by defining appropriate machining operations and sequences based on the given solid model. The machining parameters specific to each machining sequence are also specified in the VM model. Similarly, an FEA model is constructed using given boundary and loading conditions, as well as initial geometric information.

The design velocity field is then computed for sensitivity analysis and finite element mesh updates. FEA and VM are conducted to evaluate structural and machining performance measures, respectively. Machining time obtained from the VM model is important for the calculation of machining costs. Data obtained from FEA and VM models are used to evaluate objective and constraint functions. Design sensitivity analysis is conducted to compute the gradients of the objective function and constraints with respect to changes in the design variables. The gradients and values of the objective and constraint functions are passed to an optimization algorithm, which determines the design changes for the next design iteration. FEA and VM models are then updated using these changes, and the process is iterated as shown in Figure 5.40, until an optimal design is achieved.

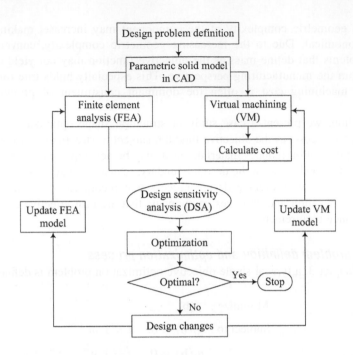

FIGURE 5.40

The design optimization process.

5.6.2.2 Manufacturing cost model

The objective function $f(\mathbf{b})$ in Eq. 5.57a is defined as follows (Edke and Chang 2006):

$$f(\mathbf{b}) = C_{mat}\gamma V(\mathbf{b}) + C_{mc}\tau(\mathbf{b}) + C_t \tag{5.58}$$

where the three terms on the right-hand side represent material cost, machining cost, and tooling cost, respectively. In Eq. 5.58, C_{mat} is the material cost rate ($/lb); γ is the specific weight of the material; $V(\mathbf{b})$ is the volume of the component that depends on design; C_{mc} is the machining cost rate ($/min); $\tau(\mathbf{b})$ is the time required to machine one component, which also depends on design; and C_t is the tooling cost ($).

The machining cost rate accounts for the cost of actual machining operations, machine shop overheads, operator wages and overheads, indirect costs such as cost of electricity, and machine depreciation (Dieter 1991). C_{mc} is given as

$$C_{mc} = \frac{1}{60}\left[\frac{M(100 + OH_M)}{100} + \frac{W(100 + OH_{OP})}{100}\right] \tag{5.59}$$

In this equation, M is the cost of machine operation per hour ($/h); OH_M is the machine overhead rate (%); W is the hourly wage for the operator ($/h); and OH_{OP} is operator overhead rate (%). The unit time $\tau(\mathbf{b})$ in Eq. 5.58 is the sum of machining time t_{mc} and idle time t_i: $\tau = t_{mc} + t_i$, in which

$$t_{mc} = \left(\frac{L}{v_f} + \frac{L'}{v_r}\right); \text{ and } t_i = t_{set} + t_{ch} + t_{hand} + t_{down} + t_{ins} \tag{5.60}$$

In Eq. 5.60, L is the total tool travel while cutting (m); L' is the tool travel during rapid traverse (m); v_f is the feed rate (m/min); v_r is the rapid traverse velocity (m/min); t_{set} is the job setup time per part (min); t_{ch} is the time for tool change (min); t_{hand} is the work-piece handling time (min); t_{down} is the machine down time (min); and t_{ins} is the time for in-process inspection (min). The cost of tooling, which includes cost of cutting tools and cost of jigs/fixtures, is given as

$$C_t = \frac{t_{mc}}{T_{tl}} \left(\frac{K_i}{n_i} + \frac{K_h}{n_h} \right) + C_f \tag{5.61}$$

where T_{tl} is the total tool life (min); K_i is cost of the insert (\$); n_i is the number of cutting edges; K_h is the tool holder cost (\$); n_h is the number of cutting edges in the tool-holder; and C_f is the cost of jigs/fixtures (\$). Machining time t_{mc} is obtained from VM. Similarly, cost models for other machining processes can be developed.

5.6.2.3 Virtual manufacturing

Virtual manufacturing is a simulation-based method that supports engineers to define, simulate, and visualize the manufacturing process in a computer environment. By using virtual manufacturing, the manufacturing process can be defined and verified early in the design process. In addition, the manufacturing time can be estimated. Material cost and manufacturing time constitute a significant portion of the product cost. The virtual machining operations, such as milling, turning, and drilling, allow designers to conduct machining process planning, generate machining tool paths, visualize and simulate machining operations, and estimate machining time. Moreover, the tool path generated can be converted into Computer Numerical Control (CNC) codes (M-codes and G-codes) (Lee 1999; Chang 2013b) and loaded to CNC machines, such as HAAS mills (www.haascnc.com), to machine functional parts as well as dies or molds for production.

Geometrically complex parts are commonly found in the automotive and aerospace industries, where molds and dies are manufactured. The time taken to manufacture molds or dies of different sizes and complexities ranges from 1200 to 3800 h (Sarma and Dutta 1997). Considerable time (49–72%) is spent in contour surface milling for the design surfaces (the surfaces to be machined) of molds and dies (Sarma and Dutta 1997). Design changes in structural geometry due to functional considerations may affect the manufacturing cost significantly. Among different types of machining processes, contour surface milling is most applicable for mold and die machining. Existing commercial CAM tools, such as Pro/MFG (www.ptc.com), support contour surface milling using various methods, including iso-parametric (Kai et al. 1995), constant curvature (Lo and Hsaio 1998), and constant scallop height (Sarma and Dutta 1997).

Typical milling operations that make molds or dies include pocket milling and surface contour milling. The quality and accuracy of the machined surfaces are largely determined by surface contour milling. In this section, surface contour milling will be briefly discussed to illustrate the connection between manufacturing time and structural geometric shape determined by performance requirements.

As discussed in Chapter 4, parametric surfaces are employed to parameterize the design boundary. The same surfaces will be assumed for machining. In general, the cutter contact (CC) curves are first generated on the design surface, as shown in Figure 5.41. The CC curves are created, for example, by uniformly splitting parameter u or w:

$$\mathbf{C}^i(w) = \mathbf{S}(u_i, w), \quad u_i \in [u_{min}, u_{max}] \tag{5.62}$$

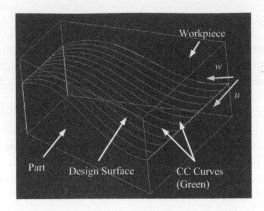

FIGURE 5.41

Design surface and CC curves.

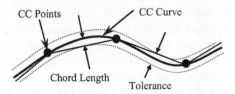

FIGURE 5.42

CC points and CC curve.

where $\mathbf{C}^i(w)$ is the ith CC curve (which is indeed a u-isoline), and $\mathbf{S}(u, w)$ is the parametric design surface. The CC points are discrete points generated along the CC curve, which form a piecewise linear approximation of the CC curve for the CNC controller to trace. The chord length between the CC curve and the polyline formed by the CC points must be less than a prescribed tolerance, as illustrated in Figure 5.42. CL (cutter location) points are then obtained by offsetting from the CC points, considering workcell and cutter shape, as illustrated in Figure 5.43.

Machining time t_{mc} can be estimated by the length of the CL polyline and a prescribed feed rate. As can be seen in Figure 5.44, scallops remaining on the design surface are not uniform. A uniform scallop (with constant scallop height) is more desirable for the follow-on grinding operations. This can be reasonably achieved by moving the cutter on the design surface with a uniform physical spacing, as illustrated in Figure 5.45.

5.6.2.4 Design sensitivity analysis

As discussed in Chapter 4, design sensitivity analysis calculates the gradients of the objective and constraint functions with respect to design variables. Shape sensitivity analysis for structural performance measures has been developed for many years (Choi and Kim 2006) and was briefly discussed in

(a)

(b)

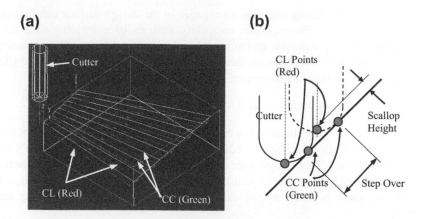

FIGURE 5.43

Offset CC for CL points: (a) CL and CC lines (3-axis mill) and (b) cutter offset and scallop.

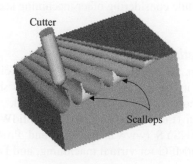

FIGURE 5.44

Uneven scallops.

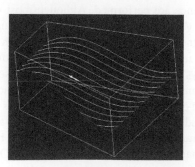

FIGURE 5.45

Uniformly spaced toolpath.

Chapter 4. Sensitivity of machining time due to changes in the design surface can be calculated for specific machining sequences. For example, sensitivity analysis of machining time for a contour surface milling using the isoparametric method can be obtained as follows. First, a new parametric surface will be generated for a small design change of the kth design variable δb_k. The CC curves on the perturbed parametric surface can be created by

$$\mathbf{C}^i\left(w; \mathbf{b} + \delta b_k\right) = \mathbf{S}\left(u^i, w; \mathbf{b} + \delta b_k\right), \quad u^i \in \left[u_{\min}, u_{\max}\right] \tag{5.63}$$

where u^i is the u-parametric coordinate of the ith CC curve, which is kept the same before and after the design perturbation. Following the procedures discussed above, CC points and CL points are calculated next, considering chord length tolerance, workcell, and cutter shape. The new machining time $t_{mc}(\mathbf{b} + \delta b_k)$ can then be calculated by multiplying the length of the polyline formed by the new CL points with the same prescribed feed rate. The sensitivity of the machining time can be approximated by

$$\frac{\partial t_{mc}}{\partial b_k} \approx \frac{t_{mc}\left(\mathbf{b} + \delta b_k\right) - t_{mc}(\mathbf{b})}{\delta b_k} \tag{5.64}$$

However, the overall finite difference method is probably more general and more straightforward to implement in CAM, especially while considering other machining sequences, such as constant scallop height surface contour milling.

5.6.2.5 Software implementation

The design optimization process is implemented using commercial CAD/CAM/FEA and design optimization tools, as illustrated in Figure 5.46. The shaded boxes show the commercial tools, while the plain boxes show software modules that need to be developed.

As a sample implementation (Edke and Chang 2006), SolidWorks and Pro/ENGINEER were selected for CAD modeling, ANSYS (www.ansys.com) and Pro/MECHANICA (www.ptc.com) for finite element modeling, Pro/MFG for virtual machining, and DOT for optimization. MATLAB and C/C++ are used to construct application programs for data transfer and mathematical computations.

Note that after a change is made in model dimensions, Pro/MFG updates the tool path for an NC sequence only after the particular NC (numerical control) sequence is run. This is when the tool path computations for that NC sequence are performed. Although most of the automation in Pro/ENGI-NEER can be achieved using Pro/TOOLKIT (e.g., for CAD model updates), Pro/TOOLKIT does not provide any function to perform the tool path computations. Tool path generation in Pro/MFG has to be done interactively by making a series Pro/ENGINEER menu and dialogue box selections. To overcome this problem, the "mapkeys" feature in Pro/ENGINEER is used.

Mapkeys are similar to the macros used in many application packages. A mapkey is a keyboard macro that maps frequently used command sequences to certain sets of keyboard keys. The approach for the recording of a mapkey and its value is shown in Figure 5.47 as an example. Once a mapkey is recorded, it is saved in a configuration file mapkey, with each macro beginning on a new line. Value of this mapkey (the string of commands) can be copied into a Pro/TOOLKIT application as a command string. Using a Pro/TOOLKIT function, the commands are loaded into a stack and are executed sequentially after the control returns to Pro/ENGINEER from the Pro/TOOLKIT application. Using

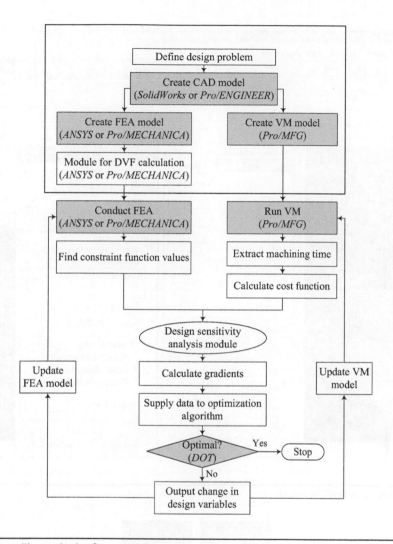

FIGURE 5.46

Optimization flow with required software modules.

these mapkeys, a Pro/TOOLKIT application is constructed to run the machining sequences and to extract total machining time.

5.6.2.6 Aircraft torque tube example

The torque tube shown in Figure 5.48 is a structural component located inside the wings of an aircraft. Loads are applied to the three brackets and the bottom face of the tube is bolted to the wing flap. The tube is made up of AL2024-T351 with yield strength of 43 ksi. Seven rectangular holes are created between fins to reduce the weight of the torque tube, as shown in Figure 5.49.

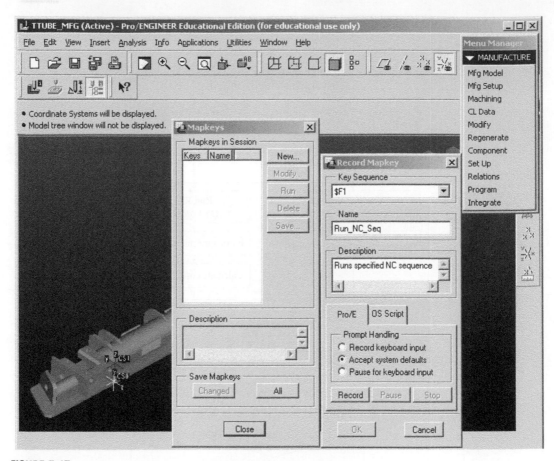

FIGURE 5.47

Recording a mapkey in Pro/ENGINEER.

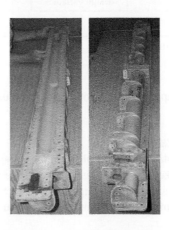

FIGURE 5.48

Airplane torque tube.

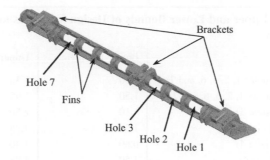

FIGURE 5.49

Torque tube with holes.

5.6.2.6.1 Problem definition

The objective of the torque tube optimization problem is to minimize the product cost subject to limits on structural performance measures. The objective function, which is similar to Eq. 5.58, is defined as:

$$\text{Minimize}: \quad f(\mathbf{b}) = C_{\text{mat}}\gamma\{V_0 - [5V_1(b_1,,b_2) - V_2(b_3,b_4) - V_3(b_5,,b_6)]\}$$

$$+\frac{1}{60}(t_{\text{mc}} + t_i)\left[\frac{M(100 + \text{OH}_\text{M})}{100} + \frac{M(100 + \text{OH}_{\text{OP}})}{100}\right] \tag{5.65a}$$

$$\text{Subject to}: \quad \sigma_{1\text{max}}(\mathbf{b}), \sigma_{2\text{max}}(\mathbf{b}), ..., \sigma_{12\text{max}}(\mathbf{b}) \leq 21.5 \text{ ksi} \tag{5.65b}$$

$$b_j^\ell \leq b_j \leq b_j^u, \quad j = 1, 6 \tag{5.65c}$$

Note that tooling cost is ignored in this problem for simplicity. The tube volume is computed by subtracting the volume of holes from its total volume. Since the five holes (1, 2, 5, 6, and 7) are grouped together, they have only two design variables in common. The maximum principal stresses at 12 locations are defined as constraint functions (see Figure 5.51 for some of the high stresses). The limit on the maximum stress is 21.5 ksi. It is apparent that changes in the hole sizes will vary the weight of the tube, may impact the structural integrity, and influence machining time of the tube. The design variable bounds are listed in Table 5.10.

5.6.2.6.2 Design parameterization

Parameterization of the torque tube holes is shown in Figure 5.50. Hole depth and half of the hole length are selected as design variables. When the length of the holes is changed, the holes either expand or contract symmetrically. Also, the position of the hole is maintained such that it always remains centered between the adjacent fins. From initial tests, it was observed that the maximum stress occurs near the middle bracket. Hence, except for the two holes adjacent to the middle bracket, the design variables of all other holes are grouped together; implying that width and length of the Holes 1, 2, 5, 6, and 7 are changed at the same amounts, respectively. This reduces the number of design variables from 14 (2 design variables per hole time 7 holes) to 6.

Table 5.10 Upper and Lower Bounds of Design Variables for the Torque Tube

	Lower Limit (in.)	Upper Limit (in.)
b1 (length of holes 1, 2, 5, 6, and 7)	0.90	2.30
b2 (depth of holes 1, 2, 5, 6, and 7)	1.50	1.95
b3 (length of hole 3)	1.10	2.50
b4 (depth of hole 3)	1.50	1.95
b5 (length of hole 4)	0.60	1.40
b6 (depth of hole 4)	1.50	1.95

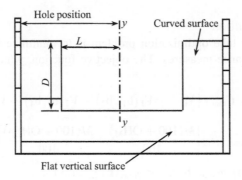

FIGURE 5.50

Parameterization of torque tube holes.

5.6.2.6.3 Finite element analysis

The finite element model constructed in Pro/MECHANICA consists of 11,502 elements. A p-convergence study is conducted at the outset to fix the polynomial level of the element shape function for the analysis. The criterion specified is 3.5% strain energy convergence. The polynomial order is fixed at 7. The FEA solves 2,242,623 equations. Because the model is constrained by fixing displacements at finite element nodes at some locations, concentrated high stresses at such nodes are neglected.

The highest maximum principal stress is located at one of the holes and the inner edge of the middle bracket, as shown in Figure 5.51. The maximum stress magnitude is 20.70 ksi, which is close to the constraint limit (21.5 ksi). Note that a safety factor of 2 is used.

The most critical information required for performing design optimization in this example is the design velocity field. Once the velocity field is obtained, the remaining optimization process becomes routine. For the torque tube, the design boundaries (hole surfaces) are plane surfaces, as shown in Figure 5.52. This simplifies the velocity field calculations, as the prescribed displacement itself is now the boundary velocity. Thus, only domain velocity field calculation is required.

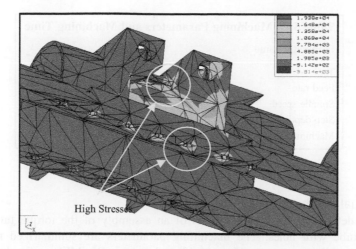

FIGURE 5.51

Finite element analysis results.

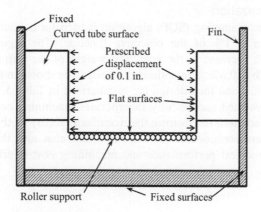

FIGURE 5.52

Velocity field computation for length design variables.

As shown in Figure 5.52, the fins and bottom surfaces are fixed. For the length design variables, a prescribed displacement of 0.1 in. is applied on the two end surfaces of the hole along the longitudinal direction. Roller boundary conditions are applied to the bottom surface of the holes. FEA is conducted to calculate the displacement of the finite element nodes, which is the design velocity field of the length design variable. To calculate the design velocity of the depth design variable, a prescribed displacement of 0.1 in. is applied to the bottom surface of the holes, and the two longitudinal end surfaces are fixed.

Table 5.11 Machining Parameters and Machining Time

Parameter name	Value
Tool	0.5 in. end mill
Feed rate	10 in./min
Spindle speed	1250 rpm
Step depth	0.2 in.
Machining time (initial design)	41.26 min

5.6.2.6.4 Virtual machining

The VM model defined in Pro/MFG consists of an assembly of the torque tube without holes and the torque tube with holes. The machining parameters are summarized in Table 5.11. A customized pocket milling sequence (Figure 5.53) is defined in Pro/MFG to simulate the machining process. The time required to machine all the holes is 41.26 min for the initial values of design variables.

5.6.2.6.5 Design optimization

The sequential quadratic programming (SQP) algorithm is used for conducting design optimization. The convergence criterion is 1% of the objective function. The algorithm converges in four iterations. There is a 2.4% decrease in the cost. The weight of the torque tube reduces by 6.1%. Machining time decreases by 10.6%. The optimization history is shown in Figure 5.54. The values of the design variables for initial and final design are summarized in Table 5.12.

The design process presented successfully incorporates machining cost into a structural shape optimization problem. In addition to ensuring the manufacturability of the optimized components, the design process delivers components with a minimum cost and the required performance. The trade-off between structural performance and machining cost is critical, as revealed in this torque tube example.

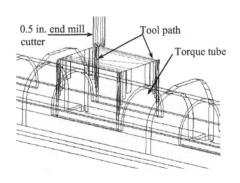

FIGURE 5.53

Virtual machining in Pro/MFG.

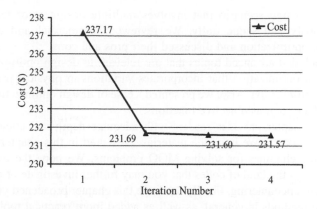

FIGURE 5.54

Optimization results for the torque tube: objective (cost $) function history.

Table 5.12 Optimization Results for the Torque Tube

	Initial Design (in.)	Final Design (in.)	% Change
b1 (length of holes 1, 2, 5, 6, and 7)	1.60	1.23	−23.1
b2 (depth of holes 1, 2, 5, 6, and 7)	1.65	1.56	−5.45
b3 (length of hole 3)	1.80	1.43	−20.5
b4 (depth of hole 3)	1.65	1.72	4.24
b5 (length of hole 4)	1.00	0.96	−4.00
b6 (depth of hole 4)	1.65	1.55	−6.06

5.7 SUMMARY

In this chapter, we discussed multiobjective design optimization, in which more than one objective is being minimized simultaneously. We started with basic concepts and Pareto optimality, and then we used a simple pyramid example to illustrate the essential points of multiobjective design optimization, especially the Pareto front in the criterion space. We introduced major solution techniques in four categories: methods with a priori articulation of preferences criteria, a posteriori articulation of preferences, no articulation of preferences, and multiobjective genetic algorithms. We pointed out the pros and cons of methods in individual categories and used simple examples for illustration. We also revisited the topic of decision-based design using decision theories, including utility theory and game theory. We reviewed the topic from the context of MOO problems and compared the decision methods with those of conventional MOO solution techniques. We pointed out that the solutions obtained using decision theories are subsets of the Pareto front. However, game theory supports a broader and practical design scenario that involves multiple designers or teams making design decisions simultaneously and sequentially. We reviewed commercial and academic codes for multiobjective design optimization and discussed their pros and cons.

In addition, we included advanced topics that are relevant to design optimization. We presented reliability-based design optimization that incorporates variations in physical parameters and design variables into design optimization. Also, we presented a case of design optimization that takes product cost, including manufacturing, as objective functions.

We hope that this chapter provides readers with adequate depth and enough breadth in these important and practical topics. We hope the materials presented in this chapter explained the basic concepts and solution techniques for solving MOO problems. We hope the overview on software programs offered ideas on the kind of codes that you may further investigate or adopt for solving the MOO problems you are encountering. Finally, we hope this chapter broadened your understanding of design theory and the methods in general, as well as added more practical tools to your toolset for engineering design.

APPENDIX A. MATLAB SCRIPTS

Script 1: Solving single-objective optimization problems of objectives A and T, respectively, in the pyramid example.

```
objfun_s.m

function f = objfun_s(x)
f = 2*x(1)*sqrt((0.5*x(1))^2+x(2)^2);

objfun_t.m

function f = objfun_t(x)
 f = 2*x(1)*sqrt((0.5*x(1))^2+x(2)^2)+x(1)^2;

confun.m
function [c, ceq] = confun(x)
% Nonlinear inequality constraints
c = 1500-x(1)^2*x(2)/3;
% Nonlinear equality constraints
ceq = [];

x0 = [10,10]; % Make a starting guess at the solution
options = optimset('Algorithm','active-set');
[x_s,fval_s] = fmincon(@objfun_s,x0,[],[],[],[],[],[],@confun, options);
[x_s,fval_s]    % Print results.
[x_t,fval_t] = fmincon(@objfun_t,x0,[],[],[],[],[],[],@confun, options);
[x_t,fval_t]    % Print results.
```

Script 2: Solving and graphing the Pareto front for the pyramid example using the generative method.

```
%Pyramid example; plot objective domain
clear all;format long
```

```
NN=1000000; %No. of random points for plotting
%-------------------------------------------------
A=1; %counting feasible points
B=1; %counting infeasible points
Smin=1000000;
aos=0;
hos=0;
Tmin=1000000;
aot=0;
hot=0;
for ii=1:NN
    a=rand*30;
    h=rand*30;
    V=(1/3)*a^2*h;
    s=sqrt((a/2)^2+h^2);
    S=2*a*s;
    T=S+a^2;
    if V>1500;
        feasi(:,A)=[S,T];
        if Smin > feasi(1,A);
            Smin = feasi(1,A);
            aos=a; hos=h;
        end
        if Tmin > feasi(2,A);
            Tmin = feasi(2,A);
            aot=a; hot=h;
        end
        A=A+1;
    else
        infeasi(:,B)=[S,T];
        B=B+1;
    end
end
figure (1)
plot(feasi(1,:),feasi(2,:),'bs','MarkerSize',3)
plot(infeasi(1,:),infeasi(2,:),'rs','MarkerSize',3)
xlabel('S'); ylabel('T')
axis equal
```

Script 3: Solving the pyramid problem using the weighted-sum method (Example 5.4).

```
objfun_pyramid_ws.m
function f = objfun_pyramid_ws(x)
ws=0.1;
wt=1-ws;
f = ws*(2*x(1)*sqrt((0.5*x(1))^2+x(2)^2))+wt*(2*x(1)*sqrt((0.5*x(1))^2+x(2)^2)+x(1)^2);

confun.m
function [c, ceq] = confun(x)
```

```
% Nonlinear inequality constraints
c = 1500-x(1)^2*x(2)/3;
% Nonlinear equality constraints
ceq = [];

x0 = [10,10]; % Make a starting guess at the solution
options = optimset('Algorithm','active-set');
[x,fval] = fmincon(@objfun_pyramid_ws,x0,[],[],[],[],[],[],@confun, options);
[x,fval]    % Print results
S = 2*x(1)*sqrt((0.5*x(1))^2+x(2)^2)
T = 2*x(1)*sqrt((0.5*x(1))^2+x(2)^2)+x(1)^2
```

Script 4: Solving Example 5.5 using the weighted min–max method.

```
objfun.m
function f = objfun(x)
f = x(3);
```

```
confun.m
function [c, ceq] = confun(x)
% Inequality constraints
c = 0.5*x(1)+0.5*x(2)+1-x(3);
% Equality constraints
ceq = x(1)^2+x(2)^2-1;
```

```
x0 = [-0.5,0,0]; % Make a starting guess at the solution
options = optimset('Algorithm','active-set');
[x,fval] = fmincon(@objfun,x0,[],[],[],[],[],[],@confun, options);
```

Script 5: Solving Example 5.7 using the NBI method.

```
objfun.m
function f = objfun(x)
f = -x(3);
```

```
confun.m
function [c, ceq] = confun(x)
% Equality constraints
a1=-0.7;
a2=-(1+a1);
c=[];
ceq(1) = a1-x(3)-x(1);
ceq(2) = a2-x(3)-x(2);
ceq(3) = x(1)^2+x(2)^2-1;
```

```
x0 = [-0.5,0,0]; % Make a starting guess at the solution
options = optimset('Algorithm','active-set');
[x,fval] = fmincon(@objfun,x0,[],[],[],[],[],[],@confun, options);
```

Script 6: Solving the pyramid example using the NBI method (Example 5.8).

```
objfun_pyramid_nbi.m
function f = objfun_pyramid_nbi(x)
f = -x(3);

confun.m
function [c, ceq] = confun(x)
w1=1;
w2=1-w1;
% Nonlinear inequality constraints
c = 1500-x(1)^2*x(2)/3;
% Nonlinear equality constraints
ceq(1)=w2*(648.5-595.8)-x(3)*(648.5-595.8)-2*x(1)*((0.5*x(1))^2+x(2)^2)^0.5+595.8;
ceq(2)=w1*(935.6-865.3)-x(3)*(935.6-865.3)-(2*x(1)*((0.5*x(1))^2+x(2)^2)^0.5+
x(1)^2)+865.3;

x0 = [10,10,10]; % Make a starting guess at the solution
options = optimset('Algorithm','active-set');
[x,fval] = fmincon(@objfun_pyramid_nbi,x0,[],[],[],[],[],[],@confun, options);
[x,fval]    % Print results
S = 2*x(1)*sqrt((0.5*x(1))^2+x(2)^2)
T = 2*x(1)*sqrt((0.5*x(1))^2+x(2)^2)+x(1)^2
```

Script 7: Solving the pyramid problem using the min–max method (Example 5.9).

```
objfun_pyramid_minmax.m
function f = objfun_pyramid_minmax(x)
f = x(3);

confun.m
% Nonlinear inequality constraints
c(1) = 1500-x(1)^2*x(2)/3;
c(2) = 2*x(1)*((0.5*x(1))^2+x(2)^2)^0.5-x(3);
c(3) = c(2)+x(1)^2;
% Nonlinear equality constraints
ceq = [];

x0 = [50,50,10]; % Make a starting guess at the solution
options = optimset('Algorithm','active-set');
[x,fval] = fmincon(@objfun_pyramid_minmax,x0,[],[],[],[],[],[],@confun, options);
[x,fval]    % Print results
S = 2*x(1)*sqrt((0.5*x(1))^2+x(2)^2)
T = 2*x(1)*sqrt((0.5*x(1))^2+x(2)^2)+x(1)^2
```

Script 8: Pareto solutions of the beam example.

```
%Beam example; plot objective domain
clear all;format long
NN=1000000; %No. of random points for plotting
L=200; rou=2.7*10^(-3); g=9806;P=1000; E=69000; Sy=27.6;
```

```
%---------------------------------------------
A=1; %counting feasible points
B=1; %counting infeasible points
Wmin=1000000;
bow=0;
how=0;
Zmin=1000000;
boz=0;
hoz=0;
for ii=1:NN
    b=rand*60;
    h=rand*80;
    W=rou*b*h*L*g*10^(-6);
    I=b*h^3/12;
    Z = P*L^3/(3*E*I);
    S=6*P*L/(b*h^2);
    if S<=Sy & 10<=b & b<=60 & 20<=h & h<=80;
        feasi(:,A)=[W,Z];
        if Wmin > feasi(1,A);
            Wmin = feasi(1,A);
            bow=b; how=h;
        end
        if Zmin > feasi(2,A);
            Zmin = feasi(2,A);
            boz=b; hoz=h;
        end
        A=A+1;
    else
        infeasi(:,B)=[W,Z];
        B=B+1;
    end
end
figure (1)
plot(feasi(1,:),feasi(2,:),'bs','MarkerSize',3)
xlabel('fw'); ylabel('fz')
```

Script 9: Solving the pressure vessel example.

```
%Vessel example; plot objective domain
clear all;format long
NN=1000000; %No. of random points for plotting
L=100; R = 30; gamma=0.283; P=4000; Sy=32000; pi = 3.1415916;
%---------------------------------------------
A=1; %counting feasible points
B=1; %counting infeasible points
Wmin=1000000;
Row=0;
tow=0;
```

```
Vmin=0;
Rov=0;
tov=0;
for ii=1:NN
    R=rand*10;
    t=rand;
    W=gamma*(pi*(R+t)^2*(L+2*t)-pi*R^2*L);
    V = -pi*R^2*L;
    if 5*t<=R & R<=8*t & t+R<=40 & 0.5<=t & t<=6 & 2.5 < R & R<=5.55;
        feasi(:,A)=[W,V];
        if Wmin > feasi(1,A);
            Wmin = feasi(1,A);
            Row=R; tow=t;
        end
        if Vmin > feasi(2,A);
            Vmin = feasi(2,A);
            Rov=R; tov=t;
        end
        A=A+1;
    else
        infeasi(:,B)=[W,V];
        B=B+1;
    end
end
figure (1)
plot(feasi(1,:),feasi(2,:),'bs','MarkerSize',3)
xlabel('Self-weight (W)'); ylabel('Negative volume (-V)')
```

QUESTIONS AND EXERCISES

1. In the design space, plot the objective function contours for the following unconstrained problem and identify the Pareto optimal set:

Minimize: $f_1(\mathbf{x}) = (x_1 - 2)^2 + (x_2 - 3)^2$
$f_2(\mathbf{x}) = (x_1 - 5)^2 + (x_2 - 8)^2$

Draw the gradients of each function at any point on the Pareto optimal set in the design space. Comment on the relationship between the two gradients.

2. Sketch the Pareto optimal set for Problem 1 in the criterion space. Find the utopia point. Is the utopia point attainable?

3. A constrained optimization of two variables and two objectives is given below:

Minimize: $f_1(\mathbf{x}) = (x_1 - 2)^2 + (x_2 - 3)^2$
$f_2(\mathbf{x}) = (x_1 - 5)^2 + (x_2 - 8)^2$

Subject to: $g_1 = -x_1 - x_2 + 10 \leq 0$
$g_2 = -10 - 2x_1 + 3x_2 \leq 0$

a. Sketch the objective function contours and constraint functions in the design space and identify a feasible set S and the Pareto optimal set.

   b. Write a MATALB script to find the Pareto front in the criterion space using the generative method. Graph the feasible and infeasible regions and identify the Pareto front in the criterion space.

4. A right circular cone shown below is considered for a redesign. The objective is to use a minimum material (surface area) to achieve a maximum volume. More specifically, we want to minimize both the lateral area S and total surface area T of the cone for a volume no less than 250 cm^3 by varying the base radius r and height h of the cone. Note that $T = S + B$, in which B is the based area $B = \pi r^2$. The size of the cone is constrained by $r \in (1, 10)$ cm and $h \in (1, 20)$ cm.

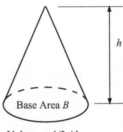

$$\text{Volume} = 1/3\ Ah$$

   a. Formulate a MOO problem for the cone design mathematically.
   b. Sketch the objective contours and constraint functions in the design space and identify a feasible set S and the Pareto optimal set.
   c. Solve the respective single-objective optimization problems to find the utopia point. Comment on the design of the cone at the utopia point.
   d. Sketch the Pareto front in the criterion space using the generative method. Graph the feasible and infeasible regions and identify the Pareto front in the criterion space.

5. Solve Problem 4 using the weighted-sum method. Find at least 10 solutions by varying the weights (including a case for $w_1 = 0.1$ and $w_2 = 0.9$). Identify those solutions in the Pareto front in the criterion space.

6. Solve Problem 4 using the weighted min–max method, assuming $w_1 = 0.1$ and $w_2 = 0.9$. Compare results with those of Problem 5. Comment on the pros and cons of the weighted min–max and the weighted-sum method based on the observation made in solving the cone problem.

7. Solve Problem 4 using the NBI methods by creating at least five points on the CHIM.

8. Solve Problem 3 using the min–max method.

9. Derive Eq. 5.21c of Example 5.7: $\beta = \frac{1}{2}\left[(\alpha_1 + \alpha_2) + \sqrt{2 - (\alpha_1^2 - \alpha_2^2)}\right]$. Hint: $(\alpha_1 - \beta)^2 + (\alpha_2 - \beta)^2 = 1$.

10. Solve Problem 4 using the NSGA II of the genetic algorithm. Follow the discussion in Section 5.3.5.3 to download the MATLAB code and write two MATLAB files for the cone design problem. Use population size $= 200$ and number of generations $= 500$.

11. Solve Problem 4 using NIMBUS software online. Comment on the advantages and disadvantages of the software from a user's perspective.

REFERENCES

Arora, J.S., 2012. Introduction to Optimum Design. Academic Press, Waltham, MA.

Branke, J., Deb, K., Miettinen, K., Slowinski, R. (Eds.), 2008. Multiobjective Optimization: Interactive and Evolutionary Approaches. Springer-Verlag, Berlin, Heidelberg.

Chang, K.H., 2013a. Product Performance Evaluation Using CAD/CAE, the Computer Aided Engineering Design Series. Academic Press, Burlington, MA. ISBN: 978-0-12-398460-9.

Chang, K.H., 2013b. Product Manufacturing and Cost Estimate Using CAD/CAE, the Computer Aided Engineering Design Series. Academic Press, Burlington, MA. ISBN: 978-0-12-401745-0.

Choi, K.K., Kim, N.H., 2006. Structural Sensitivity Analysis and Optimization 1: Linear Systems. (Mechanical Engineering Series), Springer, Dordrecht.

Das, I., Dennis, J., 1998. Normal-boundary intersection: a new method for generating pareto optimal points in multicriteria optimization problems. SIAM Journal on Optimization 8 (3), 631–657.

De Weck, O.L., 2004. Multiobjective Optimization: History and Promise, the Third China-Japan-Korea Joint Symposium on Optimization of Structural and Mechanical Systems (CJK–osm3), October 30–November 2, 2004 (Kanazawa, Japan).

Deb, K., 2001. Multiobjective Optimization Using Evolutionary Algorithms. John Wiley and Sons, Ltd, New York.

Deb, K., Pratap, A., Agarwal, S., Meyarivan, T., 2002. A fast and elitist multiobjective genetic algorithm: NSGA-II. IEEE Transactions on Evolutionary Computation 6 (2), 182–197.

Dieter, G.E., 1991. Engineering Design: A Materials and Processing Approach, second ed. McGraw Hill, New York.

Edke, M., Chang, K.H., 2006. Shape optimization of heavy load carrying components for structural performance and manufacturing cost. Journal of Multidisciplinary Structural Optimization 31 (5), 344–354.

Haug, E.J., Smith, R.C., 1990. DADS-Dynamic Analysis and Design System, Multibody Systems Handbook pp. 161–179.

Hazelrigg, G.A., 1998. A framework for decision-based engineering design. ASME Journal of Mechanical Design 120, 653–658.

Jones, D.R., 2001. A taxonomy of global optimization methods based on response surfaces. Journal of Global Optimization 21, 345–383.

Kai, T., Cheng, C.C., Dayan, Y., 1995. Offsetting surface boundaries and 3-axis gouge-free surface machining. Computer-Aided Design 27 (12), 915–927.

Kim, I.Y., de Weck, O., 2006. Adaptive weighted sum method for multiobjective optimization: a new method for Pareto front generation. Structural and Multidisciplinary Optimization 31 (2), 105–116.

Lee, K., 1999. Principles of CAD/CAM/CAE Systems. Addison Wesley Longman, Inc, Reading, MA.

Lo, C.C., Hsiao, C.Y., 1998. CNC machine tool interpolator with path compensation for repeated contour machining. Computer-Aided Design 30 (1), 55–62.

Marler, R.T., Arora, J.S., 2004. Survey of multiobjective optimization methods for engineering. Structural Multidisciplinary Optimization 26, 369–395.

Messac, A., Mattson, C.A., 2002. Generating well-distribute sets of Pareto points for engineering design using physical programming. Optimization Engineering 3, 431–450.

Miettinen, K., 1999. Nonlinear Multiobjective Optimization. Kluwer Academic Publishers, Boston.

Poles, S., Vassileva, M., Sasaki, D., 2008. In: Branke, J., Deb, K., Miettinen, K., Slowinski, R. (Eds.), Multiobjective Optimization Software, Multiobjective Optimization Lecture Notes in Computer Science, vol. 5252. Springer-Verlag Berlin Heidelberg, pp. 329–348.

Sarma, R., Dutta, D., 1997. An integrated system for NC machining of multipatch surfaces. Computer-Aided Design 29 (11), 741–749.

Sawaragi, Y., Nakayama, H., Tanino, T., 1985. Theory of Multiobjective Optimization. In: Mathematics in Science and Engineering, vol. 176. Academic Press Inc., London, UK.

Schaffer, J.D., 1985. Multiple objective optimization with vector evaluated genetic algorithms. In: Grefenstette, J.J. (Ed.), Proceedings of an International Conference on Genetic Algorithms and Their Applications. Laurence Erlbaum Associates, pp. 93–100.

Srinivas, N., Deb, K., 1994. Multiobjective function optimization using nondominated sorting genetic algorithms. Evolutionary Computation 2 (3), 221–248.

Stadler, W., 1979. A survey of multicriteria optimization, or the vector maximum problem. Journal of Optimization Theory and Applications 29, 1–52.

Steuer, R., 1985. Multiple Criteria Optimization: Theory, Computation and Application. John Wiley & Sons, New York, NY.

Yu, X., Choi, K.K., Chang, K.H., 1997. A mixed design approach for probabilistic structural durability. Journal of Structural Optimization 14, 81–90.

CHAPTER OUTLINE

Design optimization was discussed in Chapters 3 and 5 for single-objective and multiobjective problems, respectively. In these chapters, both theoretical and practical aspects of the topics were discussed. With a basic understanding of the concept, theories, and solution techniques of design optimization, we are ready to go over a few tutorial lessons on design optimization capabilities offered by commercial tools, such as SolidWorks.

In Project S5, we introduce SolidWorks design optimization capabilities and then conduct interactive design using the suite of SolidWorks computer-aided design (CAD), engineering (CAE), and manufacturing (CAM) capabilities for a multidisciplinary design problem. We include an optimization design study for a cantilever beam as a single-objective optimization. Then, we present a single-piston engine example to illustrate the interactive design process discussed in Chapter 3 involving multiple engineering disciplines. For those who are not familiar with SolidWorks and embedded CAE/CAM modules, we encourage you to go over Projects S1–S4 of the book series before moving forward with this tutorial project. Without going over these prior projects, you may get lost very quickly. Example models are available for download at the book's companion Web site.

Overall, the objective of this project is to enable readers to use SolidWorks for batch-mode design optimization and interactive design. The use of the capability for design optimization should be straightforward. Those who are interested in learning more about the batch-mode optimization may go over a few more examples offered as YouTube videos, such as those listed below, subject to their availability.

- Design study and optimization in SolidWorks finite element analysis (FEA) by Intercad (http://www.youtube.com/watch?v=TJDvfvu0WRE)
- Torsional static analysis and design study example using SolidWorks (http://www.youtube.com/watch?v=KTWEDiwZwvs)

Note that the lessons included in this project were developed using SolidWorks 2013 SP2.0 and CAMWorks 2014. If you are using a different version of SolidWorks or CAMWorks, you may see slightly different menu options or dialog boxes.

S5.1 INTRODUCTION TO DESIGN OPTIMIZATION IN SOLIDWORKS

There are two main modes for running a design study in SolidWorks: evaluation and optimization. Evaluation supports users in specifying discrete values for each variable and using sensors as constraints. The software runs the study using various combinations of the values and reports the output for each combination.

For example, consider the plate with a hole shown in Figure S5.1 for a design change. Using the evaluation capability, users may specify values of 0.2, 0.1, and 0.3 m for the length variable (currently $L = 0.2$ m); 0.1, 0.05, and 0.15 m for the height variable (currently $H = 0.1$ m); and 0.04, 0.02, and 0.06 m for the diameter variable ($\phi = 0.04$ m). You can specify a volume sensor to monitor the volume of the plate. The design study results report the volume of the plate for each combination of L, H, and ϕ—in this case, $3 \times 3 \times 3 = 27$ combinations.

Optimization, on the other hand, allows users to specify values for each variable, either as discrete values or as a range. Users use sensors as constraints and as goals. The software runs iterations of the values and reports the optimum combination of values to meet specified goal. Note that a sensor in SolidWorks monitors selected properties of parts and assemblies and alerts users when values deviate from the limits that users specify. Sensors can be used to study mass properties, dimensions, interface

FIGURE S5.1
Plate with a center hole. (a) CAD solid mode. (b) Simulation model with mesh and stress fringe plot.

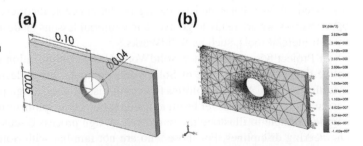

detection, proximity (available in assemblies only for monitoring the assembly for interferences be-tween components users select), simulation data, and motion data.

For example, for the plate model described earlier, users specify a range of 0.1–0.3 m for the length variable (*L*); discrete values of 0.1, 0.05, and 0.15 m for the height variable (*H*); and a range of 0.04–0.06 m for the diameter variable ($\phi = 0.04$ m). For a constraint, one specifies a stress sensor to be less than 380 MPa, and uses a mass sensor to minimize the mass of the plate. The optimization design study iterates on the values specified for *L*, *H*, ϕ, and stress, then reports the optimum combination to produce the minimum mass that satisfies the stress constraint. In this tutorial project, we focus only on the optimization of the design study.

We use a cantilever beam example for illustration of the optimization capability (batch mode) in Section S5.2. The beam example is a single-objective optimization problem of a single discipline, only involving structural performance measures. In addition, we use a single-piston engine example in Section S5.3 to illustrate the details of a multidisciplinary design problem using an interactive design process, in which structural, motion, and manufacturing are involved.

The main objective of this tutorial project is to help you, as an experienced SolidWorks user, become familiar with SolidWorks design optimization capabilities. The discussion on SolidWorks software, including motion, simulation, and CAMWorks, in this section will be brief.

S5.1.1 DEFINING AN OPTIMIZATION PROBLEM

As discussed in Chapter 3, a single-objective optimization problem can be formulated mathematically as follows:

$$\text{Minimize:} \quad f(\mathbf{x}) \tag{S5.1a}$$

$$\text{Subject to:} \quad g_i(\mathbf{x}) \leq 0, \quad i = 1, m \tag{S5.1b}$$

$$h_j(\mathbf{x}) = 0, \quad j = 1, p \tag{S5.1c}$$

$$x_k^\ell \leq x_k \leq x_k^u, \quad k = 1, n \tag{S5.1d}$$

where $f(\mathbf{x})$ is the objective function or goal to be minimized (or maximized); $g_i(\mathbf{x})$ is the *i*th inequality constraint; *m* is the total number of inequality constraint functions; $h_j(\mathbf{x})$ is the *j*th equality constraint; *p* is the total number of equality constraints; $\mathbf{x}$ is the vector of design variables, with $\mathbf{x} = [x_1, x_2, ..., x_n]^T$; *n* is the total number of design variables; and x_k^ℓ and x_k^u are the lower and upper bounds of the *k*th design variable x_k, respectively.

SolidWorks supports users in defining an optimization problem, running the optimization (using the so-called generative method, as discussed in Chapter 3), and displaying optimization results. Note that if you plan to use simulation data sensors, you must create at least one initial simulation study before creating the design study.

A design study can be initiated by choosing from the pull-down menu:

Insert > Design Study > Add.

In the Design Study Manager window shown in Figure S5.2, users define an optimization problem by entering and choosing objective (goals), constraints, and variables using the Variable View tab or Table View tab. Results view allows users to view optimization results in tables or graphs.

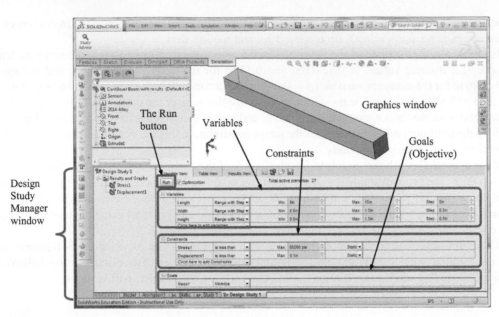

FIGURE S5.2

User interface of optimization design study in SolidWorks.

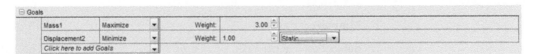

FIGURE S5.3

Example of multiple goals with weights.

Goals can be chosen from a predefined sensor. Users may add more sensors to goals. For multiple goals, users may enter the weight of each goal. For example, in Figure S5.3, mass and displacement are included in the goals with weights of 3 and 1, respectively. The weight of a goal represents its relative importance. The higher the weight of the goal, the more important it is to optimize that goal. The program normalizes the weights and uses the weight-sum method discussed in Chapter 5 to conduct multiobjective optimization.

Users may choose continuous variables to perform optimization. For each design variable, users need to define its lower and upper limits, as well as step size to vary. Dimensions of the solid model are the most common variables for optimization. In addition, parameters, such as material properties, can be chosen as design variables. Note that the model, part, or assembly must be fully parameterized to conduct a batch-mode optimization in SolidWorks, as discussed in Chapter 3. For more about design parameterization, the reader is referred to Chang (2014a).

Similar to goals, you may choose sensors to define constraints. For the condition, you may select one of the following four constraints: (1) "monitor only," monitoring the sensor value without

29 of 29 scenarios ran successfully. Design Study Quality: High

		Current	Initial	Optimal (19)	Scenario 16	Scenario 17	Scenario 18	Scenario 19	Scenario 20	Scenario 21
Length		10in	10in	5in	5in	10in	15in	5in	10in	15in
Width		1in	1in	0.5in	1.5in	1.5in	1.5in	0.5in	0.5in	0.5in
Height		1in	1in	1.5in	1in	1in	1in	1.5in	1.5in	1.5in
Stress1	< 65000 psi	99974 psi	99974 psi	42076 psi	28487 psi	67068 psi	83643 psi	42076 psi	89354 psi	1.1789e+005 psi
Displacement1	< 0.1in	0.37721in	0.37721in	0.0296in	0.03171in	0.25065in	0.84573in	0.0296in	0.22621in	0.75782in
Mass1	Minimize	1.01156 lb	1.01156 lb	0.379336 lb	0.758673 lb	1.51735 lb	2.27602 lb	0.379336 lb	0.758673 lb	1.13801 lb

FIGURE S5.4

Optimal solution reported by simulation.

History: Design Study 1

(Mass1 (lb) vs. Scenario: Set1 Set2 Set3 Set4 Set5 Set6 Set7 Set8 Set9 Set10 Set11 Set12 Set13 Set14 Set15 Set16 Set17 Set18 Set19 Set20 Set21 Set22 Set23 Set24 Set25 Set26 Set27)

— Mass1

FIGURE S5.5

Objective function history graph.

imposing any constraints; (2) "is greater than" the minimum acceptable sensor value; (3) "is less than" the maximum acceptable sensor value; and (4) "is between" the range of acceptable sensor values.

Once an optimization problem is completely defined, users may simply click the Run button above the Variable table (see Figure S5.1) to launch an optimization run.

At the end of the run, users may choose the Results View tab to review the results of the study, such as those shown in Figure S5.4. Users may click a scenario to update the model in the graphics window with the variables of that scenario. A green-colored scenario indicates the best or optimal result. A red color indicates a violation of one or more constraints by the scenario. A background color indicates the current scenario and all scenarios that are not optimal. Users may also create graphs to see optimization history, like the one shown in Figure S5.5. Note that although the graph is called "history," it is nothing but a polyline that connects results of individual scenarios.

S5.1.2 TUTORIAL EXAMPLES

Two examples are included in this tutorial project. The first example, a cantilever beam, illustrates step-by-step details of defining an optimization problem, running optimization, and showing results in table and graphs. The second example, a single-piston engine, illustrates an interactive design process of solving a multidisciplinary design problem using SolidWorks Simulation, motion, CAMWorks, the sensitivity display, and a what-if study, discussed in Chapter 3. Because the detailed steps in using individual software modules have been discussed in Projects S1–S4, we only include major steps with key information in these examples. Examples and topics to be discussed in each lesson are summarized in Table S5.1.

Table S5.1 Examples Employed in This Project

Section	Example	Solid Models	Things to Learn
S5.2	Cantilever beam design optimization		1. This in an introductory tutorial lesson, showing detailed steps in defining an optimization problem, running an optimization, and showing results in table and graphs. 2. You will learn to solve similar optimization problems using SolidWorks.
S5.3	Single-piston engine of multidisciplinary optimization		1. This example illustrates an interactive process of solving a multidisciplinary design problem using SolidWorks Simulation, motion, CAMWorks, a sensitivity display, and a what-if study for design trade-off. 2. You will learn to solve similar multidisciplinary optimization problems using SolidWorks and procedures involved.

S5.2 CANTILEVER BEAM

We use the cantilever beam example from Chapter 3 to illustrate the steps for carrying out design optimization using SolidWorks. The beam is shown again in Figure S5.6. The beam is made of cast stainless steel. In the current design, the beam length $\ell = 10$ in., and the width w and height h of the crosssectional area are both 1 in. The load is $P = 1000$ lbf acting at the tip of the beam, as shown in Figure S5.6. Our goal is to optimize the beam design for a minimum mass subject to displacement and stress constraints by varying its length as well as the width and height of the crosssection.

S5.2.1 DESIGN PROBLEM DEFINITION

The design problem of the cantilever beam example is formulated mathematically as:

$$\text{Minimize:} \quad M(w, h, \ell) = \rho w h \ell \tag{S5.2a}$$

$$\text{Subject to:} \quad g_1(w, h, \ell) = \sigma_{\text{mp}}(w, h, \ell) - 65 \text{ ksi} \leq 0 \tag{S5.2b}$$

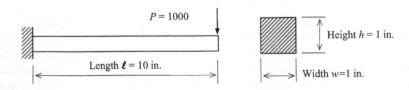

FIGURE S5.6

Cantilever beam example.

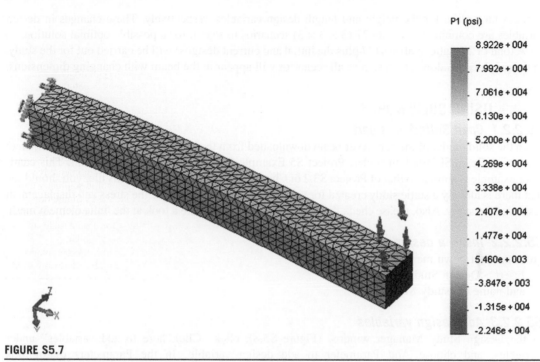

P1 (psi)

8.922e + 004

7.992e + 004

7.061e + 004

6.130e + 004

5.200e + 004

4.269e + 004

3.338e + 004

2.407e + 004

1.477e + 004

5.460e + 003

-3.847e + 003

-1.315e + 004

-2.246e + 004

FIGURE S5.7

Finite element model of the cantilever beam in a SolidWorks Simulation.

$$g_2(w, h, \ell) = \delta_y(w, h, \ell) - 0.1 \text{ in.} \leq 0 \qquad \text{(S5.2c)}$$

$$0.5 \text{ in.} \leq w \leq 1.5 \text{ in.,} \quad 0.5 \text{ in.} \leq h \leq 1.5 \text{ in.,} \quad 5 \text{ in.} \leq \ell \leq 15 \text{ in.} \qquad \text{(S5.2d)}$$

in which ρ is the mass density, σ_{mp} is the maximum principal stress, and δ_y is the maximum displacement in the Y-direction (vertical).

A static design study is created for the beam, with the boundary and load conditions shown in Figure S5.7. The material chosen is cast stainless steel from the simulation material library. The unit system chosen is IPS (in.-lbf-s). Also shown in Figure S5.7 are the finite element mesh (default median mesh) and the fringe plot of the maximum principal stress (or first principal stress).

In the current design, the maximum principal stress (g_1) is 89.2 ksi, which is greater than the required upper bound of 65 ksi (tensile strength, assumed). This performance constraint is violated (hence, design is infeasible).

The second performance constraint of maximum y-displacement (g_2) is 0.1454 in., which is greater than the required upper bound of 0.1 in. Again, this performance constraint is violated (design is infeasible).

We are changing the width, height, and length of the beam to find an optimal solution. We choose a 0.5-in. step size for varying the width and height design variables, and a 5-in. step size for the length design variable. As a result, each design variable is varied three times. For example, the width design variable is changed from its lower bound of 0.5 in. to 1 in., and then to its upper bound of 1.5 in. Similarly, three changes take place for the height and length design variables, respectively. These changes in design variables are combined to create 27 (3 × 3 × 3) scenarios to search for a possible optimal solution.

A total of 29 static analyses (27 plus the initial and current designs) will be carried out for the study. In the graphics window, the design of all scenarios will appear in the beam with changing dimensions.

S5.2.2 USING SOLIDWORKS

S5.2.2.1 *Open SolidWorks part*

Open the solid model of the cantilever beam downloaded from the book companion Web site, filename: Cantilever Beam.SLDPRT in folder: Project S5 Examples/Lesson S5.2 Cantilever Beam. This cantilever example is similar to that of Project S3.2 of Chang (2013a). After opening this file, you should see that there is already a static study created for you. You may want to browse the stress and displacement results of the study. Also, please check the unit system (IPS), and take a look at the finite element mesh.

S5.2.2.2 *Insert a design study*

From the pull-down menu, choose
    Insert > Design Study > Add
    and name the study My_Opt.

S5.2.2.3 *Add design variables*

In the Design Study Manager window (Figure S5.8), click "Click here to add variables" under Variables, and choose Add Parameter to add design variable. In the Parameters dialog box (Figure S5.9), enter Width for name, and then pick the width dimension from the solid model in the graphics window. The selected dimension should appear under Value in the Parameters dialog box (Figure S5.9). Click Apply to accept the parameter.

In the Design Study Manager window (Figure S5.10), Min, Max, and Step (0.5, 1.5, and 0.5 in., respectively) are automatically added to the length design variable. These are lower and upper bounds of the width design variable. Step: 0.5 in. implies that the design will be changed for every 0.5 in. We will stay with this default setup.

Repeat the same steps to define the height and length dimensions. All three variables should appear in the Parameters dialog box. Click OK to close the Parameters dialog box.

You may select Height and Length parameters in the Design Study Manager window, and review their respective Min, Max, and Step values.

S5.2.2.4 *Add constraints*

In the Design Study Manager window (Figure S5.8), click "Click here to add constraints" under Constraints, and choose Add Sensor to add a constraint. In the Property Manager dialog box (Figure S5.11), choose Simulation Data, Displacement, UY: Y Displacement, in, Model Max, and then Accept (checkmark). The constraint should appear in the Design Study Manager window, under Constraints. Choose "Is less than," and enter 0.1 in. for its upper bound (see Figure S5.12).

Repeat the same to create a stress constraint by choosing Simulation Data, Stress, P1: 1st Principal Stress, psi, Model Max (see Figure S5.13). The stress constraint should appear in the Design Study Manager window, under Constraints. Choose "Is less than," and enter 65,000 psi for its upper bound (Figure S5.14).

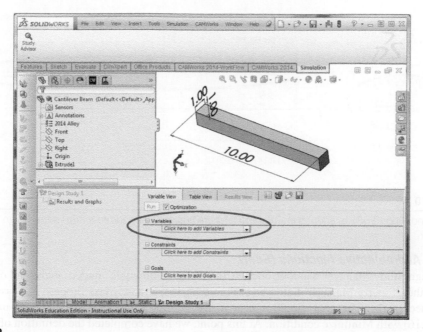

FIGURE S5.8

Add design variables.

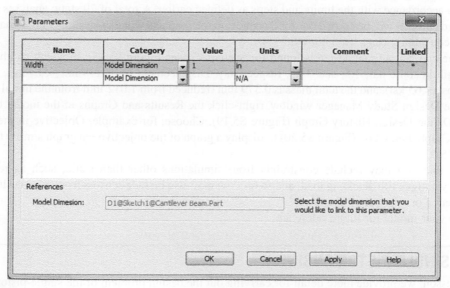

FIGURE S5.9

The selected width dimension appears under value in the parameters dialog box.

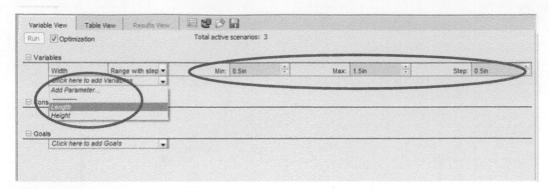

FIGURE S5.10

Min, Max, and Step (0.5, 1.5, and 0.5 in., respectively) are automatically added to the width design variable.

S5.2.2.5 Add objective functions (Goal)

In the Design Study Manager window, click "Click here to add goals" under Goals, and choose Add Sensor. In the Property Manager dialog box (Figure S5.15), choose Mass Properties, Mass, and then Accept (checkmark). The goal should appear in the Design Study Manager window, under Goals (Figure S5.16) with Minimize condition. At this point, we have completed the definition of the design problem. We are ready to carry out an optimization for this beam example.

S5.2.2.6 Run optimization

In the Design Study Manager window, click Run (Figure S5.16). Design variables will be varied one at a time in accordance with the limits and steps as defined earlier. A total of 29 static analyses (27 plus initial and current) will be carried out for the study. In the graphics window, design of all scenarios will appear sequentially (e.g., see Figure S5.17).

At the end, an optimal design is found for Scenario 7 (see Figure S5.18), in which width, height, and length become 0.5, 1.5, and 5 in., respectively. Displacement is 0.0114 in. (<0.1 in.), stress is 37,238 psi (<65 ksi), and the total mass is 0.379 lbm (reduced from 1.012 lbm from the initial design).

In the Design Study Manager window, right-click the Results and Graphs in the model tree, and choose Define Design History Graph (Figure S5.19). Choose, for example, Objective in the Design History Graph dialog box (Figure S5.20) to display a graph of the objective—a graph similar to that of Figure S5.5.

Note that you may include constraints from simulations other than static, such as resonance frequency, fatigue life, buckling load, and so on. You may also include sensors from motion simulation for objective or constraint functions.

Save your model for future reference.

S5.3 SINGLE-PISTON ENGINE

In this lesson, we provide more detail for carrying out the design problem of the single-piston engine (see Figure 1.18) discussed in Chapter 1. Recall in Section 1.5 that a cross-functional team was asked to develop a new model of the engine, with a 30% increment in both maximum torque and horsepower at the speed 1215 rpm. The design of the new engine was carried out at two interrelated levels: the system level and component level. At the system level, due to the increase in power and torque, the bore diameter of the engine case was increased from 1.416 to 1.6 in.; the crank length was increased from 0.5833 to 0.72 in.; and the connecting road length was increased from 2.25 to 2.49 in., as listed in Table 1.1. Also, the change causes the brake effective pressure P_b to increase from 140 to 180 lbf, causing peak combustion load increases from 400 to 600 lbf that act on the top face of the piston. The load magnitude and path applied to the major load-bearing components, such as the connecting rod and crankshaft, are therefore altered. Reaction forces applied to components, such as the connecting rod, increase. The increase in reaction forces raises concerns about the structural integrity of these load-bearing components in the engine. Changes may need to be made to strengthen the design of these components, which may cause an increase in manufacturing time and cost.

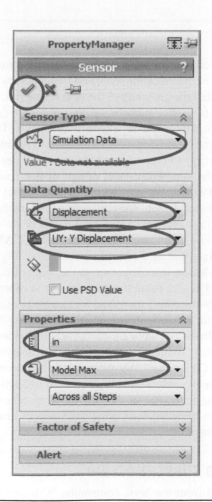

FIGURE S5.11

The Property Manager dialog box.

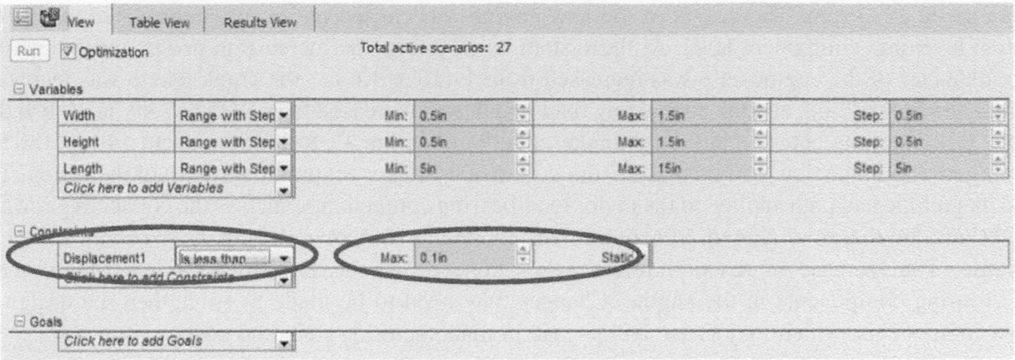

FIGURE S5.12

Displacement constraint defined.

In Section 1.5, we discuss component-level design using the connecting rod as an example. We included structural analysis for stress, buckling load, and frequency, as well as machining time for the connecting rod. We parameterized the geometry of the connecting rod, and selected three dimensions: the diameter of the large hole (d_{32}), the diameter of the small hole (d_{31}), and the thickness (d_7), as design variables (see Table 1.2). We carried out a sensitivity analysis for these structural performance measures and manufacturing time by varying these three design variables. We conducted design trade-offs and carried out a what-if study to come up with an improved design for the connecting rod.

In this lesson, we repeat the component-level design of the connecting rod using SolidWorks Motion, Simulation, and CAMWorks. We start with a motion model of the new design; that is, the changes in bore diameter, crankshaft length, and connecting rod length are made according to the new values shown in Table 1.1. We have prepared a structural model of the connecting rod with simulation results in static, frequency, buckling, and fatigue analyses. Also, a CAMWorks model is prepared. All are carried out for the new design; that is, the length of the connecting rod is 2.49 in. (increased from 2.25 in.). The objective of this lesson is to demonstrate the steps of conducting an interactive design process to achieve an improved design using the connecting rod example.

We start in Section S5.3.1 by describing the model files needed to carry out this exercise. We briefly describe the simulation results for the current design without showing the detailed steps in creating the models and running the simulations. Readers are encouraged to review tutorial projects S1–S4 for a good standing in solid modeling, motion analysis, structural analysis, and machining simulation, respectively, using SolidWorks CAD/CAE/CAM tools. With an understanding of the current design of the connecting rod, we move on to conduct sensitivity analysis in Section S5.3.2 and carry out a what-if study in Section S5.3.3. Note that some of the simulation results may not be consistent to those presented in Section 1.5, which were obtained using Mechanism, Pro/MECHANICA Structure, and Pro/MFG, with slightly different setups.

S5.3.1 MODEL FILES AND CURRENT DESIGN

The example files downloaded from the book companion Web site should contain those shown in Figure S5.21(a) under the folder named "Project S5 Examples/Lesson S5.3 Single-Piston Engine." The

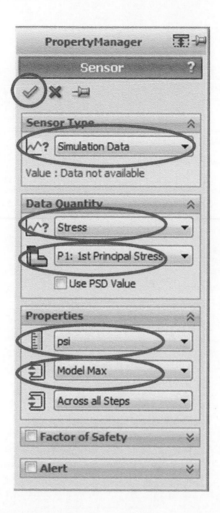

FIGURE S5.13

Defining stress constraint.

subfolder named "Connecting Rod" contains another three subfolders: Current Design, Sensitivity Analysis, and Whatif, as shown in Figure S5.21b. The "Current Design" folder contains simulation result files for the connecting rod at the current design, including static, frequency, buckling, and fatigue studies. Under "Sensitivity Analysis," there are three subfolders: Perturb d7, Perturb d31, and Perturb d32. All are empty. We will store perturbed models and simulation results for the respective three design variables for sensitivity analysis using finite difference method. In the "Whatif" folder, there is a spreadsheet file (Whatif.xlsx) prepared for carrying out a what-if study.

In the following sections, we briefly go over the models and simulations for motion and structural analyses, as well as a machining simulation.

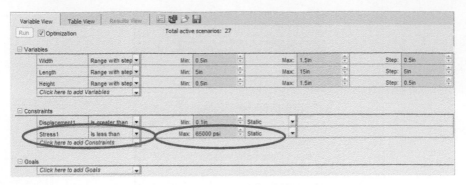

FIGURE S5.14

Stress constraint defined.

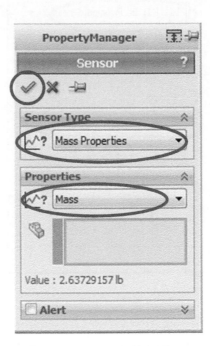

FIGURE S5.15

Defining goal.

S5.3.1.1 Motion model and simulation

Start SolidWorks and open "Engine.sldasm" under the folder "Lesson S5.3 Single-Piston Engine." You should see the engine assembly, as shown in Figure S5.22a. There are 21 assembly mates defined to create the motion model (see Figure S5.22a). One of them, Angle1, which is created to define the initial rotation angle of the spinner, is suppressed to provide the free degree of freedom. You may want

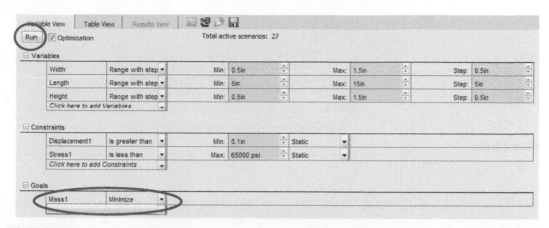

FIGURE S5.16

Goal (Minimize Mass1) defined.

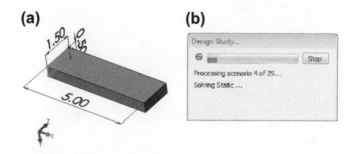

FIGURE S5.17

A total of 29 static analyses being carried out for the study. (a) Solid model changed at a scenario. (b) Design study status dialog box.

29 of 29 scenarios ran successfully. Design Study Quality: High

		Current	Initial	Optimal (7)	Scenario 1	Scenario 2	Scenario 3	Scenario 4	Scenario 5
Length		10in	10in	5in	5in	10in	15in	5in	10in
Height		1in	1in	1.5in	0.5in	0.5in	0.5in	1in	1in
Width		1in	1in	0.5in	0.5in	0.5in	0.5in	0.5in	0.5in
Displacement1	< 0.1in	0.145421in	0.145421in	0.0114011in	0.2905462in	2.3188248in	7.8263908in	0.0371238in	0.2913775in
Stress1	< 65000 psi	89224.04 psi	89224.04 psi	37238.18 psi	277730.6 psi	602408.4 psi	816796.8 psi	76076.53 psi	160168.5 psi
Mass1	Minimize	1.01156 lb	1.01156 lb	0.379336 lb	0.126445 lb	0.252891 lb	0.379336 lb	0.252891 lb	0.505782 lb

FIGURE S5.18

Results tabulated for the 29 scenarios.

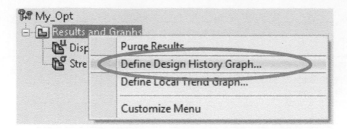

FIGURE S5.19

Define design history graph.

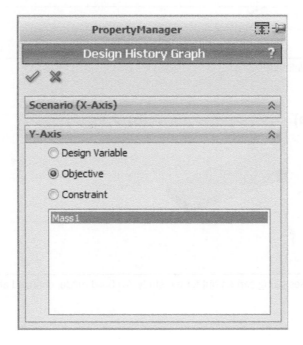

FIGURE S5.20

Choose objective for a history graph.

to browse some of the assembly mates to gain a better understanding of the assembly and motion model, especially, "Concentric7" between the connecting rod and piston pin (see Figure S5.22b). This mate is chosen to monitor the reaction force acting on the connecting rod for structural analyses.

A motion simulation called "Motion Study" has been created. An initial angular velocity is defined and a force is applied to the piston at the beginning of the simulation. The initial velocity of 1215 rpm is defined on the crankshaft, as shown in Figure S5.23. A force of 600 lbf is applied to the top face of the piston for a short duration of 0.002 s, corresponding to about 6.5° of one rotation, as shown in Figure S5.24, to mimic the engine combustion force. Note that the force is applied only once at the

(a)

Name	Type	Size
Connecting Rod	File folder	
backplate_.sldprt	SLDPRT File	67 KB
bolt516_.sldprt	SLDPRT File	58 KB
bolt716_.sldprt	SLDPRT File	58 KB
bolt1316_.sldprt	SLDPRT File	58 KB
bushing1_.sldprt	SLDPRT File	57 KB
bushing2_.sldprt	SLDPRT File	57 KB
case_.sldprt	SLDPRT File	347 KB
connectingrod_.sldprt	SLDPRT File	255 KB
crankshaft_.sldprt	SLDPRT File	66 KB
cylinderfins_.sldprt	SLDPRT File	73 KB
cylinderhead_.sldprt	SLDPRT File	112 KB
cylindersleeve_.sldprt	SLDPRT File	67 KB
drivewasher_.sldprt	SLDPRT File	61 KB
drivewasherpin_.sldprt	SLDPRT File	59 KB
Engine.sldasm	SLDASM File	18,266 KB
piston_.sldprt	SLDPRT File	67 KB
pistonpin_.sldprt	SLDPRT File	59 KB
spacer_.sldprt	SLDPRT File	60 KB
spinner_.sldprt	SLDPRT File	58 KB
spinner_pro.sldprt	SLDPRT File	170 KB

Folder that contains simulation result files for the connecting rod component

SolidWorks part and assembly files

Engine assembly Open to view motion analysis results

(b)

- Connecting Rod
  - Current Design
  - Sensitivity Analysis
    - Perturb d7
    - Perturb d31
    - Perturb d32
  - Whatif

FIGURE S5.21

Files and folder under "Lesson S5.3 Single-Piston Engine." (a) Files and subfolder and (b) subfolders included in "Connectingrod."

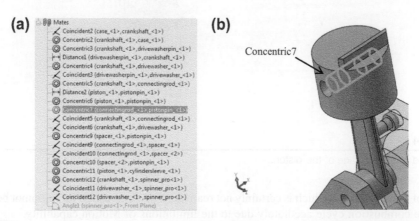

(a)

Mates
- Coincident2 (case_<1>,crankshaft_<1>)
- Concentric2 (crankshaft_<1>,case_<1>)
- Concentric3 (crankshaft_<1>,drivewasherpin_<1>)
- Distance1 (drivewasherpin_<1>,crankshaft_<1>)
- Concentric4 (crankshaft_<1>,drivewasher_<1>)
- Coincident3 (drivewasherpin_<1>,drivewasher_<1>)
- Concentric5 (crankshaft_<1>,connectingrod_<1>)
- Distance2 (piston_<1>,pistonpin_<1>)
- Concentric6 (piston_<1>,pistonpin_<1>)
- Concentric7 (connectingrod_<1>,pistonpin_<1>)
- Coincident5 (crankshaft_<1>,connectingrod_<1>)
- Coincident6 (crankshaft_<1>,drivewasher_<1>)
- Concentric9 (spacer_<1>,pistonpin_<1>)
- Coincident9 (connectingrod_<1>,spacer_<1>)
- Coincident10 (connectingrod_<1>,spacer_<2>)
- Concentric10 (spacer_<2>,pistonpin_<1>)
- Concentric11 (piston_<1>,cylindersleeve_<1>)
- Concentric12 (crankshaft_<1>,spinner_pro<1>)
- Coincident11 (drivewasher_<1>,spinner_pro<1>)
- Coincident12 (drivewasher_<1>,spinner_pro<1>)
- Angle1 (spinner_pro<1>,Front Plane)

(b)

Concentric7

FIGURE S5.22

Engine motion model. (a) Assembly mates. (b) Engine assembly with mate "Concentric7" highlighted (some parts were hidden for a clear view).

FIGURE S5.23

Initial velocity dialog box.

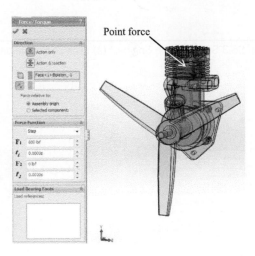

FIGURE S5.24

Force applied at the top face of the piston.

beginning of the simulation, which is certainly not realistic. The combustion force cannot be applied to every single combustion cycle accurately due to the limitations of Motion capability.

It will be more realistic if the force can be applied when the piston just starts moving downward (negative *Y*-direction) and can be applied only for a selected short period. In order to do so, we will have to define sensors that monitor the position of the piston for the combustion load to be activated.

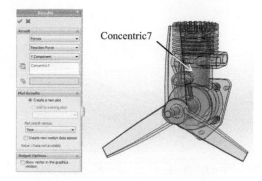

FIGURE S5.25

Result plot of reaction force in the *Y*-direction defined at Concentric7.

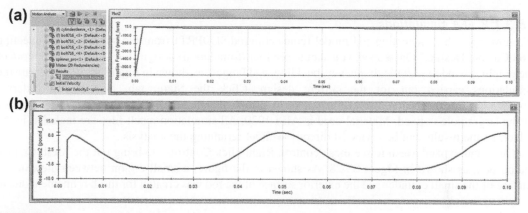

FIGURE S5.26

Reaction force at Concentric7. (a) Maximum of −600 lbf at the beginning. (b) Zoomed-in view showing small oscillations due to inertia force.

Unfortunately, such a capability is not available in Motion. Therefore, the force is simplified as a step function of 600 lbf along the negative *Y*-direction and applied for 0.002 s at the beginning of the simulation.

One result plot is created, which shows the reaction force at the upper joint of the connecting rod (Concentric7). The maximum reaction force is about −600 lbf at the beginning of the simulation. Click Results from the Motion Manager window to see an existing plot (Plot2<Reaction Force2>). Right-click Plot2 and choose Edit Feature to see the plot definition (Figure S5.25). Close the Results dialog box, right-click Plot2<Reaction Force2>, and then choose Show Plot. The graph of results appears as in Figure S5.26a. The reaction force due to inertia can be observed by zooming in the y-axis, in which the force reaches about 10 lbf, as shown in Figure S5.26b. For more information regarding the use of motion, readers are referred to Chang (2014b).

FIGURE S5.27

Simulation model and the four design studies created at the current design.

S5.3.1.2 FEA simulations

We now open the connecting rod model (connectingrod.SLDPRT) from the subfolder "Connecting Rod/Current Design." Once open, you should see four (Simulation) design study tabs at the bottom of the graphics window—static, frequency, buckling, and fatigue, as shown in Figure S5.27. These are the simulations carried out for the current design. Click the Static tab, and choose External Loads > BearingLoads-1 to display the bearing load applied at the small hole of the rod. The inner face of the larger hole is fixed. Material used is AISI 1020. All simulation result files are included in the folder, so the results can be reviewed directly without rerunning the analysis.

Click and expand Mesh to see mesh control. Right-click Control-1 to bring out the Mesh Control dialog box, as shown in Figure S5.28. As shown in the figure, a small region between the exterior surface of the small cylinder and the exterior surface of the rod was created for mesh refinement due to

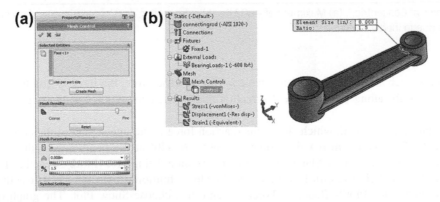

FIGURE S5.28

Finite element mesh of the connecting rod. (a) Dialog box for mesh control. (b) Small surface created for mesh refinement.

(a)   **(b)**

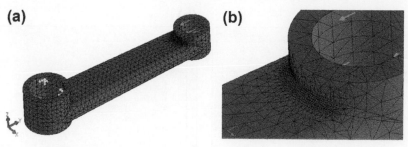

FIGURE S5.29

Finite element mesh. (a) Entire model. (b) Zoomed-in view of the refined area.

the high stress concentration found in the area. There are about 12,000 tetrahedron finite elements created for the rod with approximately 58,000 degrees of freedom. The mesh is shown in Figure S5.29.

The static analysis results show that the maximum von Mises stress occurs in the area where mesh is refined, as shown in Figure S5.30. The stress is 35.93 ksi. The yield strength of AISI 1020 is approximately 51 ksi. The safety factor for current design is about 1.42.

For frequency and buckling analyses, a fine mesh is chosen without any mesh control. The first five resonance frequencies and first five buckling loads are shown in Figures S5.31a and b, respectively. As shown in Figure S5.31a, the first (lowest) resonance frequency is 1431 Hz or 86,860 rpm, which is way higher than the operating frequency in the range of thousands rpm. Also, as shown in Figure S5.31b, the first (lowest) buckling load factor is 18.7, implying that the actual load that causes the connecting rod to buckle is $18.7 \times 600 = 11,220$ lbf, much higher than the firing load.

For fatigue analysis, we defined the 600 lbf load as a repeated load, where maximum is -600 lbf and minimum is 0 (very small inertia force in oscillation). Therefore, we choose the Zero Based (LR $= 0$) option (LR: load ratio), as shown in Figure S5.32a. Also, 1000 repetitions is defined as one

(a)   **(b)**

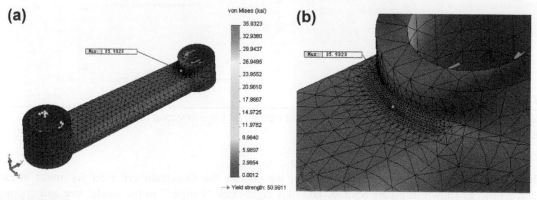

FIGURE S5.30

von Mises stress fringe plot. (a) Entire connecting rod model. (b) Zoomed-in view to the high-stress concentration area.

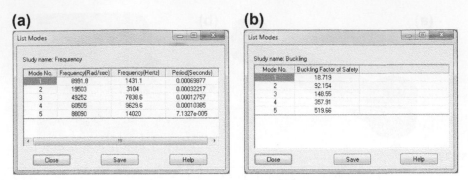

FIGURE S5.31

Simulation results. (a) First five resonance frequencies. (b) First five buckling loads.

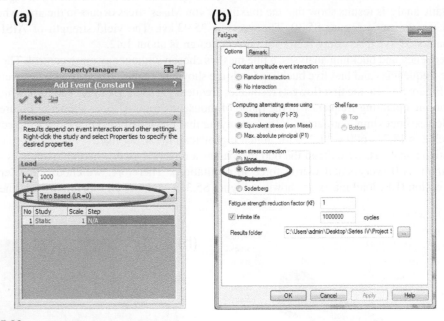

FIGURE S5.32

Setup for a fatigue analysis. (a) Fatigue (load) event defined for the connecting rod. (b) Goodman criterion chosen for mean stress correction.

block. Because the mean stress is nonzero, we choose the Goodman criterion for mean stress correction, as shown in Figure S5.32b. You may right-click "Fatigue" in the model tree and choose Properties to bring out the Fatigue dialog box of Figure S5.32b.

In addition, the S–N diagram and associated properties for AISI 1020 are shown in Figures S5.33a and b, respectively. Figure S5.33b appears by right-clicking the "connectingrod (-ASME Austenitic

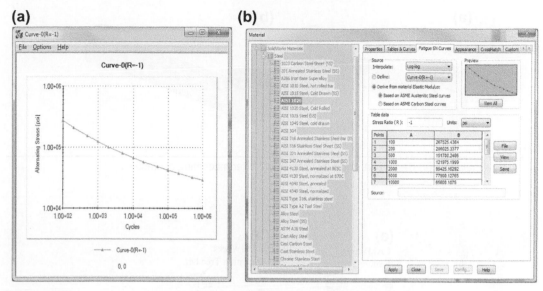

FIGURE S5.33

Material properties for fatigue analysis. (a) S–N diagram. (b) The Material dialog box.

Steel-)" in the simulation model tree and choosing Apply/Edit Fatigue Data. In the Material dialog box, click View to show the S–N diagram in Figure S5.33a. The data in the Material dialog box shows that the endurance limit is about 29 ksi (corresponding to a fatigue cycle of 1,000,000).

Note that the stress in the static analysis is 35.9 ksi, which is greater than the endurance limit of the material. If the load (therefore, stress) is fully reversed, the fatigue life would have been about 2.3×10^5, which can be estimated using the S–N diagram shown in Figure S5.33a.

The simulation results indicate that the fatigue life is infinite ($>10^6$) for the entire model because the stress is not fully reversed. For a more detailed discussion on fatigue analysis using Simulation, readers are referred to Project S3.4, which is part of Chang (2013a).

S5.3.1.3 CAMWorks machining

Manufacturing simulation was carried out using CAMWorks for the current design. We assume the machining operations start from a green part made by casting, as shown in Figure S5.34a. The machining operations remove materials on the top and bottom faces of the large and small cylinders, and drill two holes, as shown in Figure S5.34b.

Entering CAMWorks from SolidWorks is straightforward. You may click the CAMWorks Feature Tree tab or Operation Tree tab to browse existing machining entities, as shown in Figure S5.35a and b.

If you do not see the CAMWorks pull-down menu, the CAMWorks Feature Tree tab, or the Operation Tree tab, you may have not activated the CAMWorks add-in module. To activate the CAMWorks module, choose from the pull-down menu:

Tools > Add-Ins.

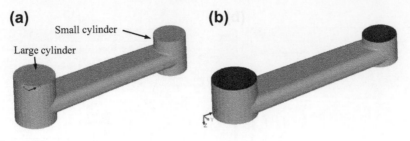

FIGURE S5.34

Machining simulation. (a) Green part made by casting. (b) After machining operations.

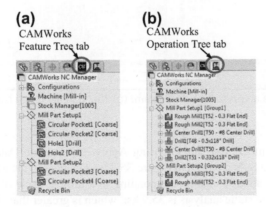

FIGURE S5.35

Virtual machining model created in CAMWorks. (a) The six machinable features under the Feature Tree tab. (b) The eight machining operations under the Operation Tree tab.

In the Add-Ins dialog box, click CAMWorks in both boxes (Active Add-Ins and Start Up), and then click OK. You should see that CAMWorks is added to the pull-down menu.

Click the Feature Tree tab. You should see that two mill part setups (Figure S5.35a), Mill Part Setup1 and Mill Part Setup2, are included for machining the faces on both sides of the connecting rod. Mill Part Setup1 consists of four machinable features: Circular Pocket1, Circular Pocket2, Hole1, and Hole2 and Mill Part Setup2 has two features, Circular Pocket3 and Circular Pocket4, as shown in Figure S5.35a.

Click the Operation Tree tab to see the eight machining operations created under Mill Part Setup1 and Mill Part Setup2, as shown in Figure S5.35b. Note that each Rough Mill operation is for machining a circular surface, whereas each hole feature will result in a center drill operation plus a follow-on hole-drilling operation. The machine zeroes for the respective setups are shown in Figures S5.36a and b. Note that the Z-axis of the machine zero of Setup1 is pointing in the negative Z-axis of the world

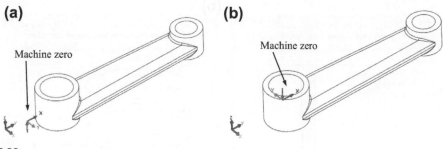

FIGURE S5.36

Machine zeroes. (a) Mill Part Setup1. (b) Mill Part Setup2.

coordinate system. For Setup2, the Z-axis is in parallel with that of the world coordinate system. In practice, jigs and fixtures must be designed and fabricated to hold the green part in support of the machining operations at shop floor. The machine zeroes may need to be adjusted according to the fixture design.

Right-click Rough Mill1 (under Mill Part Setup1) and choose Edit Definition; the Operation Parameters dialog box appears. A 3/8 EM CRB 2FL flat end cutter is chosen for all Rough Mill operations, as shown in Figure S5.37a. The machining parameters, such as spindle speed and feed per tooth, generated from CAMWorks database are under the F/S tab, as shown in Figure S5.37b. Setups for the roughing operations are under Roughing tab, as shown in Figure S5.37c. The machining time, such as Rough Mill1, which removes a thin-layer material on the bottom face of the large cylinder, is given in the Optimize tab shown in Figure S5.37d (in this case, 0.381 min). Note that this includes only the actual machining time when the cutter is in motion. Setup time, tool change time, and so forth are not included.

Note that instead of using default values, we modify the machining parameters as follows: for rough mill, feed per tooth changed from 0.006 to 0.001 in.; for center drill, the Z feed rate changed from 6 to 2 in./min; and for hole-drilling, the Z feed rate changed from 8 to 1 in./min.

You may see the complete machining operations by choosing the following from the pull-down menu:

CAMWorks > Simulate Toolpath

Then, click the Run key ($\blacktriangleright$) in the Toolpath Simulation dialog box (Figure S5.38) to play the toolpath simulation.

The overall machining time can be found by right-clicking the mill part setup, such as Mill Part Setup1, and choosing Edit Definition. In the Part Setup Parameters dialog box (Figure S5.39), choose the Statistics tab to see the machining time for the entire setup. The machining times for Setup1 and Setup2 are 2.32 and 1.81 min, respectively. Total is 4.13 min. You may easily double-check these numbers by adding the times from individual operations. For example, for Setup1, we add the machining times of the six individual operations as $0.381 + 0.206 + 0.238 + 0.768 + 0.166 + 0.567 = 2.326$ min, which is the same as the time obtained from Setup1.

A brief description of the construction of the machining model can be found in Appendix A. More detailed discussion on CAMWorks can also be found in Project S4 of Chang (2013b).

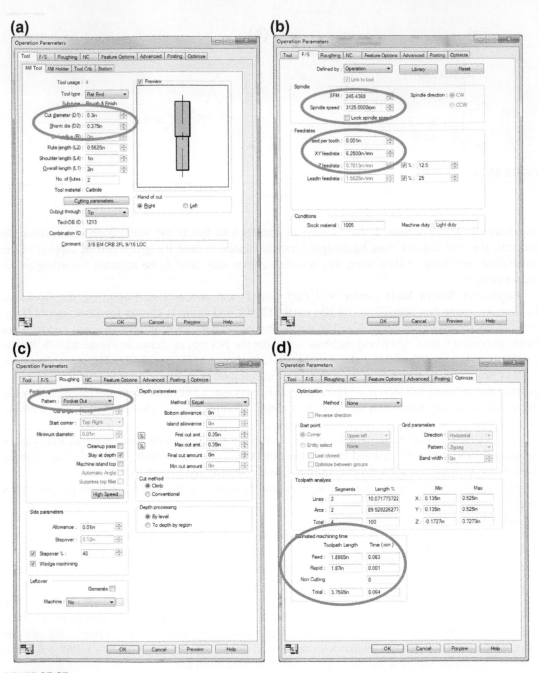

FIGURE S5.37

Definition of rough mill operations. (a) Tool chosen, (b) machining parameters, (c) roughing options, and (d) machining time.

FIGURE S5.38

The Toolpath Simulation dialog box.

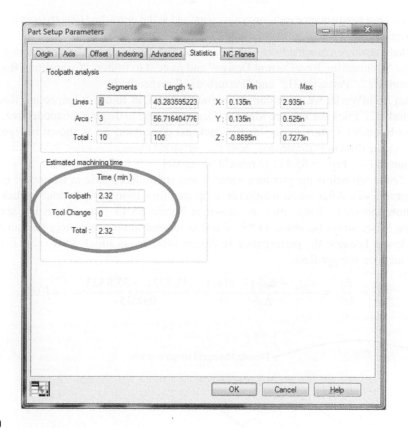

FIGURE S5.39

The Part Setup Parameters dialog box.

S5.3.2 SENSITIVITY ANALYSIS

The three design variables defined for the connecting rod are the diameter of the larger hole (d_{32}), diameter of the small hole (d_{31}), and the thickness (d_7). For the current design, the values of the design variables are, respectively, $d_{32} = 0.5$ in., $d_{31} = 0.334$ in., and $d_7 = 0.25$ in.

To carry out sensitivity analysis for the gradients of the performance measures, including von Mises stress, lowest resonance frequency, lowest buckling load, fatigue life, volume, and machining time with respect to the three design variables, we use the overall finite difference method discussed in Chapter 3.

The approach is very simple. We perturb the individual design variables, one at a time, by 1% of its design variable value. For example, we change the thickness d_7 from 0.25 to 0.2525 in., regenerate the model, and then carry out simulations to find the values of the performance measures at the perturbed design. Thereafter, the sensitivity of a performance $\psi(x)$ can be approximated as:

$$\frac{\partial \psi}{\partial x} \approx \frac{\psi(x + \Delta x) - \psi(x)}{\Delta x} \tag{S5.3}$$

where $\psi(x + \Delta x)$ is a performance measure value obtained at the perturbed design from simulation, and Δx is the design perturbation (e.g., $\Delta d_7 = 0.0025$ in.).

The detailed steps are described next. We copy the SolidWorks part file connectingrod.SLDPRT from the folder "Connecting Rod/Current Design" and paste it to folders "Connecting Rod/Sensitivity Analysis/Perturb d7," "Perturb d31," and "Perturb d32," respectively.

Then, from SolidWorks we open connectingrod.SLDPRT in folder "Connecting Rod/Sensitivity Analysis/Perturb d7," click the Feature Manager Design tree tab to display the model tree, and double-click the Boss-Extrude1 feature to display the thickness dimension d_7, as shown in Figure S5.40.

Double-click the thickness dimension, enter the perturbed dimension value 0.2525 in., and click the Regenerate button (see Figure S5.41) to rebuild the model.

Click the Static tab below the graphics window, and right-click Static in the model tree to choose Run (see Figure S5.42). After the simulation is completed (in a short period), double-click Results and Stress1 to show the stress fringe plot, as shown in Figure S5.43. The fringe plot shows that the maximum von Mises stress becomes 34.83 ksi and is located in approximately the same location as the original design because the perturbation in design variables is small. This value is plugged into Eq. S5.3 to calculate the gradient:

$$\frac{\partial \psi}{\partial x} = \frac{\partial \sigma}{\partial d_7} \approx \frac{\sigma(d_7 + \Delta d_7) - \sigma(d_7)}{\Delta d_7} = \frac{34.8321 - 35.9323}{0.0025} = -440.08$$

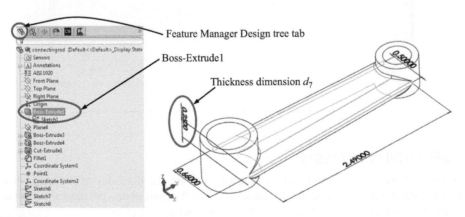

FIGURE S5.40

Perturbing design variable d_7 from 0.25 to 0.2525 in.

FIGURE S5.41

Changing the dimension value for the thickness design variable (d_7) to 0.2525 in. and rebuilding the model.

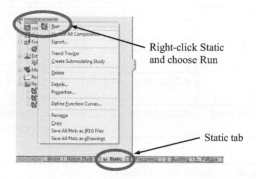

FIGURE S5.42

Rerun Static simulation for the perturbed design for $d_7 = 0.2525$ in.

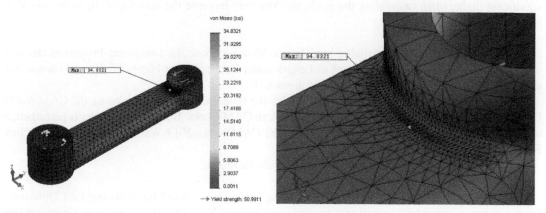

FIGURE S5.43

von Mises stress fringe plot at the perturbed design of $d_7 = 0.2525$ in.

We repeat the same steps to rerun simulations for frequency, buckling, and fatigue. The results are shown in Figure S5.44.

Note that volume of the connecting rod can be obtained by choosing from the pull-down menu: Tools > Mass Properties.

Sensitivity Matrix

	Current value	d31	d32	d7
Perturbation		3.34000E-03	5.00000E-03	2.50000E-03
von Mises Stress (psi)	3.59323E+01	3.58461E+01	3.52391E+01	3.48321E+01
Sensitivity		-2.58084E+01	-1.38640E+02	-4.40080E+02
Natural Frequency (Hz)	1.43110E+03	1.43710E+03	1.43250E+03	1.44860E+03
Sensitivity		1.79641E+03	2.80000E+02	7.00000E+03
Buckling Load Factor	1.87190E+01	1.87480E+01	1.87400E+01	1.92910E+01
Sensitivity		8.68263E+00	4.20000E+00	2.28800E+02
Fatigue Life (cycles)	1.00000E+06	1.00000E+06	1.00000E+06	1.00000E+06
Sensitivity		0.00000E+00	0.00000E+00	0.00000E+00
Volume (in^3)	4.02224E-01	4.01491E-01	3.99987E-01	4.05161E-01
Sensitivity		-2.19461E-01	-4.47400E-01	1.17480E+00
Manufacturing Time (min)	4.13000E+00	4.13000E+00	4.13000E+00	4.13000E+00
Sensitivity		0.00000E+00	0.00000E+00	0.00000E+00

FIGURE S5.44

Sensitivity spreadsheet (in file Whatif.xlsx of Whatif folder).

Volume and other mass properties are shown in the Mass Properties dialog box, as shown in Figure S5.45. Note that by default, only two or three digits are shown, which may cause a loss of significant digits when calculating the gradients. You may increase the number of digits by choosing from the pull-down menu:

Tools > Options.

In the Document Properties dialog box (Figure S5.46), choose the Document Properties tab, and click Units. Pull down a data field, such as Length under Decimals (see Figure S5.46) and choose, for example, 123456 to increase the number of digits to 6.

Now, we rerun the machining simulation. Click the Model tab below, and choose the CAMWorks Operation Tree tab above the model tree to get back to CAMWorks. Because the model is perturbed, a warning message appears (see Figure S5.47). In the CAMWorks 2014 Warning dialog box, we click Full to update CAMWorks data.

Update the toolpath by choosing from the pull-down menu:

CAMWorks > Generate Toolpath

After regenerating the toolpath, the machining time can be found by choosing Edit Definition from setups or individual operations, as discussed before. Note that the machining times are unchanged because the change in the thickness design variable d_7 is not affecting the machining operations, which remove materials at the top and bottom faces of the two cylinders and drill two holes for the respective cylinders. Therefore, sensitivity of the machining time with respect to the thickness design variable is zero.

We repeat the steps discussed earlier for design variables d_{31} and d_{32}, by changing them to $d_{32} = 0.505$ in. and $d_{31} = 0.33734$ in. We carry out simulations and toolpath generations for the two perturbed designs. The sensitivity matrix for the three design variables are calculated as shown in Figure S5.44.

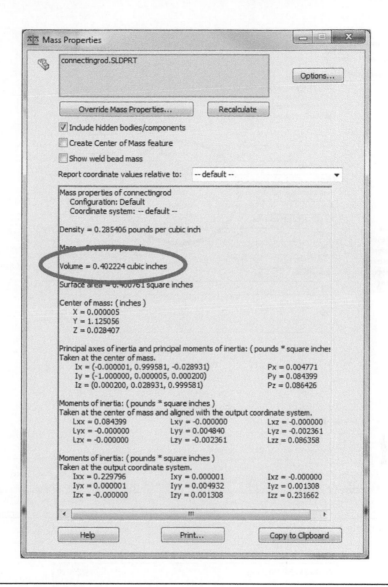

FIGURE S5.45

The Mass Properties dialog box.

In Figure S5.44, the sensitivity coefficients of the machining time for all three design variables are zero because the design perturbations did not affect the machining operations. Although the hole sizes are changed, we assume the use of drill bits of the same sizes as the holes; therefore, toolpath and machining time are unchanged. Also, the sensitivity coefficients of the fatigue life are all zero because the fatigue life is infinite before and after design perturbations. Moreover, some of the coefficients may not have enough significant digits. For example, if only four digits were chosen in the Document

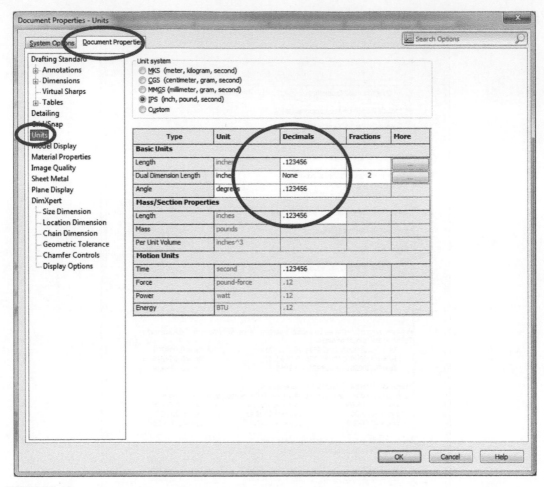

FIGURE S5.46

The Document Properties dialog box.

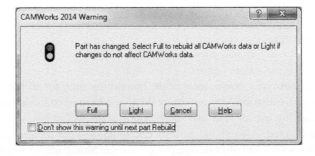

FIGURE S5.47

The CAMWorks 2014 Warning dialog box.

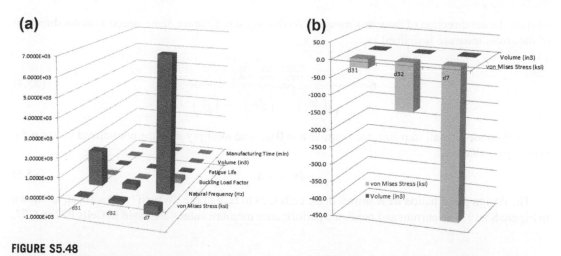

FIGURE S5.48

Bar chart of the sensitivity matrix. (a) The entire matrix. (b) Selected sensitivity coefficients: stress and volume performance measures.

Properties dialog box, the volume sensitivity of design variable d_{31} is calculated in spreadsheet as 2.70915E–05. In fact, the difference of volume before and after design perturbation is $0.4022 - 0.4015 = 0.0007$—only one significant digit. In order to increase the accuracy of the volume sensitivity coefficients, we need more digits. We did choose six decimals earlier in the Document Properties dialog box, so we should have enough significant digits. For performance measures, such as the resonance frequency, there may not be a way to increase the number of digits without extracting data from SolidWorks database via application protocol interface.

A bar chart of the entire sensitivity coefficient matrix (6×3 in this case) can be created using Spreadsheet, as shown in Figure S5.48a. The bar chart provides a global view on the influence of individual design variable to all performance measures and all design variables to a single performance measure. For example, natural frequency increases by increasing any of the three design variables. Among them, d_7 is the most significant and d_{32} is the least effective. Increasing design variable d_7 decreases stress, increases natural frequency, increases buckling load, and so on. Because the sensitivity magnitudes of different measures are different (by an order of magnitude), a common practice is normalizing these gradients by their respective maximum values for a better comparison of the effect of one variable across performance measures.

If we want to reduce stress and not increase volume much, we may graph the sensitivity coefficients for these two performance measures only, as can be seen in Figure S5.48b. The bar chart shows that the von Mises stress is reduced by increasing any of the three design variables. Among them, d_7 is the most significant and d_{31} is the least influencing. On the other hand, only a small change in volume is observed.

S5.3.3 WHAT-IF STUDY

We go over a what-if study for the goal of reducing von Mises stress and not increasing volume too much. As seen in the bar chart in Figure S5.48b, we increase all three design variables by using the steepest decent direction of the stress measure. As discussed in Chapter 3, the steepest decent direction of the stress measure is defined as:

$$\mathbf{n} = -\frac{\left[\frac{\partial \sigma}{\partial d_{31}} \quad \frac{\partial \sigma}{\partial d_{32}} \quad \frac{\partial \sigma}{\partial d_{7}}\right]^{T}}{\sqrt{\left(\frac{\partial \sigma}{\partial d_{31}}\right)^{2} + \left(\frac{\partial \sigma}{\partial d_{32}}\right)^{2} + \left(\frac{\partial \sigma}{\partial d_{7}}\right)^{2}}}$$

(S5.4)

We then use a small step size, for example $\alpha = 0.01$, and multiply it by the normalized vector $\mathbf{n}$ for design perturbations:

$$\delta \mathbf{x} = \alpha \, \mathbf{n}$$

(S5.5)

The design perturbation $\delta \mathbf{x}$ is shown in the cells C29 to E29 of the "Whatif" spreadsheet, as shown in Figure S5.49. The current and predicted performance measure values are listed in Cells F22 to G27,

	C34		f_x	=-D7/SQRT(SUMSQ(D7:F7))						
	A	B	C	D	E	F	G	H	I	J
1										
2		Sensitivity Matrix								
3										
4			Current value	d31	d32	d7				
5		Perturbation		3.34000E-03	5.00000E-03	2.50000E-03				
6		von Mises Stress (psi)	3.59323E+01	3.58461E+01	3.52391E+01	3.48321E+01				
7		Sensitivity		-2.58084E+01	-1.38640E+02	-4.40080E+02				
8		Natural Frequency (Hz)	1.43110E+03	1.43710E+03	1.43250E+03	1.44860E+03				
9		Sensitivity		1.79641E+03	2.80000E+02	7.00000E+03				
10		Buckling Load Factor	1.87190E+01	1.87480E+01	1.87400E+01	1.92910E+01				
11		Sensitivity		8.68263E+00	4.20000E+00	2.28800E+02				
12		Fatigue Life (cycles)	1.00000E+06	1.00000E+06	1.00000E+06	1.00000E+06				
13		Sensitivity		0.00000E+00	0.00000E+00	0.00000E+00				
14		Volume (in^3)	4.02224E-01	4.01491E-01	3.99987E-01	4.05161E-01				
15		Sensitivity		-2.19461E-01	-4.47400E-01	1.17480E+00				
16		Manufacturing Time (min)	4.13000E+00	4.13000E+00	4.13000E+00	4.13000E+00				
17		Sensitivity		0.00000E+00	0.00000E+00	0.00000E+00				
18										
19		What-if Study								
20										
21			d31	d32	d7	Current	Predicted	%Change	Analysis results	Accuracy
22		von Mises Stress (psi)	-2.5808E+01	-1.3864E+02	-4.4008E+02	3.59323E+01	3.13111E+01	-12.86	3.39039E+01	227.83%
23		Natural Frequency (Hz)	1.7964E+03	2.8000E+02	7.0000E+03	1.43110E+03	1.49960E+03	4.79	1.49880E+03	101.19%
24		Buckling Load Factor	8.6826E+00	4.2000E+00	2.2880E+02	1.87190E+01	2.09153E+01	11.73	2.09630E+01	97.87%
25		Fatigue Life (cycles)	0.0000E+00	0.0000E+00	0.0000E+00	1.00000E+06	1.00000E+06	0.00	1.00000E+06	NA
26		Volume (in^3)	-2.1946E-01	-4.4740E-01	1.1748E+00	4.02224E-01	4.11947E-01	2.42	4.11948E-01	99.99%
27		Manufacturing Time (min)	0.0000E+00	0.0000E+00	0.0000E+00	4.13000E+00	4.13000E+00	0.00	4.13000E+00	NA
28										
29		Design Perturbations (in.)	0.00056	0.00300	0.00952					
30		Current Value (in.)	0.334	0.5	0.25					
31		% Change	0.17	0.60	3.81					
32		New Design (in.)	0.33456	0.50300	0.25952					
33										
34		Steepest decent dir of stress	5.58475E-02	3.00007E-01	9.52301E-01					
35										

H ◀ ▶ H Sheet1 Sheet2 Sheet3

FIGURE S5.49

What-if spreadsheet (in file Whatif.xlsx of Whatif folder).

and the percentage changes between them are shown in Cells H22 to H27. The cells indicate that the changes are below 13%, which are acceptable. The stress is reduced by about 13%, with only a 2% increase in volume. The design seems to be desirable.

To carry out simulation of the new design, we first copy the SolidWorks model of the current design (located in folder: "Connecting Rod/Current Design") and paste it to folder "Connecting Rod/Whatif." We then open this model from SolidWorks and update the solid model by entering the new design variables: $d_{31} = 0.33456$, $d_{32} = 0.50300$, and $d_7 = 0.25952$. We regenerate the solid model and rerun the simulations. The simulation results are entered into the spreadsheet (Cells I22 to I27); the accuracy of the prediction and analysis results are shown in Cells J22 to J27. The predictions are accurate except for stress. This could be caused by either insufficient accuracy of the sensitivity or design perturbations that are not small enough.

In any case, the design is improved with the maximum von Mises stress reduced from 35.93 to 33.90 ksi and small increase in volume (0.4022 to 0.4119 in.^{3}). The process discussed above can be repeated if needed. For the time being, we are satisfied with the design and will stop here.

APPENDIX A. CREATING THE MACHINING MODEL IN CAMWORKS

In this appendix, we briefly discuss the key steps in creating the machining model of the connecting rod in CAMWorks. We assume readers are familiar with CAMWorks. If not, readers are referred to Project S4 of Chang (2013b) for more information before going over the discussion in this appendix.

First, the workpiece employed is an STL (stereolithography) model. To bring in an STL model as a workpiece, we first choose the CAMWorks Operation Tree tab, right-click Stock Manager, and choose Edit Definition. In the Stock Manager dialog box (Figure A.1), choose STL and click the button next to the STL path to choose the STL model and bring it in. The workpiece should appear in the graphics window.

Only the two holes are recognized automatically using Extract Machinable Features (pull-down menu: CAMWorks > Extract Machinable Features). Once extracted, they are listed under Mill Part

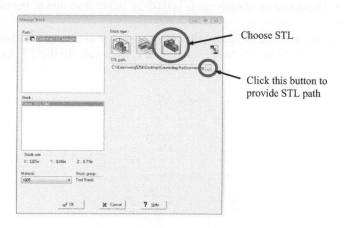

Choose STL

Click this button to provide STL path

FIGURE A.1

The Manager Stock dialog box.

FIGURE A.2

The two holes recognized and listed under Mill Part Setup1.

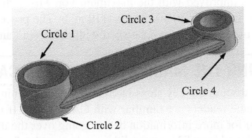

FIGURE A.3

Four circles sketched.

Setup1, as shown in Figure A.2. There is no need to do anything else. The toolpath of the center drill and follow-on hole-drilling operations can be generated automatically.

Machining features for the rough mills that remove materials on the top and bottom faces of the two cylinders cannot be extracted automatically. Sketches need to be created. Four circles are sketched, as shown in Figure A.3. The diameter of each circle is 0.05 in. larger than that of the circular edge on the respective cylinders.

Then, we create machining features manually by selecting the circles or sketches. First, right-click Mill Part Setup1 and choose New 2.5 Axis Feature (see Figure A.4). In the 2.5 Axis Feature Wizard

FIGURE A.4

Create a New 2.5 Axis Feature.

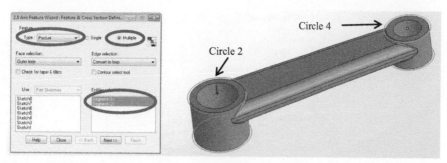

FIGURE A.5

Choosing Circles 2 and 4 for the 2.5 axis features.

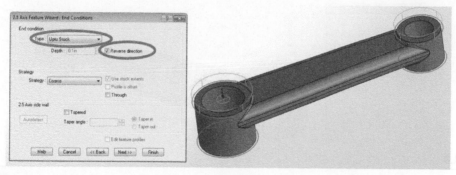

FIGURE A.6

Defining the end condition.

(see Figure A.5), select feature type Pocket, click Multiple, and then choose Circle 2 and Circle 4 in the graphics window (flip the model in the graphics window to pick the two circles). Or, simply click Sketch6 under the Use column. Click Next.

Choose upto Stock as End Condition, click Reverse direction (default), and then click Next and Finish (see Figure A.6). Close the wizard. Two newly created circular pockets (Circular Pocket1 and Circular Pocket2) are listed under Mill Part Setup1.

Next, we create another mill part setup (because part needs to be flipped before the other two faces can be machined).

Right-click Mill Part Setup1 and choose New Mill Part Setup (see Figure A.7). Rotate the model in the graphics window (if needed) and pick Face<1> as the plane to define Z-axis of machine zero, as shown in Figure A.8. Use the default machine zero.

Like before, we manually create machining features by selecting the remaining two circles (Circles 1 and 3).

First, right-click Mill Part Setup2 and choose New 2.5 Axis Feature. In the 2.5 Axis Feature Wizard (see Figure A.9), select feature type Pocket, click Multiple, and then choose Sketches 7 and 8 under the Use column. Click Next.

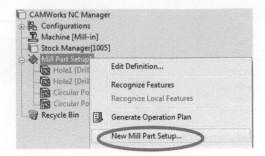

FIGURE A.7

Creating a new mill part setup.

FIGURE A.8

Setting up the machine zero for the new mill part setup.

FIGURE A.9

Choosing Sketches 7 and 8 for the 2.5 axis features.

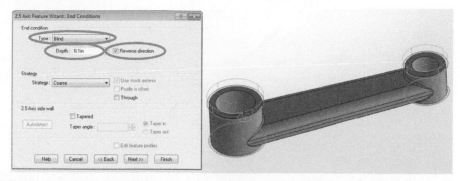

FIGURE A.10

Defining the end condition for the new circular pockets.

Choose Blind, click Reverse direction (default), and enter 0.1 in. for End Condition, and then click Next and Finish (see Figure A.10). Close the wizard. Two newly created circular pockets are listed under Mill Part Setup2.

At this point, you should be able to generate a toolpath by choosing Generate Operation Plan and Generate Toolpath from the CAMWorks pull-down menu.

QUESTIONS AND EXERCISES

S5.1 Design the following cantilever beam using optimization capability in SolidWorks Simulation.

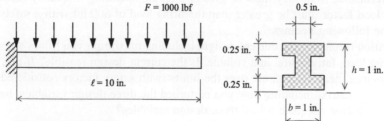

Material: 2014 Alloy
Design Problem definition:

Minimize: $\phi = \rho bh\ell$ (mass)
Subject to: $y_{max} < 0.1$ in.
$\sigma_{vm} < 30$ Ksi
0.5 in. $< b \leq 1.5$ in.
0.5 in. $< h \leq 3$ in.

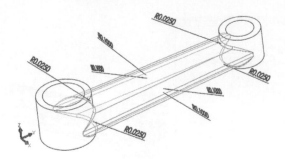

a. Create SolidWorks and Simulation models, and carry out optimization. Submit screen captures for the stress and displacement at the initial and optimal designs found by SolidWorks Simulation. Submit design history graphs for objective, constraints, and design variables.

b. Solve the same problem by adjusting the two design variables manually. Is your design better than that of SolidWorks Simulation? If so, by how much (e.g., mass %)? Submit screen captures for the stress and displacement at the improved design.

S5.2 Carry out sensitivity analysis and at least one what-if study for the connecting rod, following steps similar to that of Section S5.3. In this exercise, we use the same performance measures and three design variables: thickness (d_7, same as that of Section S5.3), and the two groups of fillets shown in the figure above. That is, fillets of radius R 0.1 in. and fillets of radius R 0.025 in. Use plain carbon steel as the material (yield strength: 32 ksi and endurance limit 30.5 ksi). When calculating the fatigue life, we assume the load is fully reversed (hence, choose load ratio $LR = -1$). The final design must be feasible; that is, maximum von Mises stress must be less than the yield strength divided by a safety factor of 1.2, fatigue life must be infinite, resonance frequency must be greater than the operating frequency (2000 rpm), and buckling load factor must be greater than the firing load of 600 lbf with a safety factor 1.2. Report the following findings:

a. Simulation results for the current design, including maximum von Mises stress, frequency, buckling load, fatigue life, and volume. Is the current design feasible? If not, identify the performance measures that are over the limits (with safety factors considered).

b. Sensitivity matrix, including how you perturbed the three design variables; for example, is 1% perturbation adequate for all three design variables?

c. Report how you came up with a feasible design.

d. Describe how you can continue to reduce the volume of the connecting rod once you are in the feasible region, if possible.

REFERENCES

Chang, K.H., 2013a. Product Performance Evaluation Using CAD/CAE, the Computer Aided Engineering Design Series. Academic Press, Elsevier Science & Technology, 30 Corporate Drive, Suite 400, Burlington, MA 01803, ISBN 978-0-12-398460-9.

Chang, K.H., 2013b. Product Manufacturing and Cost Estimate Using CAD/CAE, the Computer Aided Engineering Design Series. Academic Press, Elsevier Science & Technology, 30 Corporate Drive, Suite 400, Burlington, MA 01803, ISBN 978-0-12-401745-0.

Chang, K.H., 2014a. Product Design Modeling Using CAD/CAE, the Computer Aided Engineering Design Series. Academic Press, Elsevier Science & Technology, 30 Corporate Drive, Suite 400, Burlington, MA 01803.

Chang, K.H., 2014b. Dynamic Simulation and Mechanism Design with SolidWorks Motion 2013. Schroff Development Corporation, P O Box 1334, Mission, KS 66222, ISBN 978-1-58503-902-9.

DESIGN WITH PRO/ENGINEER

Design optimization was discussed in Chapters 3 and 5 for single-objective and multiobjective problems, respectively. In these chapters, both theoretical and practical aspects of the topics are discussed. With a basic understanding of concept, theories, and solution techniques of design optimization, we are ready to go over a few tutorial lessons on design optimization using the capabilities offered by commercial tools, such as Pro/ENGINEER.

In Project P5, we introduce Pro/ENGINEER design optimization capabilities as well as conduct interactive design using the suite of Pro/ENGINEER computer-aided design (CAD), engineering (CAE), and manufacturing (CAM) capabilities for a multidisciplinary design problem. We include an optimization design study for a cantilever beam as a single-objective optimization of single engineering discipline, including only structural performance measures. Then, we present a single-piston engine example to illustrate the interactive design process discussed in Chapter 3 involving multiple engineering disciplines. For those who are not familiar with Pro/ENGINEER and embedded CAE/CAM modules, we encourage you to go over Projects P1–P4 of the book series before moving forward with this tutorial project. Without going over these prior projects, you may get lost very quickly. Example models of the current project are available for download at the book's companion Web site.

Overall, the objective of this project is to enable readers to use Pro/ENGINEER for batch-mode design optimization and interactive design. The use of the capability for design optimization should be straightforward. Readers who are interested in learning more about the batch-mode optimization in Pro/ENGINEER may review a few more examples offered as YouTube videos, such as those listed below, subject to their availability:

- Pro/ENGINEER (Pro/E) Mechanica tutorial—Introduction to Optimization Design Study by eeprogrammer.com (www.youtube.com/watch?v=LrvvtKvse_A)
- Optimization Study in Wildfire 4.0 (www.youtube.com/watch?v=DzDlty6Vs9A)

Note that the lessons included in this current project were developed using Pro/ENGINEER Wildfire 5.0 M040. If you are using a different version of Pro/ENGINEER, you may see slightly different menu options or dialog boxes.

P5.1 INTRODUCTION TO DESIGN OPTIMIZATION IN PRO/ENGINEER

There are two main design study capabilities offered by Pro/ENGINEER—or more specifically, Pro/MECHANICA Structure (or Mechanica), which is a CAE module embedded in Pro/ENGINEER for support of finite element analysis (FEA) and design studies. These two design study capabilities are sensitivity and optimization. Sensitivity includes global and local sensitivity studies. A global sensitivity study calculates the changes in product performance measures when a design variable (or a combined change with more than one design variables) is varied over a prescribed range. Mechanica does this by calculating performance measure values at numerous designs in the design variable range(s). Design variables are perturbed a number of times (default is 10 in Mechanica) with a uniform design perturbation (or design interval). Mechanica automatically changes the solid model (provided that it is fully parameterized), updates finite element mesh, and analyzes the finite element model for each design change. This computation is usually very expensive (i.e., time-consuming) for large-scale models. Results of the global sensitivity study can be shown in a graph of a measure versus a design variable, essentially a piecewise linear approximation of the performance function ψ, as illustrated in Figure P5.1a.

A local sensitivity study calculates the gradient (slope) of a product performance measure with respect to a design variable. Mechanica calculates the slope of the performance curve $\psi(x)$ between two sample points (usually, one of them is the current design). The local sensitivity is computed using the finite difference method:

$$\frac{\partial \psi}{\partial x} \approx \frac{\psi(x + \Delta x) - \psi(x)}{\Delta x} \tag{P5.1}$$

As discussed in Chapter 3, a sensitivity study using the finite difference method could be inaccurate. Accuracy is largely dependent on size of the perturbation Δx. Note that Mechanica determines the design perturbation Δx internally.

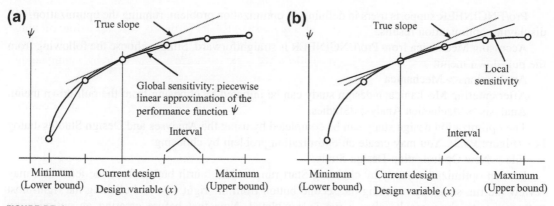

FIGURE P5.1

Sensitivity design studies: (a) global sensitivity and (b) local sensitivity.

An optimization design study, on the other hand, allows users to conduct a batch-mode optimization following gradient-based optimization approach, as discussed in Chapter 3. In this project, we focus on the optimization design study.

We use a cantilever beam example for illustration of the optimization capability (batch mode) in Section P5.2. The beam example is a single-objective optimization problem of single discipline, only involving structural performance measures. In addition, we use the single-piston engine example in Section P5.3 to illustrate details in a multidisciplinary design problem using an interactive design process, in which structural, motion, and manufacturing are involved.

The main objective of this tutorial project is to help you, as an experienced Pro/ENGINEER user, to become familiar with Pro/ENGINEER design optimization capabilities. The discussion on the basic operations in Pro/ENGINEER software, including Mechanism (for motion analysis), Mechanica, and Pro/MFG, in this section will be brief.

P5.1.1 DEFINING AN OPTIMIZATION PROBLEM

As discussed in Chapter 3, a single-objective optimization problem can be formulated mathematically as follows:

$$\text{Minimize:} \quad f(\mathbf{x}) \tag{P5.1a}$$

$$\text{Subject to:} \quad g_i(\mathbf{x}) \leq 0, \quad i = 1, \, m \tag{P5.1b}$$

$$h_j(\mathbf{x}) = 0, \quad j = 1, \, p \tag{P5.1c}$$

$$x_k^\ell \leq x_k \leq x_k^u, \quad k = 1, \, n \tag{P5.1d}$$

where $f(\mathbf{x})$ is the objective function or goal to be minimized (or maximized); $g_i(\mathbf{x})$ is the ith inequality constraint; m is the total number of inequality constraint functions; $h_j(\mathbf{x})$ is the jth equality constraint; p is the total number of equality constraints; $\mathbf{x}$ is the vector of design variables, $\mathbf{x} = [x_1, x_2,..., x_n]^T$; n is the total number of design variables; and x_k^ℓ and x_k^u are the lower and upper bounds of the kth design variable x_k, respectively.

Pro/ENGINEER supports users in defining an optimization problem, running the optimization, and displaying optimization results.

Accessing Mechanica from Pro/ENGINEER is straightforward. Simply choose the following from the pull-down menu:

Applications > Mechanica

After entering Mechanica, a design study can be initiated by choosing from the pull-down menu:

Analysis > Mechanica Analyses/Studies.

The optimization design study can be completed by using the Analyses and Design Studies dialog box (Figure P5.2). You may create an optimization problem by choosing:

File > New Optimization Design Study.

Run an optimization study by choosing Start run ⚓ (the fourth button from the left). You may review the run status by choosing the second button from the right 📧, and clicking the right-most button 🔁 to show results after a run is completed. Note that before creating an optimization study, you will have to already carry out structural analysis, such as static analysis, in which results are available for defining objective and constraint functions of the optimization problem.

In the Optimization Study Definition dialog box (Figure P5.3) that appears after choosing File > New Optimization Design Study, users may define a design optimization problem by choosing a measure as the objective function. Measures are predefined physical quantities, such as mass, maximum principal stress, and so forth, provided by Mechanica after an FEA run. Mechanica only supports single-objective optimization problems; hence, only one measure can be included in the objective.

Similar to the goal, users may choose measures to define constraints. For the condition, you may select one of the following three options: (1) ">" is greater than the minimum acceptable value, (2) "<" is less than the maximum acceptable value, and (3) "=" is equal to the given value.

Users may choose continuous variables to perform optimization. For each design variable, users need to define its lower and upper limits. Dimensions of the solid model are the most common variables for optimization. In addition, parameters, such as material properties, can also be chosen as design variables. Note that the model, part, or assembly must be fully parameterized to conduct a

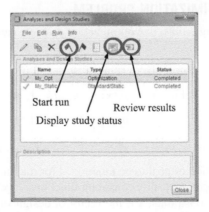

FIGURE P5.2

Analyses and Design Studies dialog box.

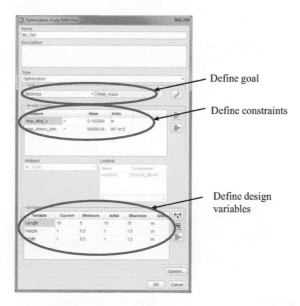

Define goal

Define constraints

Define design
variables

FIGURE P5.3

Optimization Study Definition dialog box.

batch-mode optimization in Pro/ENGINEER, as discussed in Chapter 3. More about design parameterization can be found in Chang (2014).

Once an optimization problem is completely defined, users may simply click the Start run ⬆ button (see Figure P5.2) to launch optimization. During the run, users may click Display study status button ☰ to see the status of the optimization. At the end of the run, users may choose the Review results ➡ to review the results of the study, such as a history plot of the objective function, as shown in Figure P5.4.

P5.1.2 TUTORIAL EXAMPLES

Two examples are included in this tutorial project. The first example, cantilever beam, illustrates the step-by-step details of defining an optimization problem, running the optimization, and showing results in data and graphs. The second example, a single-piston engine, illustrates an interactive design process of solving a multidisciplinary design problem using Mechanism, Mechanica, Pro/MFG, as well as the sensitivity display and what-if study discussed in Chapter 3. Because the detailed steps in using individual software modules have been discussed in Projects P1–P4, we only include the major steps with key information in these examples. Examples and topics to be discussed in each lesson are summarized in Table P5.1.

P5.2 CANTILEVER BEAM

We use the cantilever beam example shown in Chapter 3 to illustrate the steps for carrying out design optimization using Pro/ENGINEER. The beam is again shown in Figure P5.5 for illustration. The

total_mass
(lbf sec^2 / in)
Optimization Pass
Loadset:LoadSet1 : DESIGN_BEAM

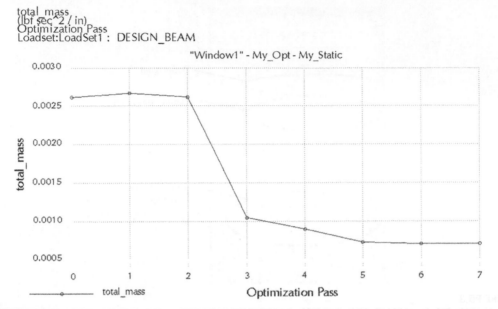

FIGURE P5.4

Objective function history graph.

Table P5.1 Examples Employed in This Project			
Section	**Example**	**Solid Models**	**Things to Learn**
P5.2	Cantilever beam design optimization		1. This in an introductory tutorial lesson, showing the steps for defining an optimization problem, running optimization, and showing results in table and graphs. 2. You will learn to solve similar optimization problems using Mechanica.
P5.3	Single-piston engine of multi-disciplinary optimization		1. This example illustrates an interactive process of solving a multidisciplinary design problem using Mechanism, Mechanica, Pro/MFG, as well as the sensitivity display and what-if study for design trade-off. 2. You will learn to solve similar multidisciplinary optimization problems using CAD/CAE/CAM capabilities in Pro/ENGINEER and the procedures involved.

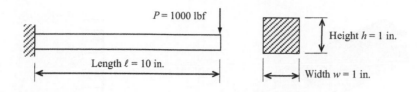

FIGURE P5.5

Cantilever beam example.

---

beam is made of AL2014. At the current design, the beam length is $\ell = 10$ in., and the width w and height h of the crosssectional area are both 1 in. The load is $P = 1000$ lbf acting vertically at the tip of the beam, as shown in Figure P5.5. Our goal is to optimize the beam design for a minimum mass subject to displacement and stress constraints by varying its length as well as the width and height of the cross-section.

P5.2.1 DESIGN PROBLEM DEFINITION

The design problem of the cantilever beam example is formulated mathematically as:

$$\text{Minimize:} \quad M(w, h, \ell) = \rho w h \ell \tag{P5.2a}$$

$$\text{Subject to:} \quad g_1(w, h, \ell) = \sigma_{\text{mp}}(w, h, \ell) - 65 \ \text{ksi} \leq 0 \tag{P5.2b}$$

$$g_2(w, h, \ell) = \delta_x(w, h, \ell) - 0.1 \ \text{in.} \leq 0 \tag{P5.2c}$$

$$0.5 \ \text{in.} \leq w \leq 1.5 \ \text{in.}, \quad 0.5 \ \text{in.} \leq h \leq 1.5 \ \text{in.}, \quad 5 \ \text{in.} \leq \ell \leq 15 \ \text{in.} \tag{P5.2d}$$

in which ρ is the mass density, σ_{mp} is the maximum principal stress, and δ_x is the maximum displacement in the X-direction (vertical).

A static design study is created for the beam with boundary and load conditions shown in Figure P5.6. The material chosen is AL2014 from Mechanica material library. The unit system chosen is IPS (in-lbf-s). Also shown in Figure P5.6 are the finite element mesh (with default settings) and the fringe plot of the maximum principal stress (or the measure: max_stress_prin of Mechanica).

In the current design, the maximum principal stress (g_1) is 71.9 ksi, which is greater than the required upper bound 65 ksi. This performance constraint is violated; hence, the design is infeasible.

The second performance constraint, the maximum X-displacement (g_2), is 0.373 in., which is greater than the required upper bound 0.1 in. Again, this performance constraint is violated (the design is infeasible).

We are changing the width, height, and length of the beam to find an optimal solution that is feasible and minimizes the mass of the beam.

P5.2.2 USING PRO/ENGINEER

P5.2.2.1 Open Pro/ENGINEER part

Open the solid model of the cantilever beam downloaded from the book's companion Web site, with the filename: beam.prt.1 in the folder: "Project P5 Examples/Lesson P5.2 Cantilever Beam." This

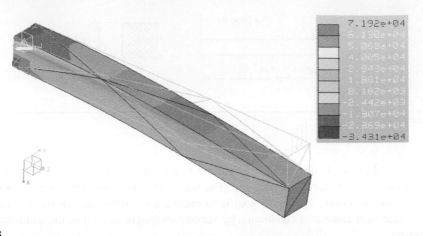

FIGURE P5.6

Finite element model and stress fringe plot of the cantilever beam in Mechanica.

cantilever example is similar to that of Project P3.2 in Chang (2013a). After opening this file, enter Mechanica and choose Mechanica Analyses/Studies. You should see that there is already a My_Static study created for you. You may want to browse the stress and displacement results of the study. Also, please check the unit system (File > Properties), and take a look at the finite element mesh. Again, the IPS Unit system is employed for this lesson.

P5.2.2.2 Create a new design study

In the Analyses and Design Studies dialog box (Figure P5.7), a My_Static study is listed. From the pull-down menu of the Analyses and Design Studies dialog box, choose:

    File > New Optimization Design Study.

    In the Optimization Study Definition dialog box (Figure P5.8), enter or choose the following:

- Enter the name of the design study as My_Opt
- Enter description (optional)
- Choose Optimization for Type (default)
- Select the total_mass as the objective function to be minimized (default).

P5.2.2.3 Add constraints

Define the performance constraint function by clicking the button 🔠 to the right of the "Design Limits" area in the Optimization Study Definition dialog box (see Figure P5.8). The Measures dialog box appears (see Figure P5.9). In the Measures dialog box, scroll down the Predefined (left) column to choose max_stress_prin, and click OK.

The max_stress_prin measure will be listed under the "Design Limits." Choose "<" (default), and enter 65,000 as the limit. Make sure the unit is lbf/in.^{2} (psi). Repeat the same steps to choose max_disp_x, choose "<," and enter 0.1 as the limit for the second constraint, as shown in Figure P5.10. The unit is inches. Note that My_Static was chosen by default and listed under Analysis (see Figure P5.10).

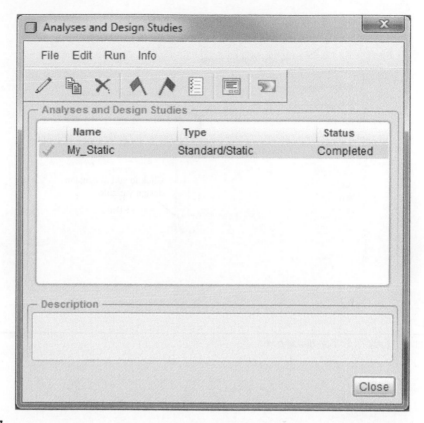

FIGURE P5.7
Analyses and Design Studies dialog box.

P5.2.2.4 Add design variables

Now, we define design variables. Click the first button ⊞ to the right of the "Variables" area in the Optimization Study Definition dialog box (see Figure P5.8), and click the beam in the graphics window (dimensions will appear like those in Figure P5.11). Pick the length dimension; the dimension will be listed under the "Variables" area in the Optimization Study Definition dialog box (see Figure P5.12). Enter LENGTH for name, and 5 and 15 for Minimum and Maximum, respectively.

Repeat the same steps for the height and width dimension variables. Enter HEIGHT and WIDTH for names. Enter 0.5 and 1.5 for Minimum and Maximum, respectively, for both variables.

Note that the Options button at the lower right corner of the Optimization Study Definition dialog box allows users to define design study options, including optimization algorithm, convergence tolerance, and number of iterations. We will use default settings for this problem (convergence tolerance: 1.0% and number of iterations: 20). Now, a design optimization problem is completely defined. Click OK in the Optimization Study Definition dialog box. The My_Opt design study is now listed in the Analyses and Design Studies dialog box. We are ready to run an optimization design study.

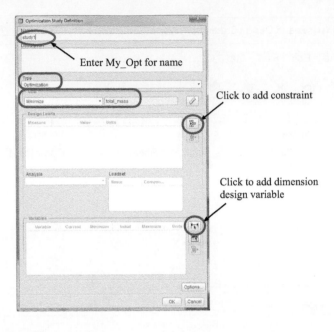

FIGURE P5.8

The Optimization Study Definition dialog box.

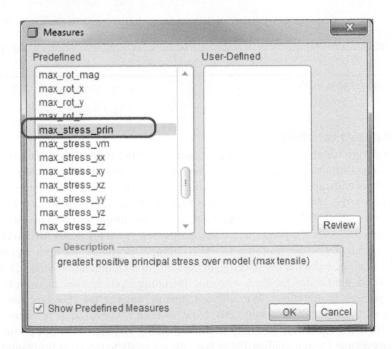

FIGURE P5.9

The Measures dialog box.

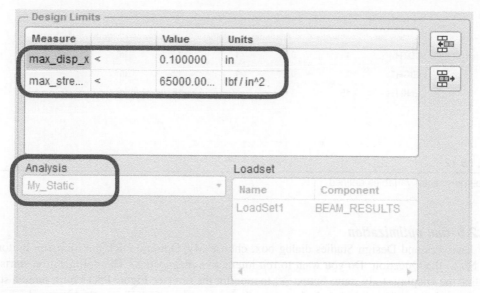

FIGURE P5.10

Constraint functions defined.

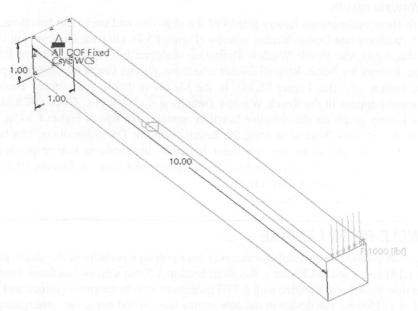

FIGURE P5.11

Pick the length dimension in the graphics window.

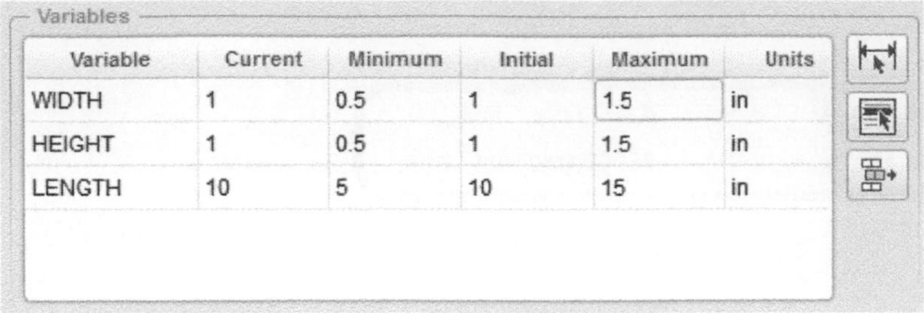

FIGURE P5.12

All three dimension variables defined.

P5.2.2.5 Run optimization

In the Analyses and Design Studies dialog box, choose My_Opt and click the Start run button ⚲. Click No to the Question: Do you want to run interactive diagnostics? The optimization starts. You may click the Display study status button 🖽 to monitor the run (see Figure P5.13 for a sample status).

At the end, an optimum is found after seven design iterations (see Figure P5.13), in which width, height, and length become 0.5, 1.08, and 5 in., respectively. Displacement is 0.0766 in. (<0.1 in.) and stress is 64.996 psi (<65 ksi). The total mass is 0.0007056 lbf-s^2/in. (or 0.2724 lbm), reduced from 0.002614 lbf-s^2/in (or 1.009 lbm) from the initial design.

P5.2.2.6 Review results

Here, we will show optimization history graphs of the objective and constraint functions.

From the Analyses and Design Studies window (Figure P5.7), click the Review result button 🗲 (first from the right), the Result Window Definition dialogue box appears (Figure P5.14). Enter Optimization_History for Name, keep all default selections, choose Graph for Display type, and click the Measure button 🖽 (see Figure P5.14). In the Measures dialog box, choose total_mass. The total_mass should appear in the Result Window Definition dialogue box. Click OK and Show. The optimization history graph for the objective function appears, like that of Figure P5.15a.

Choose Edit > Results Window to bring the Results Window Definition dialog box back. Repeat the steps to show the graphs for the constraint functions by choosing two respective measures: max_stress_prin and max_disp_x. The graphs should appear like those of Figures P5.15b and c for max_stress_prin and max_disp_x, respectively.

Save your model for future reference.

P5.3 SINGLE-PISTON ENGINE

In this lesson, we provide more details for carrying out the design problem of the single-piston engine (see Figure 1.18) discussed in Chapter 1. Recall in Section 1.5 that a cross-functional team was asked to develop a new model of the engine with a 30% increment in both maximum torque and horsepower at the speed of 1215 rpm. The design of the new engine was carried out at two interrelated levels: the system level and component level. At the system level, due to the increase in power and torque, the bore diameter of the engine case was increased from 1.416 to 1.6 in., the crank length was increased

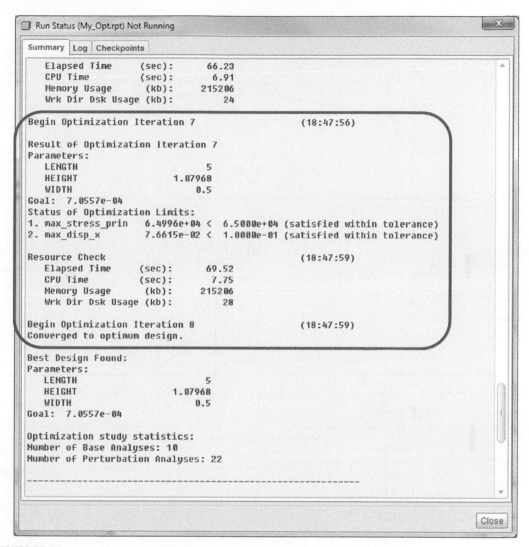

```
  Run Status (My_Opt.rpt) Not Running                                    x

  Summary  Log  Checkpoints

      Elapsed Time    (sec):        66.23
      CPU Time        (sec):         6.91
      Memory Usage     (kb):       215206
      Wrk Dir Dsk Usage (kb):          24

    Begin Optimization Iteration 7                  (18:47:56)

    Result of Optimization Iteration 7
    Parameters:
        LENGTH                      5
        HEIGHT                1.07968
        WIDTH                     0.5
    Goal:  7.0557e-04
    Status of Optimization Limits:
    1. max_stress_prin   6.4996e+04 <  6.5000e+04 (satisfied within tolerance)
    2. max_disp_x        7.6615e-02 <  1.0000e-01 (satisfied within tolerance)

    Resource Check                                  (18:47:59)
        Elapsed Time    (sec):        69.52
        CPU Time        (sec):         7.75
        Memory Usage     (kb):       215206
        Wrk Dir Dsk Usage (kb):          28

    Begin Optimization Iteration 8                  (18:47:59)
    Converged to optimum design.

    Best Design Found:
    Parameters:
        LENGTH                      5
        HEIGHT                1.07968
        WIDTH                     0.5
    Goal:  7.0557e-04

    Optimization study statistics:
    Number of Base Analyses: 10
    Number of Perturbation Analyses: 22

    -----------------------------------------------------------------

                                                                    Close
```

FIGURE P5.13

Run Status dialog box. The design converges in seven iterations.

from 0.5833 to 0.72 in., and the connecting road length was increased from 2.25 to 2.49 in., as listed in Table 1.1. Also, the change causes the brake effective pressure P_b to increase from 140 to 180 lbf, causing the peak combustion load to increase from 400 to 600 lbf acting on the top face of the piston. The load magnitude and path applied to the major load-bearing components, such as the connecting rod and crankshaft, are therefore altered. Reaction forces applied to components, such as the connecting rod, increase. The increase in reaction forces raises concerns about the structural integrity of

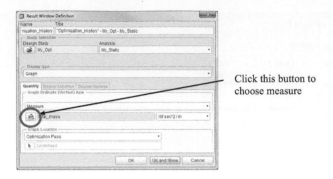

FIGURE P5.14

The Result Window Definition dialogue box.

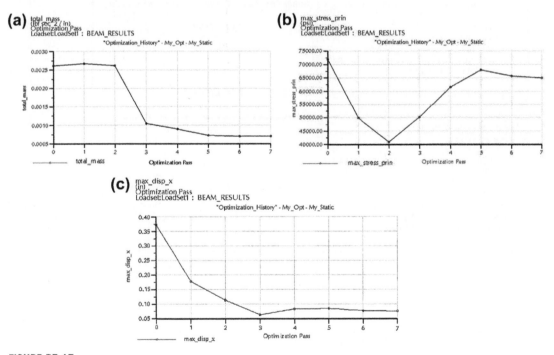

FIGURE P5.15

Optimization history graphs. (a) Objective function (total_mass). (b) Stress constraint (max_stress_prin). (c) Displacement constraint (max_disp_x).

these load-bearing components in the engine. Changes may need to be made to strengthen the design of these components, which may cause an increase in manufacturing time and cost.

In Section 1.5, we discussed component-level design using the connecting rod as an example. We included structural analysis for stress, buckling load, and frequency, as well as machining time for the connecting rod. We parameterized the geometry of the connecting rod and selected three dimensions—diameter of the large hole (d_{32}), diameter of the small hole (d_{31}), and thickness (d_7)—as design variables (see Table 1.2). We carried out sensitivity analysis for these structural performance measures and manufacturing time by varying these three design variables. We conducted design trade-off and carried out the what-if study to come up with an improved design for the connecting rod.

In this lesson, we repeat the component-level design of the connecting rod using Mechanism, Mechanica, and Pro/MFG. We start with a motion model of the new design—that is, the changes in bore diameter, crankshaft length, and connecting rod length are made according to the new values shown in Table 1.1. We have prepared the structural model of the connecting rod with simulation results in static, frequency, buckling, and fatigue analyses. Also, a Pro/MFG model was prepared. All are carried out at the new design—that is, the length of the connecting rod is 2.49 in. (increased from 2.25 in.). The objective of this lesson is to demonstrate the steps of conducting an interactive design process to achieve an improved design using the connecting rod example.

We start in Section P5.3.1 by describing the model files needed to carry out this exercise. We briefly describe the simulation results at the current design without showing the detailed steps in creating the models and running the simulations. Readers are encouraged to review tutorial projects P1–P4 for a good standing in solid modeling, motion analysis, structural analysis, and machining simulation, respectively, using Pro/ENGINEER CAD/CAE/CAM tools. With an understanding of the current design of the connecting rod, we move on to conduct sensitivity analysis in Section P5.3.2 and carry out a what-if study in Section P5.3.3. Note that some of the simulation results may not be consistent with those presented in Section 1.5, which were obtained using Mechanism, Mechanica, and Pro/MFG, with slightly different setups.

P5.3.1 MODEL FILES AND CURRENT DESIGN

The example files downloaded from the book companion site should contain those shown in Figure P5.16a under the folder named "Project P5 Examples/Lesson P5.3 Single-Piston Engine." The subfolder named "Connecting Rod" contains another four subfolders: Current Design, Machining, Sensitivity Analysis, and Whatif, as shown in Figure P5.16b. The "Current Design" folder contains simulation result files for the connecting rod at current design, including static, frequency, buckling, and fatigue studies. The "Machining" folder contains Pro/MFG files for machining simulation. Under "Sensitivity Analysis," there are three subfolders: Perturb d_7, Perturb d_{31}, and Perturb d_{32}. All are empty. We will store perturbed models and simulation results for the respective three design variables for sensitivity analysis using the finite difference method. In the "Whatif" folder, there is a spreadsheet file (Whatif.xlsx) prepared for carrying out a what-if study.

In the following, we briefly go over the models and simulations for motion and structural analyses, as well as machining simulation.

P5.3.1.1 Motion model and simulation

Start Pro/ENGINEER and open "engine.asm.1" under the folder "Lesson P5.3 Single-Piston Engine." You should see the engine assembly, as shown in Figure P5.17b. There are three subassemblies (case, propeller, and connecting rod) and one part (piston) listed in the model tree (Figure P5.17a).

Enter the Mechanism by choosing from the pull-down menu:

Applications > Mechanism.

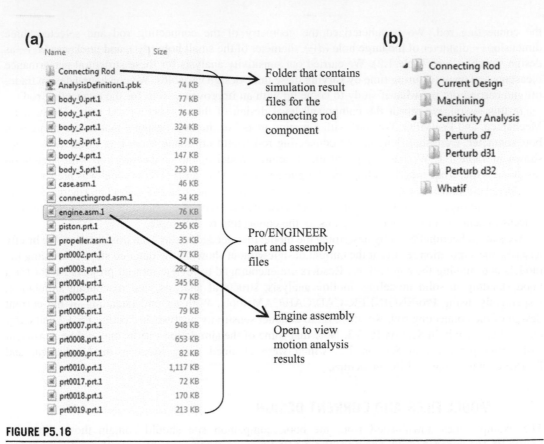

(a)

Name	Size
Connecting Rod	
AnalysisDefinition1.pbk	74 KB
body_0.prt.1	77 KB
body_1.prt.1	76 KB
body_2.prt.1	324 KB
body_3.prt.1	37 KB
body_4.prt.1	147 KB
body_5.prt.1	253 KB
case.asm.1	46 KB
connectingrod.asm.1	34 KB
engine.asm.1	76 KB
piston.prt.1	256 KB
propeller.asm.1	35 KB
prt0002.prt.1	77 KB
prt0003.prt.1	282 KB
prt0004.prt.1	345 KB
prt0005.prt.1	77 KB
prt0006.prt.1	79 KB
prt0007.prt.1	948 KB
prt0008.prt.1	653 KB
prt0009.prt.1	82 KB
prt0010.prt.1	1,117 KB
prt0017.prt.1	72 KB
prt0018.prt.1	170 KB
prt0019.prt.1	213 KB

Folder that contains simulation result files for the connecting rod component

Pro/ENGINEER part and assembly files

Engine assembly Open to view motion analysis results

(b)

- Connecting Rod
  - Current Design
  - Machining
  - Sensitivity Analysis
    - Perturb d7
    - Perturb d31
    - Perturb d32
  - Whatif

FIGURE P5.16

Files and folder under "Lesson P5.3 Single-Piston Engine": (a) files and subfolder and (b) subfolders included in "Connecting Rod."

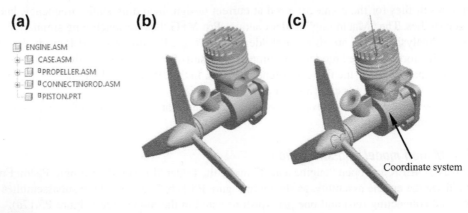

(a)

- ENGINE.ASM
  - CASE.ASM
  - PROPELLER.ASM
  - CONNECTINGROD.ASM
  - PISTON.PRT

(b)

(c)

Coordinate system

FIGURE P5.17

Engine assembly: (a) model tree, (b) assembly model, and (c) motion model in unexploded view.

The motion model appears in the graphics area with the motion entity symbols (see Figure P5.17c in an unexploded view), including connections (or joints), force, motor, and so forth. Note the coordinate system chosen is with the Y-axis pointing upward (see Figure P5.17c).

The motion model tree appearing below the model tree lists all motion entities, as shown in Figure P5.18a. There are four rigid bodies (including a Ground body), four joints (Pin1, Pin2, Slider1, and Bearing1), one motor (ServoMotor1) for kinematic analysis, one force (Firing_Load) for dynamic analysis, one initial velocity (InitCond1), and three motion simulations: FiringLoad (DYNAMICS), Constant_Angular_Speed (KINEMATICS), and Equilibrium (STATIC). Note that Equilibrium and Constant_Angular_Speed are created to verify the kinematic characteristics of the motion model. We only use the dynamic simulation (FiringLoad) results for this exercise.

You may want to browse some of the joint definitions to gain a better understanding of the assembly and motion model, especially, "Bearing1" between the connecting rod and piston (Figure P5.18b). This joint is chosen to monitor the reaction force acting on the connecting rod for structural analyses.

An initial angular velocity is defined and a force is applied to the piston at the beginning of the simulation. The initial velocity of 1215 rpm is defined on the axis of joint Pin1, as shown in Figure P5.19. In Mechanism, one of the approaches in defining the initial position and orientation is using snapshots captured in the Drag Packed Components button 🖱 at the top of the graphics window. There are two snapshots captured: S1 and S2. We use S2 as the initial condition to run the dynamic

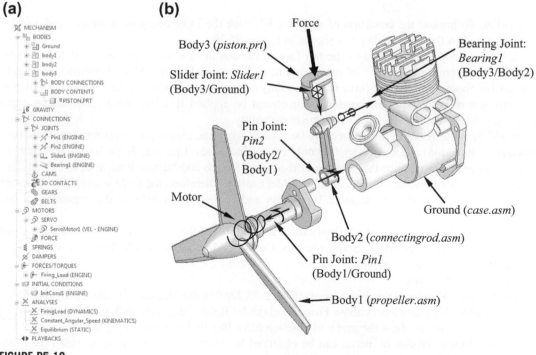

FIGURE P5.18

Engine motion model: (a) motion model tree and (b) motion model in exploded view.

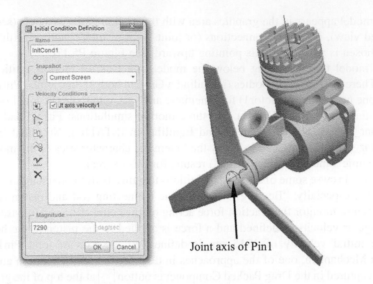

Joint axis of Pin1

FIGURE P5.19

Initial velocity dialog box.

simulation. To impose the condition of snapshot S2, click the Drag component button, and double-click S2 listed in the Drag dialog box shown in Figure P5.20.

A force of 600 lbf is applied to the top face of the piston in a short duration of 0.002 s, corresponding to approximately 6.5° of one rotation, as shown in Figure P5.21, to mimic the engine combustion force. Note that the force is applied only once at the beginning of the simulation, which is certainly not realistic. The combustion force cannot be applied to every single combustion cycle accurately due to the limitation of Mechanism capability.

It will be more realistic if the force can be applied when the piston just starts moving downward (negative Y-direction) and can be applied only for a selected short period. To do so, we will have to define measures that monitor the position of the piston for the combustion load to be activated. Unfortunately, such a capability is not available in Mechanism. Therefore, the force is simplified as a step function of 600 lbf along the negative Y-direction and applied for 0.002 s at the beginning of the simulation.

One result plot is created, which shows the reaction force at the upper joint of the connecting rod (Bearing1). To review results in a graph, click the Generate Measure Results of Analyses button ⊠ or choose from the pull-down menu:

Analysis > Measures.

The Measure Results dialog box appears (Figure P5.22a). In the Measure Results dialog box, click Reaction (under Measures) and choose FiringLoad (under Result Set), then click the Graph button ⊠ at the top left corner to show the graph of reaction force like that of Figure P5.22a.

The reaction force due to inertia can be observed by zooming in the Y-axis, in which the force reaches about 25 lbf, as shown in Figure P5.22b.

For more information regarding using Mechanism, readers are referred to Chang (2011).

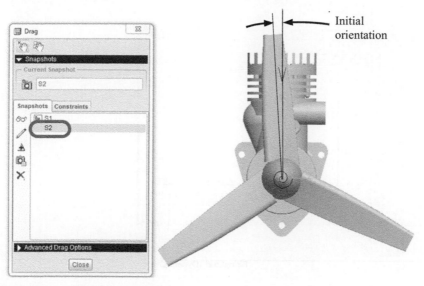

FIGURE P5.20

Snapshot S2 defining the initial orientation.

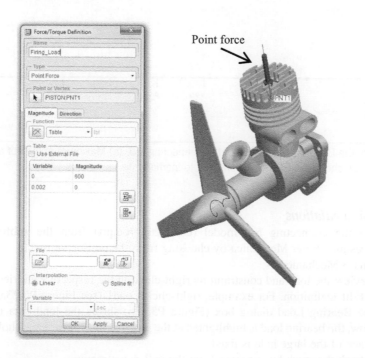

FIGURE P5.21

Force applied at the top face of the piston.

(a)

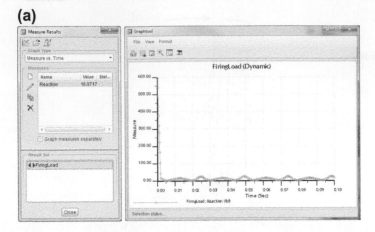

(b)

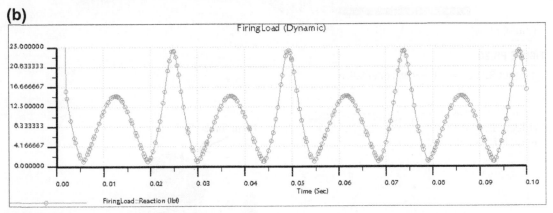

FIGURE P5.22

Reaction force of the Bearing1 joint at top of the connecting rod. (a) Maximum of 600 lbf at the beginning. (b) Zoomed-in view showing small oscillations due to inertia force.

P5.3.1.2 FEA simulations

We now open the connecting rod model (connectingrod.prt) from the subfolder "Connecting Rod/Current Design." Enter Mechanica by choosing from the pull-down menu:

Applications > Mechanica

You may review the load and constraint by right-clicking the respective entities in the model tree and choosing Edit Definition. For example, right-click Load1 (see Figure P5.23a) and choose Edit Definition. The Bearing Load dialog box (Figure P5.23b) shows the load data (−600 lbf); in the graphics window, the bearing load is highlighted at the inner surface of the small hole (Figure P5.23c). The inner surface of the large hole is fixed.

You may check the mesh by choosing from the pull-down menu:

AutoGEM > Create.

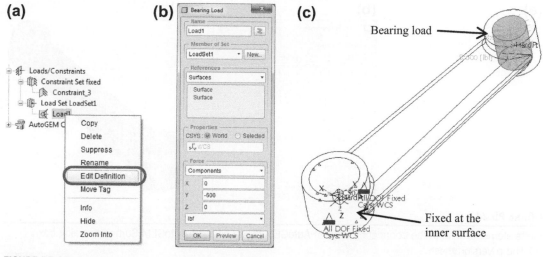

FIGURE P5.23

Mechanica model of the connecting rod. (a) Review Load1 by choosing Edit Definition. (b) Bearing Load dialog box. (c) Load and boundary conditions defined.

In the AutoGEM dialog box (Figure P5.24a), click Create. The AutoGEM Summary dialog box (Figure P5.24b) and p-version mesh appear (Figure P5.24c). There are 926 tetrahedron elements in this model. Close the dialog boxes without saving the mesh.

Choose from the pull-down menu:

Analysis > Mechanica Analyses/Studies.

In the Analysis and Study dialog box (Figure P5.25), there are four analyses listed: My_Static, My_Frequency, My_Buckling, and My_Fatigue. These are the four FEA analyses carried out at the connecting rod at current design. You may want to click an analysis and review results. For example, choose My_Static and click the Review results button 🔁 (the rightmost) to bring out a fringe plot of von Mises stress, such as that of Figure P5.26a.

The static analysis results show that the maximum von Mises stress occurs in the area of a small fillet between the outer surface of the small cylinder and the exterior surface of the rod, as shown in Figure P5.26b. The maximum stress is 21.91 ksi. The yield strength of the material is about 51 ksi. The safety factor for current design is about 2.33.

You may choose My_Frequency and click the Display study status button 📧 (second from the right) to bring out the Run Status summary dialog box, in which the first four frequencies are shown (see Figure P5.27a). Similarly, you may see results of My_Buckling analysis (see Figure P5.27b).

As shown in Figure P5.27a, the first (lowest) resonance frequency is 1,326 Hz or 79,560 rpm, which is way higher than the operating frequency in the range of thousands rpm. Also, as shown in Figure P5.27b, the first (lowest) buckling load factor is 15.3, implying that the actual load that causes the connecting rod to buckle is $15.3 \times 600 = 9,180$ lbf, which is much higher than the firing load.

(a)   **(b)**   **(c)**

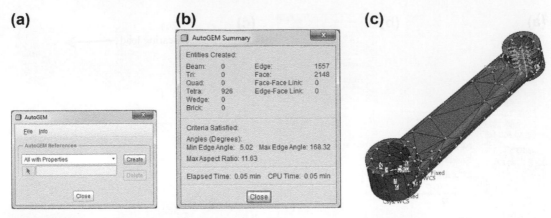

FIGURE P5.24

Finite element mesh of the connecting rod. (a) AutoGEM dialog box. (b) AutoGEM Summary dialog box.
(c) The p-version mesh.

FIGURE P5.25

The Analysis and Study dialog box.

For fatigue analysis, we defined the 600-lbf load as a repeated load, where the maximum is 600 lbf and the minimum is 0 (recall a very small inertia force in oscillation). Therefore, we choose the Zero-Peak option, as shown in Figure P5.28a. Also, 1,000,000 cycles is defined for Desired Endurance. Click the Previous Analysis tab to make sure My_Static is chosen for fatigue analysis (Figure P5.28b).

You may display a fringe plot for the fatigue life by following steps similar to those of stress fringe plot (see Figure P5.29a for the Results Window Definition dialog box). The results of fatigue life in the log scale are shown in Figure P5.29b. The lowest and highest lives are $10^{8.547}$ ($= 3.52 \times 10^8$) to 10^{20}, as seen in the color spectrum of the fringe plot of Figure P5.29c. Essentially, we see infinite fatigue life

(a)   **(b)**

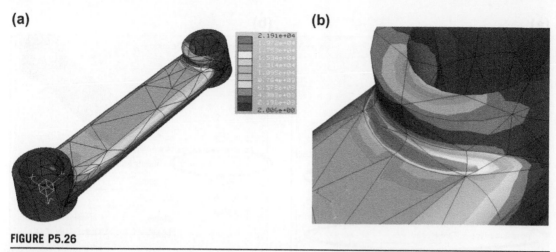

FIGURE P5.26

von Mises stress fringe plot. (a) Entire connecting rod model. (b) Zoomed-in view to the high stress concentration area.

(a)

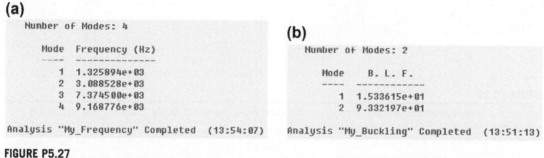

```
   Number of Modes: 4

      Mode  Frequency (Hz)
      ----  --------------
         1  1.325894e+03
         2  3.088528e+03
         3  7.374500e+03
         4  9.168776e+03

Analysis "My_Frequency" Completed  (13:54:07)
```

(b)

```
   Number of Modes: 2

      Mode   B. L. F.
      ----  --------------
         1  1.533615e+01
         2  9.332197e+01

Analysis "My_Buckling" Completed  (13:51:13)
```

FIGURE P5.27

Simulation results: (a) the first four resonance frequencies and (b) the first two buckling load factors.

in the connecting rod. For a more detailed discussion on fatigue analysis using Mechanica, readers are referred to Project P3.4, which is part of Chang (2013a).

Close the connecting rod model.

P5.3.1.3 Pro/MFG machining

Manufacturing simulation has also been carried out at the current design. We assume the machining operations start from a green part made by casting, as shown in Figure P5.30a. The machining operations remove materials on the top and bottom faces of the large and small cylinders, and drill two holes, as shown in Figure P5.30b (screen capture made in Vericut).

Change the working directory to "Connecting Rod/Machining." Open the manufacturing (assembly) file "connectingrod.asm." In the model tree, manufacturing entities are listed as shown in Figure P5.31. In the graphics window, the reference model (connectingrod.prt) and workpiece (workpiece.prt) are assembled properly, as shown in Figure P5.30a.

(a)   **(b)**

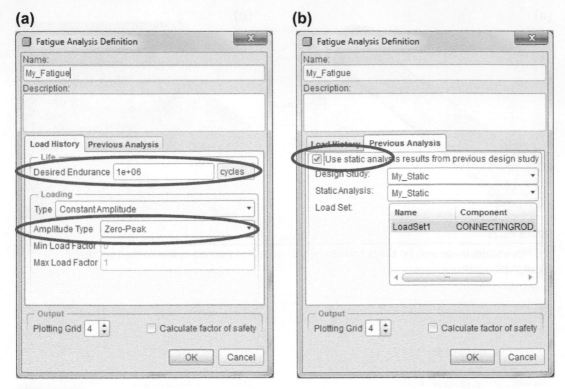

FIGURE P5.28

Setup for a fatigue analysis: (a) load history definition and (b) using stress of My_Static for the fatigue life calculation.

(a)   **(b)**   **(c)**

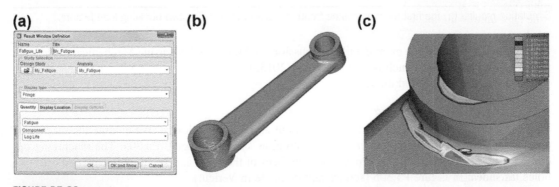

FIGURE P5.29

Displaying the fatigue life fringe plot. (a) Result Window Definition dialog box. (b) Fatigue life fringe plot. (c) Zoomed-in view.

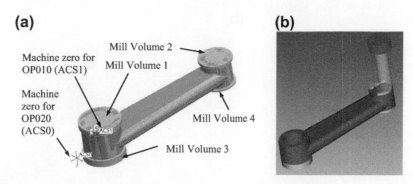

(a)

Machine zero for
OP010 (ACS1)

Machine
zero for
OP020
(ACS0)

Mill Volume 2

Mill Volume 1

Mill Volume 4

Mill Volume 3

(b)

FIGURE P5.30

Machining simulation. (a) Green part made by casting. (b) Simulation of machining operations (in Vericut).

As shown in the model tree (Figure P5.31a), there are two operations: OP010 and OP020, with respective machine zeroes ACS1 and ACS0 (see Figure P5.30a). Two volume milling NC (numerical control) sequences are created under OP010, removing materials defined in Mill Volumes 1 and 2, respectively. Another two volume milling and two hole drilling NC sequences are created for OP020, removing materials of Mill Volume 3 and 4, and drilling the two holes. In practice, jigs and fixtures must be designed and fabricated to hold the green part in support of the machining operations at shop floor. The machine zeroes may need to be adjusted according to the fixture design.

Right-click an NC sequence—for example, 1. Volume Milling—choose Edit Definition, and choose Setup to view how the sequence was defined, including the cutter (see Figure P5.32a) and machining parameters (see Figure P5.32b).

To review the toolpath, you may right-click, for example, OP010 and choose Play Path. Toolpath can be played and shown in the graphics window like that of Figure P5.31b. You may also right-click an NC sequence and choose Info > Feature to find the machining time in the Feature Info window that appears in the graphics window. Scroll down the Feature Info window to find the machining time under the caption of Manufacturing Related Information, as shown in Figure P5.33. Note that this includes only the actual machining time when the cutter is in motion. Setup time, tool change time, and so forth are not included.

The overall machining time is obtained by adding those of individual NC sequences—that is, 0.5910 (1. Volume Milling) + 0.3999 (2. Volume Milling) + 0.6031 (3. Volume Milling) + 0.3904 (4. Volume Milling) + 0.2240 (5. Holemaking) + 0.1941 (6. Holemaking) = 2.403 min.

A more detailed discussion on using Pro/MFG can also be found in Project P4 of Chang (2013b).

P5.3.2 SENSITIVITY ANALYSIS

The three design variables defined for the connecting rod are the diameter of the larger hole (d_{32}), the diameter of the small hole (d_{31}), and the thickness (d_7). For the current design, the values of the design variables are, respectively, $d_{32} = 0.5$ in., $d_{31} = 0.334$ in., and $d_7 = 0.25$ in.

To carry out sensitivity analysis for the gradients of the performance measures, including von Mises stress, lowest resonance frequency, lowest buckling load, fatigue life, mass, and machining time

(a)   **(b)**

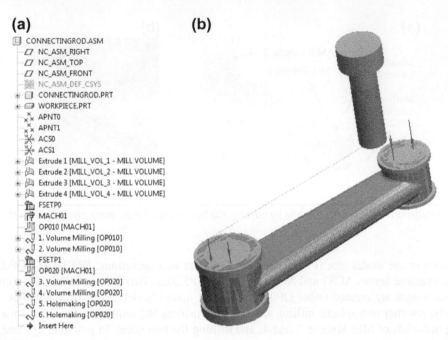

FIGURE P5.31

Machining model created in Pro/MFG: (a) model tree and (b) playing toolpath for the machining operation OP010.

(a)   **(b)**

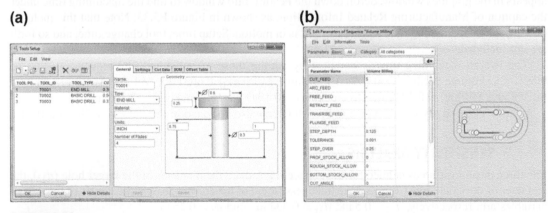

FIGURE P5.32

NC sequence setup for volume milling: (a) cutter (Tool Setup) and (b) machining parameters.

```
Manufacturing Info
|*********************************************|

|      Manufacturing Related Information      |

|*********************************************|

Feature ID       : 138

Feature Number : 18

Operation Name : OP010

Workcell Name   : MACH01

Workcell Type   : Mill

Sequence Type   : Volume Milling

Cut Mtn Feat     : 140

Axis Type          : 3 Axis

Fixture Setup   : FSETP0

Mill Volume       : MILL_VOL_1

Tool ID             : T0001

Machining time    : 0.590992 min

Retract Plane   : ACS1:F10(CSYS)(Z), Offset: 0.25

START POINT      : N/A

END POINT         : N/A

NCL File Name   : SEQ0001
```

FIGURE P5.33

Machining time of a volume milling NC sequence, 1. Volume Milling (OP010).

with respect to the three design variables, we use the overall finite difference method discussed in Chapter 3.

The approach is very simple. We perturb the individual design variables, one at a time, by 1% of its design variable value. For example, we change the thickness d_7 from 0.25 to 0.2525 in., regenerate the model, and then carry out simulations to find the values of the performance measures at the perturbed design. Thereafter, the sensitivity of a performance $\psi(x)$ can be approximated as:

$$\frac{\partial \psi}{\partial x} \approx \frac{\psi(x+\Delta x) - \psi(x)}{\Delta x} \tag{P5.3}$$

where $\psi(x+\Delta x)$ is a performance measure value obtained at the perturbed design from simulation, and Δx is the design perturbation, such as $\Delta d_7 = 0.0025$ in.

The detailed steps are described next. We copy the Pro/ENGINEER part file connectingrod.prt from the folder "Connecting Rod/Current Design" and paste it to folders "Connecting Rod/Sensitivity Analysis/Perturb d7," "Perturb d31," and "Perturb d32," respectively.

Then, from Pro/ENGINEER, we open connectingrod.prt from folder "Connecting Rod/Sensitivity Analysis/Perturb d7," right-click Protrusion id 41, and choose Edit Definition (see Figure P5.34a).

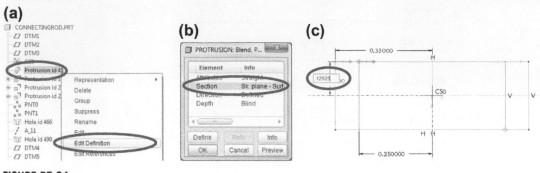

FIGURE P5.34

Perturbing design variable (half of d_7) from 0.125 to 0.12625 in. (a) Right-click Protrusion id 41. (b) Choose Section to make the change. (c) Change the height dimension to 0.12625.

In the PROTRUSION: Blend dialog box, double-click Section (see Figure P5.34b). In the sketch appearing in the graphics window, double-click the height dimension and change it to 0.12625 (half of the height, see Figure P5.34c). Click the checkmark ✔ to the right of the graphics window to accept the change. Click OK in the PROTRUSION: Blend dialog box. Choose from the pull-down menu Edit > Regenerate to rebuild the solid model.

Enter Mechanica, choose Analysis > Analysis and Design Studies. Rerun all four studies. Note that My_Static must be carried out before My_Fatigue and My_Buckling because both require the stress results from My_Static study.

Choose My_Static and click the Display study status button 🖳 to bring out the Run status window. Find the maximum von Mises stress (max_stress_vm) under Measures. The von Mises stress is now 21.440 ksi. You may create a fringe plot to locate the maximum von Mises stress, which is about the same location as the original design because the perturbation in design variable is small. This value is plugged into Eq. P5.3 to calculate the gradient:

$$\frac{\partial \psi}{\partial x} = \frac{\partial \sigma}{\partial d_7} \approx \frac{\sigma(d_7 + \Delta d_7) - \sigma(d_7)}{\Delta d_7} = \frac{21439.8 - 21906.3}{0.0025} = -1.866 \times 10^5 \text{ psi/in.}$$

We repeat the same steps to rerun simulations for frequency, buckling, and fatigue. The results are shown in Figure P5.35. The mass of the connecting rod can also be found by choosing Analysis > Model > Mass Properties from the pull-down menu.

Now, we rerun the machining simulation by simply changing the height dimension from 0.125 to 0.12625 in both the reference model and workpiece. Regenerate the models and save them. Then, we open the connectingrod.asm. The reference model and workpiece should both be updated automatically in the assembly. Regenerate the toolpath by replaying the toolpath for all six NC sequences, and then check the machining time.

Note that the machining times are unchanged because the change in the thickness design variable d_7 is not affecting the machining operations, which remove materials at top and bottom faces of the two cylinders and drill two holes for the respective cylinders. Therefore, sensitivity of the machining time with respect to the thickness design variable is zero.

Sensitivity Matrix

	Current value	d31	d32	d7
Perturbation		3.34000E-03	5.00000E-03	2.50000E-03
von Mises Stress (psi)	2.19063E+04	2.26933E+04	1.84716E+04	2.14398E+04
Sensitivity		2.35608E+05	-6.86946E+05	-1.86628E+05
Natural Frequency (Hz)	1.32589E+03	1.45738E+03	1.45458E+03	1.46921E+03
Sensitivity		3.93668E+04	2.57370E+04	5.73248E+04
Buckling Load Factor	1.53362E+01	1.92229E+01	1.92414E+01	1.92130E+01
Sensitivity		1.16370E+03	7.81046E+02	1.55075E+03
Fatigue Life (cycles)	1.00000E+06	1.00000E+06	1.00000E+06	1.00000E+06
Sensitivity		0.00000E+00	0.00000E+00	0.00000E+00
Mass (lb-s^2/in)	2.87815E-04	2.87278E-04	2.86442E-04	2.89966E-04
Sensitivity		-1.60647E-04	-2.74598E-04	8.60548E-04
Manufacturing Time (min)	2.40300E+00	2.40300E+00	2.40300E+00	2.40300E+00
Sensitivity		0.00000E+00	0.00000E+00	0.00000E+00

FIGURE P5.35

Sensitivity spreadsheet (in file Whatif.xlsx of Whatif folder).

We repeat the steps discussed above for design variables d_{31} and d_{32} by changing them to $d_{32} = 0.505$ in. and $d_{31} = 0.33734$ in. By simply double-clicking the diameter dimension of the connecting rod in the graphics window, enter the new dimension value (see Figures P5.36a and b) and regenerate the solid model. We carry out simulations and toolpath generations for the two perturbed designs. The sensitivity matrix for the three design variables are calculated as shown in Figure P5.35.

In Figure P5.35, the sensitivity coefficients of the machining time for all three design variables are zero because the design perturbations did not affect the machining operations. Although the hole sizes are changed, we assume the use of drill bits of the same sizes as the holes; therefore, toolpath and machining time are unchanged. Also, the sensitivity coefficients of the fatigue life are all zero because the fatigue life is infinite before and after design perturbations.

A bar chart of the entire sensitivity coefficient matrix (6×3 in this case) can be created using Spreadsheet, as shown in Figure P5.37a. The bar chart provides a global view on the influence of individual design variable to all performance measures and all design variables to a single performance measure. For example, natural frequency increases by increasing any of the three design variables. Among them, d_7 is the most significant and d_{32} is the least effective. Also, increasing design variable d_7 decreases stress, increases natural frequency, increases buckling load, and so on. Because the sensitivity magnitudes of different measures are different (by an order of magnitude), a common practice is normalizing these gradients by their respective maximum values for a better comparison of the effect of one variable across performance measures.

If we want to reduce stress and not increasing mass by much, we may graph the sensitivity coefficients for these two performance measures only, as can be seen in Figure P5.37b. The bar chart shows that the von Mises stress is reduced by increasing d_{32} and d_7 and reducing d_{31}. Among them, d_{32} is the most significant. On the other hand, only a small change in mass is observed.

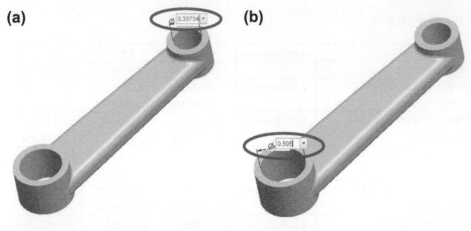

FIGURE P5.36

Perturb design variables d_{31} and d_{32}. (a) Change d_{31} to 0.33734. (b) Change d_{32} to 0.505.

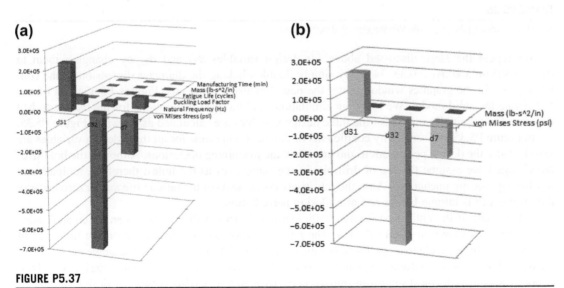

FIGURE P5.37

Bar chart of the sensitivity matrix. (a) The entire matrix. (b) Selected sensitivity coefficients: stress and mass performance measures.

P5.3.3 WHAT-IF STUDY

We go over a what-if study for the goal of reducing von Mises stress and not increasing the mass of the connecting rod too much. As seen in the bar chart in Figure P5.37b, we perturb all three design variables by using. As discussed in Chapter 3, the steepest decent direction of the stress measure is defined as:

$$\mathbf{n} = \frac{\left[\dfrac{\partial \sigma}{\partial d_{31}} \quad \dfrac{\partial \sigma}{\partial d_{32}} \quad \dfrac{\partial \sigma}{\partial d_7}\right]^T}{\sqrt{\left(\dfrac{\partial \sigma}{\partial d_{31}}\right)^2 + \left(\dfrac{\partial \sigma}{\partial d_{32}}\right)^2 + \left(\dfrac{\partial \sigma}{\partial d_7}\right)^2}}$$

(P5.4)

We then use a small step size, for example $\alpha = 0.01$, and multiply it with the normalized vector $\mathbf{n}$ for design perturbations:

$$\delta \mathbf{x} = \alpha \, \mathbf{n}$$

(P5.5)

The design perturbation $\delta \mathbf{x}$ is shown in the cells C29 to E29 of the "Whatif" spreadsheet, as shown in Figure P5.38. The current and predicted performance measure values are listed in Cells F22 to G27, and the percentage changes between them are shown in Cells H22 to H27. The cells indicate that the

| C34 | | f_x | =-D7/SQRT(SUMSQ(D7:F7)) | | | | | | | |

	A	B	C	D	E	F	G	H	I	J	K
2		**Sensitivity Matrix**									
4			Current value	d31	d32	d7					
5		Perturbation		3.34000E-03	5.00000E-03	2.50000E-03					
6		von Mises Stress (psi)	2.19063E+04	2.26933E+04	1.84716E+04	2.14398E+04					
7		Sensitivity		2.35608E+05	-6.86946E+05	-1.86628E+05					
8		Natural Frequency (Hz)	1.32589E+03	1.45738E+03	1.45458E+03	1.46921E+03					
9		Sensitivity		3.93668E+04	2.57370E+04	5.73248E+04					
10		Buckling Load Factor	1.53362E+01	1.92229E+01	1.92414E+01	1.92130E+01					
11		Sensitivity		1.16370E+03	7.81046E+02	1.55075E+03					
12		Fatigue Life (cycles)	1.00000E+06	1.00000E+06	1.00000E+06	1.00000E+06					
13		Sensitivity		0.00000E+00	0.00000E+00	0.00000E+00					
14		Mass (lb-s^2/in)	2.87815E-04	2.87278E-04	2.86442E-04	2.89966E-04					
15		Sensitivity		-1.60647E-04	-2.74598E-04	8.60548E-04					
16		Manufacturing Time (min)	2.40300E+00	2.40300E+00	2.40300E+00	2.40300E+00					
17		Sensitivity		0.00000E+00	0.00000E+00	0.00000E+00					
19		**What-if Study**									
21			d31	d32	d7	Current	Predicted	%Change	Analysis results	Accuracy	
22		von Mises Stress (psi)	2.3561E+05	-6.8695E+05	-1.8663E+05	2.19063E+04	1.44081E+04	-34.229	1.91106E+04	268.21%	
23		Natural Frequency (Hz)	3.9367E+04	2.5737E+04	5.7325E+04	1.32589E+03	1.58066E+03	19.215	1.46752E+03	179.89%	
24		Buckling Load Factor	1.1637E+03	7.8105E+02	1.5507E+03	1.53362E+01	2.26949E+01	47.983	1.98055E+01	164.65%	
25		Fatigue Life (cycles)	0.0000E+00	0.0000E+00	0.0000E+00	1.00000E+06	1.00000E+06	0.000	1.00000E+06	NA	
26		Mass (lb-s^2/in)	-1.6065E-04	-2.7460E-04	8.6055E-04	2.87815E-04	2.87946E-04	0.045	2.87930E-04	113.21%	
27		Manufacturing Time (min)	0.0000E+00	0.0000E+00	0.0000E+00	2.40300E+00	2.40300E+00	0.000	2.40300E+00	NA	
29		Design Perturbations (in.)	-0.0031422	0.0091614	0.0024890						
30		Current Value (in.)	0.334	0.5	0.25						
31		% Change	-0.94	1.83	1.00						
32		New Design (in.)	0.3308578	0.5091614	0.2524890						
34		Steepest decent dir of stress	-3.14218E-01	9.16143E-01	2.48896E-01						

| | Sheet1 | Sheet2 | Sheet3 |

FIGURE P5.38

What-if spreadsheet (in file Whatif.xlsx of Whatif folder).

maximum change is about 48% in Buckling Load Factor, which is large. The stress is reduced by about 35%, with only a 0.045% increase in mass. The design seems to be desirable.

To carry out simulation of the new design, we first copy the Pro/ENGINEER model of the current design (located in folder: "Connecting Rod/Current Design") and paste it to folder "Connecting Rod/Whatif." We then open this model from Pro/ENGINEER and update the solid model by entering the new design variables: $d_{31} = 0.3308578$, $d_{32} = 0.5091614$, and $d_7 = 0.252489$. We regenerate the solid model and rerun the simulations. The simulation results are entered to the spreadsheet (Cells I22 to I27); the accuracy of the prediction and analysis results is shown in Cells J22 to J27. The predictions are in general accurate, except for stress. This could be caused by either the insufficient accuracy of the sensitivity or a design perturbation that is not small enough.

In any case, the design is improved with the maximum von Mises stress reduced from 21.91 to 19.11 ksi and a very small increase in mass (0.00028782 to 0.00028793 lb-s^2/in.). The process discussed above can be repeated if needed. For the time being, we are satisfied with the design and will stop here.

QUESTION AND EXERCISES

P5.1 Design the following cantilever beam using optimization capability in Mechanica.

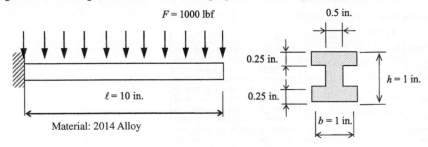

Design problem is defined as:

$$\text{Minimize:} \quad \phi = \rho b h \ell \ (\text{mass})$$
$$\text{Subject to:} \quad y_{max} < 0.1 \text{ in.}$$
$$\sigma_{vm} < 30 \text{ ksi}$$
$$0.5 \text{ in.} < b \le 1.5 \text{ in.}$$
$$0.5 \text{ in.} < h \le 3 \text{ in.}$$

a. Create Pro/ENGINEER and Mechanica models, then carry out optimization. Submit screen captures for the stress and displacement at the initial and optimal designs found by Mechanica. Submit design history graphs for the objective and constraints.

b. Solve the same problem by adjusting the two design variables manually. Is your design better than that of Mechanica? If so, by how much (e.g., mass %)? Submit screen captures for the stress and displacement of the improved design.

P5.2 Carry out sensitivity analysis and at least one what-if study for the connecting rod, following steps similar to that of Section P5.3. In this exercise, we use the same performance measures and three design variables: thickness (d_7, same as that of Section P5.3), and the two groups of fillets shown in the figure below. That is, fillets of radius R 0.04 in. (Round 3 and Round 4) and fillets of radius R 0.09 in. (Round id 918).

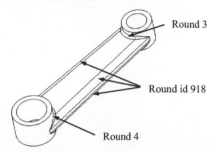

Use AL2014 as the material (assuming a yield strength of 25 ksi and endurance limit of 20.5 ksi). The final design must be feasible; that is, the maximum von Mises stress must be less than the yield strength divided by a safety factor of 1.2, fatigue life must be infinite, resonance frequency must be greater than the operating frequency (2000 rpm), and buckling load factor must be greater than the firing load of 600 lbf, with a safety factor of 1.2. Report the following:

a. Simulation results for the current design, including maximum von Mises stress, frequency, buckling load, fatigue life, and mass. Is the current design feasible? If not identify the performance measures that are over the limits (with safety factors considered).
b. Sensitivity matrix, including how you perturbed the three design variables; for example, is 1% perturbation adequate for all three design variables?
c. Report how you came up with a feasible design.
d. Describe how you continue reducing the mass of the connecting rod once you are in the feasible region, if possible.

REFERENCES

Chang, K.H., 2013a. Product Performance Evaluation Using CAD/CAE, the Computer Aided Engineering Design Series. Academic Press, Elsevier Science & Technology, 30 Corporate Drive, Suite 400, Burlington, MA 01803, ISBN 978-0-12-398460-9.

Chang, K.H., 2013b. Product Manufacturing and Cost Estimate Using CAD/CAE, the Computer Aided Engineering Design Series. Academic Press, Elsevier Science & Technology, 30 Corporate Drive, Suite 400, Burlington, MA 01803, ISBN 978-0-12-401745-0.

Chang, K.H., 2014. Product Design Modeling Using CAD/CAE, the Computer Aided Engineering Design Series. Academic Press, Elsevier Science & Technology, 30 Corporate Drive, Suite 400, Burlington, MA 01803.

Chang, K.H., 2011. Mechanism Design and Analysis with Pro/ENGINEER Wildfire 5.0. Schroff Development Corporation, P O Box 1334, Mission, KS 66222.

Index

Note: Page numbers followed by "f" denote figures; "t" tables; "b" boxes.

Printed and bound by CPI Group (UK) Ltd, Croydon, CR0 4YY

03/10/2024

01040329-0002